Root, Tuber and Banana Food System Innovations

Graham Thiele • Michael Friedmann
Hugo Campos • Vivian Polar
Jeffery W. Bentley
Editors

Root, Tuber and Banana Food System Innovations

Value Creation for Inclusive Outcomes

 Springer

Editors
Graham Thiele
CGIAR Research Program on Roots, Tubers
and Bananas, led by the International
Potato Center
Lima, Peru

Hugo Campos
International Potato Center (CIP)
Lima, Peru

Jeffery W. Bentley
Agro-Insight
Cochabamba, Bolivia

Michael Friedmann
CGIAR Research Program on Roots, Tubers
and Bananas, led by the International
Potato Center
Lima, Peru

Vivian Polar
CGIAR Research Program on Roots, Tubers
and Bananas, led by the International
Potato Center
Lima, Peru

ISBN 978-3-030-92024-1   ISBN 978-3-030-92022-7   (eBook)
https://doi.org/10.1007/978-3-030-92022-7

This Springer imprint is published by the registered company Springer Nature Switzerland AG
The registered company address is: Gewerbestrasse 11, 6330 Cham, Switzerland

Foreword

It is such a great honor to write the foreword to this landmark book volume on research and innovations in root, tuber, and banana food systems.

Root, tuber and banana crops are crucial for Africa's food security. They represent a vast number of crops, including cassava, banana, plantain, potato, sweetpotato, and yams. They all serve as major sources of calories for millions of households and consumers.

Yet, the full potentials of these crops have not been fully tapped due to low productivity, diseases and pests, poor storage systems, and limited processing and value addition to expand market demand. Compared to cereal crops, root, tuber, and banana food systems have also had more limited investment in research and development.

However, major gains have been achieved in research and development. Investments, for example, in tissue culture for propagating these crops, have helped to raise yields while tackling major pests and diseases. Recent advances in food processing with the use of cassava flour as composite flour with wheat flour in bread and confectioneries have had successes in some countries such as Nigeria. Orange-fleshed sweetpotato has seen increased use in nutritious foods to tackle malnutrition.

The CGIAR (previously Consultative Group on International Agricultural Research) has been at the forefront of breakthrough research and development on root, tuber, and banana crops. The CGIAR Research Program on Roots, Tubers and Bananas (RTB) led and implemented this work. For 10 years, from 2012 to 2021, the RTB worked in a highly participatory fashion with over 200 partners in Africa and across the world to unlock the potential of these crops.

The RTB is a highly successful program and represents an excellent model of effective and impactful collaborative research and development. It is impressive that across its entire life cycle of 10 years, RTB secured an investment of US$750 million, a greater share of which was devoted to root, tuber, and banana crops in Africa.

RTB developed several technologies and innovations that have contributed significantly to the improvement of root-, tuber-, and banana-based food systems. These include improved seed systems for delivering high yielding and disease- and pest-resistant crop varieties, improved agronomic practices, postharvest storage and handling, value addition through enhanced processing, as well as biofortification for improved nutrition.

This book compendium, written by over 70 authors, documents an impressive array of research and development innovations that have contributed to helping millions of farmers to achieve higher yields, food security, and nutrition.

The African Development Bank through its Technologies for African Agricultural Transformation (TAAT) has provided a critical platform to scale up several of these technologies to reach millions of farmers, working in strategic partnership with the CGIAR System.

I commend the authors of the book for their rich essays that offer excellent illumination on the potentials of root, tuber, and banana crops. The book is a must read for everyone interested in agricultural research and development and the power of science and innovation to tackle the challenge of food insecurity in developing countries, especially in Africa. I highly recommend the book.

Dr. Akinwumi A. Adesina
President, African Development Bank
Abidjan, Côte d'Ivoire
World Food Prize Laureate, 2017

Acknowledgments

First and foremost, we would like to thank each and every author and the organizations they are associated with for their willingness to share their most priceless asset: time. This book would just not have been possible without their generous commitment.

Drs. Gordon Prain, Jillian Lenne, and Tom Remington provided critical review of the chapters, and Christopher Butler edited and improved the manuscripts for several chapters. Thanks must also go to Zandra Vasquez for administrative support second to none, ensuring proper formatting and meticulous inclusion of the references.

We are very grateful to our editor, Joao Pildervasser, and also to Shanthini Kamaraj from Springer, for all the work behind the scenes, and specifically for the outstanding professional support as the book moved into production, publishing, and marketing.

Much of the research documented in this book was undertaken as part of the CGIAR Research Program on Roots, Tubers and Bananas (RTB) and generously supported by the CGIAR Trust Fund contributors: https://www.cgiar.org/funders/.

We would like to thank the many colleagues particularly from ABC, CIP, CIRAD, and IITA who made RTB such a success story. Especially we would like to recognize Barbara Wells, Director General of the International Potato Center (CIP), who championed RTB and sadly passed away as this book was being finalized; Rodney Cooke, Chair of the Board of Trustees of CIP; and the RTB Independent Steering Committee members, particularly present and past chairs Eugene Terry, Rupert Best, Helen Hambly, and Dunstan Spencer.

We also thank the many collaborating organizations, more than 200 partners, and all the farmers, traders, processors, and others who work with RTB crops who have made possible these achievements and have hopefully benefited in some measure with the important progress made with these hitherto neglected crops.

Some personal acknowledgments

I (Graham Thiele) am grateful to the broader "family" of RTB scientists who believe so much in the value of collaboration and for the chance to have led such a dynamic program. I especially thank my wife Ross Ana for so much love and support during the many travels and writing assignments which made this book possible.

I (Michael Friedmann) am grateful to all the scientists and research staff that contributed to the RTB program, making this book possible. Most heartfelt thanks to my wife Sigal and my daughters for their love and never-ending support through long hours both for this book and for contributing to the RTB program.

I (Hugo Campos) am grateful to the Bill & Melinda Gates Foundation, through its investment (OPP1213329) awarded to the International Potato Center (SweetGAINS). Also, to my family for putting up with yet another writing assignment, and particularly to my wife Orietta for her tireless support, for her love, and for being my most perceptive critic.

Vivian Polar is grateful to the women and men, producers, processors, traders, and consumers of RTB crops that have worked side by side with researchers to inform and support the development the multiple innovations presented across the book. She is also grateful to her husband Luis and her children who are a continuous source of inspiration and support. In loving memory of her friends and mentors, Antonio Gandarillas and Edson Gandarillas, who passed away during the Covid-19 pandemic.

I (Jeff Bentley) thank the other editors, who as always were a great team to work with. Thanks to the authors and the others at RTB, the partners, and farmers who contributed to this remarkable research program. I also thank my wife, Ana, for her patience and encouragement.

Some closing remarks

To the readers of our book, all its coauthors share the hope that you will be able to put the insights, pitfalls, and learnings gathered here to good use. That will mean this book has been a success.

Lima, Peru, January 2022

Graham Thiele, Michael Friedmann, Hugo Campos, Vivian Polar,
and Jeffery Bentley

Contents

Contributors

Adebayo Abass International Institute of Tropical Agriculture (IITA), Dar-Es-Salaam, Tanzania

George Ooko Abong Department of Food Science, Nutrition and Technology, University of Nairobi, Nairobi, Kenya

Suraju Adeyemi Adegbite Federal Institute of Industrial Research Oshodi (FIIRO), Lagos, Nigeria

Victor Attah Adejoh Synergos Nigeria, Maitama Abuja, Nigeria

Akin O. K. Adesehinwa Institute of Agricultural Research & Training, Obafemi Awolowo University, Ibadan, Nigeria

Julius Adewopo International Institute of Tropical Agriculture, Kigali, Rwanda

Laurent Adinsi Université d'Abomey-Calavi, Faculté des Sciences Agronomiques (UAC-FSA), Jericho, Benin

Beatrice Aighewi International Institute of Tropical Agriculture (IITA), Abuja, Nigeria

Susan Ajambo Alliance of Bioversity International & CIAT, Kampala, Uganda

Conny J. M. Almekinders Wageningen University, Wageningen, The Netherlands

Alfredo Augusto Cunha Alves Embrapa Mandioca e Fruticultura, Cruz das Almas, Brazil

Francis Kweku Amagloh Department of Food Science & Technology, Faculty of Agriculture, Food and Consumer Sciences, University for Development Studies, Tamale, Ghana

Tunde Amole International Livestock Research Institute, Ibadan, Nigeria

Jorge L. Andrade-Piedra International Potato Center, Lima, Peru

Frezer Asfaw International Potato Center, Addis Ababa, Ethiopia

Jacqueline A. Ashby Independent Consultant, Portland, OR, USA

Robert Asiedu International Institute of Tropical Agriculture (IITA), Ibadan, Nigeria

Marsy Asindu International Livestock Research Institute, Kampala, Uganda

Anna-Marie Ball Hartley Wintney, UK

Morufat Balogun International Institute of Tropical Agriculture (IITA), Ibadan, Nigeria

University of Ibadan, Department of Crop Protection and Environmental Biology, Ibadan, Nigeria

Luis Augusto Becerra López-Lavalle Alliance Bioversity-CIAT, Cali, Colombia

John Belalcázar Alliance Bioversity-CIAT, Cali, Colombia

Guy Blomme The Alliance of Bioversity International and the International Center for Tropical Agriculture (CIAT), Addis Ababa, Ethiopia

Sunil Bordoloi Amo Farm Sieberer Hatchery Ltd, Amo Byng Ltd, Awe, Oyo State, Nigeria

Alexandre Bouniol Université d'Abomey-Calavi, Faculté des Sciences Agronomiques (UAC-FSA), Jericho, Benin

CIRAD UMR Qualisud, Cotonou, Benin

Leon Brimer Department of Veterinary and Animal Sciences University of Copenhagen, Frederiksberg, Denmark

Hugo Campos International Potato Center (CIP), Lima, Peru

Arnaud Chapuis CIRAD, Montpellier, France

Linley Chiwona-Karltun Department of Urban & Rural Development, Swedish University of Agricultural Sciences, Uppsala, Sweden

Steven Cole International Institute of Tropical Agriculture (IITA), Dar es Salaam, Tanzania

Wilmer J. Cuellar The Alliance of Bioversity International and the International Center for Tropical Agriculture (CIAT), Cali, Colombia

Joselito da Silva Motta Embrapa Mandioca e Fruticultura, Cruz das Almas, Brazil

Luciana Alves de Oliveira Embrapa Mandioca e Fruticultura, Cruz das Almas, Brazil

Erik Delaquis Alliance Bioversity-CIAT, Rome, Italy

Elohor Diebiru-Ojo International Institute of Tropical Agriculture (IITA), Ibadan, Nigeria

Dominique Dufour CIRAD UMR Qualisud, Montpellier, France

David Eagle Mennonite Economic Development Associates (MEDA), Waterloo, Ontario, Canada

Beatrice Ekesa Alliance Bioversity International-CIAT, Kampala, Uganda

Prince M. Etwire Council for Scientific and Industrial Research, Savanna Agricultural Research Institute, Tamale, Ghana

Sarah Fernandes CGIAR Research Program on Roots, Tubers and Bananas (RTB), led by the International Potato Center, Lima, France

Michael Friedmann CGIAR Research Program on Roots, Tubers and Bananas (RTB), led by the International Potato Center, Lima, Peru

Richard Fuchs Natural Resources Institute (NRI), University of Greenwich, Kent, UK

Karen A. Garrett Plant Pathology Department and Food Systems Institute, University of Florida, Gainesville, FL, USA

Fredrick Grant International Potato Center (CIP), Kampala, Uganda

Helen Hambly University of Guelph, Guelph, ON, Canada

Simon Heck International Potato Center (CIP), Nairobi, Kenya

David P. Hughes Pennsylvania State University, Department of Biology, University Park, PA, USA

Mihiretu C. Hundayehu International Potato Center, Hawassa, Ethiopia

Paul Ilona HarvestPlus, Nigeria country office, Ibadan, Nigeria

Simon Imoro International Institute for Tropical Agriculture, Ibadan, Nigeria

Francois Iradukunda Alliance of Bioversity International & CIAT, Kampala, Uganda

Rogers Kakuhenzire International Potato Center, Addis Ababa, Ethiopia

Edward Kanju International Institute of Tropical Agriculture (IITA), Uganda, Kampala, Uganda

Regina Kapinga International Institute of Tropical Agriculture (IITA), Dar es Salaam, Tanzania

Enoch Kikulwe Alliance of Bioversity International & CIAT, Kampala, Uganda

Fleur B. M. Kilwinger Wageningen University, Wageningen, The Netherlands

Jan Kreuze International Potato Center (CIP), Lima, Peru

Apollin Fotso Kuate International Institute of Tropical Agriculture (IITA), Yaoundé, Cameroon

Peter Kulakow International Institute of Tropical Agriculture (IITA), Ibadan, Nigeria

P. Lava Kumar International Institute of Tropical Agriculture (IITA), Ibadan, Nigeria

Sanni Lateef International Institute of Tropical Agriculture (IITA), Ibadan, Nigeria

Cees Leeuwis Wageningen University, Wageningen, the Netherlands

James P. Legg International Institute of Tropical Agriculture (IITA), Dar es Salaam, Tanzania

Jan Low International Potato Center (CIP), Nairobi, Kenya

Simon Singi Lukombo International Institute of Tropical Agriculture (IITA), Kinshasa, DR, Congo

Stephen Magige Mennonite Economic Development Associates (MEDA), Dar es Salaam, Tanzania

Antonio Magnaghi Euro Ingredients Limited, Southern By-pass, Kikuyu, Kiambu County, Kenya

Jose Jackson Alliance for African Partnership, International Studies and Programs, Michigan State University, East Lansing, MI, USA

Pricilla Marimo Alliance Bioversity – CIAT, Kampala, Uganda

Norbert Maroya International Institute of Tropical Agriculture (IITA), Ibadan, Nigeria

Sarah Mayanja International Potato Center (CIP), Kampala, Uganda

Margaret A. McEwan International Potato Center, Regional Office for Africa, Nairobi, Kenya

Djana B. Mignouna International Institute of Tropical Agriculture (IITA), Cotonou, Benin

Martín Moreno Universidad del Valle (Univalle), Cali, Colombia

Mukani Moyo International Potato Center (CIP), Regional Office for Africa, Nairobi, Kenya

Kiddo Mtunda Tanzania Agricultural Research Institute (TARI), Tumbi, Tabora, Tanzania

Anna Muller Bioversity International, Rome, Italy

Lucy Mulugo Makerere University, Kampala, Uganda

Tawanda Muzhingi Department of Food, Bioprocessing and Nutritional Sciences, North Carolina State University, Raleigh, NC, USA

Leroy Mwanzia The Alliance of Bioversity International and the International Center for Tropical Agriculture (CIAT), Cali, Colombia

Janet Mwende University of Nairobi, Nairobi, Kenya

Sam Namanda International Potato Center, Kampala, Uganda

Israel Navarrete Wageningen University, Wageningen, The Netherlands

Diego Naziri International Potato Center (CIP), Hanoi, Vietnam

Robert Ndjouenkeu Ecole Nationale Supérieure des Sciences Agro-Industrielles, University of Ngaoundéré, Ngaoundéré, Cameroon

Hemant Nitturkar FAO, Riyadh, Saudi Arabia

Walter Ocimati Alliance of Bioversity International & CIAT, Kampala, Uganda

Makuachukwu Ojide Alex Ekwueme Federal University Ndufu-Alike-Ikwo (FUNAI), Ebonyi state, Nigeria

Julius J. Okello International Potato Center (CIP), Kampala, Uganda

Iheanacho Okike International Institute of Tropical Agriculture (IITA), Ibadan, Nigeria

Aman Bonaventure Omondi Alliance Bioversity-CIAT, Cotonou, Benin

Jean Pankuku Tehilah Value Chain and Food Processing, Blantyre, Malawi
Root and Tuber Crops Development Trust (RTCDT), Lilongwe, Malawi

Monica Parker International Potato Center, Nairobi, Kenya

Wolfgang Pfeiffer Alliance Bioversity-CIAT, HarvestPlus, Washington, DC, USA

Vivian Polar CGIAR Research Program on Roots, Tubers and Bananas, led by the International Potato Center, Lima, Peru

Marcelo Precoppe Natural Resources Institute (NRI), Chatham, UK

Claudio Proietti Formerly CGIAR Research Program on Roots, Tubers and Bananas, currently CIRAD, Montpellier, France

Srinivasulu Rajendran International Potato Center (CIP), Kampala, Uganda

Anandan Samireddipalle National Institute of Animal Nutrition and Physiology (NIANP), Indian Council of Agricultural Research (ICAR), Bangalore, India

Murat Sartas Wageningen University, Wageningen, the Netherlands
International Institute of Tropical Agriculture (IITA), Kigali, Rwanda

Elmar Schulte-Geldermann Bingen University of Applied Science, Bingen am Rhein, Germany

Marc Schut Wageningen University, Wageningen, the Netherlands

International Institute of Tropical Agriculture (IITA), Kigali, Rwanda

Michael Selvaraj The Alliance of Bioversity International and the International Center for Tropical Agriculture (CIAT), Cali, Colombia

Kalpana Sharma International Potato Center, Nairobi, Kenya

Josip Simunovic Department of Food, Bioprocessing and Nutritional Sciences, North Carolina State University, Raleigh, NC, USA

Issahaq Suleman International Potato Center, Kumasi, Ghana

Luis Alejandro Taborda Andrade Centro Internacional de Agricultura Tropical (CIAT), Cali, Colombia

Universidad Nacional de Colombia, Bogotá, Colombia

Béla Teeken International Institute of Tropical Agriculture (IITA), Ibadan, Nigeria

Graham Thiele CGIAR Research Program on Roots, Tubers and Bananas (RTB), led by the International Potato Center, Lima, Peru

William Tinzaara Alliance of Bioversity International & CIAT, Kampala, Uganda

Thierry Tran Alliance Bioversity-CIAT, Cali, Colombia

CIRAD UMR Qualisud, Montpellier, France

Van-Den Truong Department of Food, Bioprocessing and Nutritional Sciences, North Carolina State University, Raleigh, NC, USA

Hale Ann Tufan Cornell – GREAT, Ithaca, NY, USA

Jacob van Etten Bioversity International, Rome, Italy

Tom A. van Mourik Royal Tropical Institute (KIT), Amsterdam, The Netherlands

Seerp Wigboldus Wageningen University & Research, Wageningen, The Netherlands

Juma Yabeja International Institute of Tropical Agriculture (IITA), Dar es Salaam, Tanzania

List of Figures

List of Tables

Part I
Overview, Institutional Change and Scaling

Chapter 1
Overview

Graham Thiele (ID), **Michael Friedmann** (ID), **Vivian Polar** (ID),
and Hugo Campos (ID)

Abstract Root, tuber, and banana (RT&B) crops play a critical role in food and nutrition security in developing countries, increasingly so in sub-Saharan Africa (SSA). They have great potential to contribute to alleviate poverty, improve health and nutrition, and enhance the resilience of smallholder farmers to climate change. However, RT&Bs are characterized by unique challenges including vegetative propagation, genetic complexity, and postharvest constraints with bulkiness and perishability, compared to cereals. They are also characterized by a high yield potential and the ability to deliver micronutrients at large scale. However, until recently they have suffered from neglect in both investment and research. The CGIAR Research Program on Roots, Tubers and Bananas, which operated from 2012 to 2021, represented a novel and successful innovation model within the agricultural research for development domain not only in scientific terms but also from an organizational perspective. This program built upon the uniqueness of the RT&B crops and contributed to much of the progress reported in the book. This chapter provides an overview of challenges and opportunities facing RT&B crops in processing, marketing and distribution, enhancing productivity, and improving livelihoods. It presents underlying concepts for gender and scaling that feature prominently throughout the book, as well as an updated stance on innovation, touching base on the topic of the jobs to be done. The chapter concludes with an overview and highlights of the different sections and chapters in the remainder of the book.

G. Thiele (✉) · M. Friedmann · V. Polar
CGIAR Research Program on Roots, Tubers and Bananas (RTB), led by the International Potato Center, Lima, Peru
e-mail: g.thiele@cgiar.org; v.polar@cgiar.org

H. Campos
International Potato Center (CIP), Lima, Peru
e-mail: h.campos@cgiar.org

1.1 Introduction

Root, tuber, and banana (RT&B) crops play a critical role in food and nutrition security in developing countries, increasingly so in sub-Saharan Africa (SSA). They have great potential to contribute to alleviate poverty, improve health and nutrition, and enhance the resilience of smallholder farmers to climate change. However, RT&Bs are characterized by unique challenges and, until recently, have suffered from neglect in both investment and research (Fig. 1.1). The CGIAR Research Program on Roots, Tubers and Bananas (RTB hereafter), which operated from 2012 to 2021, represents a novel and successful innovation model within the agricultural research for development domain. This program built upon the uniqueness of RT&B crops, including vegetative propagation, genetic complexity, and postharvest constraints with bulkiness and perishability, compared to cereals, and also their high yield potential and the ability to deliver micronutrients at large scale.

Historically, RT&B crops have suffered from lower investment, and thus, this field of research features fewer researchers than other crops. In this context, RTB raised the profile of this important crop group for global food and nutrition security and served as an effective advocate for more investment. As different chapters in the book will make clear, RTB has not only focused on developing and implementing

Fig. 1.1 Value chains for bulky and perishable RT&B crops need upgrading. Market in Kampala, Uganda. (Photo credit: M. Friedmann (RTB))

particular innovations but has also addressed broader constraints for scaling innovations, since innovation only creates value when embraced and adopted by its users and beneficiaries at scale.

This book delivers an updated stance on innovations in RT&B crops, within and outside of the CGIAR program. We will touch on how to design innovations with a high likelihood of success and impact and provide several successful examples of innovations in RT&B crops while examining the scaling readiness of the innovations.

1.2 Importance of RT&B Crops

The importance of RT&B crops in developing countries has grown rapidly in the past 60 years: total production has quadrupled, and in this group, potato, cassava, and yam have experienced the highest growth. Sweetpotato declined a bit as producers in Asia who relied on sweetpotatoes as a food security crop and for animal feed switched to other crops (Table 1.1).

Table 1.1 Production and area harvested of RT&B crops and value of production in developing countries

# of countries	Crop	Production[a] (000,000 tonnes)			Area (000,000 ha)			Price USD/ Mg[b]	Value USD billions	% of total
		1961–1963	1988–1990	2016–2018	1961–1963	1988–1990	2016–2018			
86	Cassava	74.6	150.0	281.1	9.8	15.0	24.7	128	36	10.7
90	Potato	26.8	76.3	221.5	3.5	6.3	12.1	318	71	20.9
96	Sweetpotato	93.1	121.0	88.5	12.7	9.1	7.9	741	66	19.5
101	Banana	21.1	54.4	112.9	2.0	4.1	5.5	621	70	20.8
63	Yam	5.9	18.7	72.9	1.2	2.0	8.5	743	54	16.1
46	Plantain	13.4	24.5	38.7	2.5	4.2	5.5	597	23	6.9
88	Other RT&B[c]	6.9	9.6	21.0	1.4	1.7	3.1	920	17	5.2
	Total	**241.8**	**454.5**	**836.6**	**33.1**	**42.4**	**67.3**		**337.0**	**100**

Source: Scott (2020)

[a]Production and area (FAOSTAT), 2020

[b]IFPRI's IMPACT model for commodity projections utilizes average prices (USD/Mg) in 2005 of 515.99 for banana and plantain, 673.1 for sweetpotato and yam, taro, and yautia, and 116.89 and 257.85 for cassava and potato, respectively. Estimated prices for 2030 are 735.61 for banana, 684.82 for plantain, 819.28 for yam, taro, and yautia, 815.26 for sweetpotato, 384.17 for potato, and 140.25 for cassava. Prices listed above are estimates for 2017 based on a linear interpolation of prices for 2005 and 2030 and then rounded to the nearest dollar. The prices for 2005 and 2030 are based on IFPRI researchers' determinations of past and future market tendencies as well as consultations with specialized scientists in international agricultural research centers around the world

[c]Other RT&B crops include the production (in million tonnes) of taro (10.2), yautia (0.5), and not specified R&T (10.3) for 2016–2018, with a production value (USD billions) of 7.6 for taro, 0.3 for yautia, and 9.9 for not specified R&T in 2016–2018

Table 1.2 Changing share of total cropped area in developing countries in Africa

	% area in Africa		
Crops	1961–1963	1989–1991	2016–2018
Banana	26.6	34.3	36.4
Cassava	57.1	57.7	75.5
Plantain	80.5	77.6	74.8
Potato	7.8	12.4	15.1
Sweetpotato	5.2	15.2	57.4
Yam	93.0	94.9	97.9

Source: FAOSTAT, New Food Balances. Last Update: February 19, 2020

Table 1.3 Contribution of RT&B crops to food intake in selected African countries in kilocalories (kcal) per capita (2017)

Country	Population (000s)	Grand total kcal	RT&B foods kcal	% RTBs foods
Democratic Republic of Congo[a]	62,523	1605	916	57
Ivory Coast	24,437	2730	970	36
Ghana	29,121	3033	1430	47
Nigeria	190,873	2464	622	25
Rwanda	11,981	2215	1046	47
Uganda	41,167	2144	659	31

[a]DRC 2008. Source: FAOSTAT, New Food Balances. Last Update: February 19, 2020

Rapidly increasing populations in Africa, growing conditions in the humid tropics, and a strong preference for RT&B crops mean that, with the exception of plantain, an increasing share of the area under these crops is grown in this continent (Table 1.2).

The contribution of foods derived from RT&B crops to calorific needs ranges from 25% in Nigeria to 57% in the Democratic Republic of Congo. Populations are growing rapidly in SSA, particularly in urban centers. By 2050, just these six countries (Table 1.3) are projected to have a combined population of 0.81 billion (UNDP 2019). The number of people involved in RT&B-dominated agri-food systems could more than double by the end of the twenty-first century—most of it in SSA.

Globally, RT&B crops are of particular importance to the poorest households in developing countries (Wiebe et al. 2020). They enhance the resilience of food systems and increase food availability and diversity, especially during hunger periods or in the event of crop failure or damage to cereals during extreme weather events. Through biofortification, RT&B crops can contribute to reducing widespread micronutrient deficiencies of vitamin A (VA), iron (Fe), and zinc (Zn), particularly in young children and women of reproductive age.

In Asia, yields of RT&B crops mostly showed large increases, doubling or tripling (Table 1.4; Fig. 1.2). In Latin America and the Caribbean (LAC), we see general increases but more modest. However, with the partial exception of potatoes and bananas, productivity of RT&B crops in SSA remains low. Hence, production does

Table 1.4 Change in yield for RT&B crops in developing countries

Crops	Yield (t/ha)								
	Africa			Asia			LAC		
	1961–1963	1989–1991	2017–2019	1961–1963	1989–1991	2017–2019	1961–1963	1989–1991	2017–2019
Bananas	6.1	6.9	11.0	8.7	15.1	31.7	16.4	18.6	24.6
Cassava	5.7	8.0	8.7	8.1	13.1	21.7	12.2	11.8	13.0
Plantains	4.7	5.2	6.0	6.6	9.7	12.5	8.1	8.6	11.0
Potatoes	7.9	10.8	14.8	8.7	12.7	20.1	7.0	12.4	19.7
Sweetpotatoes	5.2	5.2	6.3	7.5	15.4	19.6	8.1	7.0	10.9
Yams	7.2	9.7	8.5	10.3	12.2	14.5	7.0	7.3	8.0

Source: FAOSTAT, New Food Balances. Last Update: December 22, 2020 http://www.fao.org/faostat/en/#data/FBS. FAOSTAT reports banana and plantain separately; however, no systematic criteria are used to make this separation. Note: (t) = ton; ha = hectare

Fig. 1.2 Harvesting improved cassava varieties in Dong Nai province, Vietnam. (Photo credit: G. Smith (Alliance of Bioversity & CIAT))

not satisfy basic food security needs in rural areas and does not compete well with imported staples for urban consumers, resulting in missed smallholder income opportunities. As people move to cities in SSA, value chains for RT&B crops need to be reconfigured to improve efficiency and convenience and reduce postharvest losses to compete with imports.

Outside of the humid tropics in Africa and in most of Asia and LAC, RT&B crops are generally important in rotations with cereals (Fig. 1.3) or legumes, as part of agroforestry systems, such as bananas and coffee, or as secondary crops (Fig. 1.4). In this way, they contribute resilience, increase efficiency of cropping systems, provide key micronutrients in the diet, and generate off-season income.

Fig. 1.3 The Assam Agribusiness and Rural Transformation Project (APART) project in Assam State, India, introduced potato farmers to climate resilient zero tillage with paddy straw mulch potato production. (Photo credit: CIP)

Fig. 1.4 Mr. John Ndamira in his coffee/banana plantation, with I. Arinaitwe of Mbarara District, Bushenyi, Uganda. (Photo credit: M. Friedmann (RTB))

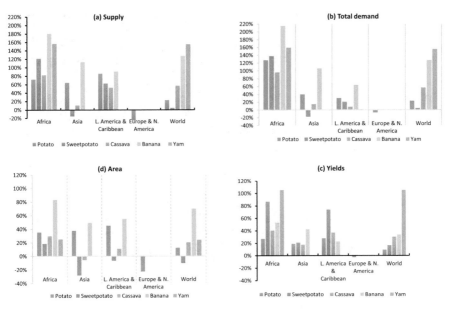

Fig. 1.5 Changes in RT&B agriculture from 2010 to 2050 under a baseline foresight scenario. (Source: Authors' calculations based on Rosegrant et al. 2017)

Using the global partial equilibrium model IMPACT, future agricultural demand and supply for RT&B crops were projected, considering future climate and socio-economic changes. These forecasts (Fig. 1.5) predicted RT&B crop production reaching 1400 million tons (of fresh produce) by 2050—a global increase of almost 50% from 2010—with Africa emerging as the world's largest producing and consuming region (Petsakos et al. 2019).

1.3 Challenges and Opportunities for RT&B Crops

Significant technological change will be needed to cope with climate change, especially in SSA, where RT&B crops are of the highest importance (Thiele et al. 2017). Therefore, a key dynamic will be crop substitution, especially in areas where RT&B crops can replace more climate-sensitive cereals and legumes. Maize is vulnerable to higher frequencies of periodic drought, whereas production of cassava and sweetpotato can be much more reliable under these conditions. In addition, there are specific traits that make RT&B crops tolerant or resistant to abiotic stresses like heat, drought, soil salinity, and waterlogging and shocks like typhoons/cyclones (Fig. 1.6). For this reason, roots and tubers are important contributors to post-disaster recovery and mitigation for their capacity for piecemeal harvesting, underground protection, and short growing cycles (Prain and Naziri 2020). Sweetpotato has played an especially important role in humanitarian relief operations combining rusticity, short crop cycles and—particularly with orange-fleshed varieties—high levels of micronutrients (Fig. 1.7).

Fig. 1.6 A field affected by typhoon with broken trees and barely damaged sweetpotato plants, Luzon, Philippines. (Photo credit: CIP)

Fig. 1.7 High β-carotene (pro-vit A) OFSP. (Photo credit: D. Gemenet (EiB, CIP))

1.3.1 Processing, Marketing, and Distribution

Because RT&B crops are more perishable and bulkier than cereals, they create opportunities for value addition and employment in postharvest and processing in rural areas, especially for women. Providing growing urban populations with

Fig. 1.8 A mechanical sieve—coarse mash retained in the wooden box and fine mash captured in the basin after separation. (Photo credit: I. Okike (IITA))

RT&B crop-based food will require extensive transformation of current technology to capture these benefits. But unless gender roles and needs are systematically considered, innovation can exacerbate gender inequality (Sarapura 2012). Increasing opportunities for women can have a powerful impact on productivity and agriculture-led development and help to reduce gender gaps in access to inputs, assets, opportunities, information, and other resources (Margolies and Buckingham 2013; FAO 2014).

RT&B crops and their residues are finding increasing use in animal feed. High-quality cassava peels (HQCP; Fig. 1.8) are being taken to scale in Nigeria (with interest elsewhere in SSA), while sweetpotato silage is now widely used in Asia and finding greater popularity in Uganda (Asindu et al. 2020).

Since cassava starch is particularly important for its specific functional properties and constitutes the largest source of starch in tropical regions, improving small- and medium-scale processing technologies is key to expanding the market in SSA (Chapuis et al. 2016).

1.3.2 Enhancing Productivity

Investment in RT&B crops compared to cereals has lagged, slowing yield growth. RT&B crops offer high potential yields, but farmers often realize less than half the potential due to poor quality planting material of limited genetic potential, biotic and abiotic constraints, and poor management practices. Seed systems may not work well because of inappropriate policies for RT&B crops or where famine and disaster-relief undermine the development of healthy seed market systems by providing large volumes of free propagation materials. These bottlenecks in markets and policy further restrain the willingness to invest in yield-increasing input intensification. Where market conditions are favorable and appropriate new technology available, yield gains have been considerable, as in Southeast Asia for cassava (Malik et al. 2020). In Africa, potato is grown increasingly as a cash crop for urban markets, showing consistent growth with the second highest rate (after banana) of yield increase of any crop over the last decade.

Recent adoption studies in SSA provide optimism that yield gaps can be closed. For instance, in Nigeria, modern varieties of cassava are grown on 39.9% of cultivated area with a yield increase of at least 60%, bringing an estimated 1.6 million people out of poverty each year (Wossen et al. 2018). However, in general, adoption rates of modern varieties of RT&B crops are below 40%; therefore, more investment in breeding and seed systems is required (Thiele et al. 2020). Seed systems for RT&B crops can be a particular challenge, as farmers can more easily share and retain planting material that discourages the interest of private seed companies. Hence, some public sector seed system investment is usually required.

Women's farm yields and incomes are typically much lower than men's, reflecting specific gender-related barriers that affect women's productivity: competing priorities (e.g., childcare), exclusion from farming decisions by men, and limited access to land, markets, and information technology (FAO 2014; Mudege et al. 2015). COVID-19 may be undoing recent progress on reducing gender gaps and compromising food and nutrition security (Doss et al. 2020).

RTB has paid explicit attention to gender in its program, and gender is a specific focus of this book, which uses the following definitions of key terms for these important discussions (Box 1.1).

1.3.3 Improving Livelihoods

With an average production of approximately 820 million tons on 64 million hectares (ha) in 2016–2018 (FAOSTAT 2020), RT&B crops represent the second most important set of crops in developing countries after cereals. The yield potential of RT&B crops is very high, providing one of the cheapest and more affordable sources of dietary energy (Lebot 2020). In 2017, RT&B crops alone provided around 10% of the daily per capita calorie intake for the 864 million people living in least developed countries (Kennedy et al. 2019).

RT&B crops can also play an important role in nutrition security. In SSA, vitamin A deficiency (VAD) is widespread, contributing to increased risks of blindness, illness, and premature death, particularly in young children and pregnant/postpartum women. Globally, 163 million children under 5 years of age are vitamin A deficient, and prevalence rates of Fe and Zn deficiencies are even higher. Orange-fleshed sweetpotato (OFSP) is a proven biofortified crop: 100 g/day can meet the vitamin A requirements of a young child (Fig. 1.9), and 6.2 million households

Box 1.1 Key Terms for Discussing Gender

Gender is not the same as sexuality. Gender is socially constructed and refers to roles and behaviors of men and women that are the intended or unintended products of social practice.

Gender analysis is a critical examination of how differences in gender roles, activities, needs, opportunities, and rights affect women, men, girls, and boys in a given situation or context.

Gender gap is the difference between women and men in terms of their levels of participation, access, rights, remuneration, or benefits.

Gender inequality is reflected in unequal access to or enjoyment of rights, remuneration, or benefits, as well as the assumption of stereotypical social and cultural roles.

Gender roles and norms are standards or expectations of the behavior of women and men, just because they are male or female. These expectations are particular for each society, culture, and community at a given point in time.

Gender relations are social relations that unite women and men as social groups in a given context, including how power and access to and/or control over resources are distributed between the sexes.

Gender power relations are how gender shapes the distribution of power at all levels of society.

Gender responsive describes the quality of a specific tool, approach, or intervention in terms of actively examining and addressing gender norms, roles, and inequalities. This type of approach often involves actions to create an environment that promotes gender equality.

Gender transformative describes the quality of a specific tool, approach, or intervention in terms of creating opportunities for individuals to actively challenge and/or transform gender norms to address power inequalities between people of different genders.

Gender intentional describes the quality of a specific program, project, or intervention in terms of purposefully addressing gender considerations in the design and/or operation. It seeks to target and benefit a specific gendered group to achieve specific goals.

Intersectionality is an analytical tool for studying, understanding, and responding to the ways that sex and gender intersect with other personal characteristics or identities and how these intersections contribute to unique experiences of discrimination.

Empowerment is the process by which individuals gain power and control over their own lives and acquire the ability to make strategic choices.

Sex-disaggregated data is data collected and tabulated separately for women and men allowing the measurement of differences between women and men in terms of various social and economic dimensions.

Source: European Institute for Gender Equality: Glossary & Thesaurus. https://eige.europa.eu/thesaurus

Fig. 1.9 Mothers and caregivers learning how to prepare dishes including orange-fleshed sweet-potato safely while retaining vitamins to improve their young children's nutrition (Photo credit: H. Rutherford, CIP)

(HH) have been reached with improved sweetpotato varieties—mostly OFSP—across 15 African countries (Low and Thiele 2020). HarvestPlus and its partners delivered cassava varieties enriched with vitamin A to more than one million farming households in Nigeria and the Democratic Republic of Congo (Ilona et al. 2017). Banana cultivars are also a significant source of vitamin A (Amah et al. 2018) and are being promoted in East Africa. Meanwhile, potato breeding has achieved significantly enhanced levels of Fe and Zn (Amoros et al. 2020). Where consumed as a staple, biofortified potatoes can contribute up to 50% of women's requirements for these micronutrients (Burgos et al. 2019). Jongstra et al. (2020) have recently found that Fe from biofortified potatoes is more bioavailable to humans than any other Fe biofortified crop developed to date.

Much current work on RT&B crops in SSA and Asia is demonstrating the power and effectiveness of these crops in alleviating poverty, fighting hunger and nutrition insecurity, especially within the current content of climate change and COVID-19 (Heck et al. 2020). RT&B crops are mostly produced, processed, and traded locally, making them less vulnerable to abrupt price fluctuations in international markets and interruptions due to epidemics. RT&B-based products (such as orange-fleshed sweetpotato puree used in baked goods) have been developed and promoted and are currently going to scale in African markets. There is growing global demand for innovative ingredients to meet growing consumer tastes and nutrition needs. Their potential is often limited, however, by the lack of preferred nutritious varieties, instability of micronutrients particularly when processed, unfavorable value chains, and weak institutional arrangements.

1.4 A Primer on Innovation and the Jobs to Be Done

It is clear that for RT&B crops to contribute more extensively, innovation is required in all the domains so far described. The RTB program put innovation and the scaling of innovations at center stage.

As part of its program design, RTB invested in a subprogram—known as a "flagship project" in CGIAR—dedicated to improving livelihoods by scaling RT&B solutions in agri-food systems. The flagship provided support to next and end users for scaling of RTB innovations from the farm to the community, the region, and beyond. We guided scientists in the program on how to better target research, design strategic youth and gender research, select partners, and develop capacity for improving livelihoods at scale. This results-oriented approach included the following:

1. Forward-looking analysis of trends
2. Decision support for the tailoring, integration, and scaling of RT&B technologies based on farmer typologies and practical investment steps
3. Identifying best possible RT&B (and other crops and livestock) technology options that contribute to sustainable livelihoods
4. Addressing constraints for RT&B technology adoption with gender and intergenerational transformations
5. Evidence base to improve scaling of RT&B agri-food system innovations with enhanced equity

The flagship developed an approach to innovation and scaling that was widely applied within the program through a dedicated fund for scaling (see Chap. 2) and is featured in several chapters of this book as a key crosscutting topic. Key definitions for innovation and scaling are provided in Box 1.2.

Several chapters in this book use the concept of Scaling Readiness to characterize different innovation packages (Fig. 1.10). "Innovation readiness" refers to the demonstrated capacity of an innovation to fulfill its contribution to development outcomes in specific locations. This concept is presented in nine stages showing progress from an untested idea to a fully mature and proven innovation. "Innovation use" indicates the level of use of an innovation or innovation package by the project members, partners, and society. This concept captures progressively broader levels

Box 1.2 Key Definitions for Innovation and Scaling

Innovations are the new ideas, products, services, and solutions capable of facilitating impact through innovation systems involving multiple partners and enablers.

Innovation systems are the interlinked sets of people, processes, assets, and social institutions that enable the introduction and scaling of new ideas, products, services, and solutions capable of facilitating impact.

Scaling of innovations is a deliberate and planned effort to enable the use of innovations to have positive impact for many people across broad geographies.

Impact is a durable change in the condition of people and their environment brought about by a chain of events to which research, innovations, and related activities have contributed.

Scaling strategy is a set of coherent activities, stakeholders, and stakeholder engagement models to enable scaling.

Innovation package is the combination of innovations that are needed for scaling in a specific location or context.

Scaling approach is an integrated set of scaling tools and procedures that can be used to design and implement scaling activities in different contexts.

Innovation readiness refers to the demonstrated capacity of an innovation to fulfill its contribution to development outcomes in specific locations. This is presented in nine stages showing progress from an untested idea to a fully mature proven innovation.

Innovation use indicates the level of use of the innovation or innovation package by the project members, partners, and society. This shows progressively broader levels of use beginning with the intervention team who develops the innovation to its widespread use by users who are completely unconnected with the team or their partners.

Scaling Readiness of an innovation is a function of innovation readiness and innovation use. Table 1.5 provides summary definitions for each level of readiness and use adapted from Sartas et al. (2020), which have been used throughout the book. Scaling Readiness also is the name of the approach to scaling described in Chap. 3.

Table 1.5 *Definition* of levels of innovation readiness and use (Sartas et al. 2020)

Stage	Innovation readiness	Innovation use
1	Idea	Intervention team
2	Basic model (testing)	Direct partners (rare)
3	Basic model (proven)	Direct partners (common)
4	Application model (testing)	Secondary partners (rare)
5	Application model (proven)	Secondary partners (common)
6	Application (testing)	Unconnected developers (rare)
7	Application (proven)	Unconnected developers (common)
8	Innovation (testing)	Unconnected users (rare)
9	Innovation (proven)	Unconnected users (common)

Responsible scaling requires ethics of co-responsibility for ensuring that the impacts from the innovation are well captured by the intended beneficiaries and minimize negative social consequences, whether these impacts are intentional or not, and whether they can be fully foreseen or not. One crucial dimension of responsible scaling is gender equity, which can be achieved through gender-responsive research. This idea was explicitly included in much RTB research and is featured in several chapters of this book as a cross-cutting topic (see Box 1.1 for definitions used in the RTB program and this book).

Source: CGIAR 2020. Scaling Brief #4: Scaling glossary. Bonn: Deutsche Gesellschaft für Internationale Zusammenarbeit: https://hdl.handle.net/10568/110632

Fig. 1.10 Scaling Readiness barrel to illustrate how innovation(s) with the lowest readiness limits an innovation package's capacity to achieve impact at scale. (Adapted from "von Liebig's barrel," after Whitson and Walster, 1912; published in Sartas et al. 2020)

of use beginning with the intervention team who develops the innovation to its widespread implementation by users who are completely unconnected with the team or their partners. "Scaling Readiness" of an innovation is a function of innovation readiness and innovation use. Table 1.5 provides summary definitions for each level of readiness and use adapted from Sartas et al. (2020).

Many innovations fall short in adoption and value creation because they lack understanding about what drives a user/beneficiary/customer to choose one good or service over another. The "jobs to be done" (JTBD hereafter) theory, developed by the late Clayton M. Christensen and colleagues at Harvard University, provides a compelling perspective about why people buy or decline what companies try to sell them, or why beneficiaries from organizations involved in international development sometimes refuse to adopt what is offered or delivered. The JTBD theory significantly increases the ability to predict the likelihood of innovation success, which is a welcome addition to the toolbox of innovators. Its main tenet is that users do not purchase goods or services. Instead, people bring things into their lives to do a specific job or to achieve progress toward a particular goal under specific circumstances. Therefore, the goods and services that users acquire are unconscious means for filling satisfaction gaps. Unless we understand the "jobs" that people want to fill, the likelihood of innovation success is severely undermined, regardless of budget size or innovator skill. Every "job" exists within a given context—the where, when, who, or what of a situation. As just one example, a female smallholder in SSA tries to provide healthy, affordable food to her family. How can RT&B crops meet that need for her?

JTBDs can be quite different from solutions, and it helps to understand why well-funded innovation efforts run by very talented people oftentimes fall short of expectations and adoption, or why their scaling up stalls. For instance, many providers of MP3 reproducers focused on the presumed solution, namely, providing music. In contrast, Apple aimed the iPod at helping customers listen to music through a seamless experience. They reconsidered the whole business around personal music management, enabling customers to acquire, organize, listen to, and share music. In addition, Apple considered social and emotional factors involved in why people want to listen to music on the go and share the experience with others. Despite being a late entrant, Apple took over the market, beating formidable competitors such as Sony and Windows, both of which had more resources, larger teams, and much more cash available.

By far, the most valuable insight from the JTBD theory is that any job has not one but rather three components:

- *Functional*, which addresses practical, technical aspects
- *Social*, which addresses behavioral aspects of users, as well as how the users take into account the effects that the action will have on those around them
- *Emotional*, which relates to deep, personal aspects of users such as feelings, sentiments, aspirations, and desires

Regrettably, since most innovation efforts are driven by their functional components from their onsets, their likelihood of success is in jeopardy since the social and emotional components are the ones that represent a strong, if not the strongest, driver of user preferences and subsequent adoption. Within the agricultural domain, innovations solely focused on productivity or technological progress but which fail to account for cultural and social aspects are equally in jeopardy. More detailed accounts about the JTBD theory and its use in agricultural contexts can be found in Christensen et al. (2016) and Campos (2021).

1.5 Layout of the Book and Key Lessons to Enable Effective Innovation in RT&B Seed Systems

The body of this book is organized by four broad topics:

1. Institutional change and scaling
2. Processing and marketing
3. Productivity
4. Improving livelihoods

In each of these sections, we identify key lessons for designing and enabling effective innovation efforts in RT&B crops.

1.5.1 Institutional Change and Scaling

In the first section on institutional change and scaling, Chap. 2 looks at the constituent institutional innovations that underpinned the RTB program and how they added value to the combined achievements of the participating centers and their partners in science and research for development outcomes (which are described in more detail in the consequent chapters). The 10-year RTB program raised the profile of RT&B crops with investors within the research and development community. Dedicated funding expanded the frontier of knowledge around several of the key constraints to expanded utilization intrinsic to vegetatively propagated crops, particularly in the areas of breeding, seed systems, pests and diseases, and postharvest utilization. Within the domain of breeding, RTB excelled at providing tools and knowledge for breeders to acquire the skills necessary to develop product profiles (a set of traits and specifications to inform breeding efforts) in such a way that improved varieties are released with a greater likelihood of adoption by smallholder farmers.

Progress by RTB was facilitated by an array of institutional innovations for collaborative governance and management, including stakeholder consultations and priority setting, a portfolio organized by aggregated innovations (i.e., clusters of activities), articulated flagship projects, incentive funds, a dynamic interactive communication capacity, programmatic embedding of strategic and integrated gender research, and purposive engagement of national partners. Overall, RTB's design, governance, and management innovations added value to the combined achievements of the participating centers in science and research for development outcomes that are described in many of the book chapters.

Another key learning evident from the book is that there were limitations in the degree to which these institutional innovations could be "scaled down," from the program level to the implementation of projects and activities. This limitation was, in part, a reflection of the predominance of bilateral funding with multiple donors under center management, which made "scaling down" outside of earmarked funding/scaling projects challenging.

RTB also pioneered support and investment in the development and implementation of scaling strategies and partnerships to catalyze the scaling of its innovations. This body of work and linked research supported the development of the Scaling Readiness approach, which is widely relevant beyond RT&B crops. Chapter 3 covers the principles and practices of Scaling Readiness, which assesses the readiness of innovations to be scaled and supports the development, implementation, and monitoring of scaling strategies at project and organizational levels. The chapter explores how the approach enabled different RT&B innovations to move to scale. The innovations described in Chaps. 4, 5, 6, 10, and 12 are described in relation to Scaling Readiness and benefited directly from RTB investments in a Scaling Fund and the application of the Scaling Readiness approach. Chapter 17, meanwhile, used the Scaling Readiness framework to assess progress in scaling biofortification across multiple RT&B crops. A key lesson from this work is that a focused scaling approach adds value particularly when dedicated funding is available. However, these investments are effective only if tested guidelines and capacity development with specialized roles for scaling in the team are also available.

1.5.2 Processing, Marketing, and Distribution

The second section addresses innovations in processing, marketing, and distribution of RT&B crop products. The RTB program had a dedicated flagship for adding value to the processing of RT&B crops and enhancing value chains, with targeted investments through earmarked funding and scaling fund grants (as described in Chap. 2). This idea was built on the insight that RT&B value chains could make more nutritious food more cheaply and widely available in cities and rural markets by improving fresh supply, by developing new RT&B food products, and by supporting different typologies of small- and medium-scale processors.

Chapter 4 examines the scaling of innovations to improve the overall efficiency of small-scale cassava industries for rural small-scale processors to access more distant and higher-value markets (Fig. 1.11). Chapter 5 investigates sweetpotato puree as a novel ingredient in the food industry, developed as an innovative value chain product to address many constraints of sweetpotato adoption and use, while delivering a highly nutritional ingredient for processed bakery goods for rural and urban populations. Together, these chapters show how investments in cassava and sweetpotato processing addressed the challenges of limited shelf life for these two crops and how they overcame bottlenecks in their conversion to value-added products.

Fig. 1.11 Toasting grated and fermented cassava is the final step in creating gari and is a task mainly done by women. (Photo credit: H. Holmes (RTB))

In the first case, improving the energy efficiency of drying cassava into flour and starch through the development and scaling out of flash dryers led to 10–15% gains in productivity and incomes for small- and medium-scale processors. In the second case, development of orange-fleshed sweetpotato puree and developments in the value chains from field to market enabled bakeries in Ghana, Kenya, Malawi, and Rwanda to substitute puree for 20–50% of the wheat flour used in cookies, donuts, and bread by some commercial bakeries. By reaching urban markets, demand for OFSP was created and nutritional outcomes improved as a result. Between the two cases, we note that cassava processors could use locally manufactured equipment while OFSP puree required imported equipment. This difference has implications for the scaling strategies and long-term sustainability—two topics that merit further analysis.

Chapter 6 looks at the conversion of waste cassava peels into a valuable animal feed ingredient while also addressing serious environmental issues created by the millions of tons of waste cassava peels generated during processing. This chapter shows how it is possible to build on existing technologies and capacities to add value by processing waste cassava peels into animal feed. However, various "soft" components of scaling needed attention here as they formed the main bottlenecks to expanding the processing of high-quality cassava peels (HQCP).

Chapters 7 and 8 look beyond the innovations in cassava of the RTB program by focusing on cassava-related technology transfer between Brazil and Africa. Chapter 7 shows that for existing and well-developed processing technologies to be transferred to potential new users, the capacity, motivation, and training of the targeted users will determine the uptake and success. Chapter 8 notes the need for renewed attention on food safety needs as processing acquires greater importance in developing countries, especially countries in Africa. While this group of chapters record technological progress, they also emphasize the need for research on processing innovations to focus on women and men operators and entrepreneurs and to avoid getting lost in the details of machinery, products, and processes. In other words, RT&B crops require more study along this theme with attention to gender in the design, adaptation, and scaling of processing technologies.

1.5.3   Enhancing Productivity

The third section addresses innovations to enhance productivity, especially around pest and disease management and the challenges of seed systems for vegetatively propagated crops.

Pests and diseases are major production constraints in vegetatively propagated crops such as RT&Bs. Chapter 9 documents how digital technologies have been developed to identify and monitor emerging pests and diseases, looking at their future potential but also the challenges, limitations, and innovative approaches taken to reach end users, particularly smallholder farmers. This work illustrates that cohesive engagement between digital tool developers, researchers, extension and advisory services, and farmers is required to define specific problems and opportunities for developing suitable digital solutions. Moreover, the use of digital tools is still in its infancy with RT&B cropping systems; thus, opportunities are available for innovation for RT&B farming systems.

Chapter 10 looks at the process, practices, challenges, lessons learned, and future policy implications associated with scaling banana Xanthomonas bacterial wilt (BXW) management practices—an endeavor that has enabled the recovery of substantial banana production areas. This work provides evidence of the effectiveness of single diseased stem removal (SDSR) to manage and control BXW (Fig. 1.12). Nevertheless, the efforts to reduce the labor intensity of the SDSR package are equally important. Action plans for training and information dissemination related to SDSR must consider gender aspects at the onset of interventions if maximum participation is to be reached with women and youths. In particular, the SDSR case in Rwanda suggests the importance of engaging policy makers at all stages of technology innovation to secure their buy-in for subsequent scaling to the intended users.

Chapters 11,12, 13, 14, and 15 expand on the work to strengthen and enable seed systems for increased access to improved varieties and for encouraging improved quality in places where informal seed systems still predominate. As seed multiplication and dissemination for RT&B crops face common and difficult constraints, RTB developed a toolkit to understand, assess, and support the design of VPC seed system interventions to improve their effectiveness. Chapter 11 reviews researcher experiences in using these tools among different crops and in different regions. Their reflections provide insights on how the toolbox has influenced our understanding of seed system performance. Work with the RTB toolbox also shows how its success depended on attracting and retaining the interest of public and private

Fig. 1.12 Application of the novel SDSR technique to control BXW: a woman in eastern DR Congo cuts a diseased stem at soil level. (Photo Credit: B. Van Schagen (Alliance Bioversity-CIAT))

sector actors to diagnose and improve VPC seed systems. Assembling the tools in one toolbox has made them more accessible, provided an intuitive structure for uninitiated users, and helped to clarify which tools and combinations of tools are most useful for addressing different types of challenges.

Chapter 12 looks at the Triple S (*S*torage in *S*and and *S*prouting) strategy to conserve and multiply sweetpotato planting material during long dry seasons, exploring evidence from Ethiopia and Ghana regarding the various communication channels used to influence the uptake of Triple S among male and female farmers. Chapter 13 examines the development and implementation of new early generation potato seed technologies, such as aeroponics and apical cuttings in East Africa, and their contribution to the dissemination of improved potato varieties. The transformation of yam seed systems is covered in Chap. 14, where the scaling of high-quality seed production technologies in seed companies in partnership with national regulatory agencies is delivering basic and certified seed yam tubers to seed entrepreneurs—a chain of activity that has resulted in localized, scalable deployment centers. The development of commercially sustainable cassava seed systems in the vast informal seed systems of Nigeria and Tanzania is explored in Chap. 15, with key lessons drawn for the further development of commercial cassava seed systems in Africa (Fig. 1.13).

Fig. 1.13 A village seed entrepreneur sells bundles of cassava stems as part of a market promotion by the BASICS program in Benue State, Nigeria (CRS)

Across the different seed system innovations, it became evident that partners have different capacities and capabilities that determine their role in contributing to the scaling process. In addition, several enabling factors related to policy, infrastructure, mindsets and behavior change, and marketing and value chains are critical for scaling innovation packages. Availability of quality seed and opportunities for commercialization were shown to have important gender implications. Continued work to understand and design gender-responsive communication channels in different sociocultural contexts is required.

Chapter 16 closes the section by exploring how to build demand-led and gender-responsive breeding programs and looks at the piloting of practical tools that enable plant breeders to examine gender differences in the targeted population of users and with regard to trait prioritization (Fig. 1.14). This chapter underscores the importance of having a good foundation for gender-responsive breeding, created by cultivating gender awareness though cross-disciplinary dialogue, gender training targeted to breeding programs, and the incorporation of a gender specialist into breeding teams.

1.5.4   Improving Livelihoods

The fourth and final section looks at improving livelihoods with biofortified RT&B crops playing a major role as an affordable source of key micronutrients in an emerging and more resilient food system in SSA. Chapter 17 examines the scaling and impacts of biofortified RT&B crops and how the lessons learned from these efforts can help illuminate pathways for having biofortified RT&B crops contribute to building back a better, climate-smart, and more nutritious food system in the

Fig. 1.14 Poundability is an important trait for West African women who pound boiled yam for traditional dishes. Portioning pounded yam, Bouaké, Côte d'Ivoire. (Photo credit: D. Dufour (CIRAD))

post-pandemic world. RT&B crops are well-positioned to move forward in the context of emergency recovery as well as gender-responsive food system transformation. As scaling efforts expand, managing the perishability and the seasonality of RT&B crops at larger scales will require greater investment in physical market chain infrastructure, storage, information systems, and enterprise development. Moreover, given the poor families' heavy dependence on staple foods, increased commitment to breeding for enhanced micronutrient and protein content is warranted (Fig. 1.15). Advocacy by the Scaling Up Nutrition movement and the recognition of biofortification by different African governments are two examples of the kind of policy engagement needed to keep biofortification and nutrition at the forefront of food policy and investment planning.

1.6 What Next for RT&B Crops and the Smallholder Farmers Relying on Them

We are writing this book during unprecedented change in the CGIAR at a time when the COVID-19 pandemic is deeply, and quickly, changing the way communications and work are done. The RTB program that generated much of the research reported

Fig. 1.15 Potatoes that are rich in iron can help alleviate the suffering of women and children who suffer from anemia. Bertha Azursa Clemente and her son with biofortified potatoes in Huancavelica, Peru. (Photo credit: S. Fajardo (CIP))

here will be transitioning into a new set of collaborative arrangements under the One CGIAR. So the book is both a reflection of what the program achieved and also a contribution to future research endeavors. We are optimistic that the increased attention, research investment, and innovations achieved bode well for the future, ultimately enhancing the livelihoods of the many millions of women and men involved in their production, processing, and consumption in RT&B food systems in developing countries.

References

Amah D, van Biljon A, Brown A, Perkins-Veazie P, Swennen R, Labuschagne M (2018) Recent advances in banana (Musa spp.) biofortification to alleviate vitamin A deficiency. Crit Rev Food Sci Nutr 59(21):1–39. https://doi.org/10.1080/10408398.2018.1495175

Amoros W, Salas E, Hualla V, Burgos G, De Boeck B, Eyzaguirre R, Zum Felde T, Bonierbale M (2020) Heritability and genetic gains for iron and zinc concentration in diploid potato. Crop Sci 60:1884–1896. https://doi.org/10.1002/csc2.20170

Asindu M, Ouma E, Elepu G, Naziri D (2020) Farmer demand and willingness-to-pay for sweetpotato silage-based diet as pig feed in Uganda. Sustainability 12(16):6452. https://doi.org/10.3390/su12166452

Burgos G, Zum Felde T, Andre C, Kubow S (2019) The potato and its contribution to human diet. In: Campos H, Ortiz O (eds) The potato crop. Springer International Publishing, Cham, pp 37–74. https://doi.org/10.1007/978-3-030-28683-5

Campos H (2021) The innovation revolution in agriculture. Springer International Publishing, Cham, 234 p. https://doi.org/10.1007/978-3-030-50991-0

CGIAR (2020) Scaling Brief #4: scaling glossary. Deutsche Gesellschaft für Internationale Zusammenarbeit, Bonn. https://hdl.handle.net/10568/110632

Chapuis A, Precoppe M, Méot J, Sriroth K, Tran T (2016) Pneumatic drying of cassava starch: numerical analysis and guidelines for the design of efficient small-scale dryers. Dry Technol. https://doi.org/10.1080/07373937.2016.1177537

Christensen CM, Hal T, Dillon K, Duncan DS (2016) Know your customers' "jobs to be done". Harv Bus Rev 96:374–382

Doss C, Njuki J, Mika H (2020) The potential intersections of Covid-19, gender and food security in Africa. AgriGender: J Gend Agric Food Secur 5(1):41–48. https://doi.org/10.19268/JGAFS.512020.4

FAO (2014) Innovation in family farming. FAO, Rome. http://www.fao.org/publications/sofa/2014/en/

FAOSTAT (2020) Crops database. http://www.fao.org/faostat/en/#data/FBS. Accessed 22 Dec 2020

Heck S, Campos H, Barker I, Okello JJ, Baral A, Boy E, Brown L, Birol E (2020) Resilient agri-food systems for nutrition amidst COVID-19: evidence and lessons from food-based approaches to overcome micronutrient deficiency and rebuild livelihoods after crises. Food Secur 12:823–830. https://doi.org/10.1007/s12571-020-01067-2

Ilona P, Bouis HE, Palenberg M, Moursi M, Oparinde A (2017) Vitamin A cassava in Nigeria: crop development and delivery. Afr J Food Agric Nutr Dev 17(2):12000–12025. https://www.ajfand.net/Volume17/No2/Biofortification%20Issue%20Chapter%209.pdf

Jongstra R, Mwangi MN, Burgos G, Zeder C, Low JW, Mzembe G, Liria R, Penny M, Andrade MI, Fairweather-Tait S, Zum Felde T, Campos H, Phiri KS, Zimmermann MB, Wegmüller R (2020) Iron absorption from iron-biofortified sweetpotato is higher than regular sweetpotato in Malawian women while iron absorption from regular and iron-biofortified potatoes is high in Peruvian women. J Nutr 150(12):3094–3102. https://doi.org/10.1093/jn/nxaa267

Kennedy G, Raneri J, Stoian D, Attwood S, Burgos G, Ceballos H, Ekesa B, Johnson V, Low JW, Talsma EF (2019) Roots, tubers and bananas: contributions to food security. In: Encyclopedia of food security and sustainability, vol 3. Elsevier, Oxford, pp 231–256. https://doi.org/10.1016/B978-0-08-100596-5.21537-0

Lebot V (2020) Tropical root and tuber crops: cassava, sweet potato, yams and aroids. CABI, United Kingdom. https://www.cabi.org/bookshop/book/9781789243369/

Low JW, Thiele G (2020) Understanding innovation: the development and scaling of orange-fleshed sweetpotato in major African food systems. Agric Syst 179(102770):15. https://doi.org/10.1016/j.agsy.2019.102770

Malik AI, Kongsil P, Nguyễn VA, Ou W, Srean P, Becerra López-Lavalle LA et al (2020) Cassava breeding and agronomy in Asia: 50 years of history and future directions. Breed Sci 70(2):145–166. https://doi.org/10.1270/jsbbs.18180

Margolies A, Buckingham E (2013) The importance of gender in linking agriculture to sustained nutritional outcomes. In: Agriculture and Nutrition Global Learning and Evidence Exchange (AgN-GLEE), Joint USAID & SPRING conference, Guatemala City, Guatemala from 5–7 March. https://www.spring-nutrition.org/sites/default/files/2.5-a_the_importance_of_gender_amy_margolies_elizabeth_buckingham.pdf

Mudege NN, Chevo T, Nyekanyeka T, Kapalasa E, Demo P (2015) Gender norms and access to extension services and training among potato farmers in Dedza and Ntcheu in Malawi. J Agric Educ Ext 22(3):291–305. https://doi.org/10.1080/1389224X.2015.1038282

Petsakos A, Prager SD, Gonzalez CE, Gama AC, Sulser TB, Gbegbelegbe S, Kikulwe EM, Hareau G (2019) Understanding the consequences of changes in the production frontiers for roots, tubers and bananas. Glob Food Sec 20:180–188. https://doi.org/10.1016/j.gfs.2018.12.005

Prain G, Naziri D (2020) The role of root and tuber crops in strengthening agri-food system resilience in Asia. A literature review and selective stakeholder assessment. International Potato Center, Lima. isbn:978-92-9060-539-3, 66 p. https://cgspace.cgiar.org/handle/10568/106669

Rosegrant MW, Kulakow P, Mason-D'Croz D, Cenacchi N, Nin-Pratt A, Dunston S, Zhu T, Ringler C, Wiebe K, Robinson S, Willenbockel D, Xie H, Known H-Y, Johnson T, Walker TS, Wimmer F, Schaldach R, Nelson GC, Willaarts B (2017) Quantitative foresight modeling to inform the CGIAR research portfolio. Washington, International Food Policy Research Institute (IFPRI)

Sarapura S (2012) Gender analysis for the assessment of innovation processes: the Case of Papa Andina in Peru. Agricultural innovation systems: an investment sourcebook. World Bank, Washington, DC, pp 211–230. 598–602. http://documents1.world-bank.org/curated/pt/140741468336047588/pdf/672070PU-B0EPI0067844B09780821386842.pdf

Sartas M, Schut M, van Schagen B, Thiele G, Proietti C, Leeuwis C (2020) Scaling readiness: concepts, practices, and implementation. CGIAR Research Program on Roots, Tubers and Bananas (RTB). January 2020, 217 p. https://doi.org/10.4160/9789290605324

Scott G (2020) A review of root, tuber and banana crops in developing countries: past, present and future. Int J Food Sci Technol 56:1093–1114. https://doi.org/10.1111/ijfs.14778

Thiele G, Khan A, Heider B, Kroschel J, Harahagazwe D, Andrade M, Bonierbale M, Friedmann M, Gemenet D, Cherinet M, Quiroz R, Faye E, Dangles O (2017) Roots, tubers and bananas: planning and research for climate resilience. Open Agric 2(1):350–361. https://doi.org/10.1515/opag-2017-0039

Thiele G, Dufour D, Vernier P, Mwanga RO, Parker ML, Schulte Geldermann E, Teeken B, Wossen T, Gotor E, Kikulwe E, Tufan H, Sinelle S, Kouakou AM, Friedmann M, Polar V, Hershey C (2020) A review of varietal change in roots, tubers and bananas: consumer preferences and other drivers of adoption and implications for breeding. Int J Food Sci Technol. https://doi.org/10.1111/ijfs.14684

United Nations, Department of Economic and Social Affairs, Population Division (UNDP) (2019) World Population Prospects 2019, Online Edition. Rev. 1. https://population.un.org/wpp/Download/Standard/Population/

Wiebe K, Sulser TB, Dunston S, Rosegrant MW, Fuglie K, Willenbockel D, Nelson GC (2020) Modeling impacts of faster productivity growth to inform the CGIAR initiative on Crops to End Hunger. https://doi.org/10.31235/osf.io/h2g6r

Wossen T, Alene A, Abdoulaye T, Feleke S, Rabbi IY, Manyoung V (2018) Poverty reduction effects of agricultural technology adoption: the case of improved cassava varieties in Nigeria. J Agric Econ 70(2):392–407. https://doi.org/10.1111/1477-9552.12296

Chapter 2
Innovation Models to Deliver Value at Scale: The RTB Program

Helen Hambly (ID), **Michael Friedmann** (ID), **Claudio Proietti** (ID), **Vivian Polar** (ID), **Sarah Fernandes** (ID), **and Graham Thiele** (ID)

Abstract Collaborative programs that facilitate innovation to deliver value at scale require attention to effective program design, management, governance, and leadership. The CGIAR has experimented with different collaborative program design options over its 50-year history, most recently with the CGIAR Research Programs (CRPs) implemented from 2012 to 2021. This chapter examines the structure and processes of the CGIAR Research Program on Roots, Tubers and Bananas (RTB). It unpacks the constituent institutional innovations that underpinned the RTB program, their key design principles, how they evolved over the 10 years of the program, the innovations achieved, and the outcomes to which they contributed. Turbulence and transformations in the CGIAR system influenced the CRPs' emergence, design, and delivery. In this chapter, we discuss the RTB approach to collaborative governance and management as complex institutional innovations operating within this broader, dynamic system. This includes attention to opportunities, limitations, and other contextual factors influencing RTB's work. Institutional innovations include stakeholder consultations and priority setting, a portfolio organized by aggregated innovations, or clusters of activities, articulated flagship projects, incentive funding, a dynamic interactive communication ability, and programmatic embedding of strategic and integrated gender research. RTB's design, governance, and management innovations added value to the combined achievements of the participating centers in science and research for development outcomes, described in the following chapters.

H. Hambly (✉)
University of Guelph, Guelph, ON, Canada
e-mail: hhambly@uoguelph.ca

M. Friedmann · V. Polar · S. Fernandes · G. Thiele
CGIAR Research Program on Roots, Tubers and Bananas (RTB), led by the International Potato Center, Lima, Peru
e-mail: M.Friedmann@cgiar.org; V.Polar@cgiar.org; S.Fernandes@cgiar.org; g.thiele@cgiar.org

C. Proietti
Formerly CGIAR Research Program on Roots, Tubers and Bananas, currently CIRAD, Montpellier, France
e-mail: claudio.proietti@cirad.fr

© The Author(s) 2022
G. Thiele et al. (eds.), *Root, Tuber and Banana Food System Innovations*,
https://doi.org/10.1007/978-3-030-92022-7_2

2.1   Introduction

For 10 years, the CGIAR Research Program on Roots, Tubers and Bananas (RTB) advanced a significant body of scientific knowledge and practice on the clonally propagated staple food crops: banana (principally cooking banana and plantain), cassava, potato, sweetpotato, and yam. Root, tuber, and banana (RT&B) crops have been historically neglected, even until recently (Krishna Bahadur et al. 2018; Almekinders et al. 2019; Scott 2020). RTB's vision was to tap the potential of these crops across the world to improve food security, nutrition, income, and the climate change resilience of smallholders, especially women and youth. RTB brought together the work on RT&B crops across four CGIAR centers (Bioversity, CIAT, CIP, and IITA) and French research partners represented by CIRAD. It sought to strengthen the agri-food system of these crops in partnership with numerous research and scaling organizations. RTB was part of an integrated portfolio of CGIAR Research Programs (CRPs) established in 2012. This chapter explains how RTB developed an integrated set of institutional innovations shaped by a historical context of change in the CGIAR system.

In this chapter, we first briefly address the historical context of the CGIAR and then present the institutional memory of RTB from its emergence in the "New CGIAR" in 2011 to its transition within the "One CGIAR" (2022–2030). RTB had two 5-year phases of program design and management (2012–2016 and 2017–2021). In both phases, the institutional innovation by RTB was critical in developing adaptive capacity and resilience in the context of emerging priorities and resource disruptions. The final sections of the chapter draw lessons of broader relevance for designing international agricultural research for development programs. Highlighted examples of strategic and integrated science outcomes of the RTB program will be covered in more detail in subsequent chapters of this book.

2.2   Context

International agricultural research plays an important role in advancing agricultural science to improve the lives of billions of people in the world, especially the food insecure who face exigencies of poverty and environmental and climate crises. Since the 1970s, coordinating agricultural research for development and its coordination has changed substantially (Dalrymple 2008; Feldman and Biggs 2012; Immonen and Cooksy 2019). Linear models that implicate top-down problem-solving, for example, from scientists to farmers, have shifted toward demand-driven solutions. The inequitable distribution of the benefits of new technologies, for example, seed that is high yielding but requires costly inputs that disadvantage resource-poor farmers, has led to greater recognition of the socio-technical complexity of knowledge plurality and the need for bridging cross-sectoral and disciplinary boundaries (Pigford et al. 2018).

There are advantages to co-produced knowledge and co-innovation with diverse stakeholders in the agricultural system (Pant and Hambly 2009). Approaches to improving innovation have included participatory design process, reflective practice in science management, regional hubs and innovation platforms, learning alliances for scaling, adaptive collaborative management that strengthens local capacity, and partnership projects to foster collective knowledge assets (Horton et al. 2009; Ekboir 2009; Pigford et al. 2018; Berthet et al. 2018). Recently, the COVID-19 pandemic has drawn attention to the role of science and trust in expert knowledge, informing not only research in human diseases but also agri-food systems, climate change, and biodiversity loss (Barrett et al. 2020). The current transition known as One CGIAR (Coffman et al. 2020) is both a response to system changes and a contributor to innovation that can unleash the creativity needed to respond to shocks such as COVID-19 that prioritize donor funding to health.[1]

2.3 CGIAR Collaborative Programs

Created in 1971, the CGIAR is the world's largest group of publicly funded, agricultural research for development organizations. Its creation involved vested geopolitical interests and other powerful forces, creating the context that was to shape the CRPs (Hardin and Collins 1974; Baum and Lejuene 1986; Harwood and Kassam 2003; Ozgediz 2012; McCalla 2014, 2017; Byerlee and Lynam 2020). Early international agricultural research programs, largely funded by the Rockefeller and Ford Foundations, generated positive social returns on investment from partnerships that developed high-yielding crop varieties. From the 1960s onward, growth in and modernization of smallholder agriculture were predicated on new Green Revolution technologies (high-yielding crop varieties, fertilizers, and pesticides), guaranteed commodity prices, subsidized inputs, research, extension, training services, and infrastructure (e.g., irrigation). The evolving mandates of the United Nations Food and Agriculture Organization (FAO), the United Nations Development Programme (UNDP), and the World Bank generated influential global agricultural policy analyses and assessments that led to the CGIAR's creation (Feldman and Biggs 2012). Increased multidisciplinary scientific exchange, civil society debate, and structural adjustment policies of the 1970s–1980s combined to reassess the priorities of the Green Revolution, which was argued to have typically benefited wealthier, larger-scale, and male farmers. This led to greater attention to equity, including gender and interventions directed toward less well-off smallholder farmers (Glaeser 1987; Feldman and Biggs 2012).

By the early 1990s, the CGIAR's first four decades were strained by the complexity of taking on issues from the global agenda set out in the World Commission on Sustainable Development (WCSD). This included adding new CGIAR centers,

---

[1] See also https://www.cultivation.hps.cam.ac.uk/CGIAR-histories.

which grew from 4 in 1971 to 12 in 1979. Limited funding led to competitive stress and coordination challenges. The CGIAR centers had, up to the 2009 reform era, two forms of funding: unrestricted (referred to as core funding provided jointly by the World Bank and other donors) and restricted funds (bilateral funding from specific donors for projects). The former supported centers' management costs, allowing operational flexibility when new cost items emerged. The latter covered donor-specified costs in projects and centers. Unrestricted/core funding declined relative to restricted and special project funding, from upwards of 80% in the first decades of the CGIAR (Operations Evaluation Department 2004) to about 30% of the system funds by its fourth decade (Ozgediz 2012). Consequently, some centers faced financial difficulties, as they competed for project funds and had little appetite for the unfunded transaction costs of system-wide initiatives (McCalla 2014, 2017). Added to this was a degree of protectionist behavior in the system that resisted further merging or a reduction in the number of centers.[2]

New priorities reflected in the WCSD led the CGIAR to give increasing attention to the environment, biodiversity, and sustainability. Against this backdrop, reform sought a strategy of unification respecting centers' autonomy with large, coordinated multicenter research programs to tackle these global issues (Fig. 2.1). This led to the creation from 1993 of system-wide and eco-regional and CGIAR initiatives involving attention to greater interdisciplinary farming systems research, farmer participatory research, location-specific research, and new centers focused on natural resource management and biodiversity (Greenland 1997). By 2002, a mechanism to invite collaborative proposals on large, system cross-cutting research programs emerged. These 5 Challenge Programs (CPs) aimed to improve CGIAR business processes while shifting the consultative group of centers toward a

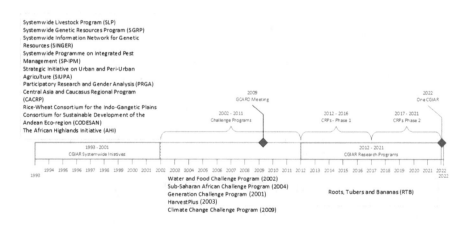

Fig. 2.1 Timeline for CGIAR collaborative programs with examples

---

[2] ILCA and ILRAD merged to become ILRI. INIBAP (banana and plantain) joined IPGRI (later Bioversity). ISNAR closed with some activities folded into IFPRI.

"consortium" model of setting system goals and regional priorities and enabling co-funding to reduce growing dependency on restricted funding (TAC/Science Council 2003).[3] One key element to engage partners in each CP was independent priority setting and management oversight bodies independent from the centers (Woolley et al. 2011).

This formative period from 1981 to 2010 prepared the ground for what would be the future cross-cutting CRPs (Fig. 2.1). Nevertheless, despite the desire for integration, reflected in new system-wide initiatives, the CGIAR center-based model of collaborative international agricultural research for development not only survived but also grew, with scientific work by the centers having a high impact.[4]

Seeking further changes toward more coherent responses to the United Nations' Sustainable Development Goals (SDGs) reduced fragmentation of the centers, and to avoid duplication of efforts among them and to optimize available funding, CGIAR funders and centers came together for a major reform effort in 2009. In 2010, delegates of the Global Conference on Agricultural Research for Development (GCARD), in Montpellier, France, helped shape a *Road Map to Transform the Agricultural Research for Development (AR4D) System for Greater Global Impacts* (GCARD 2011). The conference finalized a new Strategic Research Framework (SRF) for the CGIAR integrating the work of all the centers in a results-oriented research for development system with an integrated programmatic structure (Consortium Board 2011).

2.4 Design and Evolution of CRPs

The proposed building blocks of the SRF were a set of (initially) 15 interdependent CGIAR Research Programs (CRPs). The CRPs were envisaged as multicenter, interdisciplinary, and collaborative results-oriented programs whose impact was expected to be greater than the sum of their parts, because of the gains from synergies and system-wide cooperation (Consortium Board 2011). Donors responded favorably to the SRF, creating pooled funding through the mechanism of the "CGIAR Trust Fund," designed primarily to finance centers' collaboration and AR4D synergies (Renkow and Byerlee 2010) (Table 2.1).

---

[3] The World Bank initially hosted the CGIAR Secretariat, and the FAO hosted the Technical Advisory Council (TAC) in Rome, which held science impact, shared research service platforms, and other accountability functions. Under the consortium, the System Office was a virtual entity, and later the Consortium Office was based in Montpellier. TAC was renamed the Science Council retaining its advisory and evaluative functions.

[4] The impact of the CGIAR is summarized in two reports: at its 31-year milestone (Operations Evaluation Department 2004) and at 40 years (Ozgediz 2012). For example, the latter report stated that every dollar invested in CGIAR research meant $9 worth of additional food was produced in developing countries.

The Fund was intended to pledge stable, long-term financing for collaboration and integration of activities across CGIAR centers. W2 gave funders the discretion to invest in particular CRPs or platforms. However, an internal stabilization mechanism allowed for compensating allocations through W1 when there was a reduction in W2 allocation for any CRP. This created more stable funding from year to year but somewhat frustrated the funders' intentions to support those research areas they deemed of the highest relevance.

Initially, with the enthusiasm for the new reform, funder investments in the CGIAR rose markedly, climbing above $1 billion in 2014. However, as difficulties appeared, funding fell to $800 million in 2019. With this promising start, pooled funding (W1&2) reached 35% of total CGIAR funding but then fell sharply through 2015–2017 to around 20% of the total (Fig. 2.2). The CRPs would face challenges responding to funding-related turbulence and the late confirmation of funding within each financial year.

Even midway through SRF Phase 1 in 2015, there was a growing feeling among donors that the portfolio was too complex, inhibiting the desired collaboration. A decision was taken to close the system's CRPs and reduce from the original 15 to 11 CRPs (CGIAR 2015a, 2016). Building on strong prior collaboration with RTB, much of the CRP on Integrated Systems for the Humid Tropics (CRP-Humidtropics 2016) that was led by IITA, including the work in innovation platforms and their budgets, was integrated into RTB. This incorporation made RTB the largest Phase 2 CRP in the system portfolio by 2017.

The CGIAR portfolio was redesigned in preparation for a Phase 2 beginning in 2017. This standardized the programmatic structures for each CRP, based on flagship projects (FPs), and established a clear framework for cross-CRP integration. The design process had two stages: (a) pre-proposals and (b) full proposals, each evaluated by the Independent Science and Partnership Council (ISPC). This significantly strengthened CRP and system-level consistency across the CGIAR portfolio compared to Phase 1, with common design principles and carefully crafted theories of change to which the FPs contributed (ISPC 2012). The new portfolio was

Table 2.1 CGIAR funding windows, 2012–2021

Funding window	Purpose
Window 1 (W1)	Portfolio investments: funding allocated to the entire CGIAR portfolio of approved system-wide investments, prioritized and allocated by funders collectively through the System Council – supporting CGIAR as a whole
Window 2 (W2)	Program investments: funding allocated by funders individually to any component (e.g., CRP, Platform) of the system-wide portfolio as prioritized, defined and approved by the funders collectively through the System Council
Window 3 (W3)	Project investments: funding allocated by funders individually to projects that are defined by the funders themselves (with partners) and that are aligned with system-wide investments

Source: CGIAR website

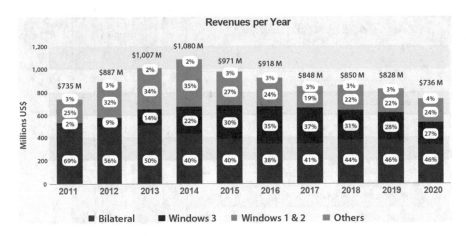

Fig. 2.2 CGIAR revenues per year

comprised of seven agri-food system programs and four interlocking globally integrating programs (Fig. 2.3).

2.5 RTB Program

The planning and evolution of the RTB program is part of the long history of CGIAR reform described above, comprising three phases of activity: a preparatory design phase (2009–2011), Phase 1 (2012–2016), and Phase 2 (2017–2021), summarized in Fig. 2.4.

During its design and Phase 1, RTB was conceived as one of the commodity CRPs, which aimed at modernizing crop breeding programs, creating synergies based on the commonalities of clonal crops and linking breeding to seed systems; these domains made up about 40% of total investment. In the design phase, a group of scientists from four CGIAR centers,[5] CIP (as the lead center), with Bioversity, CIAT, and IITA, met up to develop a scope of work and identify research components of the participating CGIAR centers (CRP-RTB 2011).

To design its program impact pathway and ensure consistency with the system-level outcomes, RTB sought to link relevant center activities at multiple levels, from farm households to wider production systems, and across national, international, and regional levels with diverse public and private sector stakeholders. A central part of the impact pathway focused on programmatic integration across CGIAR

---

[5] Following a scoping analysis, led by the RTB Steering Committee, other strategic science partners and CIRAD (a non-CGIAR center representing the French research partners) were subsequently invited to join the consortium.

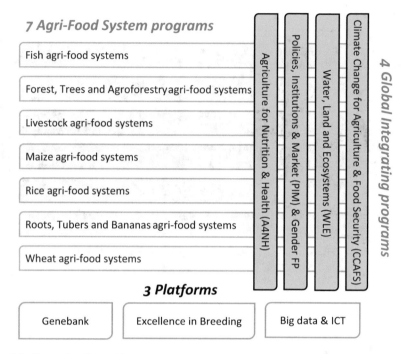

Fig. 2.3 Comprehensive and integrated CGIAR portfolio for Phase 2

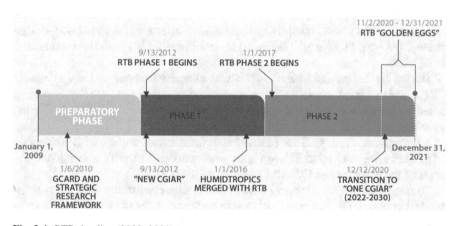

Fig. 2.4 RTB timeline (2009–2021)

core skills, adding value and capturing synergies across the five prioritized[6] crops (banana, cassava, potato, sweetpotato, and yam) and across the five research centers.

---

[6] RTB also included the minor root and tuber crops such as taro and several Andean crops. However, these were not prioritized for pooled funding despite their importance in some localities.

During implementation in Phase 1, new AR4D partnerships were created through RTB's strategic use of the W1/W2 funds. One outstanding example that began in the first phase was for work on metabolomics across all crops with Royal Holloway, University of London (Price et al. 2020). RTB also linked with other CRPs, in particular to Humidtropics, which was noted in its 2016 annual report:

> Tremendous success was made towards the incorporation and mainstreaming of components of Humidtropics research within some AFS-CRPs. The biggest success in this regard was in relation to partnership with CRP Roots, Tubers and Bananas (RTB) Agri Food Systems. RTB was open to partnership with Humidtropics and took on a number of research activities including Innovation Platforms and place-based systems research operations in a number of locations. Humidtropics contributed in the mainstreaming of systems approaches and the development of a Livelihoods Flagship within RTB.[7]

Against this backdrop of continuous reform, RTB developed as a complex institutional innovation that made purposeful change in governance and program management, drawing on "soft skills" (Woodhill 2010) such as communication, trust building, networking, and leadership with clear goal orientation. This enabled enhanced collaboration by cultivating and pursuing collective action against shared goals, which is highly challenging if institutions and organizations work alone (Roberts and Bradley 1991; Arena et al. 2017). Early on, RTB management recognized that collaborative practices were key to driving innovation and responding strategically to shifts in the external environment. The main institutional innovations that facilitated collaborative practices, described in this section, are the following:

1. Collective action in management, leadership, and associative governance
2. Stakeholder consultation and participatory design
3. Priority setting to guide investments and build/adapt the portfolio
4. Portfolio organized by aggregated innovations
5. Programmatic embedding of strategic and integrated gender research
6. Internal funding mechanisms and incentives
7. Dynamic interactive communication capability to build a shared vision and stakeholders' support
8. Purposive engagement of national partners

---

[7] See https://cgspace.cgiar.org/bitstream/handle/10568/89312/HUMIDTROPICS-Annual-Report-2016.pdf?sequence=1&isAllowed=y.

2.5.1   Collective Action in Management, Leadership, and Associative Governance

Early on, during the design phase, the centers involved in RTB opted for an associative style of governance. In terms of a "hard contract," CIP is the lead center, and its Board of Trustees holds fiduciary responsibility for RTB. Yet with an equally important "soft contract" approach, through the management committee, participating centers in RTB were given an equal weight in taking decisions with transparent access to information, particularly on finances. The different management and governance levels in RTB recognized that mechanisms were needed to support collective action, which could not be achieved by any of the centers acting separately (Horton et al. 2009). An associative style of governance was both a reason for and a result of collective action in RTB. Each participating center in RTB was an important contributor to the overall impact of clonally propagated crops, and bananas and cassava had mandates shared by two centers.

In Phase 1, RTB governance involved a steering committee comprised only of center DGs and a science advisory committee known as the PAC (Program Advisory Committee) whose membership included subject and regional specialists from Non-Governmental Organizations (NGOs), universities, and research institutes around the world (Fig. 2.5). The PAC played a role as part of the system of checks and balances by looking out for collective interests above those of each of the centers.

Operationalizing collaboration was facilitated by the full-time RTB Program Director (PD) supported by the Project Management Unit (PMU) in close communication with the RTB Management Committee composed of apex research managers in the participating CGIAR centers (usually Deputy Director General of Research). The PD led the Program Management Unit (PMU) that included five to seven full-time support staff responsible for grants, finance, communications, planning and reporting, and gender research, with a science officer included from 2015.

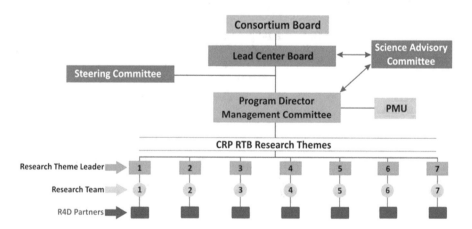

Fig. 2.5 RTB organogram Phase 1

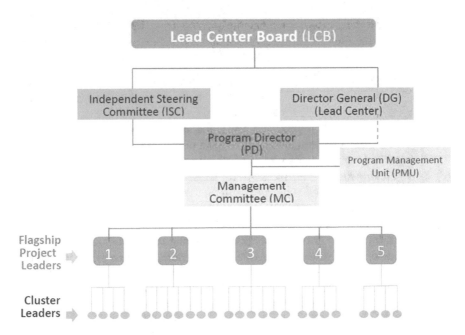

Fig. 2.6 RTB organogram for Phase 2 showing governance and oversight functions

The PD, guided by the Management Committee, oversaw FP leaders and the cluster leaders with a programmatic reporting line (Figs. 2.5 and 2.6). Reflecting the soft contract approach, FP and cluster leaders were drawn in a balanced way from all five participating centers.

During Phase 1, the RTB collaborative governance structure evolved further, following system guidelines for all CRPs, into the Independent Steering Committee (ISC, formerly the Program Advisory Committee) where the center DGs are represented, but most independent members are selected according to needed competencies for program oversight while ensuring some level of regional and gender representation. The ISC Terms of Reference specified that it had advisory input to the RTB to ensure oversight of the strategic alignment of the RTB with the SRF at crucial periods such as the design of CRP-level program proposals, draft annual Plan of Work and Budget (PoWB), and Annual Reports. The Chair of ISC reported annually to the CIP Board of Trustees, which had fiduciary responsibility for RTB. The Program Committee of the CIP Board received regular reports on program progress and approved the RTB annual PoWB. RTB developed a comprehensive and coordinated governance function with collaborative practices such as collective review of the PoWB and the review of FPs' progress and plans as a central feature of ISC annual meetings.

RTB's associative style of governance and collaborative management was further reinforced following a recommendation from the Independent Evaluation Arrangement (IEA) review (2016) to create an "alliance compact" as a trust-built, soft contract among the participating centers to bolster the legal agreements established between the lead center and the other centers. The ISC played a key role in translating the alliance compact into a specific set of partnership statements, signed by the DGs of the centers in 2017, as follows:

(a) *RTB Partnership Collaboration* – RTB is a shared asset that should be jointly promoted and nurtured in a collaborative way to support collective action and add value to members in the long term.
(b) *Inclusive Partnership* – Openness, trust, and mutual respect and learning lie at the core of effective partnerships in support of the RTB goals, recognizing the different and complementary roles of all members.
(c) *Strengthening Business Partnerships* for RTB for cross-cutting, multi-crop, collective action research for development that would not be possible by each of the members acting separately.
(d) *Donor Relations* – Promoting joint stewardship to maintain the engagement of the existing set of Window 2 donors and sharing responsibilities for resource mobilization intelligence for potential new Window 2 funding for RTB. Centers agree to pursue a policy of minimal reciprocal overheads for pass-throughs that relate to transfers across the members. For Window 3 or bilateral funding connected to the RTB Program, members agree to map the funding and results into the RTB Program so that the program as a whole could benefit.
(e) *Talent Management* that flows across centers for RTB management positions such as FP and cluster leaders is a key part of the overall compliance mechanism.
(f) *Communications* – Ensure that communication/public relations activities accurately reflected collaborative efforts and the contribution of each member.

In summary, this associative governance style supported major transitions within the first and second phases of RTB, including after 2016, the incorporation of parts of the Humidtropics CRP. The soft contract of the "alliance compact" recommended by IEA was acted upon, and a collective approach to management and governance was adopted.

2.5.2 Stakeholder Consultation and Participatory Design

Stakeholder consultation was an institutional innovation supporting the design process, particularly in the preparatory phase of RTB. There was a short timeframe between the start of proposal writing (late June 2010) and the deadline for submitting the proposal (the first week of September 2010); 255 stakeholders, about half from developing-country national agricultural research systems (NARS) and universities, were consulted using surveys. This resulted in an initial RTB structure with seven disciplinary themes, each with a mix of existing, expanded, and new product lines as well as cross-cutting activities. To create its initial structure, more than 25 researchers from Bioversity, CIAT, CIP, and IITA participated in a 3-day workshop, held at CIP's headquarters in 2010, to define and organize a strategy for developing the proposal for a CRP. Writing teams were formed across topics to

Table 2.2 RTB stakeholders' aggregated score of importance assigned to seven themes in the proposal

Theme	Regional survey	Global survey
Theme 2: Accelerating the development, delivery, and adoption of varieties with stable yields, stress resistance, and high nutritional value	4.60	4.55
Theme 6: Enhancing postharvest technologies and adding value in markets	4.58	4.22
Theme 4: Promoting sustainable systems for clean planting material for farmers	4.51	4.38
Theme 1: Conserving and accessing genetic resources	4.42	3.81
Theme 3: Managing priority pests and diseases and beneficial microbial communities	4.29	4.24
Theme 7: Enhancing impact through partnerships	4.33	4.00
Theme 5: Developing tools for more productive, ecologically robust crops	4.12	4.05

Note: Themes were scored on a 0–5 scale (0 meant "not important" and 5 was "very important")

encourage cross-center collaboration, and writing responsibilities were assigned. A proposal development schedule was developed and agreed upon, as was a protocol for writing, editing, and managing the draft sections. The inter-center workshop proposed seven core themes for RTB (Table 2.2).

The seven CRP-RTB themes were ratified by the stakeholders (Table 2.2). The stakeholder surveys during the design phase confirmed previous findings that RTB crops are generally absent from government rural development strategies (Woolley et al. 2011). Following the design workshop and stakeholder consultation, RTB was organized by these seven themes. Each theme was ordered by crop. Each theme leader was drawn from a different center, and each theme included cross-cutting work in gender, biophysical research, and processes and partnerships for scaling innovations.

By 2014, with the implementation of results-based management in the CGIAR, and more consideration of its theory of change, the PMU realized that the first thematic structure (labelled RTB 1.0) had limitations for creating a compelling and viable theory of change and for organizing science teams. RTB 1.0 neatly arranged work by different themes but lacked a process to bridge the different themes, e.g., connecting varietal development with seed system development for each crop.[8] Working groups in the 2014 RTB annual meeting in Kampala considered options for a new program structure (referred to as RTB 2.0). Before the start of Phase 2, which began in 2017, the RTB portfolio had been reorganized into five FPs and clusters of activities (CoA) that encompassed five crops, four CGIAR centers, and by this point a non-CGIAR center, CIRAD (CGIAR-IEA 2015a, b, 2016). The tran-

---

[8] The analogy was that RTB 1.0 was like a closet with separate groups of clothes on hangers, but no way to combine them into complete and matching outfits.

sition toward the RTB 2.0 structure, which replaced seven themes with five FPs, was gradually completed in 2016.

The proposal for Phase 2 (RTB 2016a) including a scope of work and a theory of change for each of the five FPs was prepared by design teams drawn from RTB scientists from each FP (see Sect. 2.5.5). Key staff from the Humidtropics CRP joined RTB on these writing teams, particularly around scaling innovation (FP5), which included researchers from Wageningen University & Research (WUR). Additional teams worked on cross-cutting topics of gender-responsive research, capacity development, and partnerships (Fig. 2.7). This proposal combining flagships and cross-cutting topics was reviewed and approved by the ISPC in a two-stage submission process. RTB received one of the highest ratings of any CRP by the ISPC (2016).

In summary, RTB 1.0 became RTB 2.0 by supporting the interaction between innovations, which is essential for wider adoption and scaling (see Sect. 2.5.4). At the same time, there was extensive CGIAR consultation during 2014–2016 that restructured all CRPs with shared design and management principles around FPs and clusters for Phase 2.

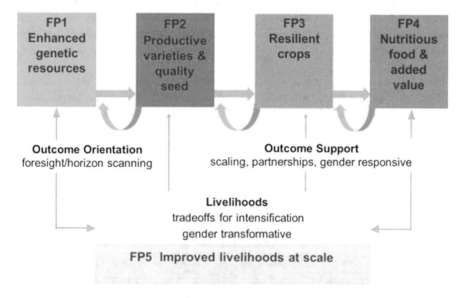

Fig. 2.7 Program structure organized by FPs, 2016–2021 – RTB 2.0

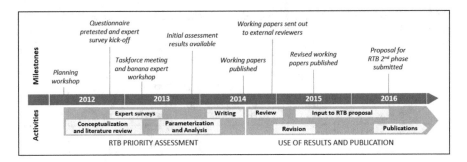

Fig. 2.8 Timeline and key activities of RTB priority assessment. (Source: Pemsl et al. 2022)

2.5.3 Priority Setting to Guide Investments and Build the Portfolio

Reflecting on the stakeholder consultations in 2010, RTB responded to a direct request from donors and the ISPC to conduct a rigorous priority assessment as a precondition for funding. The steps of the priority assessment method ran in parallel with the participatory design/redesign of the portfolio and the comprehensive framework, including the impact pathways described above (Fig. 2.8).

Priority setting in RTB covered five crops (cassava, banana, potato, sweetpotato, and yams) using a harmonized method that drew on the approach developed earlier at CIP (Fuglie and Thiele 2009). The assessment included four key steps: (1) elicitation of major production constraints and research opportunities through global and regional expert surveys; (2) identification of priority research interventions by crop; (3) ex ante estimation of costs and benefits for two adoption scenarios using partial equilibrium economic surplus models with poverty impact simulations; and (4) publication and use of findings. For the first step, to identify key constraints and research opportunities, 1709 experts were consulted, from a wide variety of disciplinary backgrounds, mostly scientists from national agricultural research organizations.

Specific research options were selected based on overall and regional scores from the expert survey (step 2) in consultation with experts and center scientists, considering the scope of RTB's research activities to ensure a good match of options with the program portfolio.

The results (Table 2.3) showed large benefits for all potential research investments and provided useful outcome and impact indicators (adoption area, number of beneficiaries, net present value, internal rate of return, and poverty reduction). In Table 2.4, results of computed performance indicators suggest two key findings: (1) expected adoption areas are large, ranging from several hundred thousand to almost four million hectares, which translates into high numbers of beneficiaries (estimated to be as high as 36 million persons), and (2) all options have large positive net present values. There was a high level of congruence between the research options of highest priority in the assessment and the crop-specific clusters in RTB's

Table 2.3 Results of RTB priority assessment – adoption, beneficiaries, economic benefits, and poverty impacts (lower adoption scenario)

Top ranked research options by crop	Adoption area (million ha)	Number of beneficiaries (million HH)	(million persons)	Net present value (NPV) (US$ million)	Internal rate of return (IRR) (%)	Poverty reduction (million persons)
Banana						
Banana Xanthomonas wilt (BXW) management cultural practices	0.64	3.22	15.67	1982	72	1.61.
Recovery from banana bunchy top virus (BBTV)	0.40	2.02	9.67	1337	61	0.64
Resistant plantain (RELEASE)	0.45	1.70	7.57	1111	64	0.25
Cassava						
High-quality planting material production and distribution systems for improved varieties	3.38	6.73	33.08	7585	416	2.10
Sustainable crop and soil fertility management	3.27	6.43	31.72	8284	210	2.66
High yielding, drought-tolerant varieties and increased water-use efficiency	3.99	7.89	36.49	3025	61	2.00
Potato						
Late blight resistance	0.77	6.73	33.08	7585	416	2.10
Virus-resistant varieties	0.36	6.43	31.72	8284	210	2.66
Bacterial wilt-resistant varieties	0.64	1.72	7.85	253	29	0.20
Sweetpotato						
Orange-flesh sweetpotato (OFSP)[a]	0.67	3.00	14.60	563	35	0.48
Weevil-resistant varieties	0.72	2.94	14.11	363	41	0.36
Sweetpotato virus disease (SPVD)-resistant varieties	0.48	1.96	9.41	673	116	0.34
Yam						
Clean planning materials and agronomic practices	0.68	2.39	17.72	570	37	0.18

(continued)

Table 2.3 (continued)

Top ranked research options by crop	Adoption area (million ha)	Number of beneficiaries (million HH)	(million persons)	Net present value (NPV) (US$ million)	Internal rate of return (IRR) (%)	Poverty reduction (million persons)
Improved varieties with complementary ICM	0.43	1.58	11.74	3026	60	0.66
Yam pest and disease management options	0.43	1.60	11.85	412	43	0.10

Source: Pemsl et al. (2022)
[a]Including health benefits from the adoption of OFSP (DALY method) substantially increases benefits: NPV: US$1298 million, IRR: 51% (lower adoption scenario)

Table 2.4 RTB portfolio organized by clusters or "innovation packages"

Discovery	Delivery		
FP1: Enhanced genetic resources	FP2: Productive varieties and quality seed	FP3: Resilient crops	FP4: Nutritious food and added value
DI1.1 Breeding CoP **DI1.2** Next generation breeding **DI1.3** Game-changing traits **DI1.4** Genetic diversity	**CC2.1** Access to quality seeds/varieties **BA2.2** User-preferred banana cultivars/hybrids **CA2.3** Added value cassava varieties **PO2.4** Seed potato for Africa **PO2.5** Potato varieties for Asia **SW2.6** User-preferred sweetpotato varieties	**CC3.1** Pest/disease management **CC3.2** Crop production systems **BA3.3** Banana fungal and bacterial wilts (Foc/BXW) **BA3.4** Banana viral diseases (BBTD) **CA3.5** Cassava biological constraints, Asia/Americas **CA3.6** Cassava biological threats, Africa	**CC4.1** Post-harvest innovation **CA4.2** Cassava processing **CA4.3** Biofortified cassava **SW4.4** Nutritious sweetpotato
FP5: Improved livelihoods at scale			
CC5.1 Foresight and impact assessment **CC5.2** Sustainable intensification and diversification for improvement resilience, nutrition, and income **CC5.3** Gender equitable development and youth employment **CC5.4** Scaling RTB agri-food system innovations			

programmatic structure for FPs 2, 3, and 4 where most delivery research is located (Pemsl et al. 2022). Hence, the exercise gave coherence to the selection of crop clusters for Phase 2, discussed below. The findings informed the development of RTB's research portfolio and were critical for facilitating continued program funding.

In summary, RTB responded to a direct request from donors and the ISPC to conduct a rigorous priority assessment as a precondition for funding. The entire exercise brought social scientists, breeders, agronomists, and other disciplines from different centers together around a shared task and method to provide consistent metrics across crops. Establishing such a community of practitioners was an institutional innovation, which provided a strong, collaborative base and set an example for cross-center cooperation.

2.5.4 RTB 2.0: Portfolio Organized by Aggregated Innovations with Linked Impact Pathways

After 2016, RTB 2.0 was used as a comprehensive framework for planning, implementing, reporting, and learning. FPs were organized into 25 clusters of activities comprising the full RTB CRP portfolio. The three types of clusters were the following:

- Discovery clusters that included the upstream work, feeding into the crop-specific clusters.
- Crop-specific delivery clusters generating direct impact
- Cross-cutting clusters that synthesized, linked, and supported work across the crop-specific clusters

The cross-cutting clusters are a key institutional innovation as they contribute to establishing communities of practice across different crops and centers. Few other CRPs had such clusters (Jill Lenné, personal communication).

Each cluster was designed with a lead or core innovation (lead product) and an array of complementary innovations (linked products). The clusters sought to include multidisciplinary expertise from the RTB themes. Each cluster was designed by a science team that was responsible for developing the innovations and preparing a business case for the cluster that demonstrated the value added (Table 2.4).

The lead innovation[9] is the centerpiece of a work package that also consists of complementary, linked, or enabling products and includes a theory of change with quantified indicators (RTB 2013b). A lead innovation was defined as:

1. A significant measurable and time-bound product (including knowledge, technology, and organizational and institutional models) that results from a research activity or set of related activities attributable to RTB
2. Used by a well-defined group of next users who may be researchers or development actors, with strong evidence of demand-pull from these users
3. Near market-ready set of ideas, technologies, or science products that generate excitement among researchers and other users
4. With potential for contributing to large-scale impact

RTB 2.0 recognizes that a single innovation cannot be used at scale on its own as it requires complementary innovations for broader use (Sartas et al. 2020). This is illustrated by the seed potato for Africa cluster (Fig. 2.9). "Business models for

---

[9]The lead innovation was referred to as a flagship product or flagship products. We use the term "lead innovation" here so as not to be confused with FPs.

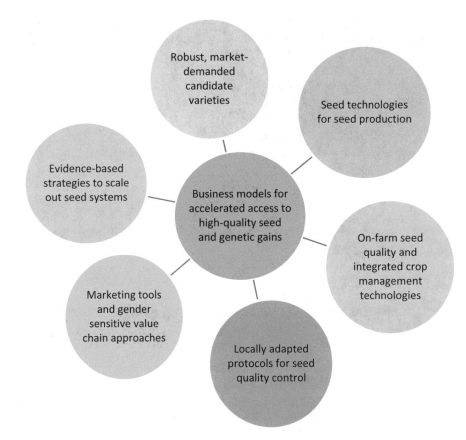

Fig. 2.9 Lead innovation and complementary innovations for the cluster of seed potato for Africa

fostering farmers' access to seed" was the lead innovation that could facilitate farmer access to market-demanded varieties and would require new seed production and quality management technologies, gender-sensitive approaches for value chain development, and scaling strategies. Thus, by organizing each cluster as aggregated innovations, connections could be made to the theory of change developed for the FP, potentially bringing together the bilateral projects mapped under the cluster of activities.

RTB 2.0 paid attention to impact pathways linked to the Intermediate Development Outcomes (IDOs) as set out in the CGIAR Strategic Research Framework (SRF) (CGIAR 2015b). Impact pathways were not drafted by one individual at a desk but were designed with the FP and cluster teams and articulated to the SRF IDOs. Where possible, RTB consulted with stakeholders, using participatory and reflective tools like pathway visualization (Fig. 2.10) to develop impact pathways (Fig. 2.11). During 2014 and 2015, as part of the piloting of CGIAR results-based management, RTB held workshops with a broad group of stakeholders for a detailed validation and co-construction of four cluster impact pathways. Some clusters identified

Fig. 2.10 Stakeholders co-constructed the impact pathway for potato quality seed in Africa during the workshop in Kenya in 2014. (Photo: G. Thiele)

quantified performance indicators to guide results-based management based on the impact pathway (Fig. 2.12). RTB did not, however, have sufficient resources to implement the monitoring and evaluation (M&E) system as initially planned with all of the cluster teams, considering the intensive convening and facilitation required.

RTB's 2.0 comprehensive framework also supported institutional innovations for monitoring and reporting. A major improvement introduced in Phase 2 was the use of the Monitoring, Evaluation, and Learning (MEL) software platform co-developed among several CGIAR centers. MEL allowed systematic tracking of progress and aggregation of results across the RTB and other CRP portfolios. Figure 2.13 illustrates the indicators and case studies included at the output, outcome, and impact level. MEL used a generic description of the program's scope of control (output), scope of influence (research outcomes with next users), and scope of interest (development outcomes with end users) developed by the CGIAR System Office. MEL enabled reporting through a CGIAR-wide dashboard.[10]

Initially, MEL had limitations for comprehensive results-based management. First, it did not show the progress of innovations moving from one state of readiness to the next in successive years. Second, it did not adequately capture the aggregated innovations within a cluster. Third, there was an issue of synchronicity, as the

---

[10] See the system dashboard at https://www.cgiar.org/food-security-impact/results-dashboard/.

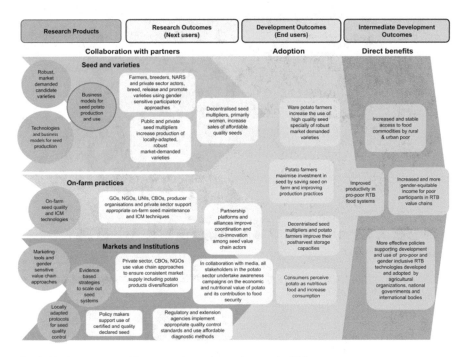

Fig. 2.11 Co-constructed impact pathway for seed potato in SSA

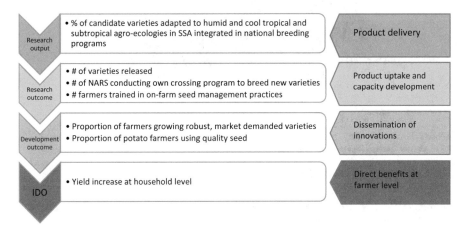

Fig. 2.12 Variables initially proposed for tracking along the impact pathway for seed potato in SSA

outcome cases and the adoption and impact studies reported in any year related to innovations developed earlier, sometimes much earlier. MEL was improved, and from 2018, in coordination with the CGIAR System Office, the readiness of each innovation was assessed, and the progress of the innovations was tracked as they moved from one level of readiness to the next. This permitted a visualization of

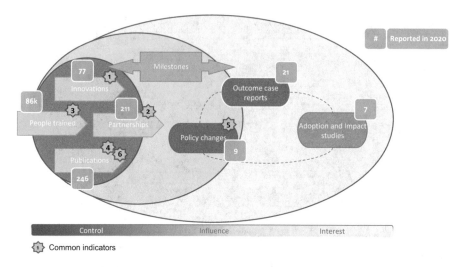

Fig. 2.13 Systematic reporting of deliverables, research outcomes, and development outcomes with MEL in 2020

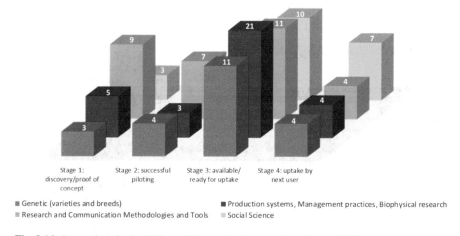

Fig. 2.14 Innovations in the RTB portfolio and their stage of readiness (2019)

aggregated innovations by type and level of readiness across the program and for each flagship (Figs. 2.14 and 2.15).

In summary, aggregated innovations made up the 25 clusters in the RTB portfolio in Phase 2. RTB's comprehensive framework for the portfolio also provided the basis for determining and linking groups of lead and complementary innovations. Aggregated innovations were embedded in a FP that included a theory of change with quantified indicators. Although RTB did not have sufficient resources to implement the M&E system as initially planned with all the cluster teams, where they were developed, co-constructed impact pathways helped to identify policy changes,

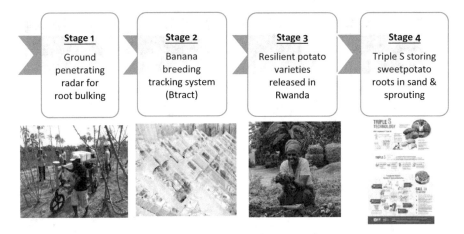

Fig. 2.15 FP 2: innovations reported (2017–2019)

outcomes, and milestones. The MEL software was effectively used in RTB monitoring and evaluation and adapted to better assess innovation readiness. This contributed to CGIAR-wide CRP reporting and performance assessment by the System Office.

2.5.5 Programmatic Embedding of Strategic and Integrated Gender Research

A key institutional innovation in RTB was to differentiate and embed two different, complementary types of research: (a) integrated gender research on specific technologies where gender scientists worked with biological and other social scientists to address specific gender constraints and opportunities for that technology and (b) strategic research addressing the knowledge gaps on how gender roles and norms affect the uptake of RTB technologies generally (Fig. 2.16).

During its first phase, RTB adopted a strategy to integrate gender in technical areas across the entire CRP and to conduct strategic gender research. The gender team comprised one gender specialist in each center, a gender coordinator based in the PMU, and part-time support from a senior gender specialist at CIP. The RTB gender team linked up with other gender scientists, in particular as key members of the CGIAR global study on Enabling Gender Equality in Agriculture and Natural Resource Management (GENNOVATE). RTB gender scientists wrote 15 case studies in target countries including Uganda (4), Malawi (2), Burundi (1), Nigeria (2), Colombia (4), Bangladesh (2), and Vietnam (2). GENNOVATE was a strategic research endeavor that explicitly sought to understand how household and community power relations and self-perceptions of personal power shape innovation decisions. A key learning from this strategic research was that when interventions do not

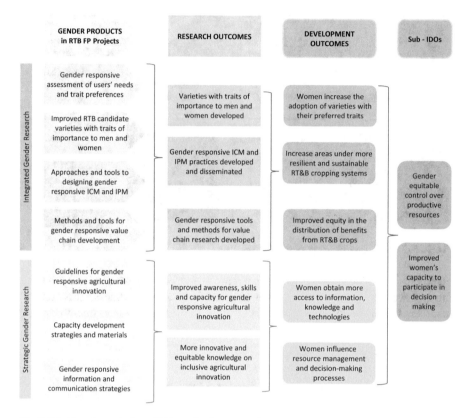

Fig. 2.16 Key elements from RTB gender impact pathway

consider underlying social structures and gender norms related to household decision-making and control of income, agriculture research may not benefit women.

Besides the strategic research in GENNOVATE, there was integrated gender research. For example, in Phase 1, a gender specialist in a seed potato project helped the team to realize that women did not have access to quality potato seed because they lacked access to credit and training and did not control household income to purchase seed (Mudege et al. 2015). Research on long shelf-life banana and potato ambient storage technology in Uganda reached similar conclusions.

During Phase 2, increased emphasis was given to integrated gender research supported by specific earmarked funding grants for gender research in some technological clusters. For example, in the breeding community of practice cluster, earmarked funding supported gender specialists and breeders to develop tools for gender-responsive breeding. Similarly, earmarked funding for gender research was provided to clusters working on seed systems and pest and disease management. By 2021, gender knowledge resources were organized in the Gender Responsive AR4D

Portal and shared with other CRPs its work on Gender Plus (G+) tools for gender-responsive breeding.

In summary, RTB adopted an approach to strengthen its gender research capacity, fostering feedback loops between strategic and integrated gender research. This required leadership of the RTB PMU, with support from the ISC, and earmarked funding directed to technological clusters and to the gender and youth cluster in FP5 that supported strategic research. This programmatic embedding generated concrete outputs to help address the challenges of gender and social inclusion for innovation design and scaling.

2.5.6 Internal Funding Mechanisms and Incentives

In Phase 2, total funding across all sources averaged around $80 million a year, making RTB the largest of all CRPs and the largest single program of any sort in the CGIAR. About 80% of RTB funding was managed directly by the centers (W3 and bilateral funding, or non-pooled funding[11]). Only about 20% was directly managed by RTB through contracts with CIP as the lead center (W1 and W2, or pooled funding[12]). This pooled funding was managed by the PMU, with the management committee and under the guidance of the ISC and, ultimately, the CIP Board of Trustees. Nevertheless, the proposals for both phases, the theories of change, and the milestones and deliverables were planned and reported for both pooled and non-pooled funding. Pooled funding was the smaller part of the overall budget, but it was carefully planned to add value and complement non-pooled funding to ensure the greatest impact.

Internal funding mechanisms for pooled funding evolved over time as PMU acquired more experience but broadly fell into three categories:

1. Earmarked funding (Table 2.5) represented about 31% of the budget allocated each year by RTB and was targeted to areas with the greater scope for synergistic value addition. This was related to multicenter investments that ordinarily were not covered through bilateral funding grants. Much of this was linked to multicenter work in the discovery and cross-cutting clusters. Additionally, some of the earmarked funding was linked to multicenter work in cassava and banana clusters where the mandate was shared between several centers. Each earmarked funding request was organized as a project with specific deliverables proposed in funding submission and then tracked over a 3-year period.

---

[11] Bilateral funding was where the donor did not cover the 2% charge made by the center to the system.

[12] Pooled funding was of two types: W1 funding, which was received from the system from donors contributing to all system entities with no restrictions, and W2 funding, which was awarded by donors specifically to RTB, sometimes with conditions on which flagship it should be used in.

Table 2.5 Earmarked funding awards in Phase 2 – 2017–2021

FP	Name of earmarked funding	Center/partners	Duration	Funding (USD, 000)
FP1	Breeding platform	Bioversity – CIAT – CIP – IITA	2017–2021	2184
FP1	Next-generation breeding	Bioversity – CIP – IITA – CIRAD	2017–2018	436
FP1	Game-changing traits	Bioversity – CIAT – CIP – IITA	2017–2021	1410
FP1	Genetic diversity	Bioversity – CIAT – CIP – IITA – CIRAD	2017–2021	1376
FP2	Quality seeds and access to improved varieties	Bioversity – CIAT – CIP – IITA – WUR	2017–2021	4343
FP2	GBI gender tool	IITA	2020	25
FP3	Pest/disease management	Bioversity – CIAT – CIP – IITA	2017–2021	3587
FP3	Crop production systems	Bioversity – CIAT – CIP – IITA	2017–2018	268
FP3	Banana fungal and bacterial diseases/ FOC/ BXW	Bioversity – CIAT – IITA – CIRAD	2017–2021	1355
FP3	Banana viral diseases/BBTV	Bioversity – IITA – CIRAD	2017–2021	1531
FP3	Cassava biological constraints, Asia/ Americas	CIAT – IITA	2019–2020	170
FP4	Postharvest innovation and nutrition improvement	Bioversity – CIAT – CIP – IITA	2017–2021	1128
FP4	Cassava processing	CIAT – IITA – CIRAD – NRI	2017–2021	1942
FP5	Foresight, impact, monitoring, and co-learning	Bioversity – CIAT – CIP – IITA – CIRAD – WUR	2017–2021	3663
FP5	Sustainable intensification and diversification for improved resilience, nutrition, and income	Bioversity – CIP – IITA – WUR	2017–2018	252
FP5	Gender equitable development and youth employment	Bioversity – CIP – IITA	2017–2021	1099
FP5	Institutional innovation, scaling	Bioversity – CIAT – CIP – IITA – CIRAD – WUR	2017–21	3075

2. Non-earmarked funding, about 57% of the annual budget, was available for centers to invest on specific deliverables in alignment with the proposal for the phase and theories of change and to complement Window 3 investments but was not projectized and therefore more flexibly managed.
3. Scaling funds, about 12% of the yearly budget, were allocated through competitive calls. These funds supported the scaling of innovations that had higher levels of readiness.

Earmarked funding was awarded and renewed through an internal competitive process with submissions of project proposals evaluated by the PMU based upon criteria developed collaboratively with the Management Committee and ISC. After feedback, many submissions received some funding, and the strongest ones received the most. This created a space for constrained competition, ensuring that most areas of synergistic investment received some funding.

During Phase 2, an institutional innovation called the RTB Scaling Fund was introduced to foster the scaling of innovations, generate an evidence base of their scalability, and improve scaling strategies (Sartas et al. 2020). As discussed above, scaling was a critical limitation identified in Phase 1. In 2017, RTB held a workshop with some stakeholders to review best bet scalable innovations. Later, RTB clusters with innovation packages in advanced stages of readiness were invited to apply for the RTB Scaling Fund. These awards were made for 2 years in order to move innovation packages to a higher level of scaling readiness. This funding was competitive with two external reviewers and one PMU member evaluating submissions against clearly defined criteria (Table 2.6).

The innovation packages receiving scaling fund support and the allocations are shown in Table 2.7.

In summary, RTB's use of W1, W2, and W3 funding mechanisms enabled maximum and strategic collaboration. The RTB Scaling Fund was a significant innovation to address scaling more comprehensively.

Table 2.6 Scaling Fund selection criteria (for concept notes and full proposals)

Criteria	Score Max (100)
1. Relevance Is the proposal relevant to the purpose of the fund? Is the contribution of the proposal in moving the innovation along the scaling readiness levels convincing? Are the objective and outcomes significant and realistic (number of actual beneficiaries already reached and number of expected beneficiaries in the coming 2 years)?	25
2. Partnerships Does the concept note provide solid elements to build the partnership and scaling strategies? Is the description of stakeholders' roles and synergies clear and convincing?	20
3. Scaling strategy Does the scaling strategy identify key opportunities and bottlenecks to achieve the project outcomes? Does the strategy clearly present how scaling of the proposed innovations will contribute to achieving project outcomes?	20
4. Strength of the proposed multidisciplinary team (it may include partners and staff funded through different sources)	15
5. Level of co-investment by key government, public or private scaling partners	20

Table 2.7 Scaling fund awards in Phase 2

Flagship	Innovation package	Center/ partners	Duration	Funding (USD, 000)
FP3/ FP5	Broadening the scaling of BXW management in East and Central Africa (Chap. 10)	Bioversity – IITA – WUR	2018– 2019	700
FP2/ FP5	Scaling Sweetpotato Triple S PLUS – gender-responsive options for quality planting material, higher yields and extended shelf life for storage roots (Chap. 12)	CIP – WUR	2018– 2019	701
FP4/ FP5	Scaling the transformation of wet cassava peels into high quality animal feed ingredients (Chap. 6)	IITA – CIAT – WUR	2018– 2019	404
FP4/ FP5	Scaling approach for flash drying of cassava starch and flour at small scale (Chap. 4)	CIAT – IITA – CIRAD – WUR	2019– 2020	903
FP2/ FP5	Market-driven scaling up and adoption of potato in Africa through a technology package combining climate resilient, novel potato varieties with a seed system innovation (Chap. 13)	CIP – WUR	2019– 2020	1013
FP4/ FP5	Orange Fleshed Sweetpotato (OFSP) Puree for Safe and Nutritious Food Products and Economic Opportunities for Women and Youths in Kenya, Uganda and Malawi (Chap. 5)	Bioversity – CIP – WUR	2019– 2020	977
FP3/ FP5	Scaling AKILIMO, a digital fertilizer recommendation service	CIP – IITA – WUR	2020– 2021	1021
FP1/ FP5	Scaling RTB crop variety validation and diffusion using farmer citizen science in Ghana and Rwanda	Bioversity – CIP – IITA – WUR	2020– 2021	1000

2.5.7 Dynamic Interactive Communication Capability to Build a Shared Vision and Engage Stakeholders

Using a strategic communication approach, RTB pursued shared goals that would not have been achievable if the centers had worked alone. RTB developed its goal-oriented strategy through a SWOT analysis in 2013 (RTB 2013a). The PMU hired a full-time specialist to develop the communication strategy, using social media campaigns and joint blogs and coordinating with focal points in the RTB centers. Building on the SWOT analysis, RTB communications served multiple functions, including the following:

(a) To advance the *image of RT&B crops* as relevant for investment and research for nutrition, food security, and women farmers' income generation. RTB communicated the importance of RT&B crops to policymakers, donors, and researchers.

(b) To promote the program externally as *a globally recognized leader on knowledge and research about RT&B crops* with its own branding. RTB developed branding guidelines, publication and acknowledgment guidelines, and an "about us" statement to ensure the brand was understood internally and could be differentiated from other CRPs and centers externally. RTB had its own website and published several blogs monthly about scientific achievements and development impact. RTB developed an illustrated annual report for stakeholders that complemented more technical reporting (CGIAR Research Program on Roots, Tubers and Bananas, 2021). This showcased the breadth of the program, its integrated vision, and specific focus on the collective assets and collaborative innovations. Communication products included videos and podcasts on social media (Facebook, Twitter, LinkedIn, YouTube). Activities and achievements were summarized in blogs and in a quarterly newsletter. RTB supported several international and regional events each year, in particular meetings of the International Society for Tropical Root Crops (ISTRC) where national partners participated.

(c) To support *internal communications* on RTB vision and goals. The communication function targeted RTB scientists, emphasizing why the program added value to what the centers and their scientists did and the demonstrated value of collaboration. RTB held community meetings to report progress and plan cross-cutting collaboration about every 18 months; these were organized around the thematic subthemes and then FPs. The RTB communication specialist supported networking among researchers and encouraged FP and cluster teams to communicate in the ways they were more comfortable with.

(d) *To identify and nurture collective knowledge assets.* As discussed in Sect. 2.5.6, RTB earmarked funding and the cross-cutting clusters that created multicenter communities of practice and generated collective knowledge assets. With the end of RTB in 2021 and the creation of new initiatives within One CGIAR, there was an urgent need to ensure that RTB's legacy and collective assets would find a place in future initiatives. The collective knowledge assets at risk in the transition were designated "golden eggs" (Fig. 2.17). The landing page of the RTB golden eggs and writing descriptions of each golden egg and promoting them through social media was a communications goal in 2021. The CAS Secretariat (2020) evaluation considered it critical for "RTB to develop and expand these packages to inclusively cover the full program achievements." The CGIAR as a whole subsequently adopted this legacy initiative.

In summary, RTB effectively used communications to manage program complexity by drawing on goal-oriented internal and external communications and using communications strategically to draw attention to RT&B crops. As an institutional innovation, the RTB collective knowledge assets or golden eggs identify the legacy of the RTB CRP, and the close link between scaling innovations and communications informs new initiatives in One CGIAR.

Fig. 2.17 RTB golden eggs, key collective knowledge assets

2.5.8 Purposive National Partner Engagement

RTB works with national partners, especially within the national agricultural research systems (NARS) through principal program participants, and the CGIAR centers. Consequently, RTB's higher-level, coordinating role is not always visible to those partners. Nevertheless, RTB had a strategy for engaging national partners, for instance, during stakeholder consultations and in some cases while preparing theories of change. However, RTB also opted to use existing events where national partners came together, rather than creating its own dedicated partnership platforms. RTB supported three such partnership platforms, the most important of which was the ISTRC (http://www.istrc.org/).

RTB supported both the general meeting of the ISTRC and of its Africa branch (ISTRC-AB). These meetings brought ownership of and participation by national scientists and extensionists, with an increasing participation of African members. The ISTRC meetings came the closest to a generalized platform for the RT&B crops, as all are included with the exception of bananas. RTB also co-organized its own meetings to follow up on these events. This strengthened the ISTRC meetings by enhancing attendance and knowledge sharing. On multiple occasions, RTB organized presentations of the program and of each flagship on the final day of ISTRC meetings, preceding the RTB meeting, where national scientists shared their experiences. RTB also engaged national partners at the Africa Potato Association and the Global Cassava Partnership for the 21st Century. For all these meetings, RTB made small travel grants to encourage young female and male scientists to present their work. This national partner engagement created a broader awareness of RTB's contributions and its role in partnerships and advocacy on RT&B crops.

RTB also had two experiences where the PMU directly managed projects in Uganda with the RTB-ENDURE project (Bentley et al. 2021) and in Nigeria with BASICS (Bentley et al. 2020a, b). These two experiences created deeper collaborative networks with national partners in these two countries and helped to capitalize on experiences from the wider RTB community to benefit national partners.

2.6   Program Outcomes

RTB's institutional innovations facilitated various outcomes across centers (Table 2.8). These outcomes are concentrated in the areas of breeding, seed systems, pests and diseases, and postharvest, all of which were built on significant bilateral investment that complemented the earmarked funding for cross-center collaboration. Agronomy was a high priority for RTB, but in the absence of significant bilateral investment, its cross-cutting outcomes were limited. This changed in 2019 with the cassava agronomy decision support tool AKILIMO (https://www.akilimo.org/) led by IITA. This digital application and database could incorporate other RTB crops and therefore scale to other contexts and uses. Table 2.8 does not include other notable center-specific outcomes such as breeding and scaling of the orange-fleshed sweetpotato (Low and Thiele 2020).

As discussed in Sect. 2.5, the RTB 2.0 portfolio identified impact pathways that could be linked directly to the Intermediate Development Outcomes (IDOs) as set out in the CGIAR Strategic Research Framework (SRF). Figure 2.18 shows how RTB tracked contributions to system-level outcomes linked to the United Nations' Sustainable Development Goals, particularly SDG1, SDG2, and SDG13.

2.7   Lessons Learned from RTB

One of the difficulties of a complex program like RTB is to grasp how the partnership contributed to research and development outcomes and to document evidence of value at scale (Horton et al. 2009). RTB's complex institutional innovations (Sect. 2.5) facilitated the outcomes presented in Table 2.8 and Fig. 2.16. RTB made value at scale possible because there was:

- A common, shared vision and purpose and realistically defined goals
- Support for the partnership from participating organizations
- Equitable sharing of resources, responsibilities, and benefits
- Transparent governance and decision-making
- Creation of genuine respect and trust between the partners
- Achievement of higher-level outcomes beyond the partnership itself
- Committed leadership in the RTB Management Committee and PMU, notably the role of the Program Director with deep RTB knowledge and many years of experience working in the CGIAR and its partnership networks.

RTB was considered a strong model of good partnering within the CGIAR and by its stakeholders with transparent and equitable decisions about the use of funding mechanisms and the program direction (CGIAR-IEA 2016; ISPC 2016; CAS Secretariat 2020). Collaboration and partnering were articulated with an effective structuring of its portfolio around cross-cutting clusters or innovation packages that facilitated reciprocal learning exchange across crops and among partners. This

Table 2.8 Eight key RTB research outcomes based upon multicenter collaboration

Cluster/partners	Research outcome	How/when
DI1.1 Breeding community of practice, with Excellence in Breeding (EiB)	Researchers use more focused design of breeding products to meet farmer and consumer demands and improved management of breeding product pipelines	2013: RTB brought breeders together 2016: Creation of a Breeding Community of Practice with earmarked funding, comparing breeding strategies and developing shared methods, including the Tricot citizen science approach for varietal testing 2019: Product profiles registered (47) for the main targets of RTB breeding registered with CGIAR EiB Platform 2020: Tools to incorporate gender into product profile development are tested and adapted in collaboration with EiB 2020: Hackathon, breeders, social scientists, gender specialists, and food scientists peer-review and improve market segment definitions and variety product profiles for four breeding programs
DI1.2 Next-generation breeding, with Royal Holloway, University of London (RHUL)	Scientists at IITA, CIAT, and CIP incorporate design, metabolite extraction, and interpretation and use metabolomic data for all crops	2013: Theme 2 leader (CIAT) puts together an earmarked funding proposal across all centers and crops to begin this work 2018–2020: Ten peer-reviewed publications published; see overview in Friedmann et al. (2019) 2020: Compound database and concentration range for metabolites detected in the major RT&B crops available for breeding programs
CC2.1 Seed systems (with WUR and U Florida)[a]	Improved seed systems. Seed system toolbox validated. Thirteen tools developed and web accessible, for improving the design and execution of seed-system interventions and the management of seed degeneration	2014: Researchers from CGIAR centers, Wageningen University & Research (WUR), and Kansas State University completed and analyzed 12 case studies of RT&B seed systems in Africa and Latin America using common framework (RTB 2016b). Improved models for seed degeneration management developed based on field trials and integrated seed health strategy. 2016: Community of Practice Cluster CC2.1 funded with earmarked funds 2017: Cluster CC2.1 collaborates to incorporate tools for understanding RTB seed systems into a single toolbox using a standard format. 2020: Toolbox validated in 14 projects in Asia, Africa, and South America across all major RTB crops 2021: Online version of the Toolbox available for use by government agencies, NARS, NGOs, and donors

(continued)

Table 2.8 (continued)

Cluster/partners	Research outcome	How/when
CC3.1 Pests and diseases (with various NARS and ARIs)[b]	Researchers, agricultural ministry officials, national plant protection organizations, and extension agents use tools and strategies to manage major pests and diseases	2012: Joint RTB and CCAFS workshop on management of critical pests and diseases through enhanced risk assessment and surveillance and understanding of climate impacts through enhanced modeling 2015: In a cross-crop consultative intervention of all RTB centers, key regional target pests and diseases were identified for pest risk analysis (PRA) 2017: Cassava Disease Surveillance (CDS) virtual network supports accurate diagnosis and offers solutions for prevention and management. CGIAR and national partners strengthen capacities to perform PRAs predicting risk of insect-transmitted viruses and generation of georeferenced risk maps 2018: ICT tools used to identify major diseases in the field for surveillance and plant disease management 2020: Digital alliance for pests and diseases as a golden egg RTB ensures gender is addressed across topics in the webinars for the International Year of Plant Health
CA3.5 Cassava biological constraints, Asia/Latin America, with 21 NARS, national plant protection organizations (NPPOs, U of Queensland)	Scientists use response plan developed to contain emerging cassava mosaic disease (CMD) in SE Asia	2015: CMD detected in Cambodia. Network of experts to monitor and manage cassava mealybug and cassava witches' broom disease in Asia, progress in helping farmers combat pests, with local partners 2018: Regional workshop supported by the Global Cassava Partnership for the Twenty-first Century and CIAT led to the joint development of a response plan, with IITA participation under the RTB umbrella 2018: Adoption of biological control for cassava mealybug using a host-specific parasitoid provided by IITA involving collaboration from CIAT, contributed to restoring the cassava yield that had dropped by 27% after the arrival of the cassava mealybug (first reported in 2008) 2019: Improved capacities of NPPOs and cassava farmers to contain CMD and implement management strategies in Cambodia, Laos, Thailand, and Vietnam following a multi-pronged approach (Siriwan et al. 2020)

(continued)

Table 2.8 (continued)

Cluster/partners	Research outcome	How/when
BA3.4 Banana viral diseases (with 14 NARS, FAO, U Queensland)	Multinational, multi-stakeholder Alliance coordinated action to halt the expansion of banana bunchy top disease (BBTD) and recover banana production in disease-affected areas, especially in sub-Saharan Africa	2014: RTB learning alliance to contain BBTD and help farmers recover from it in eight African countries; established field sites and capacity building for researchers from each participating country 2015: BBTD Alliance launched by RTB. Fifteen pilot sites in eight African countries set up to implement strategies for recovering banana production (https://www.bbtvalliance.org/) 2019: Training workshops in Togo, Nigeria, and Cameroon 2021: Gender roles analysis (Nkengla-Asi et al. 2021) 2021: Alliance developing capacity for disease recognition and control options, training farmers and entrepreneurs in field inspection, rogueing infected plants and production of clean planting material 2021: Training African scientists to continue surveillance on the extent of disease incidence to update the spatial distribution map of BBTV spread in SSA 2021: Location-specific clean banana production and distribution systems are being established to replenish banana plants that have been eradicated
CC4.1 Postharvest innovation, led by CIRAD with eight NARS and Cornell U	Framework and tools to breed for quality traits in 11 food products from RTB crops in five target countries	2015 RTB supported work on end-user preferences through initiatives with national partners and the Natural Resources Institute (NRI, UK) 2018: RTBfoods project funded by Bill & Melinda Gates Foundation (BMGF) (https://rtbfoods.cirad.fr/) 2020–2021: Special journal issue (Dufour et al. 2021) published with 17 articles on end-user preferences of RT&B crops and method to measure quality traits

(continued)

Table 2.8 (continued)

Cluster/partners	Research outcome	How/when
CC5.4 Scaling, with WUR[c]	Scaling readiness approach implemented across the RTB portfolio, in collaboration with the scaling experts of CC5.4	2017: Scaling Readiness Approach and RTB Scaling Fund developed and piloted 2018–2021: Eight projects supported by RTB Scaling Fund 2019: Scaling Readiness web portal (www.scalingreadiness.org) and principles (Sartas et al. 2020) published 2020–2021: Other CRPs, CGIAR centers, and the CGIAR System Office expressed interest in RTB's scaling work and using the Scaling Readiness approach 2021: RTB gender research, tools, and methods are compiled and shared for broader use in collaboration with the Gender Platform

[a]See Chap. 11 of this book
[b]See Chap. 9 of this book
[c]See Chap. 3 of this book

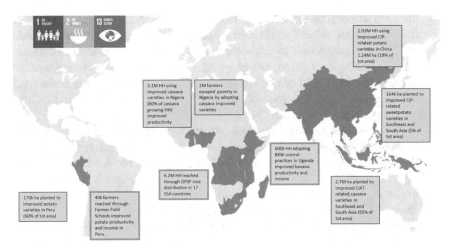

Fig. 2.18 RTB's contributions to system level outcomes, 2017–2020. (Source: CGIAR Research Program on Roots, Tubers and Bananas, 2018, 2019, 2020, 2021)

value addition by RTB aligned with institutional innovation around scaling that was a strong feature of RTB 2.0.

One challenge with RTB closing at the end of 2021 is that the organizing principles of the new initiatives in the One CGIAR (2022–2030) are not organized by crop type (CGIAR System Organization 2021). There should be significant work continuing on the RT&B crops, but the RTB program and its clusters will be redistributed and work will be organized in different ways. The RTB golden eggs provide some continuity of innovation packages, but not necessarily the synergies among the teams. At the time of writing, there is interest in keeping the RT&B crops together as a work package in the One CGIAR seed initiative, but it is less clear what happens to other RTB work in One CGIAR.

In summary, various lessons emerge from the RTB experience that are relevant for future multi-organizational programs for agricultural research for development and for One CGIAR.

(a) *Build on the unique value proposition of the collaborative program.* For RTB, this relates to the commonalities of clonally propagated crops, with program design to add value to them through the cross-cutting clusters and their communities of practice on topics of breeding, seed systems, pests and diseases, and postharvest, all of which had dedicated funding. A recommendation for One CGIAR would be to keep clonally propagated crops together as a group within key research areas to further enhance synergies and achievements.

(b) *Use strategic communications to build broad-based stakeholder support around the value proposition.* RTB did this with stakeholder consultations and communications demonstrating progress.

(c) *Ensure that programmatic design can bring together the different innovations required for impact.* For RTB, this happened at three different levels. First, at the level of program architecture, the crop-specific clusters were purposively designed as aggregated innovations (technical, organizational, institutional) and contributed together to a systemic change. Second, at the level of context-specific scaling fund grants, where specific innovation packages were enabled to move to a high level of readiness and greater impact. And third, at the level of purposively engaged partners who could scale results further and created further systemic change.

(d) *Embed gender in technical research and engage with and support researchers to mainstream gender in biological sciences.* Without effort to resource and monitor the progress of gender in A4RD, the impact of socio-technical innovations can be limited or eroded. RTB's targeted support helped to achieve impact in gender in breeding and the uptake of G+ tools for breeding.

(e) *Establish structural incentives for integration among initiatives and across crops* if cross-CGIAR contributions to the IDOs and SDGs are to be fully captured. This was an area where RTB and other CRPs underachieved compared to the promise at program design. One of the primary reasons for this was insecurity and late award of funding, which made it more challenging to secure strategic partnerships.

(f) *Link theories of change with flexible and utilization-focused M&E systems* to strengthen adaptive management and reflexivity at different management levels. This can enhance planning and reporting, including milestones and indicators used to plan and report at the level of the CGIAR system, which were excessively rigid for CRPs.

(g) *Recognize and incorporate key partnerships* that add value to the program participants. These need to respect good partnering principles including identification of the value added from the partnership, transparency and access to funding, shared responsibilities, and clarity of the role of independent advisory groups to adjudicate any conflicts in the partnership. When the program is closed, attention needs to be paid to ensure that relationships (corporate, technical, and personal) built up by RTB and other CRPs are retained. For this reason, this book intentionally documents RTB experiences.

(h) *Develop mechanisms to promote continuity of key innovations and teams of collaborators associated with them (inside and outside of CGIAR) as the program is closed.* RTB, for example, used the concept of golden eggs with resourced and linked nurturing plans.

2.8 Conclusion

RTB demonstrates that institutional innovations involving collaborative program design and management can enable comprehensive research and development outcomes aligned to program goals. Over its 10 years, RTB evolved as a unique, global collaborative program with connections among multiple centers and stakeholders bridging upstream research, translational research, and innovation processes to deliver value at scale. RTB developed pragmatic institutional innovations to manage the program effectively, ensuring research outcomes and contributing to the global Sustainable Development Goals. Being embedded within the CGIAR has required RTB to adapt its capacity to respond to system change. One CGIAR could draw on the RTB approach to collaboration as complex institutional innovations organized by a portfolio of aggregated innovations that led to concrete achievements. These innovations include the strategic use of funding mechanisms, co-constructed impact pathways, stakeholder consultations and priority setting, goal-oriented communications, committed and effective program leadership, and governance and partnerships for strategic interventions to realize cross-cutting priorities, such as strategic and integrated gender research. The following chapters in this book describe RTB's legacy in greater detail.

Acknowledgments The research reported here was undertaken as part of, and funded by, the CGIAR Research Program on Roots, Tubers and Bananas (RTB) and supported by CGIAR Trust Fund contributors (https://www.cgiar.org/funders/) as well as numerous other donors.

The authors thank the following individuals who provided input and feedback on earlier drafts of this chapter: Rupert Best, Rodney Cooke, Jill Lenné, Gordon Prain, Oscar Ortiz, Selcuk Ozgediz, and Yvonne Pinto. We greatly appreciate Zandra Vasquez's assistance with references and formatting.

References

Almekinders CJM, Walsh S, Jacobsen KS, Andrade-Piedra JL, McEwan MA, de Haan S, Kumar L, Staver C (2019) Why interventions in the seed systems of roots, tubers and bananas crops do not reach their full potential. Food Security 11:23–42. https://doi.org/10.1007/s12571-018-0874-4

Arena M, Cross R, Sims J, Uhl-Bien M (2017) How to capitalize innovation in your organization. MIT Sloan Manag Rev. https://sloanreview.mit.edu/article/how-to-catalyze-innovation-in-your-organization

Barrett CB, Benton Y, Fanzo J, Herrero M, Nelson RJ, Bageant E, Buckler E, Cooper K, Culotta I, Fan S, Gandhi R, James S, Kahn M, Lawson-Lartego L, Liu J, Marshall Q, Mason-D'Croz D, Mathys A, Mathys C, Mazariegos-Anastassiou V, Miller A, Misra K, Mude AG, Shen J, Majele Sibanda L, Song C, Steiner R, Thornton P, Wood S (2020) Socio-technical innovation bundles for agri-food systems transformation, report of the international expert panel on innovations to build sustainable, equitable, inclusive food value chains. Cornell Atkinson Center for Sustainability and Springer Nature, Ithaca and London

Baum WC, Lejuene M (1986) Partners against hunger: the consultative group on international agricultural research. The World Bank for the CGIAR, Washington, DC

Bentley J, Nitturkar H, Friedmann M, Thiele G (2020a) BASICS phase I final report. 23 Dec 2020. 78

Bentley J, Nitturkar H, Friedmann M, Thiele G (2020b) Is there a space for medium-sized cassava seed growers in Nigeria? J Crop Improv. https://doi.org/10.1080/15427528.2020.1778149p

Bentley JW, Naziri D, Prain G, Kikulwe E, Mayanja S, Devaux A, Thiele G (2021) Managing complexity and uncertainty in agricultural innovation through adaptive project design and implementation. Dev Pract 31(2):198–213. https://doi.org/10.1080/09614524.2020.1832047

Berthet ET, Hickey GM, Klerkx L (2018) Opening design and innovation processes in agriculture: insights from design and management sciences and future directions. Agric Syst 165:111–115

Byerlee D, Lynam J (2020) The development of the international center model for agricultural research: a prehistory of the CGIAR. World Dev 135:105080. https://doi.org/10.1016/j.worlddev.2020.105080

CAS Secretariat (CGIAR Advisory Services Shared Secretariat) (2020) CGIAR research program 2020 reviews: roots, tubes and bananas (RTB). CAS Secretariat Evaluation Function, Rome. https://cas.cgiar.org/evaluation/publications/crp-2020-review-roots-tubers-and-bananas-rtb

CGIAR (2015a) Research program portfolio report 2015 https://hdl.handle.net/10947/4480

CGIAR (2015b) CGIAR strategy and results framework 2016–2030: overview. https://hdl.handle.net/10947/4069

CGIAR (2016) Research program portfolio report 2016 https://hdl.handle.net/10568/89018

CGIAR Research Program on Roots, Tubers and Bananas (2018) RTB Annual report 2017. Phase II – year 1. Lima (Peru). https://hdl.handle.net/10568/97476

CGIAR Research Program on Roots, Tubers and Bananas (2019) RTB Annual report 2018. Phase II – year 2. Lima (Peru). https://hdl.handle.net/10568/103623

CGIAR Research Program on Roots, Tubers and Bananas (2020) RTB Annual report 2019. Phase II – year 3. Lima (Peru). https://hdl.handle.net/10568/108931

CGIAR Research Program on Roots, Tubers and Bananas (2021) RTB Annual report 2020: innovation and impact. Phase II – year 4. Lima (Peru). https://cgspace.cgiar.org/handle/10568/114576

CGIAR System Organization (2021) CGIAR 2030 research and innovation strategy: transforming food, land, and water systems in a climate crisis. CGIAR System Organization, Montpellier. https://hdl.handle.net/10568/110918

CGIAR-IEA (2015a) Evaluation of CGIAR Research Program on Roots, Tubers and Bananas (RTB), vol I. Independent Evaluation Arrangement (IEA) of the CGIAR, Rome. http://iea.cgiar.org/

CGIAR-IEA (2015b) Evaluation of CGIAR Research Program on Roots, Tubers and Bananas (RTB), vol II. Independent Evaluation Arrangement (IEA) of the CGIAR, Rome. http://iea.cgiar.org/

CGIAR-IEA (2016) CGIAR Research Program on Roots, Tubers, and Bananas. Evaluation brief. https://cas.cgiar.org/sites/default/files/pdf/Evaluation-Brief-RTB.pdf

Coffman WR, Acevedo M, Evanega SD, Porciello J, Tufan HA, McCandless L (2020) Viewpoint: five recommendations for an inclusive and collaborative One CGIAR. Food Policy 91:101831. https://doi.org/10.1016/j.foodpol.2020.101831

Consortium Board (2011) A strategy and results framework for the CGIAR. https://storage.googleapis.com/cgiarorg/2011/08/CGIAR-SRF-Feb_20_2011.pdf

CRP-Humidtropics (2016) Annual performance report. https://cgspace.cgiar.org/bitstream/handle/10568/89312/HUMIDTROPICS-Annual-Report-2016.pdf?sequence=1&isAllowed=y

CRP-RTB (2011) CRP-RTB 3.4. Roots, tubers, and bananas for food security and income. Revised proposal. 170 p. https://hdl.handle.net/10568/83375

Dalrymple D (2008) International agricultural research as a global public good: concepts, the CGIAR experience and policy issues. J Int Dev 20:347–379. https://doi.org/10.1002/jid.1420

Dufour D, Hershey C, Hamaker B, Lorenzen J (2021) Integrating end-user preferences into breeding programmes for roots, tubers and bananas. Int J Food Sci Technol 56(3):1071–1075. https://doi.org/10.1111/ijfs.14911

Ekboir J (2009) The CGIAR at a crossroads: assessing the role of international agricultural research in poverty alleviation from an innovation systems perspective. ILAC Working Paper 9, Institutional Learning and Change Initiative, Rome

Feldman S, Biggs S (2012) International shifts in agricultural debates and practice: an historical view of analyses of global agriculture. In: Campbell W, López Ortíz S (eds) Integrating agriculture, conservation and ecotourism: societal influences. Issues in agroecology – present status and future prospectus, vol 2. Springer, Dordrecht. https://doi.org/10.1007/978-94-007-4485-1_2

Friedmann M, Becerra LA, Fraser PD (2019) Metabolomics in CGIAR Research Program on Roots, Tubers and Bananas (RTB). Lima (Peru). CRP RTB. RTB working paper 2019-2. ISSN 2309-6586. 11 p. https://doi.org/10.4160/23096586RTBWP20192

Fuglie K, Thiele G (2009) Research priority assessment at the international potato center (CIP). In: Raitzer DA, Norton GW (eds) Prioritizing agricultural research for development. CABI, Cambridge

GCARD (2011) The GCARD road map: transforming agricultural research for development (AR4D) systems for global impact. http://www.fao.org/docs/eims/upload/290017/the_gcard_road_map_finalized.pdf

Glaeser B (1987) The green revolution revisited. Allen & Unwin, London

Greenland DJ (1997) International agricultural research and the CGIAR system- past, present, and future. J Int Dev 9(4):459–482

Hardin LS, Collins NR (1974) International agricultural research: organising themes and issues. Agric Adm 1(1):13–22. https://doi.org/10.1016/0309-586X(74)90024-7

Harwood R, Kassam AH (2003) Research towards integrated natural resources management. Examples of research problems, approaches and partnerships in action in the Rome: CGIAR. Consultative Group on International Agricultural Research (CGIAR); Food and Agriculture Organization of the United Nations (FAO), 168 p

Horton D, Prain G, Thiele G (2009) Perspectives on partnership: a literature review. Working paper 2009-3. CIP, Lima, p 111

Immonen S, Cooksy L (2019) Role and use of independent evaluation in development-oriented agricultural research: the case of CGIAR, an agricultural research network. Outlook Agric 48(2):94–104. https://doi.org/10.1177/0030727019850835

Interim Science Council (TAC) (2003) Eighty-fourth meeting, June 9–13, 2003: end of meeting report. https://hdl.handle.net/10947/654

ISPC (2012) Strategic overview of CGIAR research programs, Part I. Theories of change and impact pathways. Independent Science and Partnership Council, CGIAR, Rome

ISPC (2016) ISPC assessment of the roots, tubers and bananas (RTB) CRP II revised proposal (2017–2022). Independent Science and Partnership Council, CGIAR, Rome

Krishna Bahadur KC, Dias GM, Veeramani A, Swanton CJ, Fraser D, Steinke D et al (2018) When too much isn't enough: does current food production meet global nutritional needs? PLoS One:16. https://doi.org/10.1371/journal.pone.0205683

Low J, Thiele G (2020) This understanding innovation: the development and scaling of orange-fleshed sweetpotato in major African food systems. Agric Syst 179(102770). https://doi.org/10.1016/j.agsy.2019.102770

McCalla AF (2014) CGIAR reform – why so difficult? Review, reform, renewal, restructuring, reform again and then "The New CGIAR" – so much talk and so little basic structural change-why? Working paper no. 14-001. UC Davis. p 55

McCalla AF (2017) The relevance of the CGIAR in a modernizing world. In: Pingali P, Feder G (eds) Agriculture and rural development in a globalizing world: challenges and opportunities, 1st edn. Routledge. https://doi.org/10.4324/9781315314051

Mudege NN, Kapalasa E, Chevo T, Nyekanyeka T, Demo P (2015) Gender norms and the marketing of seeds and ware potatoes in Malawi. J Gender Agric Food Secur 1(2):18–41

Nkengla-Asi L, Eforuoku F, Olaosebikan O, Ladigbolu TA, Amah D, Hanna R, Kumar PL (2021) Gender roles in sourcing and sharing of banana planting material in communities with and without banana bunchy top disease in Nigeria. Sustainability 13(6):3310. https://doi.org/10.3390/su13063310

Operations Evaluation Department (2004) The CGIAR at 31: an independent meta-evaluation of the Consultative Group on International Agricultural Research. Revised Edition. The World Bank, Washington, DC. http://hdl.handle.net/10986/15041

Ozgediz S (2012) The CGIAR at 40. Institutional evolution of the world's premier agricultural research network. CGIAR Fund Office, Washington, DC. http://hdl.handle.net/10986/23845

Pant LP, Hambly HO (2009) The promise of positive deviants: bridging divides between scientific research and local practices in smallholder agriculture. Knowl Manag Dev J. https://doi.org/10.1080/18716340903201504

Pemsl DE, Staver C, Hareau G, Alene AD, Abdoulaye T, Kleinwechter U, Labarta R, Thiele G (2022) Prioritizing international agricultural research investments: lessons from a global multi-crop assessment. Res Policy 51(4):104473. https://doi.org/10.1016/j.respol.2022.104473

Pigford AE, Hickey G, Klerkx L (2018) Beyond agricultural innovation systems? Exploring an agricultural innovation ecosystems approach for niche design and development in sustainability transitions. Agric Syst 164:116–121. https://doi.org/10.1016/j.agsy.2018.04.007

Price EJ, Drapal M, Perez-Fons L, Amah D, Heider B, Rouard M, Swennen R, Becerra Lopez-Lavalle LA, Fraser PD (2020) Metabolite database for root, tuber and banana crops to facilitate modern breeding in understudied crops. Plant J 103(5):1959–1959. https://doi.org/10.1111/tpj.14649

Renkow M, Byerlee D (2010) The impacts of CGIAR research: a review of recent evidence. Food Policy 35:391–402. https://doi.org/10.1016/j.foodpol.2010.04.006

Roberts N, Bradley R (1991) Stakeholder collaboration and innovation: a study of public policy initiation at the state level. J Appl Behav Sci 27:209–227. https://doi.org/10.1177/0021886391272004

RTB (2013a) RTB media and communication strategy for 2013–2014. CIP, Lima

RTB (2013b) Planning for greater impact: RTB current thinking 2016–2024. CIP, Lima

RTB (2016a) CGIAR research program on root, tubers and bananas. 2016. Full proposal 2017–2022. CIP, Lima. http://hdl.handle.net/10945/53701

RTB (2016b) User's guide to multistakeholder framework for intervening in RTB systems. CGIAR Research Program on Roots, Tubers and Bananas RTB Working Paper 2016–1

Sartas M, Schut M, Proietti C, Thiele G, Leeuwis C (2020) Scaling readiness: science and practice of an approach to enhance the impact of research for development. Agric Syst 183(102874)

Scott G (2020) A review of root, tuber and banana crops in developing countries: past, present, and future. Int J Food Sci Technol. https://doi.org/10.1111/ijfs.14778

Siriwan W, Jimenez J, Hemniam N, Saokham K, Lopez-Alvarez D, Leiva AM, Martinez A, Mwanzia L, Becerra Lopez-Lavalle LA, Cuellar WJ (2020) Surveillance and diagnostics of the emergent Sri Lankan cassava mosaic virus (Fam. Geminiviridae) in Southeast Asia. Virus Res 285:197959., ISSN 0168-1702. https://doi.org/10.1016/j.virusres.2020.197959

Woodhill J (2010) Capacities for institutional innovation: a complexity perspective. IDS Bull 41:47–59. https://doi.org/10.1111/j.1759-5436.2010.00136.x

Woolley J, Johnson VB, Ospina B, Lemaga B, Jordan T, Harrison G, Thiele G (2011) Incorporating stakeholder perspectives in international agricultural research: the case of the CGIAR Research Program for Roots, Tubers, and Bananas for Food Security and Income. International Potato Center (CIP), Lima, Peru. Social sciences working paper 2011-3. p 92

Chapter 3
Scaling Readiness: Learnings from Applying a Novel Approach to Support Scaling of Food System Innovations

Marc Schut (ID), **Cees Leeuwis** (ID), **Murat Sartas** (ID),
Luis Alejandro Taborda Andrade (ID), **Jacob van Etten** (ID), **Anna Muller** (ID),
Thierry Tran (ID), **Arnaud Chapuis** (ID), and **Graham Thiele** (ID)

Abstract Scaling of innovations is a key requirement for addressing societal challenges in sectors such as agriculture, but research for development programs struggles to make innovations *go to scale*. There is a gap between new

M. Schut (✉) · M. Sartas
Wageningen University, Wageningen, the Netherlands

International Institute of Tropical Agriculture (IITA), Kigali, Rwanda
e-mail: marc.schut@wur.nl; m.schut@cgiar.org; murat.sartas@wur.nl

C. Leeuwis
Wageningen University, Wageningen, the Netherlands
e-mail: cees.leeuwis@wur.nl

L. A. Taborda Andrade
Centro Internacional de Agricultura Tropical (CIAT), Cali, Colombia

Universidad Nacional de Colombia, Bogotá, Colombia
e-mail: latabordaa@unal.edu.co

J. van Etten · A. Muller
Bioversity International, Rome, Italy
e-mail: j.vanetten@cgiar.org; a.muller@cgiar.org

T. Tran
Alliance Bioversity-CIAT, Cali, Colombia

CIRAD UMR Qualisud, Montpellier, France
e-mail: thierry.tran@cirad.fr

A. Chapuis
CIRAD, Montpellier, France
e-mail: arnaud.chapuis@cirad.fr

G. Thiele
CGIAR Research Program on Roots, Tubers and Bananas (RTB), led by the International
Potato Center, Lima, Peru
e-mail: g.thiele@cgiar.org

complexity-aware scientific theories and perspectives on innovation and practical approaches that can improve strategic and operational decision-making in research for development interventions that aim to scale innovations. To bridge this gap, Scaling Readiness was developed. Scaling Readiness is an approach that encourages critical reflection on how ready innovations are for scaling in a particular context for achieving a particular goal and what appropriate actions could accelerate or enhance scaling to realize development outcomes. Scaling Readiness provides decision support for (1) characterizing the innovation and innovation system; (2) diagnosing the current readiness and use of innovations; (3) developing strategies to overcome bottlenecks for scaling; (4) facilitating multi-stakeholder negotiation and agreement; and (5) navigating the implementation process. This chapter explains how Scaling Readiness was used in the CGIAR Research Program on Roots, Tubers and Bananas (RTB) and describes how Scaling Readiness informed the design and management of the RTB Scaling Fund, an instrument for identifying and nurturing scaling-ready innovations. We introduce the key principles and concepts of Scaling Readiness and provide a case study of how Scaling Readiness was applied for scaling a cassava flash dryer innovation in different countries in Africa and Central America. The chapter concludes with a reflection and recommendations for the further improvement and use of Scaling Readiness.

3.1   Scaling of Innovation and Scaling Readiness

Innovation and the use of innovations at scale form an important element for achieving the Sustainable Development Goals (SDGs). The international community invests significant resources in the design and testing of innovations to overcome key challenges such as food insecurity, malnutrition, and environmental degradation. Many of those innovations start as bright ideas that are consequently developed in controlled conditions, to be tested with farmers and other end users. Although initial results and testing are often promising, relatively few lead to the desired positive impact at scale (Woltering et al. 2019). The agricultural research for development (AR4D) sector, in which the CGIAR is an important player, has been struggling with the question of how to best nurture impactful innovation and scaling pathways (Leeuwis et al. 2018).

One of the main reasons why innovations do not lead to impact at scale is that ideas about how scaling happens are not realistic. The notion of "find out what works and do more of the same" (Wigboldus et al. 2016) does not take into account the complex and diverse biophysical, socioeconomic, and political contexts that shape agriculture across the globe and limit the effectiveness of one-size-fits-all approaches (Hammond et al. 2020). Furthermore, research organizations often focus on technological innovations and pay less attention to the behavioral, organizational, and institutional changes that are needed to enable the effective use of technology (Schut et al. 2016). In addition, we observe that R4D interventions often see scaling as something that happens at the end of a short-term project or program as opposed to long-term systemic change processes and that scientists have limited capacities to shape impactful processes and partnerships needed for the scaling of innovations (Schut et al. 2020).

Revealing misconceptions and bottlenecks to the scaling of innovations, and supporting the development and implementation of effective strategies to overcome them, inspired the development of Scaling Readiness. In this chapter, the term "Scaling Readiness" (capitalized) is used as a brand name for the decision-support process that we have developed and as a key concept and metric that scores the maturity and scalability of an innovation (not capitalized).

The notion of "readiness" refers to whether an innovation has been tested and validated for the role it is intended to play in a specific context. The concept resonates with levels of technology readiness that have been proposed by the National Aeronautics and Space Administration (NASA) of the United States, the European Commission (EU), and scholars in technology studies who assess advancements in technology development, commercialization, and transition pathways (Verma and Ramirez-Marquez 2006; European Commission 2014; Kobos et al. 2018).

Scaling Readiness builds on the key principles of Agricultural Innovation Systems (AIS) thinking (Spielman et al. 2009; Hall and Clark 2010; Hounkonnou et al. 2012). Innovation systems are the interlinked set of people, processes, assets, and social institutions that enable (or constrain) the development and scaling of new technologies, products, practices, services, and solutions to deliver impact. A key lesson from conducting innovation systems research in the AR4D sector was that complexity-aware approaches (such as AIS) need to be operationalized or translated into simple tools that can be overseen and managed by program and project teams to guide their practice (Schut et al. 2015). Without such operationalization, AIS approaches can easily be perceived as fuzzy with the risk of people abandoning their willingness to engage with systems approaches altogether.

One of the aims of Scaling Readiness is to support a complexity-aware decision-making process that assists R4D interventions in designing, implementing, and monitoring scaling strategies in a structured and evidence-based way. To this end, Scaling Readiness proposes an iterative cycle of five steps that builds on key principles and concepts that will be further discussed and illustrated in this chapter (Fig. 3.1).

This book chapter has four main objectives:

1. To briefly introduce the key principles and concepts of Scaling Readiness.
2. To explain how Scaling Readiness was used in the CGIAR Research Program on Roots, Tubers and Bananas (RTB) Scaling Fund.
3. To present a case study that illustrates how Scaling Readiness can support the development of better-informed scaling strategies for R4D interventions.
4. To present lessons and recommendations for the further development and use of Scaling Readiness.

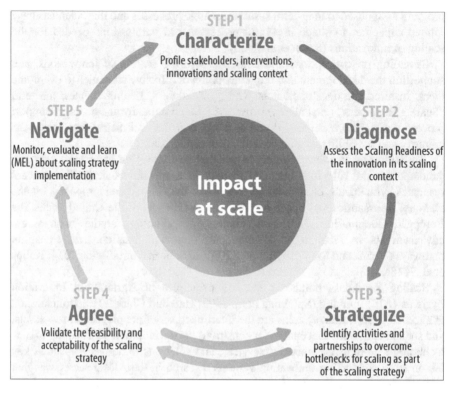

Fig. 3.1 Scaling Readiness proposes a stepwise approach to operationalize AIS thinking in support of the development, implementation, and monitoring of better-informed scaling strategies (Sartas et al. 2020b)

3.2 Scaling Readiness in the CGIAR Research Program on Roots, Tubers and Bananas Scaling Fund

In an attempt to close the gap between the *science* and the *practice* of scaling innovations, RTB developed an institutional innovation to support the scaling of RTB innovations: the *Scaling Fund*. Scaling Readiness was used in two distinct ways in the Scaling Fund: (1) to identify and select scaling-ready RTB innovations and (2) to nurture and support the design, implementation, and monitoring of strategies to scale those RTB innovations. Both will be explained in more detail in the below sections.

It is worthwhile mentioning that the RTB Scaling Fund provided an opportunity for Scaling Readiness not only to develop scaling strategies but also to test and improve the various tools, processes, and workstreams that Scaling Readiness offers.

3.2.1 Identifying and Selecting Scaling-Ready RTB Innovations

In 2017, a first call for Scaling Fund project proposals was announced and elicited 12 submissions that were assessed by an independent panel. Five of the proposals were selected to submit full proposals. The evaluation was based on the following assessment criteria:

1. Define and provide *evidence on the scaling readiness* of the selected RTB innovation (referred to as the *core innovation*).
2. Define *site-specific complementary innovations* or enabling conditions that are needed to scale the *core innovation*.
3. Scaling strategy in the proposal is *congruent with existing projects and public and private partners' initiatives*.

From the initial batch of five proposals, the three with the highest scores were awarded a total investment of approximately USD two million to further improve and implement their scaling strategies with their partners. In 2018 and 2019, five additional Scaling Fund projects were funded (Table 3.1).

Table 3.1 Overview of the eight Scaling Fund projects awarded and implemented between 2017 and 2021

RTB Scaling Fund batch	RTB Scaling Fund projects		
2018–2019	1. Single diseased stem removal (SDSR) for BXW banana disease in Burundi, eastern DR Congo, Rwanda, and Uganda	2. Triple S storage process for conserving sweetpotato roots to produce planting material in Ethiopia and Ghana	3. A technology for turning cassava peels into an ingredient of animal feed in Nigeria
2019–2020	1. Orange-fleshed sweetpotato (OFSP) puree for safe and nutritious food products and economic opportunities for women and youths in Kenya, Uganda, and Malawi	2. Approach for flash drying of cassava starch and flour at small scale in Nigeria, DR Congo, and Colombia	3. Rooted apical cuttings in Kenya
2020–2021	1. RTB crop variety validation and diffusion using farmer citizen science in Ghana and Rwanda (TRICOT)	2. A digital fertilizer recommendation service (AKILIMO) in Nigeria, Tanzania, and Rwanda	

3.2.2 RTB Scaling Strategy Design, Implementation, and Monitoring

After the initial selection of the Scaling Fund projects, a kick-off and capacity development workshop was organized for each batch of projects (Fig. 3.2). For 2–3 days, the Scaling Fund project teams were trained in the basics of Scaling Readiness and discussed how these would be applied and useful to their own Scaling Fund projects. At the end of the workshop, each of the project teams had a road map for the implementation of their Scaling Fund projects.

To ensure sufficient capacity within the projects to manage scaling processes and implement Scaling Readiness, a key requirement was to assign several people with designated functions, including scaling champions and Scaling Readiness monitors. The scaling champion was primarily responsible for the implementation of scaling strategies and stakeholder engagement plans. They were usually people with a good understanding of the innovation and the local partnership dynamics. Their role was to broker and network for key partners to work together and make scaling happen. The Scaling Readiness monitors were mainly responsible for applying the Scaling Readiness tools to collect and analyze data with the objective to influence decision-making and strategy development with the broader scaling project team. They were usually research assistants with good data collection and analysis skills.

Two different scaling consultants were recruited to backstop the scaling champions and Scaling Readiness monitors and to ensure cross-project learning.

Fig. 3.2 Group photo of participants representing the three 2019–2020 Scaling Fund project teams during the kick-off workshop in Nairobi, Kenya

During the two years of implementation, the scaling champions, Scaling Readiness monitors, and scaling consultant worked closely together with the Scaling Fund project teams.

One of the 2019–2020 Scaling Fund projects (a scaling approach for flash drying of cassava starch and flour at small scale in Nigeria, DR Congo, and Colombia) is used as a case study in this book chapter to explain how Scaling Readiness was used. The cassava flash drying project was selected because of its systematic use and documentation of the Scaling Readiness approach.

3.2.3 *Introduction to the Scaling Fund Cassava Flash Dryer Case Study*

Cassava is a starchy root crop that is a major staple food for people in developing countries. It is grown in tropical regions of the world because of its ability to withstand difficult growing conditions. Cassava in sub-Saharan Africa is generally a subsistence crop, but there is increased commercial interest here in processing cassava flour and for starch production. Rapid perishability of roots requiring agile and efficient processing is one of the greatest challenges facing smallholder cassava farmers and small-scale cassava processors. The most common practice is sun-drying cassava roots to make flour, which is challenging during the extended rainy seasons in the tropics and affects the overall quality of the starch. As a result, farmers and small-scale processors face difficulties to offer their produce to industries that need regular, all-year-round supply and consistent quality for flour and starch production (IITA 2016).

Flash drying, compared to sun drying, enables substantial gains in product quality and productivity by reducing the drying time from between 10 and 48 hours to a few seconds and providing constant drying conditions. Flash dryers are used mostly by large-scale processors (production capacity of >50 tons of starch/day) in countries such as Brazil and Thailand, which have highly developed commercial starch production. Small-scale flash dryers (production capacity of between 1–3 tons of flour/day) are not widely used due to a combination of factors including high energy consumption and production costs. Since 2013, more reliable methods to design energy-efficient flash dryers, based on numerical modelling, have been developed and successfully tested in small-scale pilot flash dryers (Fig. 3.3) that have proven to achieve the same energy efficiency as large-scale industrial flash dryers.

The Scaling Fund Cassava Flash Drying project focusses on three countries where the scaling of small-scale flash dryers has potential: Colombia, DR Congo, and Nigeria. In Nigeria, between 2006 and 2016, prior to the Scaling Fund project, 157 cassava processors had invested in first-generation small-scale flash dryers to produce cassava flour. However, 50% of these are not in use anymore because of the low energy efficiency and subsequent high costs of operation. In the DR Congo, the long rainy season (1300–1900 mm/year) led to strong demands for cost-effective drying solutions such as flash drying. In Colombia, labor-intensive sun drying is

Fig. 3.3 Example of a small-scale cassava flash drying system in Nigeria

costly, motivating cassava processors to seek other drying solutions to increase their production capacity and reduce costs. In each of these countries, partners who were willing to co-invest were identified and brought on board. An initial scaling strategy was proposed in 2019 to train scaling partners (equipment manufacturers and cassava processors) on theoretical and practical aspects of building and operating energy-efficient small-scale flash dryers. In addition, the Scaling Fund project would provide technical support to enable scaling partners to upgrade their existing flash dryers or to invest in new ones. During the project, the Scaling Readiness approach was used to identify bottlenecks and adjust the scaling strategy.

3.3 Principles, Concepts, and Case Study Application of Scaling Readiness

This section introduces the main Scaling Readiness principles and concepts and describes how these were applied in the cassava flash dryer case study following the five Scaling Readiness steps (Fig. 3.1).

3.3.1 Scaling Readiness Step 1: Characterize

During step 1, the project team characterizes the innovation, innovation package, and scaling contexts to explore interdependencies related to the scaling ambitions and aspired impacts.

3.3.1.1 Scaling Readiness Step 1: Principles and Concepts

Scaling Innovation Requires Context-Specific Approaches

A key starting point for Scaling Readiness is that scaling is contextual. Whether something goes to scale and supports the achievement of desired outcomes or impacts depends, for example, on the specific institutional setting (including cultural values, market arrangements, legal frameworks, and policy conditions), on agroecological conditions, and on the interactions that take place within and between networks of interdependent actors and stakeholders (Klerkx et al. 2010; Schut et al. 2015). This implies that an innovation may be scalable in one context but not in another and that scaling strategies successful in one situation may not be effective elsewhere or at another point in time (Baur et al. 2003; Sartas et al. 2019). Similarly, the outcomes of scaling may vary across contexts.

Innovations Never Scale in Isolation

There has been a tendency in both theory and practice to focus on the scaling of a particular – often technological – innovation (Rogers 2003). However, research has shown that the scaling of one particular innovation (e.g., a hybrid seed variety) depends on the simultaneous uptake or enhancement of other practices and services (e.g., seed multiplication, input provision, reorganization of labor, pro-poor credit models, etc.) and/or the downscaling of preexisting practices (e.g., use of open pollinated seed). All of these require attention for successful scaling.

In Scaling Readiness, we consider all innovations or changes that need to take place, including products, technologies, services, and institutional arrangements, and distinguish between "core" and "complementary" innovations. The *core innovation* refers to the initial innovation that an R4D intervention or project aims to develop or scale in order to achieve an assumed societal benefit, for example, the cassava flash dryer. *Complementary innovations* are additional advances or changes in technology, capacity, or policy on which the scaling of the core innovation depends. Together these are labelled the "innovation package."[1] In view of the contextual nature of scaling, the composition of a viable and meaningful innovation package is likely to differ across contexts. That is, the package of core and complementary innovations that is advocated needs to be tailored to different contexts, different target beneficiaries (e.g., specific gender or age groups), and may also need to change over time in view of changing conditions.

On target beneficiaries specifically, there is evidence that different groups in society may face diverse challenges and opportunities in having awareness of, having access to, being able to use, and/or benefitting from innovations. If, for example, market information is provided through a mobile phone-based SMS

---

[1] See also Bundling innovations to transform agri-food systems. Nature Sustainability 3(12): 973

service, this may benefit men who typically have more control over the household's mobile phone than women. Additional innovations that provide the same market information through a different channel (e.g., printed information provided at a community health center where women regularly visit) may be considered while trying to address some of the underlying inequalities with regard to control over communication assets. This shows that different combinations of core and complementary innovations as part of an innovation package need to be considered for achieving a specific objective or outcome. By being explicit about target beneficiaries during step 1, different types of innovation packages can be considered and designed for different groups of beneficiaries.

3.3.1.2 Scaling Readiness Step 1 in the Cassava Flash Dryer Case Study

During step 1 of Scaling Readiness, the project team characterized the innovation and its context and formulated context-specific innovation package(s) for the three countries. In the flash dryer case study, all core and complementary innovations were defined by the project team and its partners with backstopping from the scaling consultant. A total of 15 core and complementary innovations (Table 3.2) were characterized. The innovations were classified under different innovation types including technologies (i.e., the subcomponent of the machinery), products, services, and institutional arrangements. Since the contexts are different, complementary innovations necessary to scale the flash dryer also differed among the countries.

3.3.2   Scaling Readiness Step 2: Diagnose

During step 2, the project team assesses the current readiness and use of the various core and complementary innovations in the innovation package with the aim of identifying the main bottlenecks toward scaling.

3.3.2.1 Scaling Readiness Step 2 Principles and Concepts

The Scaling Readiness of an Innovation Is a Function of Innovation Readiness and Innovation Use

The notion of "innovation readiness" refers to the demonstrated capacity of an innovation to fulfill its promise or contribute to specific development outcomes. The level of innovation readiness increases as innovations progress from an untested idea to something that has been validated to work in an artificial setting (e.g., a laboratory or controlled project environment) all the way to settings where the innovation has fully matured and has been proven to work under uncontrolled

Table 3.2 Description of the flash dryer core and complementary innovations in the country-specific innovation packages

Core/ complementary innovation	Innovation type	Description	Geographical relevance		
			Nigeria	DR Congo	Colombia
Core innovation	Technology	The **Efficient Flash Dryer** is a pneumatic-conveying dryer that reduces processing cost due to innovative design.	x	x	x
Core innovation sub-component	Technology	Innovation in **the starch/ flour feed system** to the dryer that improves the homogeneity (particle size) of the supply	x	x	x
Core innovation sub-component	Technology	Technological proposal of **mechanical pre-treatment of the raw material** to reach +/- 35% humidity prior to flash drying. Options considered are press and/or centrifuge to remove water	x	x	x
Core innovation sub-component	Technology	**Hot air generator designs** adapted to the requirements and particular conditions of each country to optimize energy consumption and production costs. Depends on type of fuel available (e.g., diesel, gas, agricultural residues/biomass, etc.) and type of burner technology and heat exchanger technology available	x	x	x
Core innovation sub-component	Technology	**New fans/blowers** to achieve sufficient air velocity and flow rate, which improves production capacity	x	x	
Core innovation sub-component	Technology	Adaptation of **drying technology** to the production of sour cassava starch by testing the expansion quality of flash dried sour starch compared to sun-dried			x
Complementary innovation	Product	Preparing a **business plan template** for estimation of costs of investment and operations; estimation of revenues generated; business plans and testing them with manufacturers and processors	x	x	x
Complementary innovation	Product	A spreadsheet **template for assessing availability of raw material and energy** at acceptable cost in the target locations for a cassava starch/flour factory	x	x	x
Complementary innovation	Service	Stimulation of the cassava flour market and **promotion** through social networks; creation of new linkages between actors in the cassava flour value chain and exploration of new domestic and international markets	x		x
Complementary innovation	Service	Information from banks about the conditions and **support** to provide to access investments **loans**	x		
Complementary innovation	Service	**Capacity building** on installing and operating flash dryers	x	x	x
Complementary innovation	Service	Technical forum through facilitated **WhatsApp group** (English) and **technical support** through visits to construction sites	x	x	x
Complementary innovation	Institutional arrangement	A big physical gathering, **forum**, to bring together and promote multi-stakeholder dialogue between entrepreneurs, processors, eqpt manufacturers, funders, government agencies, etc.	x	x	x
Complementary innovation	Institutional arrangement	Assess the feasibility of establishing **cooperatives** of cassava producers (possible support by central bank loans) to ensure sufficient supply of cassava roots to the proposed starch or flour factories	x	x	x
Complementary innovation	Institutional arrangement	**Contracts** between processors, equipment manufacturers, and project teams in order to define the responsibilities, commitments (financial and otherwise), and expected benefits of all parties	x	x	x

Technologies are presented in green, products in blue, services in yellow, and institutional arrangements in orange

conditions (Table 3.3). In contrast to the notion of "technology readiness" that is used by NASA and EU, we use the term "innovation readiness" to signal that the framework can also be applied to measure the maturity of non-technological innovations.

Table 3.3 Innovation readiness and innovation use levels, short names, and basic descriptions of each term (Sartas et al. 2020a)

Level	Innovation readiness		Innovation use	
	Short name	Basic description	Short name	Basic description
0	Idea	Genesis of the innovation. Formulating an idea that an innovation can meet specific goal	None	Innovation is not used for achieving the objective of the intervention in the specific spatial-temporal context where the innovation is to contribute to achieving impact
1	Hypothesis	Conceptual validation of the idea that an innovation can meet specific goals and development of a hypothesis about the initial idea	Intervention team	Innovation is only used by the intervention team who is developing the R4D intervention
2	Basic model (unproven)	Researching the hypothesis that the innovation can meet specific goals using existing basic science evidence	Effective partners (rare)	Innovation has some use by effective partners who are involved in the R4D intervention
3	Basic model (proven)	Validation of principles that the innovation can meet specific goals using existing basic science evidence	Effective partners (common)	Innovation is commonly used by effective partners who are involved in the R4D intervention
4	Application model (unproven)	Researching the capacity of the innovation to meet specific goals using existing applied science evidence	Innovation network (rare)	Innovation has some use by stakeholders who are not directly involved in the R4D intervention but are connected to the effective partners
5	Application model (proven)	Validation of the capacity of the innovation to meet specific goals using existing applied science evidence	Innovation network (common)	Innovation is commonly used by stakeholders who are not directly involved in the R4D intervention but are connected to the effective partners
6	Application (unproven)	Testing of the capacity of the innovation to meet specific goals within a controlled environment that reflects the specific spatial-temporal context in which the innovation is to contribute to achieving impact	Innovation system (rare)	Innovation has some use by stakeholders who work on developing similar, complementary, or competing innovations but who are not directly connected to the effective partners

(continued)

Table 3.3 (continued)

Level	Innovation readiness		Innovation use	
	Short name	Basic description	Short name	Basic description
7	Application (proven)	Validation of the capacity of the innovation to meet specific goals within a controlled environment that reflects the specific spatial-temporal context in which the innovation is to contribute to achieving impact	Innovation system (common)	Innovation is commonly used by stakeholders who are developing similar, complementary, or competing innovations but who are not directly connected to the effective partners
8	Incubation	Testing the capacity of the innovation to meet specific goals or impact in natural/real/uncontrolled conditions in the specific spatial-temporal context in which the innovation is to contribute to achieving impact with support from an R4D	Livelihood system (rare)	Innovation has some use by stakeholders who are not in any way involved in or linked to the development of the R4D innovation
9	Ready	Validation of the capacity of the innovation to meet specific goals or impact in natural/real/uncontrolled conditions in the specific spatial-temporal context in which the innovation is to contribute to achieving impact without support from an R4D	Livelihood system (common)	Innovation is commonly used by stakeholders who are not in any way involved in or linked to the development of the R4D innovation

However, the maturity of an innovation along the innovation readiness scale is not the only factor that is important for understanding and assessing the scalability of an innovation or innovation package in a specific context. There are many examples of innovations with a high level of readiness that were never used at scale. Similarly, there are also examples of innovations that go to scale even if their performance is limited, contested, or poorly documented. Scaling Readiness, therefore, assumes that scalability also depends on the networks in which innovations are embedded and through which their use is supported and advocated (Geels and Schot 2007; Leeuwis and Aarts 2011; Hermans et al. 2017). For example, it makes a difference whether an innovation is only being used by directly incentivized R4D project partners or whether there are partners or beneficiaries that use or promote the innovation independently from the R4D intervention. To capture the degree to which an innovation has penetrated networks, we have introduced the notion of "innovation use" (Table 3.3). The concept also measures the relative magnitude of use (e.g., rare vs. common) to indicate both the scalability potential and actual innovation use at scale. Innovation use is measured using a network analysis approach (Sartas et al. 2018; Sartas et al. 2020a). Scaling readiness, then, must be seen as the function of innovation readiness and innovation use.

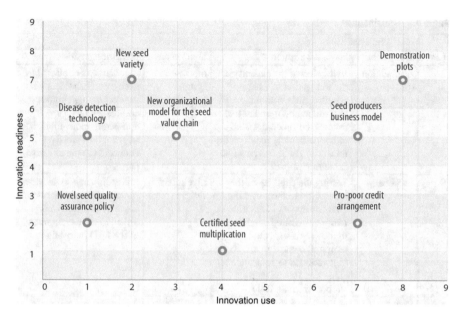

Fig. 3.4 Stylized example of an innovation package (with 8 innovations) that have been assessed for their innovation readiness (y-axis) and innovation use (x-axis) specific to space, time, and goals (Sartas et al. 2020a)

In step 2 of Scaling Readiness, these three concepts are used for diagnostic purposes. With the help of survey techniques, each innovation in a package is assessed for its innovation readiness and innovation use, and evidence of the proclaimed assessment is provided. The scaling readiness of a particular core or complementary innovation results can be found by multiplying the two scores. If the innovation readiness of a particular innovation in the package is at level 3 and innovation use at level 2, the scaling readiness for that innovation is 6 (Fig. 3.4).

3.3.2.2   Scaling Readiness Step 2 in the Cassava Flash Dryer Case Study

Based on the innovation packages defined for the flash dryer scaling work in Colombia, DR Congo, and Nigeria, the project team assessed the innovation readiness and innovation use. The first step to determine the innovation readiness and innovation use level was to collect background information via a short desktop study. The Scaling Readiness monitors gathered information about the available evidence on the readiness and use of the innovation package core and complementary innovations from academic and technical databases and repositories. To complement the desktop study, the project team also collected new data. For the innovations categorized as *core innovation subcomponents*, small-scale processors of cassava flour and starch and flash dryer manufacturers from Colombia, DR Congo, and Nigeria were visited at each location to collect technical information

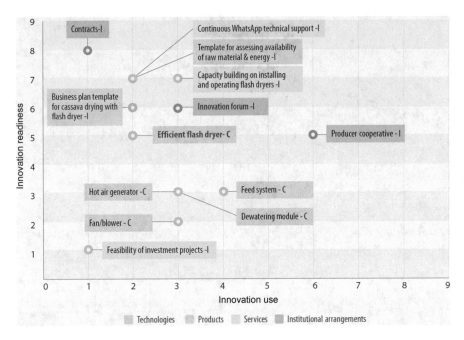

Fig. 3.5 Assessment of the innovation readiness and innovation use of the cassava flash dryer innovation package in Nigeria. (The boxes in Figs. 3.5, 3.6, and 3.7 that include "C" are the core innovation subcomponents of the flash dryer, and "I" refers to complementary innovations. Technologies are presented in green, products in blue, services in yellow, and institutional arrangements in orange.)

used in the analysis. For innovations categorized as *services* and *institutional arrangements* (Table 3.2), information was collected through surveys with different value chain actors, such as bankers, cassava producers, processors, and representatives of government organizations, among others. These surveys were administered during stakeholder meetings and forums in each location.

The information collected via the desktop review, field measurements, and survey results were processed by the Scaling Readiness monitor to determine the innovation readiness and innovation use level for each of the innovations in the innovation package for the three country contexts. A Microsoft Excel template was used to plot the Scaling Readiness graph. The template enabled selection of innovation readiness and innovation use levels from a drop-down list and automatically generated the graph (see Figs. 3.5, 3.6, and 3.7). To validate the results, the Scaling Readiness monitor presented the template to the project team and collaborators and to other key project partners who made their contributions.

The innovation readiness levels ranged between 1 and 8, while the use scores ranged from 1 to 7. These indicated that innovation packages included some new ideas (e.g., organizing an innovation forum in Colombia). Some of the subcomponents of the flash dryer were at the conceptual model stage and not yet validated by the existing applied literature (e.g., fan/blower in Nigeria and DR Congo), while other

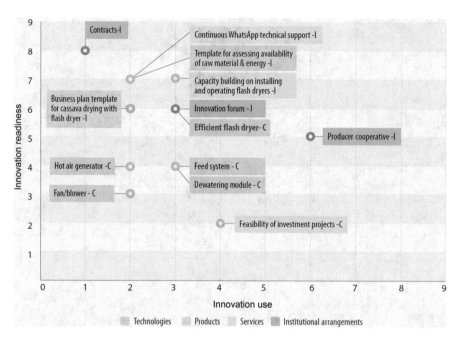

Fig. 3.6 Assessment of the innovation readiness and innovation use of the cassava flash dryer innovation package in DR Congo

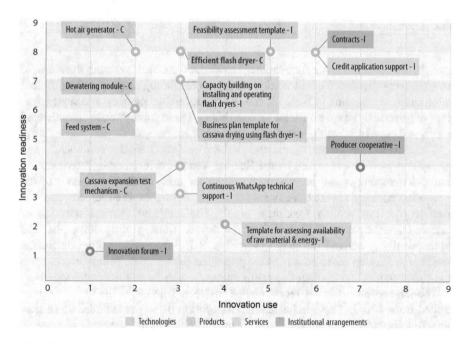

Fig. 3.7 Assessment of the innovation readiness and innovation use of the cassava flash dryer innovation package in Colombia

innovations (e.g., continuous WhatsApp technical support mechanism) were shown to work without a support from R4D interventions in real conditions. Also, the readiness and use levels of some of the components varied among different locations. For example, contracts between the cassava producers and processors were not used in DR Congo and Nigeria beyond those contracts of the project partners, while contracts are commonly used in Colombia.

The diagnosis of the innovation packages indicated that "strengthening the *feasibility of investment projects* through market promotion campaigns for small-scale cassava flour" was the key bottleneck in Nigeria. There was no campaign design that could guide the flash dryer marketing (innovation readiness level 1), and the idea of having flash dryer market promotion campaigns was still under development by the project team (innovation use level 1). In addition, some technical subcomponents of the flash dryer (i.e., hot air generator, dewatering module, and fan/blowers) were assessed at the lower readiness and use, thus needing a strategy for improvement.

In Colombia, the lack of an "innovation forum" was identified as the key bottleneck (Fig. 3.7). "Continuous WhatsApp technical support" for cassava flash dryer installation and use and a "cassava expansion testing" mechanism for the flash dryer were prioritized as the other bottlenecks.

To identify the key partners to overcome the bottlenecks, the flash dryer team used social network analysis. A survey was administered to potential partners, and results were used to characterize stakeholders and partners. Due to the limitations caused by COVID-19, only a few identified partners could be feasibly reached, and those reached did not occupy the most strategic positions in the network. Findings were captured in the stakeholder engagement reports written for DR Congo and Nigeria (Taborda et al. 2020a, 2020b).

3.3.3   Scaling Readiness Step 3: Strategize

During step 3, the project team considers different options and strategies that may be used to address the main bottlenecks to scaling for each innovation package.

3.3.3.1   Scaling Readiness Step 3 Principles and Concepts

Bottlenecks for Scaling Can Be Identified by Assessing Innovation Readiness and Innovation Use

When the core and complementary innovations have been assessed for their level of innovation readiness and innovation use, it becomes pertinent to think about strategies to enhance the readiness of the package as a whole. Scaling Readiness directs most attention to the innovations in the package with the lowest levels of readiness and use, labelled "bottleneck innovations," as they are most likely to limit the scaling of the innovation package. Unless bottlenecks have been addressed, the

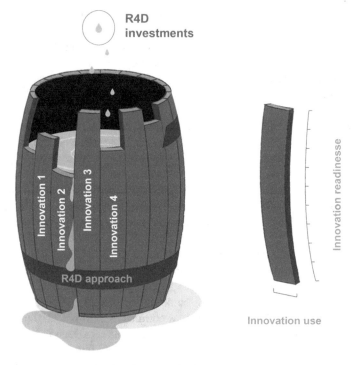

Fig. 3.8 Scaling Readiness barrel to illustrate how innovation(s) with the lowest readiness limits an innovation package's capacity to achieve impact at scale. (Adapted from "von Liebig's barrel," after Whitson and Walster 1912; published in Sartas et al. 2020a, 2020b)

value added to the effort in core or complementary innovations that already have a relatively high innovation readiness and innovation use is low. This point is illustrated in Fig. 3.8, where one can observe that R4D investments (symbolized as water drops) are wasted as they leak away from the lowest stave in the barrel, which symbolizes the bottleneck in the innovation package.

Bottleneck Innovations Can Be Overcome Through Different Strategic Options

Scaling Readiness distinguishes strategic options (i.e., innovation management options) that may be used that address a bottleneck. The choice of an appropriate strategy may be informed by available time, financial and human resources, and organizational mandates and capacities, considering what is feasible and resource efficient (derived from Sartas et al. 2020a):

1. *Substitute:* Can the bottleneck be replaced by another innovation with higher readiness and/or use in the given context?

2. *Outsource*: Are there any organizations or external experts that can more efficiently improve the Scaling Readiness of the bottleneck?
3. *Develop*: Can the intervention team improve the readiness and/or the use by investing available intervention capacities and resources?
4. *Relocate*: Can the intervention objectives be realized more effectively if the intervention is implemented in another location where innovations have higher readiness and use levels?
5. *Reorient*: Can the objective or outcome of the intervention be reconsidered if addressing the bottleneck is not possible and relocation is not an option?
6. *Postpone*: Can scaling the innovation package be achieved at a later point in time?
7. *Stop*: If none of the above strategic options are feasible, should the team consider stopping the intervention?

The strategic options are ranked according to their resource intensity, starting with the least demanding option. The options effectively imply reconsideration of the innovation package and/or the objectives and context of scaling. While we realize that existing project frameworks, budget allocations, and partnership configurations may pose limits to choosing the most sensible and efficient option, Scaling Readiness assumes that considering all options enhances discussion and critical reflection in project teams and thus contributes to the prioritization of relevant and feasible strategies to overcome bottlenecks for scaling. Clearly, the options chosen have further practical implications in terms of who are relevant partners to work with.

3.3.3.2 Scaling Readiness Step 3 in the Cassava Flash Dryer Case Study

From June 2019 onward, the flash dryer project team, scaling champions, and Scaling Readiness monitors explored strategic options for each country. Each bottleneck was discussed, and the most viable options were explored in consultation with key stakeholders and experts. Further information on the strategies is provided in Sect. 3.3.4.2.

3.3.4 Scaling Readiness Step 4: Agree

During step 4, the proposed scaling strategies are shared and discussed with relevant stakeholders to work toward effective collaboration with partners relevant to scaling.

3.3.4.1 Scaling Readiness Step 4 Principles and Concepts

Implementing Scaling Strategies Requires Multi-stakeholder Agreement
and Coalition Building

The scaling of an innovation package inherently requires the involvement and cooperation of the various stakeholders. Depending on the package and bottlenecks, these may include policy makers, value chain parties, farmer organizations, community leaders, and/or service (e.g., extension and credit) providers. While AR4D projects depend on the collaboration of such partners to realize their scaling ambition, these parties may not necessarily agree with the proposed scaling strategy, nor may they be ready to take effective action (Sahay and Walsham 2006; Wigboldus et al. 2016). Thus, it is important that initiatives are taken to align interdependent actors and work toward agreement and accommodation on, for example, objectives, strategies, task division, timelines, and investment of resources to enable scaling.

In essence, the process of aligning stakeholders amounts to building an effective coalition that supports change in a particular direction, even if the rationales and interests of stakeholders may only partially overlap (Biggs and Smith 1998; Aarts and Leeuwis 2010). Reaching the necessary degree of accommodation and consensus is far from automatic and often requires active facilitation of learning and negotiation (Leeuwis and Aarts 2011).

3.3.4.2 Scaling Readiness Step 4 in the Flash Dryer Case Study

From June 2019 onward, the flash dryer project team began engaging partners and broader stakeholders in multiple countries to discuss the proposed strategies to improve the scaling readiness of the flash dryer innovation package. In August 2019, combined with a training workshop on small-scale flash drying, the cassava flour processors and equipment manufacturers partners were presented the strategies, and their feedback was collected.

Based on the consultations with partners and key experts and feedback from the workshop, a final strategy for scaling the flash dryer was formulated for the three countries (Table 3.4). To enhance the commitment of the partners to the new strategy, partners were requested to provide their consent and support in writing clearly specifying their intention to participate in the implementation of the scaling action plans. This took the form of an umbrella *participation agreement* explaining the roles, responsibilities, and commitments of the partners (processors, equipment manufacturers) and of the project team to accomplish the goals of the project.

Table 3.4 Scaling Readiness strategic options selected to overcome the bottlenecks for scaling the cassava flash dryer innovation in the different project countries

Bottlenecks (see Figs. 3.5, 3.6, and 3.7)	Bottleneck description	Location	Strategic option	Strategy description
Strengthening the feasibility of investment projects through market promotion campaigns for small-scale cassava flour	Initially, the campaign was conceived to address the lack of markets for small-scale flour producers. However, *the agree step* showed that the dominant majority of the processors did not think increasing the markets is viable in the short term due to the huge efficiency gap and lack of the implementation of the existing local production incentives	Nigeria	Reorient	Since yam prices are higher and yam dryer has higher profitability, the team decided to explore options to use flash dryer for yam
			Postpone	To capitalize a possible favorable change in cassava markets, a demonstration flash dryer was built for promotion and training at the R&D institution FIIRO (although an impact was not anticipated until the end of the project)
Fan/blower, hot air generator, and dewatering subcomponent (only Nigeria)	The capacity of fans to achieve adequate air velocity was too low, resulting in low production capacity. Since the flash dryer team had advanced engineering capabilities, the team chose to optimize the fan/blower designs, share those designs with the manufacturers, and help processors to install them	DR Congo/ Nigeria	Develop	The team has worked on developing the fan and calibrating this tool for efficient drying. Improved fans/blowers were developed and installed (see Figs. 3.5 and 3.6)
Innovation forum	The flash dryer team initially strategized that an innovation forum could increase the awareness of the cassava processors and match them with manufacturers of flash dryers, creating business opportunities	Colombia	Develop and outsource	The team started preparations for the forum and engaged with several organizations to co-organize with. Several activities were planned for outsourcing with other organizations

(continued)

Table 3.4 (continued)

Bottlenecks (see Figs. 3.5, 3.6, and 3.7)	Bottleneck description	Location	Strategic option	Strategy description
Continuous WhatsApp technical support	At the strategizing step, based on the cassava project experience in Africa, the flash dryer team identified that a WhatsApp group could be an instrument for building capacity with the manufacturers and processors	Colombia/ DR Congo/ Nigeria	Develop	The team has opened the WhatsApp group and invited the manufacturers and processors from Africa and Latin America who attended the august workshop in Cali. The information traffic in this network and number of members increased continuously
Cassava expansion mechanism	In Colombia, cassava starch is used for special breads, which requires the expansion. The team has strategized that it can further develop the flash dryer	Colombia	Develop	The team conducted an experiment to measure the effect of various expansion options

3.3.5   Scaling Readiness Step 5: Navigate

During step 5, the project teams monitor the unfolding dynamics in relation to the implementation of agreed-upon scaling strategies and scaling action plans and signal whether major changes in the innovation package configuration or scaling context require a new cycle of Scaling Readiness assessment.

3.3.5.1   Scaling Readiness Step 5 Principles and Concepts

Scaling Projects Need Capacity to Adjust to Emergent Dynamics

When implementing scaling strategies, partners and project teams are likely to meet with unforeseen developments and unintended effects (Hall and Clark 2010; Paina and Peters 2012). This is because scaling contexts are ever-changing and, therefore, can never be fully anticipated (Schot and Geels 2008). It is quite conceivable, for example, that scaling partners meet with new constraints and challenges in their efforts to enhance the Scaling Readiness of a package, or that successful scaling appears to have unwanted side effects for the environment or for specific segments in farming communities. Thus, AR4D interventions require mechanisms to capture and navigate such emergent dynamics. Thus, project teams need to continue to invest in learning and critical reflection when scaling strategies and action plans are implemented.

Relatedly, Scaling Readiness distinguishes between short- and long-term learning cycles and feedback loops (Sartas et al. 2020b). In short-term learning and feedback loops, the focus is on monitoring how the agreed-upon action plans for addressing the bottlenecks are being implemented and on whether plans must be adapted. The long-term learning and feedback loop actually involves a new round of going through the Scaling Readiness cycle, starting with reiterating the Characterize Step 1 and Diagnose Step 2 (See Fig. 3.1). Here, the emphasis is on assessing whether the scaling context has changed and on whether the implementation of scaling strategies has yielded the desired effects. Insights derived from such assessments may result in a reconfiguration of the innovation package, the identification of new bottleneck innovations, and subsequent adaptation of agreed-upon scaling strategies.

3.3.5.2   Scaling Readiness Step 5 in the Cassava Flash Dryer Case Study

The flash dryer team has implemented the agreed-upon strategies presented in Table 3.4 and closely monitored the activities and whether they resulted in the desired improvements in the project. However, travel restrictions due to the COVID-19 pandemic and related closure of businesses necessitated changes in the strategies.

Short-Term Learning and Feedback Loops

In Colombia, the innovation forum was initially postponed (and later cancelled altogether) when the project team realized that the COVID-19 travel restrictions would last much longer than expected. Furthermore, the partners that would co-invest in the flash dryer suspended their commitments to the project and their investment plans. In addition, the development efforts of the flash dryer by the team for the cassava expansion mechanism did not result in desired improvements. As a result, the flash dryer team has decided to stop activities in Colombia and revisited the Scaling Readiness strategic options, adopting a dual strategy:

1. *Relocate and outsource*: Initial consultations with processors in the Dominican Republic showed that there is large interest for the flash dryer; thus, the team decided to relocate there. Since the Dominican Republic was not one of the initial project countries and because the organizational partners of RTB do not have implementation capabilities, the team also decided to outsource the work there to a company called Angavil. The project provided technical support to Angavil to develop its investment plans in flash drying technology for production of cassava flour in the Dominican Republic.

2. *Reorient the flash dryer toward production of high-grade cassava starch* for bio-plastics and support an initiative led by the Universidad del Cauca to develop a start-up company in Colombia, funded by Colciencias, the national agency for scientific development. The flash dryer project team decided to provide technical support to this initiative.

In Nigeria, the main bottlenecks were market options for cassava flash drying and inefficient fans/blowers that increased the cost of drying (Fig. 3.5). The customers of cassava flour (e.g., millers, brewers) required much higher volumes than could be supplied by small-scale producers. The buyers typically demand 30 or 60 tons per order, whereas some of the small-scale cassava flour factories can produce up to 1–2 tons per day. This situation led to underutilization of the flash drying capacity since some of the flour producers were too small to be economically viable suppliers for the large-scale buyers: Only 32 flash dryer businesses (out of 64 known flash dryers) were viable users for the cassava flash drying. The fans/blowers of the drying system presented another bottleneck in Nigeria. The team has developed the fans/blowers of the drying system by designing improvements and testing them in the flash drying producers' workshops (Table 3.4).

In the DR Congo, like Nigeria, the main bottleneck was the fans/blowers (Fig. 3.6).

The capacity of fans to achieve adequate air velocity was too low, resulting in low production capacity. Equipment manufacturers acknowledged that they did not have enough experience to build larger fans (due to balancing issues with the rotor) and that they did not know the methods to determine the efficiency of the fan (e.g., air velocity measurements). To address this need, the complementary innovation of flash dryer fans/blowers was added to the innovation package. The team has worked on developing the fan and is calibrating it for efficient drying (Table 3.4).

In both Nigeria and DR Congo, the *heat exchanger* was developed with the goal of driving specific modifications to existing (diesel) heat exchanger designs as well as promoting the manufacture of a new, more efficient heat exchanger design. Out of eight initial private sector scaling partners, three had adopted this innovation by the end of 2019 and increased their processing capacity by 23–50% and profitability by 8–10% – which corresponds to an extra USD 10,000 per year per processor. Cassava producers also benefit from a higher processing capacity, which increases the demand for cassava roots and, hence, economic opportunities for farmers in the regions around cassava factories. Since the most commonly used fuel in Nigeria and DR Congo is diesel, the partners were recommended to change their heating systems to liquid propane gas (LPG), as long as the price of LPG stays competitive in the region. This is a more cost-effective solution (because there is no need to manufacture a heat exchanger) and approximately 10% more efficient with respect to the use of diesel. In 2020, two partners in DR Congo invested in this innovation.

In the Dominican Republic, a scaling action plan was agreed to in early 2020 between the cassava processors, equipment manufacturers and R4D team. The plan was then revised several times through short-term feedback loops. The main feedback was related to the definition of responsibilities for the investment risk, as the cassava processor wanted a guaranteed return on investment, while the equipment manufacturer or the R4D team could not take responsibility for this guarantee due to the novelty of the innovation. Each of the partners reviewed their expectations until the scaling action plan was revised and agreed upon.

Long-Term Learning and Feedback Loops

The short-term learning and changes in the strategies were complemented by an annual assessment of activities. The teams initiated a second Scaling Readiness characterization Step 1, but due to the COVID-19 pandemic, the collection of necessary qualitative information was difficult and this step was suspended.

More generally speaking, in all countries, the COVID-19 pandemic affected most of the programmed activities. For example, two project partners in Latin America (from Colombia and the Dominican Republic) postponed their planned investment in the flash dryer due to the economic impact of the pandemic and a 17% increase in the price of the flash dryer due to the depreciation of the local currency.

The pandemic also caused new challenges and/or bottlenecks to emerge. For example, providing remote technical support for dryer manufacturing, installation, and testing was slower and more complicated than doing it on-site. Some instructions and recommendations provided by video conference were misinterpreted, leading to a need to repeat the work and consequent extra costs. Additionally, in some places in Nigeria and DR Congo, Internet service and electricity supply are patchy, which hindered effective communication with project partners. One strategy to provide efficient remote technical support during the pandemic was the development of protocols and video tutorials for the project partners (e.g., step-by-step assessment methodologies for drying efficiency).

In Nigeria, the implementation of scaling action plans was slower than expected as most project partners delayed their decisions to invest in the flash dryer innovations. Consultation with the different actors in the high-quality cassava flour value chain revealed an emerging bottleneck that had not been identified during the initial Scaling Readiness characterization and diagnosis steps, namely, the rising cost of cassava roots in 2020 that reduced the profit margin of processing. A detailed economic analysis revealed the limited use of high-yielding varieties and good agronomic practices as two of the underlying causes for the limitations.

Although the flash dryer Scaling Fund project did not have the time nor resources to contribute directly to the improvement of cassava production yields scenario, exploring *synergies with other projects* (see Fig. 3.5) became an immediate priority, and a cooperation with the African Development Bank-funded Technologies for African Agricultural Transformation (TAAT) program was established. This program has pursued an objective to provide technical assistance for efficient cassava root production in several African countries, including Nigeria. Overcoming this bottleneck will be key to reducing the cost of cassava roots and related bottle-necks on investment in processing of high-quality cassava flour.

Toward the end of the project, due to the persistence of the COVID-19-related limitations, the team decided to create a virtual platform to scale the flash dryer. Having realized that returning to previous offline work is unlikely, digital solutions were considered one of the best options for advancing the scaling work.

3.4   Reflection on the Use of Scaling Readiness in the RTB Scaling Fund

3.4.1   Reflections on the Use of Scaling Readiness by the Flash Dryer Case Study Team

One of the strengths of the Scaling Readiness approach is that innovation packages are formulated and diagnosed for different scaling contexts. This was acknowledged by the flash dryer project team and was generally appreciated by scaling project teams who used Scaling Readiness. Since each country has its particular context and related bottlenecks, strategies must be adapted to each context to define the most appropriate way to achieve scaling. For example, the teams appreciated analysis required to identify the most appropriate heat generation system for each country to reduce fuel consumption and contribute to the energy efficiency of cassava drying.

To determine the degree of maturity or readiness of the innovations and the level of innovation use, it is necessary to have a deep knowledge of the context. In this regard, one of the lessons learned from the flash dryer case was that it is not enough to carry out this analysis only at the beginning of the project, but that periodic diagnoses must also be carried out, since the context is dynamic and changing, and *emerging* bottlenecks can arise.

Collecting information to design strategies and monitor scaling progress was a challenge in the case study project. Scaling Readiness collects this information through electronic surveys, but it was found that most of the project partners were unresponsive. Some were very busy or did not believe they had the capacity to complete an electronic survey. The challenge will be to develop mechanisms that capture as much information as possible while being user-friendly for the project partners. Collecting information during project meetings or as part of workshops seemed to be more promising in terms of response rates and data quality.

In the course of project implementation, differences emerged between countries in terms of the distribution of responsibilities between processors, equipment manufacturers, and the project team. This dynamic also was felt at the level of co-investment realized and the distribution of financial risks. In DR Congo, cassava processors were confident in the market for cassava flour, which is a staple food product in this country, and, therefore, were more willing to invest in flash dryers and take the financial risks without having a formal agreement. In Nigeria, the market for cassava flour is not functioning smoothly due to a combination of factors, such as low availability of cassava roots at competitive prices for flour production, high processing costs due to low energy efficiency of current flash dryers, and a mismatch between production capacity of cassava processors and demand from large buyers. Consequently, cassava processors were less confident to bear investment risks, with the majority preferring to wait for successful implementation of the flash drying innovation before investing themselves. In Latin America, investment costs (and financial risk) were significantly higher due to higher labor costs and other constraints. In addition, the market for cassava flour is not mature yet. Consequently, cassava processors were not willing to fully take on the investment risks and required that equipment manufacturers, or the flash dryer Scaling Project, offer guarantees against construction cost overruns and potential financial underperformance of the flash dryer system. This led to negotiations and written agreements in the form of a sales contract between the cassava processor and the equipment manufacturer.

These examples underscore the idea that scaling projects always entail financial risk-taking, considering that innovations, by nature, are not yet fully proven with guaranteed return on investment. Therefore, a key bottleneck is finding agreement between project partners who will take on responsibility for these risks. One option to manage this is to identify and select early in the project private partners who are in a position to accept the risks. That is, partners with financial capacity for investment, confidence in the benefits of the innovation, and access to technical expertise to remedy emerging challenges before and after the construction and delivery of the equipment.

3.4.2 Reflections on the Use of Scaling Readiness in the RTB Scaling Fund

We offer four main reflections:

1. When the first batch of Scaling Fund projects was selected and approved – early 2018 – the Scaling Readiness approach was still under development. Although the basic principles and concepts of Scaling Readiness were defined, tested, and validated, there were no clear guidelines and workflows that supported its application with partners in controlled conditions – the Scaling Fund projects. Those guidelines and capacity development materials were developed in parallel to Scaling Fund project implementation, which sometimes resulted in confusion (e.g., What is an innovation? What is an innovation package? How to measure and document innovation readiness and innovation use? How to deal with gender and diversity among beneficiaries?). This lack of development also meant a steep learning curve between the Scaling Readiness team and Scaling Fund project teams. The second and especially third batches of Scaling Fund projects

benefitted from those learnings, resulting in a more organized and tailored application of Scaling Readiness. A very concrete spin-off of such learning is the development of a gender-responsible scaling tool for identifying relevant diversity in relation to scaling ambitions (which is currently being designed and tested for use in combination with Scaling Readiness).

2. Scaling of innovation (increasing innovation use) is very different from processes related to designing, testing, and validating the innovation through basic and applied research (improving innovation readiness). For doing the scaling, different skills and competencies, language, organizational space, and incentive structures are required. Many of these skills are very different from those that scientists obtain during their PhD trajectories and require competencies related to being opportunistic, taking risks, and negotiating with scaling partners. After the first year of Scaling Fund implementation, we decided that projects had to identify dedicated scaling champions and Scaling Readiness monitors to ensure that scaling projects were not treated and organized in the same way as science projects. Having dedicated scaling champions and Scaling Readiness monitors clarified the division of tasks and responsibilities in the RTB scaling projects. In addition, capacity development on innovation and scaling processes was very much appreciated by the Scaling Fund project teams.

3. Having senior staff in charge of scaling project design, implementation, and decision-making was not always compatible with best practices for managing scaling projects. The time and responsibilities of senior staff are often fragmented, meaning they need to juggle to a broad variety of science, management, and leadership demands and expectations. Furthermore, scientists are often not on the ground in the context where scaling is desired. The environments in which the cassava flash dryer and other RTB Scaling Fund projects operate are very dynamic and require ongoing *navigation* and *re-strategizing*, which requires operational knowledge. One of the opportunities we see here is to decentralize management and decision-making in scaling projects so that on-the-ground scaling champions can act in a flexible manner based on the analysis and data provided by Scaling Readiness monitors. We have seen that Scaling Fund projects where such a decentralized model for decision making was applied seemed more successful in capitalizing on emerging opportunities and navigating change.

4. When starting Scaling Fund implementation, it was expected that in the first months of the projects, research and scaling partners would go through a cycle of Scaling Readiness steps and start implementing and monitoring their scaling strategies and action plans. This turned out to be very different in actual practice. First, the process of sense-making and capacity development took much longer than expected. Many of the project teams had very different or unclear ideas about their "innovations" and struggled to think critically about what scaling pathways and mechanisms would be required to actually make their products, services, or tools available to end users. Second, working with co-investing scaling partners was essential but also difficult. Scaling partners – especially when they are co-investing – are very deliberate on whether and how

to engage and need to see added value in investing their time and resources in the partnership. As in the flash dryer case, scaling partners may propose risk-sharing strategies – or pull out altogether – if the investment conditions change. It made us realize that the 2-year Scaling Fund projects were essentially about finding common ground between research and scaling partners and creating the space for negotiation, adaptation, and integration that is needed before the actual scaling can happen.

3.5   An Outlook on the Broader Use of Scaling Readiness

The CGIAR Research Program on Roots, Tubers and Bananas has pioneered support and investment in the development and implementation of scaling strategies and partnerships to catalyze the scaling of its innovations. The consequent investment in the Scaling Fund has been timely. The entire CGIAR (2020) is reorganized around an impact-oriented approach, and *scaling* will figure prominently among the different parts of the organization.

Several of the Scaling Fund and Scaling Readiness principles could be embedded in a new way of doing business in CGIAR.

1. *Keep track of innovation readiness and innovation use.* Tracking these elements can support monitoring, prioritization, and resource allocation. Doing this in an evidence-based and structured way can increase transparency, facilitate decision-making, support resource mobilization, and demonstrate return on investment at both the innovation package and portfolio levels. By portfolio level, we mean the management of a broad number of innovation packages and making decisions on which ones to prioritize.
2. *Combine innovation readiness and innovation use in one framework.* With this idea, international organizations, such as the CGIAR, can better link research and development as part of its mandate. Within such a framework, science and applied research focusses on improving innovation readiness in close collaboration with expected beneficiaries and innovation partners, and – once proven to work – scaling can focus on improving innovation use with scaling partners.
3. *Capitalize on the promise of the Scaling Fund co-investment model where research and scaling partners jointly commit funds and capacities to preparing the innovation.* This recommendation would provide a more level playing field between partners and create a higher likelihood that innovations are adapted to become of real value to scaling partners. During the initial stages of sense-making and finding agreement on what the innovation package looks like and which bottlenecks should be prioritized, a *safe incubation space* – such as provided in the Scaling Fund – serves to reduce risk and incentivize partners to find common ground.

4. *Work to create an ecosystem in which the rules and cultures of scaling are different than what people are accustomed to.* This made us realize that the appropriate use of novel approaches such as Scaling Readiness and working in an impact-oriented manner needs to go hand in hand with organizational culture change, capacity development and new incentive structures that reward project teams to prioritize work on bottlenecks in innovation packages. Similarly, strategic options such as reorientation, postponing, relocating, or stopping an intervention when innovation and scaling bottlenecks cannot be overcome should be encouraged, rather than be labelled as a failure as it can avoid wasting of valuable R4D resources.

5. *Ongoing efforts to make Scaling Readiness more sensitive to gender and social differentiation will need to continue.* It seems promising to explore whether and how innovation packages, and scaling strategies can be tailored to groups that are at risk of being excluded.

Acknowledgments This research was largely undertaken as part of, and funded by, the CGIAR Research Program on Roots, Tubers and Bananas (RTB) and supported by CGIAR Trust Fund contributors. The authors thank the following individuals who provided input and feedback on earlier drafts of this chapter: Tom Remington, Jill Lenne, and Christopher Butler. We greatly appreciate Zandra Vasquez's assistance with references and formatting and Claudio Proietti for his support in effectively managing the RTB Scaling Fund.

References

Aarts N, Leeuwis C (2010) Participation and power: reflections on the role of government in land use planning and rural development. J Agric Educ Ext 16(2):131–145

Baur H, Poulter G, Puccioni M, Castro P, Lutzeyer HJ, Krall S (2003) Impact assessment and evaluation in agricultural research for development. Agric Syst 78(2):329–336

Biggs S, Smith G (1998) Beyond methodologies: coalition-building for participatory technology development. World Dev 26(2):239–248

CGIAR (2020) CGIAR performance and results knowledge Hub. Retrieved 10 May 2021, from https://sites.google.com/cgxchange.org/performance/hom

European Commission (2014) Horizon 2020 work program 2014–2015 general annexes revised. Extract from Part 19 - Commission Decision C(2014)4995. ec.europa.eu. European Commission. Retrieved 11 November 2019

Geels FW, Schot J (2007) Typology of sociotechnical transition pathways. Res Policy 36:399–417

Hall A, Clark N (2010) What do complex adaptive systems look like and what are the implications for innovation policy? J Int Dev 22(3):308–324

Hammond J, Rosenblum N, Breseman D, Gorman L, Manners R, van Wijk MT, Sibomana M, Remans R, Vanlauwe B, Schut M (2020) Towards actionable farm typologies: scaling adoption of agricultural inputs in Rwanda. Agric Syst 183:102857

Hermans F, Sartas M, van Schagen B, van Asten P, Schut M (2017) Social network analysis of multi-stakeholder platforms in agricultural research for development: opportunities and constraints for innovation and scaling. PLoS One 12(2):e0169634

Hounkonnou D, Kossou D, Kuyper TW, Leeuwis C, Nederlof ES, Röling N, Sakyi-Dawson O, Traoré M, Van Huis A (2012) An innovation systems approach to institutional change: smallholder development in West Africa. Agric Syst 108(5):74–83

IITA (2016) Making large scale-cassava drying technologies work for small-scale processors. http://bulletin.iita.org/index.php/2016/04/17/making-large-scale-cassava-drying-technologies-work-for-small-scale-processors/

Klerkx L, Aarts N, Leeuwis C (2010) Adaptive management in agricultural innovation systems: the interactions between innovation networks and their environment. Agric Syst 103(2010):390–400

Kobos PH, Malczynski LA, Walker LTN, Borns DJ, Klise GT (2018) Timing is everything: a technology transition framework for regulatory and market readiness levels. Technol Forecast Soc Chang 137:211–225

Leeuwis C, Aarts N (2011) Rethinking communication in innovation processes: creating space for change in complex systems. J Agric Educ Ext 17(1):21–36

Leeuwis C, Klerkx L, Schut M (2018) Reforming the research policy and impact culture in the CGIAR: integrating science and systemic capacity development. Glob Food Sec 16:17–21

Paina L, Peters DH (2012) Understanding pathways for scaling up health services through the lens of complex adaptive systems. Health Policy Plan 27(5):365–373

Rogers EM (2003) Diffusion of innovations. Free Press, New York

Sahay S, Walsham G (2006) Scaling of health information systems in India: challenges and approaches. Inf Technol Dev 12(3):185–200

Sartas M, Schut M, Hermans F, van Asten P, Leeuwis C (2018) Effects of multi-stakeholder platforms on multi-stakeholder innovation networks: implications for research for development interventions targeting innovations at scale. PLoS One 13(6):e0197993

Sartas M, van Asten P, Schut M, McCampbell M, Awori M, Muchunguzi P, Tenywa M, Namazzi S, Sole AA, Thiele G, Proietti C, Devaux A, Leeuwis C (2019) Factors influencing participation dynamics in research for development interventions with multi-stakeholder platforms: a metric approach to studying stakeholder participation. PLoS One 14(11):e0223044

Sartas M, Schut M, Proietti C, Thiele G, Leeuwis C (2020a) Scaling readiness: science and practice of an approach to enhance the impact of research for development. Agric Syst 183:102874

Sartas M, Schut M, van Schagen B, Velasco C, Thiele G, Proietti C, Leeuwis C (2020b) Scaling readiness: concepts, practices, and implementation, CGIAR Research Program on Roots, Tubers and Bananas (RTB). 2020, 217 pp. Available at www.scalingreadiness.org and www.rtb.cgiar.org

Schot J, Geels FW (2008) Strategic niche management and sustainable innovation journeys: theory, findings, research agenda, and policy. Tech Anal Strat Manag 20:537–554

Schut M, Klerkx L, Rodenburg J, Kayeke J, Raboanarielina C, Hinnou LC, Adegbola PY, van Ast A, Bastiaans L (2015) RAAIS: rapid appraisal of agricultural innovation systems (part I). A diagnostic tool for integrated analysis of complex problems and innovation capacity. Agric Syst 132(2015):1–11

Schut M, van Asten P, Okafor C, Hicintuka C, Mapatano S, Nabahungu NL, Kagabo D, Muchunguzi P, Njukwe E, Dontsop-Nguezet PM, Sartas M, Vanlauwe B (2016) Sustainable intensification of agricultural systems in the central African highlands: the need for institutional innovation. Agric Syst 145:165–176

Schut M, Leeuwis C, Thiele G (2020) Science of Scaling: understanding and guiding the scaling of innovation for societal outcomes. Agric Syst 184(102908):1–10

Spielman DJ, Ekboir J, Davis K (2009) The art and science of innovation systems inquiry: applications to sub-Saharan African agriculture. Technol Soc 31(4):399–405

Taborda LA, Adegbite S, Sartas M (2020a) Stakeholder engagement strategy for Scaling small scale flash drying system for cassava starch and flour production in Nigeria, CIAT and IITA

Taborda LA, Lukombo SS, Sartas M (2020b) Stakeholder engagement strategy for Scaling small scale flash drying system for cassava starch and flour production in Democratic Republic of Congo, CIAT and IITA

Verma D, Ramirez-Marquez J (2006) From TRL to SRL: the concept of systems readiness levels. Presented at the Systems Engineering Research, www.boardmansauser.com

Whitson AR, Walster HL (1912) Soils and soil fertility. St. Paul, MN: Webb. 73 p. OCLC 1593332. 100. Illustration of limiting factors

Wigboldus S, Klerkx L, Leeuwis C, Schut M, Muilerman S, Jochemsen H (2016) Systemic perspectives on scaling agricultural innovations. A review. Agron Sustain Develop 36(3):1–20

Woltering L, Fehlenberg K, Gerard B, Ubels J, Cooley L (2019) Scaling – from "reaching many" to sustainable systems change at scale: a critical shift in mindset. Agric Syst 176:102652

Part II
Processing, Marketing and Distribution

Chapter 4
Cost-Effective Cassava Processing: Case Study of Small-Scale Flash-Dryer Reengineering

Thierry Tran ⓘ, Adebayo Abass, Luis Alejandro Taborda Andrade ⓘ, Arnaud Chapuis ⓘ, Marcelo Precoppe ⓘ, Laurent Adinsi ⓘ, Alexandre Bouniol ⓘ, Makuachukwu Ojide ⓘ, Suraju Adeyemi Adegbite ⓘ, Simon Singi Lukombo ⓘ, Murat Sartas ⓘ, Béla Teeken ⓘ, Apollin Fotso Kuate ⓘ, Robert Ndjouenkeu ⓘ, Martín Moreno ⓘ, John Belalcázar ⓘ, Luis Augusto Becerra López-Lavalle ⓘ, and Dominique Dufour ⓘ

Abstract The development and scaling out of flash-dryer innovations for more efficient, small-scale production of high-quality cassava flour (HQCF) and starch is described. The diagnoses of cassava-processing SMEs (small and medium enterprises) revealed their energy expenditures for drying were considerably higher than those of large-scale industrial companies, which was mostly due to suboptimal design of flash-drying systems. As a result, small-scale production of cassava starch and HQCF often incurs high production costs, incompatible with market prices of final products. Taking stock of this situation, RTB scientists have developed several innovations to optimize energy efficiency and costs, including a longer drying pipe, reengineered heat exchanger, larger blower for higher air velocity, and a higher product/air ratio. This was based on numerical modelling to determine the key design features of energy-efficient flash dryers, followed by construction and demonstration of a pilot-scale prototype. As a result, improved small-scale flash dryers are now being scaled out to the private sector in various countries, using the Scaling Readiness framework and achieving 10–15% gains in productivity and incomes. A method for diagnosis of process efficiency is also described, to identify technical bottlenecks and to document and measure the outcomes and impacts during the implementation of scaling-out projects.

T. Tran (✉)
Alliance Bioversity-CIAT, Cali, Colombia

CIRAD UMR Qualisud, Montpellier, France
e-mail: thierry.tran@cirad.fr

A. Abass
International Institute of Tropical Agriculture (IITA), Dar-Es-Salaam, Tanzania
e-mail: a.abass@cgiar.org

G. Thiele et al. (eds.), *Root, Tuber and Banana Food System Innovations*,
https://doi.org/10.1007/978-3-030-92022-7_4

4.1 Introduction

Postharvest processing is a key link between cassava crops and end products, determining the quality and food safety for consumers. Postharvest processing also contributes to food security by stabilizing and increasing the shelf life of

L. A. T. Andrade
Centro Internacional de Agricultura Tropical (CIAT), Cali, Colombia
Universidad Nacional de Colombia, Bogotá, Colombia
e-mail: latabordaa@unal.edu.co

A. Chapuis · D. Dufour
CIRAD, Montpellier, France
e-mail: arnaud.chapuis@cirad.fr; dominique.dufour@cirad.fr

M. Precoppe
Natural Resources Institute (NRI), Chatham, UK
e-mail: M.Precoppe@greenwich.ac.uk

L. Adinsi
Université d'Abomey-Calavi, Faculté des Sciences Agronomiques (UAC-FSA), Jericho, Benin

A. Bouniol
Université d'Abomey-Calavi, Faculté des Sciences Agronomiques (UAC-FSA), Jericho, Benin

CIRAD UMR Qualisud, Cotonou, Benin
e-mail: alexandre.bouniol@cirad.fr

M. Ojide
Alex Ekwueme Federal University Ndufu-Alike-Ikwo (FUNAI), Ebonyi state, Nigeria

S. A. Adegbite
Federal Institute of Industrial Research Oshodi (FIIRO), Lagos, Nigeria

S. S. Lukombo
International Institute of Tropical Agriculture (IITA), Kinshasa, DR, Congo
e-mail: S.Lukombo@cgiar.org

M. Sartas
Wageningen University, Wageningen, Netherlands

International Institute of Tropical Agriculture (IITA), Kigali, Rwanda
e-mail: M.Sartas@cgiar.org; murat.sartas@wur.nl

B. Teeken
International Institute of Tropical Agriculture (IITA), Ibadan, Nigeria
e-mail: B.Teeken@cgiar.org

A. F. Kuate
International Institute of Tropical Agriculture (IITA), Yaoundé, Cameroon
e-mail: A.Fotso@cgiar.org

R. Ndjouenkeu
Ecole Nationale Supérieure des Sciences Agro-Industrielles, University of Ngaoundéré, Ngaoundéré, Cameroon

M. Moreno
Universidad del Valle (Univalle), Cali, Colombia
e-mail: martin.moreno@correounivalle.edu.co

J. Belalcázar · L. A. Becerra López-Lavalle
Alliance Bioversity-CIAT, Cali, Colombia
e-mail: l.a.becerra@cgiar.org

perishable crops such as cassava and other roots, tubers, and bananas. Beyond product quality and safety, processing is also crucial for improving the sustainability of cassava value chains: by optimizing processing technologies, it is possible to reuse product waste while reducing energy and water consumption, product losses, production costs, and the overall environmental footprint of cassava industries. Processing is a crucial way for many small and medium farmer-processors to add value to their crops. From a gender perspective, especially in Africa, women are important in processing roles, so improving processing technologies can potentially improve their working conditions and increase their incomes. Women can benefit from processing innovations that match their dominant type of technology, i.e., small and medium scale (Taiwo and Fasoyiro 2015; Teeken et al. 2018, 2021; Ndjouenkeu et al. 2021; Thiele et al. 2021). Processing innovations may also influence gender roles if new technologies change the processing scale or the level of involvement of different stakeholders in the value chain (Forsythe et al. 2015, 2016).

Postharvest processing of cassava involves several steps, called unit operations, to transform fresh roots into several finished products. Some unit operations are common to most cassava products, including washing, peeling, rasping/grating, and dewatering (pressing). Other unit operations are specific to certain products, particularly the cooking or drying operations such as toasting for gari, steaming for attiéké, sun drying for fufu and conventional cassava flour, and flash drying for high-quality cassava flour (HQCF) and starch. Lactic fermentation is also a specific unit operation used to preserve and confer a desirable sour taste to the end products (gari, fufu, chikwangue). Cassava processing generates large amounts of by-products (e.g., peels, fiber bagasse, and wastewater), which typically accumulate around the processing sites or pollute local water systems (Tran et al. 2015). Expansion of postharvest technologies must therefore include strategies for by-product management and processing to reduce environmental impacts and to create additional income streams for processors, most commonly animal feed, but also biomass to energy, such as solid fuel and biogas (Okudoh et al. 2014; Ozoegwu et al. 2017; Patrizi et al. 2020; Yank et al. 2016; Zvinavashe et al. 2011).

Over the past 20 years, research for development projects has identified the high potential of interventions in postharvest processing to improve working conditions and reduce inefficiencies in cassava value chains across Africa, Latin America, and Asia (Abass et al. 2013, 2017, 2018; Awoyale et al. 2017; Kuye et al. 2011; Dufour et al. 2002). Cassava processing is mainly a small-scale operation, in particular in Africa, at factories handling less than 5 tons of fresh roots per day or at household level. Current small-scale technologies are often suboptimal, resulting in high use of resources (energy, water, firewood, product losses) and high production costs (Adenle et al. 2017; Da et al. 2013; Kitinoja et al. 2011; Nzudie et al. 2020). Improving the efficiency of small-scale equipment, their design, safety, and ergonomics is therefore essential for the competitiveness of cassava value chains (Abass et al. 2018; Edeh et al. 2020; Oni and Oyelade 2014; Taiwo 2006; Nweke 1994), together with the ability of small-scale processors to access investment capital (Taborda 2018). The potential impacts of such interventions are to empower small-scale processors and farming communities to process more of their crops close to the production areas, resulting in more affordable products and increased food security and resilience against fluctuations in the prices of imported foods.

Improving cassava processing also encompasses food safety. African food products tend to have high microbiological and chemical contamination levels exceeding regulatory limits. Relative to industrialized countries, various deficiencies at government, sector, retail, and company levels affect performance of food safety management systems. Collective efforts needed to address food safety include developing stringent certification standards and product specifications, improving hygiene and control of raw materials, enhancing monitoring systems, developing quality assurance and supportive administrative structures, developing risk-based legislative frameworks, and strengthening food safety authorities (Kussaga et al. 2014).

As a staple crop, cassava cultivation and processing tend to expand in lockstep with growing populations in developing countries. From this perspective, improving the efficiency of cassava processing technologies is crucial to minimize resource consumption and environmental footprint and to avoid a business-as-usual scenario whereby current inefficient technologies multiply to handle increasing volumes of cassava, leading to unsustainable levels of pollution and resource depletion in the face of climate change and other environmental risks. This issue also has important gender ramifications: "An analysis of distributive impacts of the environment on human well-being cannot ignore features such as gender. [...] Women and girls often carry a disproportionate burden from environmental degradation compared to men" (UNEP 2007, p 15).

In recent years, researchers federated under CGIAR's Research Program on Roots, Tubers and Bananas (RTB) have made vital contributions to a range of products and to the links between processing, product quality, and consumer expectations (Escobar et al. 2021; Adinsi et al. 2019; Alamu et al. 2019; Bouniol et al. 2021; Luna et al. 2021). This chapter presents the reengineering and scaling out of flash-drying technology for small-scale cassava processing, supported by the development of a methodological framework for R&D on postharvest processing of cassava.

4.2 Optimization of Energy Efficiency for Drying of Starch and Flours

In countries with industrial production of cassava starch, such as Thailand, Vietnam, Brazil, and Paraguay, large factories produce 50–400 tons of starch per day. These factories use large, highly efficient pneumatic-conveying dryers known as flash dryers (Brennan 2011; Sriroth et al. 2000; Aichayawanich et al. 2011). Even so, drying often has the highest processing cost due to the energy needed for heating. In contrast, in most cassava-producing countries, less intensive cassava production, limited transport infrastructure, and a tradition of small-scale household or family-based working units predominantly using manual labor do not permit the development of large factories. As a result, most cassava processing into flour and starch for food consumption or industrial applications is done at small scale, with an average production capacity of 1–3 tons/day (Adegbite et al. 2019). Such processing systems are beneficial in providing jobs for many in the community, including women, thus contributing to social stability and reducing migration to cities. From that perspective, proposed technical innovations and related efficiency gains imply a trade-off between the risk of job losses and the prospect of lower production costs and consequently higher incomes for processors and lower food prices for consumers. At any rate, small-scale processing is often affected by low investment capacity and limited repair-and-maintenance services, slowing down the adoption of technical innovations.

For drying, many processors still rely on sun drying, which is low cost but has limitations, including being subject to the weather, the limited availability of flat surfaces, and contamination from animals, dust, or microbes (Precoppe et al. 2020; Alonso et al. 2012). This is a particular disadvantage to access larger markets that need regular, all-year-round supply and consistent quality (Dziedzoave et al. 2006). Attempts to solve these limitations in the past 20 years have involved scaling down the design of large-scale flash dryers. However, so far, there has been little use of the resulting models of locally built, small-scale flash dryers due to high energy consumption related to suboptimal design and consequently high production costs incompatible with market prices of final products. For instance, in Nigeria between 2006 and 2016, 157 cassava processors have invested in "first-generation" flash dryers to produce HQCF, most of which are no longer in use in spite of some improvements (Ojide et al. 2021).

Taking stock of this situation and given the strategic importance of efficient post-harvest processing for the sustainable expansion of cassava value chains, research initiated in 2013 by RTB identified several critical points to optimize energy efficiency and minimize the operating costs of small-scale flash dryers. This research was based on numerical modelling of flash drying (Chapuis et al. 2017) and validated through the construction and testing of a pilot-scale dryer (capacity, 100 kg/h) at the International Center for Tropical Agriculture (CIAT, Colombia) that achieves the same energy efficiency as large-scale industrial flash dryers. The Scaling Readiness framework was then applied to scale out this innovation to cassava processors.

4.3   What Is an Efficient Dryer? Definition and Expression of Drying Efficiency and Affecting Factors

Drying of flour and starch consists of removing water by thermal treatment, inducing a phase change, generally from liquid to gas. Flash dryers use convective drying, i.e., applying a hot air stream to the product. Evaporation of liquid water requires an important quantity of energy, called latent heat of vaporization. Its value of 2500 kJ/

kg of water is thermodynamically incompressible and sets the minimum energy requirement for drying. In real drying systems, energy losses are unavoidable, so an energy consumption of 3000–4000 kJ/kg water is considered efficient.

In practice, energy is delivered to the dryer by heating the air flowing into the system. Gas burners are convenient because they produce combustion gases clean enough to be in contact with the product, even food. Otherwise, heating systems using fuels such as diesel, fuel oil, or biomass require a burner coupled to a heat exchanger. These systems are more common because they allow using cheaper and widely available fuels, thus reducing operating costs.

During drying, the hot air releases heat to evaporate water and cools down as it absorbs water vapor. Efficient drying, i.e., maximum water evaporation, requires *good exchange area and contact time* between the air and the product and *good mixing (turbulence)* to promote heat exchanges. Many drying technologies are available depending on the product to dry. Flash dryers are among the most efficient for granular materials (Crapiste and Rotstein 1997), as they maximize the exchange area between air and suspended solids (Saravacos and Kostaropoulos 2016). The capacity of air to hold water vapor is limited by phase equilibrium and increases with temperature (hot air can hold more water). Therefore, the *higher the initial temperature* of the drying air, the better, without damaging the product. As an illustration, air heated to 200 °C can absorb 60 g water/kg air, three times more than at 80 °C (20 g water/kg air). Under optimum drying conditions, the exhaust air is saturated with moisture, and the product has reached the target moisture for a long shelf life, i.e., 12–13% wet basis (wb) in the case of starch and flour.

Finally, mechanical dewatering to reduce the water content before drying is also important, as it requires much less energy and is more cost-effective than drying by evaporation (Mujumdar 2006). As an illustration, drying HQCF to 12% moisture content from an initial 40% (after good dewatering) requires 1860 kJ/kg of final product, assuming a dryer using 3500 kJ/kg of water evaporated. If starting from an initial 50% moisture content (after inefficient dewatering), drying requires markedly more energy at 3020 kJ/kg of product.

4.4 Key Design Components for Efficient Small-Scale Flash Drying

4.4.1 Surveys on Flash-Dryer Designs and Energy Efficiency

The first action of the RTB initiative to improve energy efficiency of small-scale flash drying consisted of a survey of processors owning flash dryers in various countries (Thailand, Vietnam, Tanzania, Nigeria, Colombia, and Paraguay) to identify common design features underpinning energy efficiency, such as dimensions and operating conditions (Tran et al. 2015a, b, c; Saengchan et al. 2015). The physical properties of cassava particles before and after flash drying were also studied

(Romuli et al. 2017). While the drying principle and key components remained the same across all flash dryers (Fig. 4.1), important variations in designs, dimensions, and operating conditions were observed (Fig. 4.2, Table 4.1). Large-scale dryers had longer drying pipes (20–57 meters) than small-scale ones (7–15 m). The quantity of air used for drying varied more than tenfold between dryers, with air-to-product ratio ranging from 6 to 75 kg of air/kg of product (Table 4.1). Small-scale dryers tended to use more hot air per kg of product, resulting in higher energy use and drying costs.

Standardized drying costs ranged from 23 USD up to 400 USD/t of product (Table 4.1), assuming for comparison purpose that all the flash dryers used diesel fuel (energy density, 40 MJ/L; price, 1 USD/L). In reality, flash dryers are powered by various fuels, renewable or nonrenewable, including wood, biogas, agricultural biomass (palm oil kernel, cashew nut kernels), liquefied petroleum gas (LPG), heavy fuel oil, diesel, black oil (a mixture of kerosene and used motor oil), and coal. Processors owning flash dryers with high fuel consumption manage to keep drying operations profitable by using cheaper fuels than diesel. Nevertheless, flash dryers with high fuel consumption reduce the profitability of processing and contribute to small-scale factories abandoning production of HQCF or starch, often at a significant loss on their investment. In Nigeria, a recent survey of 41 HQCF processors to

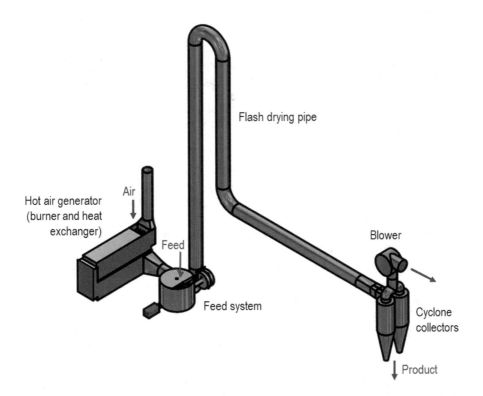

Fig. 4.1 Key components common to all flash-dryer designs

Fig. 4.2 Diversity of flash-dryer designs in Thailand, Nigeria, Paraguay (dryer model from Brazil), Paraguay (dryer model from Sweden), and Argentina

ascertain current perceptions on flash dryers found that 50% of them have stopped producing in the past 10 years and that most see their market prospects as limited (Ojide et al. 2021). Processors who remain in operation have done so through cost-saving measures: being near the source of raw materials and the end-product market, thus cutting transportation costs; replacing petroleum-based fuels with more economical coal or agricultural biomass; and diversification to produce gari or fufu when there is no demand for HQCF.

Given the small profit margins in HQCF production, small-scale processors currently select cheaper fuels, without considering the environmental costs of more polluting fuels. As a result, nonrenewable hydrocarbon fuels are still widely used. Cleaner renewable fuels such as biogas may become viable options for small-scale processing in the future, as long as the greater use of renewable energies does not also incentivize the burning of firewood, which would exacerbate deforestation. Renewable fuels are possible, and investments by large-scale cassava starch factories to generate biogas from factory wastewater can substantially reduce both drying costs and environmental impacts (Hansupalak et al. 2016).

Table 4.1 Diversity of dimensions and operating conditions of flash dryers in selected countries

	Unit	C-III	C-IV	TH-1	TH-2	TH-3	TH-4	N-2	N-3	T-1
Country		Paraguay	Paraguay	Thailand	Thailand	Thailand	Thailand	Nigeria	Nigeria	Tanzania
Capacity	t prod. wb/h	4.57	4.36	7.54	14.9	8.30	11.2	0.04	0.05	0.2
Total pipe length	m	33	20.7	46.0	57.7	45.0	44	12.5	7.7	13.6
Pipe diameter	m	1.2	1.5	1.16	1.50	1.14	1.25	0.38	0.75	0.36
Product input moisture	% wb	34	40	34	36	35	35	39	48	–
Product output moisture	% wb	13	13	13	13	13	13	5	10	–
Air flow	t air/h	69.9	52.2	49.2	114.9	63.8	66.2	2.2	3.8	2.3
Air/product ratio	kg/kg	15.1	12.2	6.4	7.5	7.5	5.8	56.1	75.3	11.6
Input temperature	°C	130	155	173	177	170	200	222	137	275
Output temperature	°C	70	62	52	55	54	59	62	86	
Energy use	MJ/t prod.	1503	1534	933	1109	1077	996	16,087	8212	3055
Energy use[a]	MJ/t water	4613	3541	3080	3087	3182	3035	15,393	11,238	4377
Drying costs[b]	USD/t prod.	37.6	38.4	23.3	27.7	26.9	22.6	402	205	76

[a]Energy used in the drying pipe only. The energy losses at the burner and heat exchanger are not considered
[b]For comparison purposes, the drying costs presented above have been standardized assuming diesel as fuel (40 MJ/L diesel) costing 1 USD/L

The surveys revealed the absence of a consensus design among flash dryers. These variations in design and ultimately in energy efficiency and drying costs suggest that there is a huge potential for optimization, with benefits in terms of production costs and environmental impacts (fossil fuel consumption and greenhouse gas emissions). The surveys were thus followed by an in-depth study of the technical characteristics needed to achieve energy-efficient flash drying.

4.4.2 Numerical Modeling of Flash Drying Provided Design Guidelines for Energy Efficiency

Several physical phenomena govern drying kinetics of a particle in suspension: heat transferred from the air to the particle causes water to evaporate from the surface, which, in turn, causes water within the particle to diffuse toward the surface. At the same time, the particle is entrained by the air flow along the drying pipe. These phenomena are described using conservation equations of mass, momentum, and heat (Mujumdar 2006). In the case of starch and flour, the slowest phenomenon is the diffusion of water within the particles; therefore, the size of the particles and their residence time in the dryer are critical for quick drying.

Applying this theoretical framework to flash drying, a numerical model was developed to simulate various designs and predict energy efficiency and final moisture content of cassava flour and starch (Chapuis et al. 2017). The technical data collected during the surveys (Sect 4.4.1) were used to validate that the model gave accurate drying predictions. Deviations between predicted data and actual data were satisfactorily low (less than 10%) for moisture content of the end product, air temperature after drying, and energy consumption.

The numerical model was applied to investigate optimum dryer designs that minimize energy consumption and maximize the quantity of water evaporated per unit mass of hot air (Sect. 2.1). The main challenge was to identify a configuration where the hot air carried just enough energy to dry the product flow rate (thermodynamics design objective) and the particles stayed in contact with the hot air long enough for complete heat and water transfers (kinetics design objective). Numerical simulations showed that such configuration could be attained through the following key design components:

(i) The target production capacity of the flash dryer defines the flow rate of water to evaporate, or *drying rate* (kg water evaporated/h):
     Drying rate = MassFlowProductIn*MoistureIn - MassFlowProductOut*MoistureOut, where *MassFlowProductIn = MassFlowProductOut*(1-MoistureOut)/(1-MoistureIn)*.

(ii) The drying rate in turn defines the minimum energy input for drying and hence the required quantity of hot air to enter the dryer, or *air mass flow*. For efficient drying, the minimum energy input (or specific energy use) should aim for 3000 kJ/kg of water evaporated. Then.

HeatInput (kW) = SpecEnergyUse (kJ/kg)*DryingRate (kg/h)/3600 and

AirMassFlow (kg/h) = 3600*Heatinput / (HeatCapacity*(TempAirIn-TempAirAmb)), where HeatCapacity = 1.01 kJ/K/kg.

For instance, drying from 38% to 12% moisture, using initial air temperature of 180 °C, requires about 10 kg of air per kg of final product.

(iii) *Air velocity* must remain high enough to keep the particles in suspension. For particles of typical size distribution (10 μm up to 2 mm), the threshold for suspension is 8–9 m/s. Air velocity should be set above the threshold, typically 12 m/s, as a safety margin. A long drying pipe (20 m or longer) gives flexibility to use higher air velocities to increase the dryer capacity without affecting energy efficiency significantly.

(iv) Air mass flow and air velocity in turn define the *section and diameter of the pipe*:

PipeSection (m²) = AirMassFlow/(3600*AirDensity*AirVelocity), with AirMassFlow in kg/h and AirVelocity in m/s. AirDensity (kg/m³) depends on temperature (0.77 kg/m³ at 180 °C).

(v) The *residence time* for the product to fully dry depends on the particle size, initial moisture content, initial air temperature, and air velocity. For typical drying conditions of cassava starch or HQCF, a residence time of 2 s is recommended.

(vi) The *length of the drying pipe* is defined by residence time and air velocity, but also input moisture content and initial air temperature. A length of 17 meters is a minimum, and 20 meters or longer is recommended for energy efficiency and stability of operations. A dryer with a longer pipe is more versatile for handling variations in product moisture, air temperature, and air velocity while delivering stable moisture content in the final product.

(vii) Finally, initial *air temperature* should be as high as possible without damaging the product (Sect. 2.1), with optimum between 180 and 200 °C.

These findings helped to interpret the surveys, according to which longer pipe and lower air/product ratio achieve better energy efficiency (Table 4.1 and Fig. 4.3). Approaches to improve the less efficient, small-scale flash dryers were identified: First, all small-scale flash dryers surveyed had short drying pipes (7.7–13.6 m), which should be extended to 20 m to increase residence time and water evaporation. Second, with a longer drying pipe, the air/product ratio can be reduced by increasing the feed rate of the product.

To validate these conclusions, the next step was to build and test a prototype of a small-scale flash dryer with adjustable pipe length and adjustable air velocity.

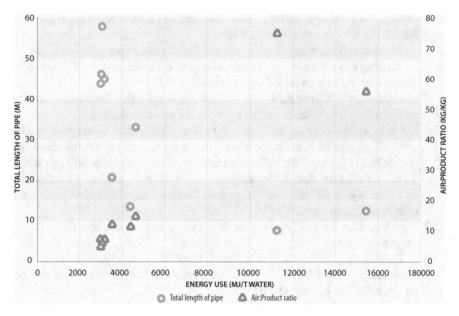

Fig. 4.3 High energy use by flash dryers is correlated with shorter pipe and high air/product ratio. Each point represents a different model of flash dryer (Table 4.1)

4.4.3 Prototype Flash Dryer Confirmed That High Energy Efficiency at Small Scale Is Achievable

Numerical simulations indicated that energy-efficient flash drying at small scale was achievable. To validate these results through actual experiments, a prototype flash dryer (100 kg/h capacity) was designed and built at CIAT (Cali, Colombia) in 2017 (Fig. 4.4). In addition to integrating the key design components (Sect. 4.4.2), the following elements were also essential for operating the dryer:

– *Regulated feed system:* Wet cassava flour or starch is a semi-sticky powder with a tendency to form clumps during handling. Accordingly, the feed system was composed of a cylindrical hopper equipped with stirring paddles to prevent caking and powder bridging and discharging into an endless screw conveyor leading to the drying pipe. A pin mill located between the screw and the drying pipe disaggregated clumps, to minimize particle size and maximize the exchange surface. The speed of the endless screw was regulated by a feedback loop maintaining the temperature of the air stable after drying (setpoint 55 °C). This regulation maintained a stable moisture of the end product, maximized energy efficiency, and reduced fuel consumption and costs. This control strategy can be implemented by manually reading the air temperature after drying and by manually adjusting the feeding rate; however, an automatic temperature controller (PID) is more effective.

Fig. 4.4 Flash-dryer prototype installed at CIAT

- *Conveying system and collection system of the dry product:* Inside the drying pipe, air and product are kept in movement with a blower, which can be located at the beginning or at the end (positive or negative pressure, respectively). Negative pressure is recommended because the blower is located after the cyclone that collects the dry product, which requires less power as the air is cooler (55 °C). Moreover, after the cyclone, the air is free of solid particles, which reduces wear and breakdowns and allows using a blower with more efficient curved blades (as opposed to straight blades). On the other hand, negative pressure requires that the bottom output of the cyclone be closed with a rotary airlock valve to prevent the product from being re-entrained into the cyclone.
- In a positive-pressure system, the blower is located at the beginning of the pipe after the feed system and needs to be bigger to move the same quantity of air as an equivalent negative-pressure system due to the higher temperature (150–180 °C) and lower air density. The power consumption of a positive-pressure blower is thus 50–100% higher. Another drawback of positive pressure is that the product passes through the blower, which requires more maintenance and allows only less efficient straight blades to avoid accumulation of product on the blades. On the other hand, in positive-pressure dryers, a single flap valve at the bottom of the cyclone is sufficient instead of a rotary airlock valve.

– *Hot air generator (burner and heat exchanger):* Common fuels such as diesel or biomass generate fumes and particles, which must be kept separated from the product that is being dried. For small-scale dryers, this is achieved through an air-air heat exchanger installed with the burner. Current models are mostly single pass concentric pipe type, with sub-optimum efficiency. The RTB team developed an improved heat exchanger integrating the following:

  (i) Counter-current flow of the combustion fumes and fresh air
 (ii) Increased heat exchange surface through a bundle of thin smoke pipes for the combustion fumes, around which fresh air circulates in a series of chambers with chicanes
(iii) Increased turbulence of the flow of fresh air forced to pass through the chicanes (Fig. 4.5). This design increased air outlet temperature while reducing fuel consumption

Alternatively, if a distribution network of LPG is available, the heat exchanger system can be replaced with a direct-combustion LPG gas burner, which reduces fuel consumption (by 10–15%) and investment costs compared to a system with a heat exchanger.

The performance of the prototype flash dryer was assessed under diverse operating conditions (Table 4.2). High energy efficiency was achieved with a pipe longer than 17 m, with specific energy consumption in the range 3300–4000 kJ/kg of water evaporated (Table 4.2). These figures are slightly higher than for large-scale dryers due to higher surface-to-volume ratio and consequently higher heat losses in smaller equipment (Kemp 2012). The best energy efficiency was obtained with a 20-m-long pipe (3268 kJ/kg water) and high air velocity (24 m/s).

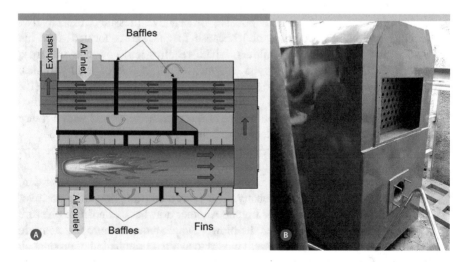

Fig. 4.5 (**a**) Cut view of the counter-current improved air-air heat exchanger developed by RTB; (**b**) one of the six heat exchangers built during the scaling-out phase of the project. (Source: Authors and Agrimac Ltd, DR Congo)

Table 4.2 Performance of the prototype small-scale flash dryer built at CIAT (Cali, Colombia) under various operating conditions (fixed pipe diameter, 16 cm)

	Unit	Trial #							
		1	2	3	4	5	6	7	8
Pipe length	m	17.2	17.2	17.2	17.2	20.3	20.3	27.2	27.2
Inlet air velocity[a]	m/s	11	11	23	23	11	24	22	22
Inlet air temperature	°C	140	180	180	180	140	180	140	140
Outlet air temperature	°C	47.5	50.6	53.6	56.5	47.1	52.6	51.9	52.0
Product initial moisture	% wb	31.6	31.4	36.8	36.9	34.0	38.5	36.7	36.4
Product final moisture	% wb	13.4	13.3	13.5	11.9	14.3	13.6	12.9	13.3
Product feed rate	kg/h	77	99	164	152	67	160	122	124
Specific energy consumption	kJ/kg water	3563	3610	3340	3486	3979	3268	3399	3417

[a]Air velocity is provided at the corresponding inlet air temperature

The numerical model predicted that high air velocity would increase specific energy use due to shorter residence time and consequently incomplete drying. On the contrary, actual performance tests indicated that increasing air velocity is not necessarily detrimental (Table 4.2). A possible explanation is that higher air velocity also reduces particle size (a factor not included in the numerical model) through increased collisions, thus intensifying the drying. Consequently, when the drying pipe is long enough (>20 m), increasing air velocity may substantially increase drying capacity, without affecting energy efficiency.

Design guidelines and tools resulting from this RTB initiative are available from the authors upon request.

4.5   From Prototype to Technology Transfer: Optimization of Feeding Rate for Energy Efficiency of a Small-Scale Flash Dryer in Ghana

4.5.1   Introduction

After demonstrating the energy efficiency of the small-scale flash-dryer prototype, the RTB research team moved to transfer these findings to the private sector. Processors from several countries expressed interest (DR Congo, Nigeria, Ghana, Cameroon, Uganda, Tanzania, Colombia, Brazil, and the Dominican Republic) to reduce operating costs and/or increase production of HQCF or starch. Reasons cited included growing demand for HQCF for industry or to replace sun-dried cassava flour, which consumers increasingly see as lower quality due to contamination during drying. Underpinning these trends was ongoing economic development with the emergence of a larger middle-class and more urban consumers with higher incomes. In 2018, a small-scale energy-efficient flash dryer was developed and put into commercial use in the Central Region of Ghana. This provided an opportunity to further evaluate the effect of feeding rate on energy efficiency, following previous findings by Precoppe et al. (2016).

4.5.2   Materials and Methods

Drying procedure and processing equipment   At the partner processing center in Ghana (Tropical Starch Company Ltd.), cassava roots are peeled and washed manually and mashed using a mechanical grater. The mash is dewatered with a screw-operated press. The resulting press cake is pulverized into wet grits using another mechanical grater. The wet grits are fed into a pneumatic dryer, and the resulting dried grits are milled into flour. The pneumatic dryer was developed with joint funding from RTB and the Cassava: Adding Value for Africa project (CAVA II). CAVA-II partners (equipment manufacturers and Tropical Starch Company Ltd.) were given training on the design and construction of small-scale pneumatic dryers, including the key design components described in Sect. 4.4.2. The equipment was constructed with stainless steel and thermally insulated with 50-mm-thick mineral wool, shielded with aluminum sheeting (Fig. 4.6b). The drying duct had a diameter of 0.18 m and a length of 24.84 m. To reduce the height of the equipment, the drying duct was divided into seven vertical meandering sections (Fig. 4.6a). Air flow was forced by a centrifugal 7.5-kW blower located at the start of the drying duct, just after the feeding point (positive-pressure system). The design point used for the

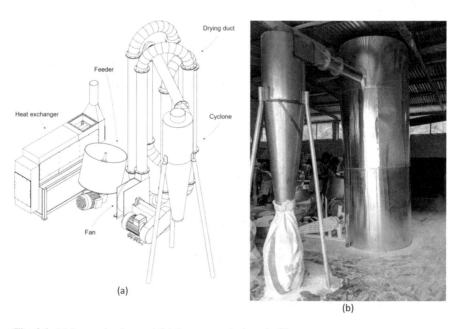

Fig. 4.6 (**a**) Isometric view and (**b**) the pneumatic dryer in Ghana

blower construction was an air mass flow rate (dry basis, db) of 550 kg/h and a static pressure of 10.2 hPa. Air was heated by a 70-kW diesel burner (model B14; Bairan, Wenling, China) and the improved RTB heat exchanger (Fig. 4.5). Drying air temperature was thermostatically controlled with the sensor placed at the dryer inlet, between the heat exchanger and the feeding point.

Experimental design The energy performance of the dryer was first evaluated following the processing center's drying procedure, with no interventions other than sampling material and recording drying conditions. Based on the data collected, the feeding rate was adjusted to optimum value, and the data collection was repeated. Data were collected over six consecutive days (recording 5 hours/day), the first 3 days with the original feeding rate and the subsequent 3 days with the optimum feeding rate. Data collection started after the dryer had been in operation for 1 hour to ensure steady-state conditions.

Statistical analyses Statistical analyses were performed with SAS 9.4 (SAS Institute Inc., Cary, NC, USA) software, following a comparative experimental design (Precoppe et al. 2016) with three replicates and two treatments: before and after the adjustment on the feeding rate. One-way analysis of variance (ANOVA) was performed, and to determine whether the pairwise difference comparisons were significantly different, Fisher's Least Significant Difference (LSD) was used at a 5% level of significance.

Data collection Temperatures, relative humidity, pressures, and air velocities were measured at various points on the dryer (Fig. 4.7) with suitable sensors connected to a wireless data logger (LOG-HC2-RC, Rotronic) recording values at 1-minute intervals. Mass flow rates of wet and dry cassava grits were measured using a digital balance (AWB120; Avery Weigh-Tronix, Smethwick, UK) and a chronometer. Samples of wet and dried cassava grits were collected at 1-hour intervals for moisture content analysis (3 h at 103 ± 1 °C in a convection oven) according to AOAC 935.29 (AOAC 1998).

4.5.3 Calculations

Energy performance Psychrometric calculations used the formulas provided by British Standard (2004). Energy performance was calculated as described by Precoppe et al. (2016). The *solid mass flow rate* ($\dot{m}_{dm}$, db) was calculated based on $\dot{m}_{ws}$ and the moisture content of the wet cassava grit. *Heat input rate to the dryer* (Q_{in}) was calculated based on air temperature, relative humidity, and air flow rate ($\dot{m}_{air}$) at the dryer inlet. *Water evaporation rate* ($\dot{m}_w$) was calculated using $\dot{m}_{ws}$ and the difference in moisture content between the wet and dried cassava grits. *Specific heat consumption* (q_s) was calculated dividing Q_{in} by $\dot{m}_w$, and energy efficiency (η) was calculated dividing the heat used for moisture evaporation ($Q_{w)\,by}$ Q_{in}. Finally,

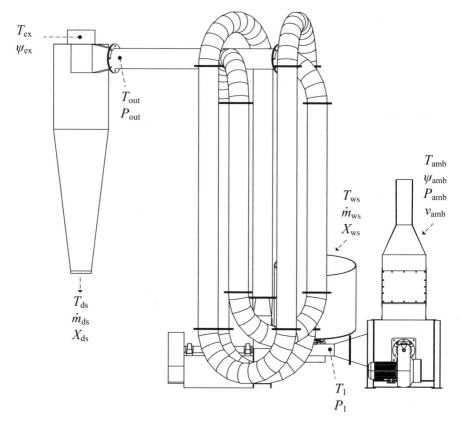

Fig. 4.7 Measurements on the pneumatic dryer in Ghana. T, temperature; ψ, relative humidity; P, pressure; v, air velocity; $\dot{m}$, mass flow rate; X, moisture content; ws, wet cassava grits; ds, dry cassava grits; amb, 1, out, and ex refer to air characteristics at various points of the dryer

specific heat utilization (q_u), i.e., energy consumption per kg of dried product, was calculated dividing Q_{in} by $\dot{m}_{dm}$.

Optimum feeding rate The dryer's original feeding rate was 65.7 ± 11.4 kg.h^{-1}. This rate was adopted by trial and error by the operators of the flash dryer over a period of 1 year of use. With this feeding rate, T_{out} was 67.3 ± 4.9 °C, and ψ_{out}, calculated using T_{ex} and ψ_{ex}, was $37.6 \pm 8.7\%$. At this temperature and relative humidity, the enthalpy at the dryer outlet (h_{out}) was 249.9 ± 15.5 kJ.kg^{-1}. Based on these values, the lowest allowable air temperature at the dryer outlet was $T^*_{out} = 60$ °C (Precoppe et al. 2016). Keeping h_{out} unchanged $\left(h_{out} = h^*_{out}\right)$ and reducing T^*_{out} to 60 °C, the highest allowable relative humidity at the dryer outlet $\left(\psi^*_{out}\right)$ can be raised to $\psi^*_{out} = 53\%$. Keeping the other settings unchanged and monitoring the air at the dryer outlet, the feeding rate was gradually increased until T_{out} approached T^*_{out} and ψ_{out} approached ψ^*_{out}. At this point, the optimum feeding rate was determined to be 98.6 kg.h^{-1}, a 50% increase compared to the original configuration.

4.5.4 Results and Discussion Regarding the Small-Scale Flash Dryer in Ghana

Dryer operating conditions Adjusting the dryer to optimum feeding rate increased the output of dried cassava grits from 42.2 ± 7.3 kg.h^{-1} to 65.0 ± 5.5 kg.h^{-1}, a 54% increase in productivity without increasing energy consumption (Table 4.3). Input parameters independent from the feeding rate, i.e., air temperature at inlet and moisture content of wet cassava grits, were unchanged between the treatments. The moisture content of the dried cassava grits (0.14 kg.kg^{-1}) was unchanged as well, confirming that it is possible to increase the feeding rate to its optimum without altering the quality of the end product.

The T_1 values seemed high for food drying (Kudra 2009); nevertheless, short residence times in pneumatic dryers allow the use of high temperatures without jeopardizing product quality (Pakowski and Mujumdar 2014). When drying cassava, the temperature of the product (T_{ds}) must remain below 56 °C to avoid starch gelatinization (Breuninger et al. 2009), which was realized both before and after adjusting the feeding rate.

Energy performance The water evaporation rate is driven by heat input to the dryer and by the amount of material (Kudra 2009). Increasing the feeding rate to optimum value increased the evaporation rate, which significantly improved the energy performance of the dryer (Table 4.3). Heat input rate is independent from the feeding rate and hence remained unchanged.

Energy efficiency at optimum feeding rate (72.1%) was close to the top of the range reported for pneumatic dryers (50–75%, Strumiłło et al. 2014) but still lower than the values reported by Sriroth et al. (2000) for large-scale pneumatic dryers. Small-scale dryers inevitably have an unfavorable surface-to-volume ratio that results in higher heat losses (Kemp 2012). The higher feeding rate increased the solid loading ratio of the drying air; nevertheless, the conveying mode remained in the dilute phase, and pneumatic transport was not jeopardized.

4.5.5 Conclusions: The Small-Scale Flash Dryer in Ghana

This case study in a commercial setting confirmed that small-scale flash dryers for HQCF can achieve high energy efficiency similar to large-scale dryers. Adjusting the feeding rate to its optimum value also improved energy performance significantly and is easier to implement than adjusting the air flow rate as it does not require hardware changes.

Table 4.3 Operating conditions and energy performance of a pneumatic dryer processing cassava in Ghana before and after adjustment to the feeding rate

Adjustment to the feeding rate	Unit	Before	After
Solid mass flow rate (m_{dm})	kg.h^{-1}, db	37.3^a ± 6.1	56.0^b ± 5.1
Air temperature at dryer inlet (T_1)	°C	236.6^a ± 5.1	238.4^a ± 9.5
Wet cassava grits moisture content (X_{ws})	kg.kg^{-1}, db	0.76^a ± 0.03	0.75^a ± 0.06
Dried cassava grits moisture content (X_{ds})	kg.kg^{-1}, db	0.14^a ± 0.04	0.14^a ± 0.06
Dried cassava grits temperature (T_{ds})	°C	55.9^a ± 3.3	54.4^a ± 2.2
Air temperature at dryer outlet (T_{out})	°C	67.3^a ± 4.9	60.7^b ± 3.5
Relative humidity at dryer outlet (ψ_{out})	%	37.6^a ± 8.7	55.3^b ± 9.2
Heat input rate	kW	31.4^a ± 1.0	32.4^a ± 5.2
Water evaporation rate	kg.h^{-1}	26.4^a ± 4.4	33.6^b ± 5.6
Specific heat consumption	kJ.kg^{-1} water evaporated	4388^a ± 716	3509^b ± 527
Energy efficiency^a	%	57.8^a ± 9.0	72.1^b ± 9.9
Specific heat utilization	kJ.kg^{-1} dried product	2746^a ± 437	1798^b ± 278
Solid loading ratio	g.kg^{-1}	70.1^a ± 11.5	105.7^b ± 9.5

In each line, means with different superscript letters are significantly different by Fisher's Least Significant Difference test at 5% level of significance
^{a}% of the heat input to the dryer that was used for water evaporation (not including the energy efficiency of the air heating system)

4.6 Applying Scaling Readiness to Scale Out Flash-Drying Innovations to DR Congo and Nigeria: Successes and Lessons Learned

4.6.1 Scaling Readiness Framework

Scaling Readiness is a stepwise approach for analyzing the characteristics of innovations from a scaling-out perspective, diagnosing the issues (bottlenecks) that hinder scaling out, developing optimum scaling strategies, building common agreements with the key stakeholders on the strategy, monitoring and learning from the implementation of the agreed strategies, and finally updating the strategies accordingly (Sartas et al. 2020a). Scaling Readiness analyzes the core innovation by breaking it down into complementary innovation components. "Innovation Readiness" refers to the demonstrated capacity of the innovation to fulfill its contribution to development outcomes in specific locations. This is presented in nine stages showing progress from an untested idea to a fully mature, proven innovation. "Innovation Use" indicates the level of use of the innovation or innovation package by the project members, partners, and society. This shows progressively broader levels of use beginning with the intervention team who develops the innovation to its widespread use by users who are completely unconnected with the team or their partners. "Scaling Readiness" of an innovation is a function of innovation readiness and innovation use. Table 4.4 provides summary definitions for each level of readiness and use, adapted from Sartas et al. (2020a).

Table 4.4 Summary definitions of levels of innovation readiness and use (Sartas et al. 2020a)

Stage	Innovation readiness	Innovation use
1	Idea	Intervention team
2	Basic model (testing)	Direct partners (rare)
3	Basic model (proven)	Direct partners (common)
4	Application model (testing)	Secondary partners (rare)
5	Application model (proven)	Secondary partners (common)
6	Application (testing)	Unconnected developers (rare)
7	Application (proven)	Unconnected developers (common)
8	Innovation (testing)	Unconnected users (rare)
9	Innovation (proven)	Unconnected users (common)

Readiness and use levels are evaluated using conceptual, applied, and experimental evidence. Scaling Readiness helps to identify the gaps in the design of the innovations and to prioritize the research and engineering work to address these gaps (Sartas et al. 2020b). The Scaling Readiness framework is discussed in more detail in Chap. 3.

Building on previous steps (Sects. 4.4 and 4.5), the RTB research team applied Scaling Readiness to support the scaling out of the flash-dryer innovation in the private sector in DR Congo and Nigeria, as a first case example of applying Scaling Readiness to a postharvest processing industry (Table 4.5).

4.6.2 Identification of the Innovation Package and Determination of the Innovation Readiness and Innovation Use of Each Innovation Component

In the RTB flash-dryer case, the complementary innovations were identified collaboratively with stakeholders. Small-scale processors of cassava flour (1–3 tons of flour/day) and equipment manufacturers from Nigeria and DR Congo were interviewed to generate a baseline of the technical and socioeconomic conditions of potential scaling project partners. In Nigeria, the cassava processors were selected based on being active in the cassava processing business or continuous processing operations in their factories. Equipment manufacturers were pre-selected based on experience in the commercial manufacture and sale of flash dryers. For sustainability of the flash-dryer innovation, the final selection gave priority to relatively younger manufacturers willing to learn new technology innovations. In DR Congo, flash-dryer manufacturing and use are new but with high potential to improve the quality and food safety of conventional sun-dried cassava flour. Only two local equipment manufacturers had exposure to flash dryers, and three processors were using flash dryers. All three processors and one equipment manufacturer were selected. The scaling partners explained the constraints of the current flash-dryer

Table 4.5 Levels of innovation readiness (adapted from Sartas et al. 2020a) applied to flash drying

Stage	Innovation readiness	Flash-dryer example	Evidence proving completion of the level
1	Idea	Flash drying can increase the energy efficiency of drying HQCF and starch	Crapiste and Rotstein (1997); Saravacos and Kostaropoulos (2016)
2	Basic model (testing)	An energy-efficient flash dryer has five key components (hot air generator, feed system, flash-drying pipe, blower, cyclone collector)	Figure 4.1
3	Basic model (proven)	An updated version of the basic model increased the efficiency of cassava flash drying, as validated with numerical simulations covering multiple scenarios	Chapuis et al. (2017); Sect. 4.4
4	Application model (testing)	A prototype based on the proven basic model and numerical model findings was designed	(Available upon request)
5	Application model (proven)	Multiple versions of the prototype designs were assessed by RTB researchers and finalized by technical drawings with specifications	Figure 4.4 3D drawing, Figs. 4.5a and 4.6a
6	Application (testing)	CIAT prototype was built based on the technical drawings in Colombia	Figure 4.4 photos; Figs. 4.5b and 4.6b
7	Application (proven)	CIAT prototype assessment was shown to improve the efficiency of flash drying significantly	Table 4.2
8	Innovation (testing)	Commercial small-scale, energy-efficient flash dryer was built in Ghana by RTB and CAVA II	Section 4.5, Tables 4.3 and 4.4, Fig. 4.7
9	Innovation (proven)	Testing of flash dryers in commercial setups is ongoing	N.A.

models (in particular high fuel consumption) and the expected performance improvements.

Based on on-site visits and interviews with stakeholders of the cassava value chains, scaling partners, and researchers, 16 innovation components were identified, of which ten were essential for the scaling process: four core technical innovations and six complementary innovations related to capacity building and the socioeconomic context of scaling out. The Scaling Readiness assessment showed varying levels of innovation readiness and use depending on the country (Fig. 4.8).

In both Nigeria and DR Congo, the levels of use and readiness of the technical components were relatively low (2–4 and 3–6, respectively). Nigeria had slightly higher levels of use of the feed system, hot air generator, and blower components. On the other hand, the readiness of the core innovation, i.e., efficient flash dryer, was higher in DR Congo. The main reason for this disparity was because at the beginning of the project, Nigeria had markedly more first-generation flash dryers in operation (higher use), albeit less efficient (lower readiness) than DR Congo. The levels of use and readiness of the complementary innovations were similar in both

Scaling readiness diagnosis of flash drying partners – DCR – May 2019

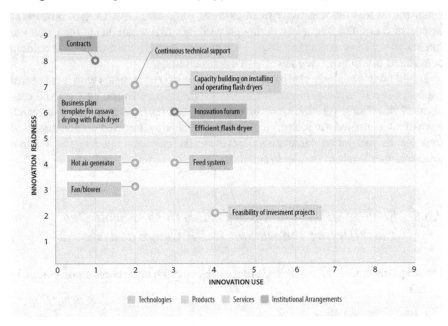

Scaling readiness diagnosis of flash drying partners – Nigeria – May 2019

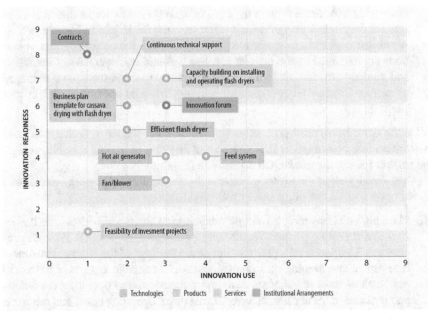

Fig. 4.8 Innovation Readiness and Innovation Use of the cassava flash-dryer innovation package in DR Congo and Nigeria after the Scaling Readiness assessment (May 2019). Core innovation components (technologies) are presented in green and complementary innovations in blue (products), yellow (services), and orange (institutional arrangements)

countries, except the feasibility of investment projects: in DR Congo, processors evaluated feasibility frequently (although with informal tools) due to the perceived market potential of producing HQCF, whereas in Nigeria, investment was seldom considered due to perceived adverse market conditions.

Based on this analysis, bottlenecks, i.e., innovation components with low use and readiness levels, were technical in DR Congo (hot air generator, blower) and economical in Nigeria (feasibility of investment). The strategy of the RTB research team then consisted of identifying and implementing targeted interventions to improve the innovation components and move the innovation package as a whole toward higher levels of readiness and use.

4.6.3   Scaling Strategy and Key Partnerships to Scale Out Flash-Drying Innovations to DR Congo and Nigeria

Project partners were all small-scale processors but with differences among them, in terms of:

(i) Investment capacity.
(ii) Factory setup, i.e., different models of machines and management of operations.
(iii) Quality of the final product expected by customers, which requires different processing techniques prior to drying. For example, lactic fermentation is a key step for fufu in DR Congo, which is not required for HQCF in Nigeria.
(iv) Administrative and labor management: eight of the ten project partner operations are managed by the owners and two by hired managers. Most managers and workers are men, and only two out of ten project partners were managed by women. Female labor is used for specific, low-paid operations such as root peeling.

On the other hand, most processors shared common concerns, such as the need to decrease fuel consumption (DR Congo and Nigeria) or finding a stable, consistent market for small-scale HQCF production (Nigeria).

After identifying the innovation bottlenecks, the scaling strategy developed and agreed upon with the scaling partners followed four key steps:

(i) Planning and delivering a training workshop to share and address the bottlenecks with scaling partners (equipment manufacturers and HQCF/starch processors). This included a presentation of the findings of the Scaling Readiness assessment and training on technical and socioeconomic aspects of flash drying, such as tools to calculate optimum dimensions and operating conditions and business plans to estimate investments costs, operating costs, and return on investment.
(ii) Implementation of the business plans by the scaling partners willing and able to do so. The process involved back-and-forth exchanges between cassava processors and the project team to develop business plans adapted to each partner's

circumstances. Realistic investment costs were provided by the equipment manufacturers who participated in the workshop. Some of the processors then used the resulting business plans to present their investment projects to banks or private investors. For those who invested their own money, the business plan contributed to accurate evaluation and of planning of the investment.

(iii) Design, construction, and commissioning of flash dryers by the scaling partners with continuous support from project scientists, through on-line technical consultations and regular on-site visits.

(iv) End-of-project debriefing workshop and evaluation of outcomes and lessons learned.

4.6.4   Results and Outcomes of Scaling Out Flash-Drying Innovations to DR Congo and Nigeria

By its completion in December 2020, the RTB Scaling project on flash-drying innovations fostered investment in energy-efficient, small-scale flash dryers by seven cassava processors: five in DR Congo and two in Nigeria. In addition, two demonstration flash dryers were built for promotion and training at two R&D institutions: the Federal Institute of Industrial Research Oshodi (FIIRO) in Nigeria and the International Institute of Tropical Agriculture (IITA) in Eastern Congo. Five equipment manufacturers were involved in constructing these improved flash dryers and are now well positioned with practical experience to build further dryers as new investors come forward in the future.

In terms of efficiency gains, performance evaluation of the improved flash dryers that have reached commercial operation indicated a 23 to 50% increase in production capacity (from 10 to 12–15 tons of flour/month) and at the same time a 30 to 33% reduction in fuel (diesel) consumption per kg of product, compared to the situation before the project. Feedback from the flash-dryer operators in DR Congo after incorporating the RTB innovations resulted in much improvement in drying efficiency: A processor reported that re-setting the heat exchanger burner and air inlet decreased the fuel consumption by 30% while other adjustments suggested by the project team resulted in an increase in the flour output from 150 to 250 kg/h. As a result, cassava processors reported an increase in net profitability of 8% to 10%.

Through end-of-project evaluation and learning events, the innovation package was evaluated a second time, showing the changes achieved during the project (Fig. 4.9). As intended initially, technical components moved toward higher levels of use and readiness, which reflects the focus of the research team and scaling partners to construct more energy-efficient flash dryers and prioritize the success of the investments. Complementary innovations also progressed toward higher levels of use and readiness, reflecting the positive contribution of the various events organized during the project and the continuous engagement with manufacturers and processors. Some components did not move, either because they were already at

Scaling readiness diagnosis of flash drying partners – DCR – Dec. 2020

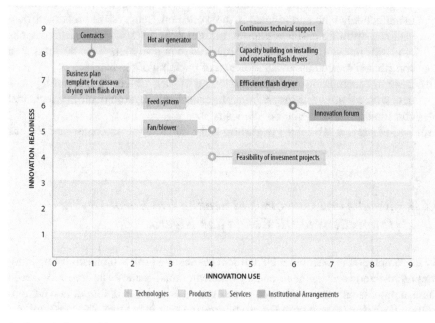

Scaling readiness diagnosis of flash drying partners – Nigeria – Dec. 2020

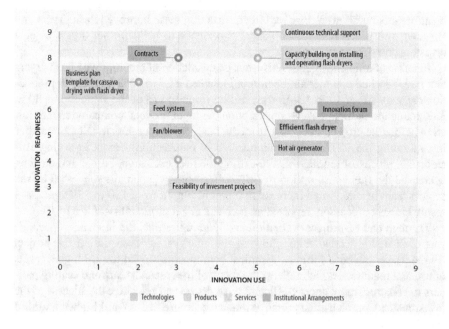

Fig. 4.9 Innovation Readiness and Innovation Use of the cassava flash-dryer innovation package in DR Congo and Nigeria at the end of the RTB Scaling flash-drying project (December 2020). Core innovation components (technologies) are presented in green and complementary innovations in blue (products), yellow (services), and orange (institutional arrangements)

sufficiently high levels of use and readiness or because of logistical delays in implementing activities, in part compounded by Covid restrictions.

Technical components reached higher levels of readiness in DR Congo compared to Nigeria. In DR Congo, the expanding market demand for HQCF to replace conventional sun-dried cassava flour boosted the determination of scaling partners to invest and to rapidly move energy-efficient flash dryers into commercial use. In contrast, in Nigeria, investment was impeded by the following:

(i) Mismatch between production capacity of HQCF factories (~1 t HQCF/day) and the purchase needs of large food companies (30 or 60 t per order, i.e., one or two trailer trucks).
(ii) Projected production costs of HQCF after adoption in flash-drying innovations may still be too high to compete easily with flour from imported wheat.
(iii) Cassava flour is not a major staple food in Nigeria, so there is limited preexisting market where HQCF could replace a traditional cassava flour by offering better quality or food safety.

During project implementation, several constraints to the development of cassava processing were identified. These included access to investment capital, stable access to cassava roots at cost-effective prices, and availability of engineering skills to conduct maintenance and repairs in a timely manner.

Over several years, RTB has made significant investments in cassava postharvest processing, first in R&D and then scaling-out activities. The outcomes of the scaling project for efficient small-scale flash drying confirm that this initiative has started to bear fruits, with flash-dryer innovations now reaching the stage of independent adoption and dissemination by the private sector in several countries.

Postharvest processing is an important part of sustainable cassava value chains to reduce losses of perishable cassava roots through transformation into food products with a long shelf life. The successful approach presented above for the efficiency of small-scale flash drying can be extended and replicated to optimize other unit operations and reduce processing costs. At the same time, more efficient processing can reduce the environmental footprint of cassava agro-industries through lower fuel and water consumption and lower product losses during processing.

4.7 Performance Diagnosis of Small-Scale Processes to Support Scaling Out of Innovations for Cassava-Based Products

Many of the activities for scaling-out innovations in cassava processing must start with an accurate comprehensive diagnosis including processing parameters, socioeconomic feasibility, and environmental impacts as a basis for decision-making for all subsequent actions. Comprehensive diagnosis is also useful at the end of scaling-out projects, as part of Monitoring, Evaluation, and Learning (MEL) to document

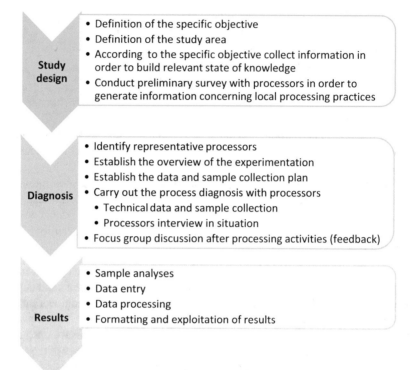

Fig. 4.10 Steps of the method for diagnosis of small-scale cassava processing units

and measure outcomes and impacts. The diagnosis contributes to improvements in product quality through the optimization of existing processes and the introduction of new operations or equipment. In this section, we lay out a versatile diagnosis method (Fig. 4.10) developed from several case studies (Adinsi et al. 2019; Escobar et al. 2018; Bouniol et al. 2017a, b, 2018, 2019, 2020) for processing of cassava into gari, fufu, and other products in West Africa, with support from RTB.

4.7.1 Design of the Diagnosis Study: Specific Objective, Study Area, and State of Knowledge

The diagnosis of small-scale processes can focus on various aspects, such as the following:

(i) Influence of cassava variety and processing on product quality.
(ii) Resource efficiency (energy, water, raw materials) in relation to production costs and environmental impacts.

(iii) Comparing different technologies for processing the same product. The first step is therefore to clearly define the specific objective(s) (Fig. 4.10), which in turn will guide the diagnosis methods to be selected.

Traditional, small-scale processing of cassava follows diverse processing pathways linked to different technologies and know-how. In turn, this variability results in different end-product quality, usually matching different consumer preferences. To collect representative data about a given process, it is therefore necessary to define where to carry out the diagnosis. These can be the areas where the target product is most frequently processed (usually areas with high cassava production), or the project's region of intervention. The diagnosis may target user segments to explicitly be socially inclusive or to take gender into account, besides other criteria: size of towns or cities, production capacity, distances to markets, etc.

The diagnosis design may be based on a review literature about the product, process, and value chain and preliminary interviews with experts, extension officers, and processors to gather information on supply, availability, and quality criteria of raw materials. The design should include a description of each of the process steps and their importance for the quality of the product, as well as gender aspects, business environment (access to infrastructures such as energy, water, road network), and market aspects.

4.7.2 Diagnosis

Conducting a process diagnosis requires time, qualified people, and funding, so pre-planning is important (Sect. 3.1) to coordinate the collection of samples and data and to ensure the quality of the results.

Identify representative processors The first step of the diagnosis is to identify in the study area a panel of processors representative of current practices, according to the following criteria:

- Main economic activity based on the studied product.
- Recognized know-how and ability to produce a specific quality of finished products matching the expectations of consumers.
- Gender division of labor among the selected processors that is representative of actual gender labor distribution.
- Technological level among the selected processors that is representative of the local most common technology. If most processors are small scale, the focus should be on them to mitigate negative social consequences of scaling technologies and to ensure more users of the technology from a social inclusive perspective.
- Evaluate the volumes processed (by month/week/year) and the type of market targeted (retail, wholesale, town, village, etc.).

Overview of the experimentation A *schematic overview of the experimentation* is necessary to clearly delimit the scope of the diagnosis according to the specific objective, to let the experimenters understand where they intervene in the overall process, and to refine the diagnosis by ensuring that the workflow and experimental design are suitable to collect scientifically valid data (representativeness of the sampling, number of repetitions). Several approaches are possible:

(i) Replicate the entire transformation process from the same batch of raw cassava several times. Tens or hundreds of kilograms are often necessary.

(ii) Repeat the experimentation over several years. As much as possible, use the same cassava varieties harvested at the same age and processed under the same conditions.

(iii) If appropriate, repeat only specific unit operations, for example, in the case of a series of complex unit operations where all variables may be difficult to control at the same time.

Data and sample collection plan The *collection plan* (e.g., Table 4.6) maps all the data and samples to collect at each step of the process, which may include temperature, weight, flow rate, dry matter, relative humidity, time, pH, and pressure. This helps to manage the complexity of the diagnosis by checking in advance with the processors the feasibility of the planned measurements and by keeping track of data and sample collection during the diagnosis.

Carry out the process diagnosis with processors

Technical data collection The same batch of *raw materials* (e.g., cassava roots) should be used for all the replications and in all the locations of the diagnosis to

Table 4.6 Example of data collection plan: diagnosis of "bâtons de manioc" (cassava sticks) in study from Cameroon in 2015

Data and sample acquisition	Weight; material balance	Sample collect. / Lab. analyses	Dry matter	Time	Work force	Water, pH, energy	Temperature	Losses of dry material	Remarks
Raw material	X Roots		X						Photos of outside and inside of roots
Peeling	X Roots & peels			X	X				Photos
Washing				X	X	X Volume water			
Steeping / Rinsing	X Initial weight / X Final weight	X Water end / X Filtered water	X Whatman / X pulp	X		X Volume water / X Final water	X Whatman filter / X Filtered water		Daily observations with operator advisories. Photos
Draining / Grinding	X Final weight	X	X	X	X				Photos
Crushing	X	X Crushed pulp		X	X	X Type of motor, power			Principle & dimensions of crushing system. Rotation speed. Photos
Shaping	X Leaves / X Bâtons	X Leaves							Sampling sheets for drying. Photos
Cooking	X End cooking	X kinetics / X		X	X	X Weight wood or gas	X DAQ (Almemo) / X + cooling		Principle & dimensions of cooking system. Photos
Bâtons	X	X							Photos

facilitate comparisons between technologies. This may not always be possible, however, due to the short shelf life of fresh roots and transportation distances. In this case, raw materials are sourced locally in each location, and the raw material is considered as one variable of the experimentation. Raw materials then need to be described in detail, including photos of the fresh roots (before peeling), size measurements, and any appropriate laboratory analyses.

All *equipment, instruments, and data collection sheet* necessary to carry out the experiments, to collect samples, and to record data are prepared in advance, based on the collection plan. The following characterization protocol is then applied: Before the first unit operation, the research team records the weight of the raw materials that will be processed into the product under study. Then after each unit operation (peeling, washing, etc.), the duration is recorded and the intermediate product (e.g., peeled product) and residues (e.g., peels, stems, soil) are weighed for later calculations of yields and mass balance. Samples can also be collected for further laboratory analyses (dry matter, starch content, fibers, pH, etc.). The following diagnosis parameters are then calculated for each unit operation and the process as a whole:

- *Yield*, defined as the quantity of product recovered after each unit operation and expressed as percentage (wet basis) of the quantity of raw material.
- *Material balance* evaluated by checking that the weights of all the inputs (raw materials, water, etc.) and outputs (final product, peels, fibers, wastewater, etc.) are equal.
- *Productivity*, defined as the quantity of raw material (in kg) processed per hour and per processor. This is an important indicator of efficiency and also of drudgery.

Interview of processors and participative approach To capture processors' expertise, at each step of the process, interviews can elicit the details that processors use to recognize that the raw materials or intermediate products will give a good final product and how they may adjust their process to ensure the best possible final product. In the case of experiments comparing the processing potential of several varieties, processors can be asked to rank them from good to poor. This ranking can be repeated with intermediate products during processing.

4.7.3   Exploitation of Results

In addition to regular reporting of results collected for each unit operation, the process as a whole can also be analyzed with the following:

- The *process flowsheet* (e.g., Fig. 4.11) describes the sequence of the unit operations, their duration (productivity), inputs and outputs (raw materials, water, electricity, thermal energy, etc.), and material balance based on the yields of the intermediate products.

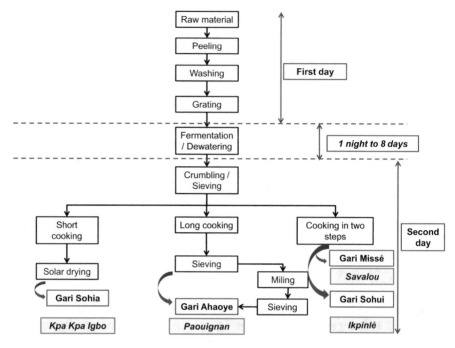

Fig. 4.11 Example of a process diagram for gari processing in four locations (shown in blue, italics) (study from Benin in 2015)

– *Yields and productivity of intermediate products*: The yields of intermediate products reveal which unit operations most impact the overall yield of the process (e.g., Fig. 4.12). Productivity of each unit operation reveals bottlenecks, i.e., which operations slow down the overall process. Yields and productivity can also depend on the processing ability of different cassava varieties, as some varieties can be harder to process, e.g., more difficult to peel, thicker peels, lower dry matter, or containing fibers that need to be manually removed. Yield and productivity are important acceptability criteria that influence the adoption of improved technologies and/or new varieties, in addition to acceptability criteria of the end product such as visual appearance, texture, and taste.

4.7.4 Conclusions

Process diagnosis is most effective when implemented with a formal methodological framework such as the one presented above. Combined with socioeconomic and environmental surveys, process diagnosis is a prerequisite to scaling-out innovations for several purposes: to provide a baseline to monitor the progress and outcomes of scaling-out activities; to reveal local constraints and expectations for the

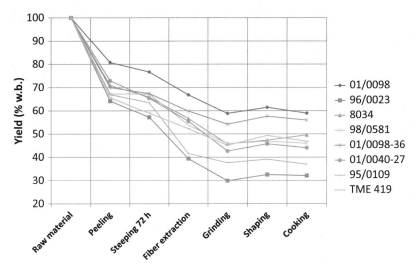

Fig. 4.12 Example of the yields of the unit operations for processing gari. Eight cassava varieties revealed the effect of variety on process yields (study from Cameroon in 2015)

innovations with respect to local traditions and gender, employment, and environmental and economic criteria; and to identify the innovation components package and establish the scaling readiness diagram when applying the Scaling Readiness framework. A well-executed diagnosis provides not only reliable information on the process and product under study but also builds dialog with processors on the benefits and constraints of processing technologies and improved cassava varieties to facilitate the adoption of innovations.

4.8 Conclusions and Perspectives: Ongoing Research and Strategic Areas for Future Research on Cassava Processing

Since 2013, RTB has fostered several initiatives to improve cassava processing and federated a team of multidisciplinary researchers and engineers from IITA, CIAT, CIRAD, NRI, academic partners, and private partners in several countries. One notable outcome has been the reengineering of small-scale flash dryers for better energy efficiency and lower production costs. The flash-drying results presented in this chapter are useful to illustrate in a practical, step-by-step way the overall approach to reengineering postharvest technologies, from the initial state-of-the-art surveys to the stages of design, proof of concept, pilot testing, and scaling out with private partners. This work also provided a relevant case study for the application of Scaling Readiness, showing the progression of the project through the levels of innovation readiness and use (Table 4.5).

This experience shows that R&D interventions at postharvest processing level have an important part to play in advancing sustainable value chains of RTB crops, especially when combined with scaling-out activities in partnership with private small-scale processors and equipment manufacturers. In addition to flash dryers, several cassava products and unit operations have been investigated by RTB: cabinet drying (Precoppe et al. 2017), attiéké (Alamu et al. 2020), gari (Escobar et al. 2018, 2021; Dahdouh et al. 2021), as well as links between processing and human health (Parmar et al. 2019; Bede-Ojimadu and Orisakwe 2020). Applying the same reengineering approach to other unit operations such as peeling, grating, dewatering, and milling will improve the overall efficiency of small-scale cassava industries and over time should increase incomes and employment along the whole cassava value chain (Escobar et al. 2021; Ezeocha et al. 2019; Dahdouh et al. 2021; Dou et al. 2020). Apart from interventions on postharvest processing, RTB researchers are currently also working on mobile cassava processing units to reduce the time and distance from the field to final products, so as to address the barrier of poor transportation infrastructure in West Africa.

The perspectives for cassava value chains over the next 10–20 years are an overall expansion to meet the food security and nutritional needs of growing populations. In this context, gains in processing efficiency will be crucial to improving both product quality and production capacity, which in turn are a prerequisite to connect rural small-scale cassava processors with higher-value and more distant markets, in particular large cities and international exports. Integrating processing innovations with the current scale of processing is important to provide a comparative advantage from the bottom up, as focusing on larger-scale processing can result in scale advantages and also put smaller-scale processors out of business with the risk of increasing inequality, social instability, and adverse gender effects as women are often highly represented in small-scale processing.

Acknowledgments This research was undertaken as part of, and funded by, the CGIAR Research Program on Roots, Tubers and Bananas (RTB) and supported by CGIAR Trust Fund contributors. Funding support for this work was provided by CAVA-II, CIRAD, and Technologies for African Agricultural Transformation (TAAT) Cassava Compact. The authors thank all the data providers, in particular cassava processors and equipment manufacturers, who have kindly participated to this study, as well as the companies Nutripro, Agrimac, Layuka, and Ecosac (DR Congo); Open Door System, Hickman Ventures, Deban Faith, Lentus Food, and El-Rasheed Farms (Nigeria); Angavil (Dominican Republic); Codipsa (Paraguay); and Deriyuca (Colombia). The participation of Charlène Lancement, Aymeric Delafosse, Francisco Giraldo, Andrés Escobar, Jhon Larry Moreno, Jorge Luna, and María Alejandra Ospina in support of project activities is gratefully acknowledged.

References

Abass AB, Mlingi N, Ranaivoson R, Zulu M, Mukuka I, Abele S, Bachwenkizi B, Cromme N (2013) Potential for commercial production and marketing of cassava: experiences from the small-scale cassava processing project in East and Southern Africa. IITA, Ibadan

Abass AB, Amaza P, Bachwenkizi B, Alenkhe B, Mukuka I, Cromme N (2017) Adding value through the mechanization of postharvest cassava processing, and its impact on household poverty in north-eastern Zambia. Appl Econ Lett 24(9):579–583

Abass AB, Awoyale W, Alenkhe B, Ndavi M, Asiru BW, Manyong V, Sanginga N (2018) Can food technology innovation change the status of a food security crop? A review of cassava transformation into "bread" in Africa. Food Rev Intl 34(1):87–102

Adegbite SA, Abass AB, Olukunle OJ, Olalusi AP, Asiru WB, Awoyale W (2019) Mass and energy balance analysis of pneumatic dryers for cassava and development of optimization models to increase competitiveness in Nigeria. Curr J Appl Sci Technol 32(3):1–11. https://doi.org/10.9734/CJAST/2019/46215

Adenle AA, Manning L, Azadi H (2017) Agribusiness innovation: a pathway to sustainable economic growth in Africa. Trends Food Sci Technol 59:88–104. https://doi.org/10.1016/j.tifs.2016.11.008

Adinsi L, Akissoé N, Escobar A, Prin L, Kougblenou N, Dufour D, Hounhouigan DJ, Fliedel G (2019) Sensory and physicochemical profiling of traditional and enriched gari in Benin. Food Sci Nutr 7(10):3338–3348

Aichayawanich S, Nopharatana M, Nopharatana A, Songkasiri W (2011) Agglomeration mechanisms of cassava starch during pneumatic conveying drying. Carbohydr Polym 84:292–298

Alamu EO, Ntawuruhunga P, Chibwe T, Mukuka I, Chiona M (2019) Evaluation of cassava processing and utilization at household level in Zambia. Food Secur 11:141–150

Alamu EO, Abass A, Maziya-Dixon B, Diallo TA, Sangodoyin MA, Kolawole P, Tran T, Awoyale W, Kulakow P, Parkes E, Kouame KA, Amani K, Appi A, Dixon A (2020) Report on the status of Attiéké production in Côte d'Ivoire. International Institute of Tropical Agriculture (IITA), Ibadan, p 42

Alonso L, Viera MA, Best R, Gallego S, García JA (2012) Artificial cassava drying systems. In: Ospina Patiño B, Ceballos H (eds) Cassava in the third millennium: modern production, processing, use, and marketing systems. Centro Internacional de Agricultura Tropical (CIAT); Latin American and Caribbean Consortium to support Cassava Research and Development (CLAYUCA); Technical Center for Agricultural and Rural Cooperation (CTA), Cali, pp 427–441

AOAC (1998) Moisture in malt (method 935.29). In: Official methods of analysis. Association of Official Analytical Chemists (AOAC), Gaithersburg

Awoyale W, Abass AB, Ndavi M, Maziya-Dixon B, Sulyok M (2017) Assessment of the potential industrial applications of commercial dried cassava products in Nigeria. J Food Meas Charact 11:598–609

Bede-Ojimadu O, Orisakwe OE (2020) Exposure to wood smoke and associated health effects in Sub-Saharan Africa: a systematic review. Ann Glob Health 86(1):32

Bouniol A, Prin L, Hanna R, Fotso A, Fliedel G (2017a) Assessment of the processability of improved cassava varieties into a traditional food product ("baton" or "chikwangue") in Cameroon. CGIAR Research Program on Roots, Tubers and Bananas (RTB), Lima, Peru. RTB working paper, p 28

Bouniol A, Ospina MA, Fotso A, Hanna R, Dufour D (2017b) Assessment of the processing ability of improved cassava varieties into a traditional food product (fufu) in Cameroon. CGIAR Research Program on Roots, Tubers and Bananas (RTB), Lima, Peru. RTB working paper, p 24

Bouniol A, Escobar A, Adinsi L, Prin L, Fliedel G, Tran T, Dufour D (2018) Standard Operating Procedure (SOP) to establish a diagnosis of a process with processors. CGIAR Research Program on Roots, Tubers and Bananas (RTB), Lima, Peru. RTB working paper, p 25

Bouniol A, Adinsi L, Padonou SW, Hotegni F, Gnanvossou D, Akissoé N, Fliedel G, Tran T, Dufour D (2019) Assessment of the processing ability of improved cassava varieties into a traditional food product (lafun) in Benin. CGIAR Research Program on Roots, Tubers and Bananas (RTB), Lima, Peru. RTB working paper, p 29

Bouniol A, Adinsi L, Hotegni F, Delpech A, Akissoé N, Fliedel G, Tran T, Dufour D (2020) Assessment of the processing ability of improved cassava varieties into a traditional food product (gari) in Benin. CGIAR Research Program on Roots, Tubers and Bananas (RTB), Lima, Peru. RTB working paper, p 34

Bouniol A, Adinsi L, Padonou SW, Hotegni F, Gnanvossou D, Tran T, Dufour D, Hounhouigan DJ, Akissoé N (2021) Rheological and textural properties of lafun, a stiff dough, from improved cassava varieties. Int J Food Sci Technol. https://doi.org/10.1111/ijfs.14902

Brennan JG (2011) Evaporation and dehydration. In: Brennan JG, Grandison AS (eds) Food processing handbook, 2nd edn. Wiley-VCH, Weinheim, pp 77–130

Breuninger WF, Piyachomkwan K, Sriroth K (2009) Tapioca/cassava starch: production and use. In: BeMiller J, Whistler R (eds) Starch: chemistry and technology. Academic Press, San Diego, pp 541–568

Chapuis A, Precoppe M, Méot JM, Sriroth K, Tran T (2017) Pneumatic drying of cassava starch: numerical analysis and guidelines for the design of efficient small-scale dryers. Dry Technol 35:393–408

Crapiste GH, Rotstein E (1997) Design and performance evaluation of dryers. In: Valentas KJ, Singh RP, Rotstein E (eds) Handbook of food engineering practice. CRC Press, Boca Raton, pp 121–162

Da G, Dufour D, Giraldo A, Moreno M, Tran T, Vélez G, Sánchez T, Le Thanh M, Marouzé C, Maréchal PA (2013) Cottage level cassava starch processing systems in Colombia and Vietnam. Food Bioprocess Technol 6(8):2213–2222. https://doi.org/10.1007/s11947-012-0810-0

Dahdouh L, Escobar A, Rondet E, Ricci J, Fliedel G, Adinsi L, Dufour D, Cuq B, Delalonde M (2021) Role of dewatering and roasting parameters on the quality of handmade gari. Int J Food Sci Technol 56(3):1298–1310. https://doi.org/10.1111/ijfs.14745

Dou GY, Wang XT, Zhao BC, Yuan XA, Pan CX, Tran T, Zellweger H, Zhu KS, Guo YJ, Wu H, Yin J, Bai YY (2020) The transformation and outcome of traditional cassava starch processing in Guangxi, China. Environ Technol. https://doi.org/10.1080/09593330.2020.1725647

Dufour D, O'Brien GM, Best R (2002) Cassava flour and starch: progress in research and development. Centre de Coopération Internationale en Recherche Agronomique pour le Développement (CIRAD), Montpellier, France and Centro Internacional de Agricultura Tropical (CIAT), Cali, 409 pp

Dziedzoave NT, Abass AB, Amoa-Awua WKA, Sablah M (2006) In: Adegoke GO, Brimer L (eds) Quality management manual for the production of high quality cassava flour. International Institute of Tropical Agriculture (IITA), Ibadan, 68 p

Edeh JC, Nwankwojike BN, Abam FI (2020) Design modification and comparative analysis of cassava attrition peeling machine. AMA Agric Mech Asia Africa Latin America 51(1):63–71

Escobar A, Dahdouh L, Rondet E, Ricci J, Dufour D, Tran T, Cuq B, Delalonde M (2018) Development of a novel integrated approach to monitor processing of cassava roots into Gari: macroscopic and microscopic scales. Food Bioprocess Technol 11:1370–1380

Escobar A, Rondet E, Dahdouh L, Ricci J, Akissoé N, Dufour D, Tran T, Cuq B, Delalonde M (2021) Identification of critical versus robust processing unit operations determining the physical and biochemical properties of cassava-based semolina (gari). Int J Food Sci Technol 56(3):1311–1321. https://doi.org/10.1111/ijfs.14857

Ezeocha CV, Ihesie LC, Kanu AN (2019) Comparative evaluation of toasting variables and the quality of gari produced by different women in Ikwuano LGA, Abia State, Nigeria. J Food Process Preserv 43(9):e14060. https://doi.org/10.1111/jfpp.14060

Forsythe L, Martin AM, Posthumus H (2015) Cassava market development: a path to women's empowerment or business as usual? Food Chain 5(1–2):11–27. https://doi.org/10.3362/2046-1887.2015.003

Forsythe L, Posthumus H, Martin AM (2016) A crop of one's own? Women's experiences of cassava commercialization in Nigeria and Malawi. J Gender Agric Food Secur 1(2):110–128

Hansupalak N, Piromkraipak P, Tamthirat P, Manitsorasak A, Sriroth K, Tran T (2016) Biogas reduces the carbon footprint of cassava starch: a comparative assessment with fuel oil. J Clean Prod 134(part B):539–546

Kemp IC (2012) Fundamentals of energy analysis of dryers. In: Tsotsas E, Mujumdar AS (eds) Modern drying technology: energy savings, vol 4. Wiley-VCH, Weinheim, pp 1–45

Kitinoja L, Saran S, Roy SK, Kader AA (2011) Postharvest technology for developing countries: challenges and opportunities in research, outreach and advocacy. J Sci Food Agric 91(4):597–603. https://doi.org/10.1002/jsfa.4295

Kudra T (2009) Energy aspect in food dehydration. In: Ratti C (ed) Advances in food dehydration. CRC Press, Boca Raton, pp 423–445

Kussaga JB, Jacxsens L, Tiisekwa BPM, Luning PA (2014) Food safety management systems performance in African food processing companies: a review of deficiencies and possible improvement strategies. J Sci Food Agric 94(11):2154–2169. https://doi.org/10.1002/jsfa.6575

Kuye A, Ayo DB, Sanni LO, Raji AO, Kwaya EI, Otuu OO, Asiru WB, Alenkhe B, Abdulkareem IB, Bamkefa B, Tarawali G, Dixon AGO, Okechukwu RU (2011) Design and fabrication of a flash dryer for the production of high quality cassava flour. Cassava Enterprise Development Project, International Institute of Tropical Agriculture, Ibadan, 50 pp

Luna J, Dufour D, Tran T, Pizarro M, Calle F, Garcia Dominguez M, Hurtado IM, Sanchez T, Ceballos H (2021) Postharvest physiological deterioration in several cassava genotypes over sequential harvests and effect of pruning prior to harvest. Int J Food Sci Technol 56(3):1322–1332. https://doi.org/10.1111/ijfs.14711

Mujumdar AS (2006) Handbook of industrial drying, 3rd edn. CRC Press, Taylor & Francis Group, Boca Raton, 1312pp

Ndjouenkeu R, Kegah FN, Teeken B, Okoye B, Madu T, Olaosebikan OD, Chijioke U, Bello A, Osunbade AO, Owoade D, Takam-Tchuente NH, Njeufa EB, Nguiadem-Chomdom ILN, Forsythe L, Maziya-Dixon B, Fliedel G (2021) From cassava to gari: mapping of quality characteristics and end-user preferences in Cameroon and Nigeria. Int J Food Sci Technol 56(3):1223–1238. https://doi.org/10.1111/ijfs.14790

Nweke FI (1994) Cassava processing in Sub-Saharan Africa: the implications for expanding cassava production. Outlook Agric 23(3):197–205. https://doi.org/10.1177/003072709402300307

Nzudie HLF, Zhao X, Tillotson MR, Zhang F, Li YP (2020) Modelling and forecasting roots & tubers losses and resulting water losses in sub-Saharan Africa considering climate variables. Phys Chem Earth 120:102952. https://doi.org/10.1016/j.pce.2020.102952

Ojide M, Abass A, Adegbite S, Tran T, Taborda LA, Chapuis A, Lukombo S, Totin E, Sartas M, Schut M, Becerra Lopez-Lavalle LA, Dufour D (2021) Processors' experience in the use of flash dryer for cassava-derived products in Nigeria. Frontiers in Sustainable Food Systems. https://www.frontiersin.org/articles/10.3389/fsufs.2021.771639/abstract

Okudoh V, Trois C, Workneh T, Schmidt S (2014) The potential of cassava biomass and applicable technologies for sustainable biogas production in South Africa: a review. Renew Sustain Energy Rev 39:1035–1052. https://doi.org/10.1016/j.rser.2014.07.142

Oni KC, Oyelade OA (2014) Mechanization of cassava for value addition and wealth creation by the rural poor of Nigeria. AMA Agric Mech Asia Africa Latin America 45(1):66–78

Ozoegwu CG, Eze C, Onwosi CO, Mgbemene CA, Ozor PA (2017) Biomass and bioenergy potential of cassava waste in Nigeria: estimations based partly on rural-level garri processing case studies. Renew Sustain Energy Rev 72:625–638. https://doi.org/10.1016/j.rser.2017.01.031

Pakowski Z, Mujumdar AS (2014) Drying of pharmaceutical products. In: Mujumdar AS (ed) Handbook of industrial drying. CRC Press, Boca Raton, pp 681–701

Parmar A, Tomlins K, Sanni L, Omohimi C, Thomas F, Tran T (2019) Exposure to air pollutants and heat stress among resource-poor women entrepreneurs in small-scale cassava processing. Environ Monit Assess 191:693

Patrizi N, Bruno M, Saladini F, Parisi ML, Pulselli RM, Bjerre AB, Bastianoni S (2020) Sustainability assessment of biorefinery systems based on two food residues in Africa. Front Sustain Food Syst 4:522614. https://doi.org/10.3389/fsufs.2020.522614

Precoppe M, Tran T, Chapuis A, Müller J, Abass A (2016) Improved energy performance of small-scale pneumatic dryers used for processing cassava in Africa. Biosyst Eng 151:510–519

Precoppe M, Chapuis A, Müller J, Abass A (2017) Tunnel dryer and pneumatic dryer performance evaluation to improve small scale cassava processing in Tanzania. J Food Process Eng 40:1–10

Precoppe M, Komlaga GA, Chapuis A, Müller J (2020) Comparative study between current practices on cassava drying by small-size enterprises in Africa. Appl Sci 10:7863

Romuli S, Abass A, Müller J (2017) Physical properties of Cassava Grits before and after pneumatic drying. J Food Process Eng 40:e12397

Saengchan K, Tran T, Faye M, Cantero-Tubilla B (2015) Technical and economic assessment of cassava starch processing in Paraguay. CGIAR Research Program on Roots, Tubers and Bananas (RTB), Lima, Peru. RTB working paper, p 52

Saravacos G, Kostaropoulos AE (2016) Food evaporation equipment. In: Handbook of food processing equipment. Springer, Cham, pp 367–419

Sartas M, Schut M, Proietti C, Thiele G, Leeuwis C (2020a) Scaling readiness: science and practice of an approach to enhance the impact of research for development. Agric Syst 183:102874

Sartas M, Schut M, van Schagen B, Velasco C, Thiele G, Proietti C, Leeuwis C (2020b) Scaling readiness concepts, practices and implementation. International Potato Center on behalf of RTB, 217 p

Sriroth K, Piyachomkwan K, Wanlapatit S, Oates CG (2000) Cassava starch technology: the Thai experience. Starch-Stärke 52(12):439–449

Standard B (2004) Guide to the measurement of humidity (Standard BS 1339-3:2004). British Standards Institution, London

Strumiłło C, Jones PL, Żyłła R (2014) Energy aspects in drying. In: Mujumdar AS (ed) Handbook of industrial drying. CRC Press, Boca Raton, pp 1077–1100

Taborda LA (2018) Determinación y análisis integral de impactos de la agroindustria rural de almidón de yuca en Cauca, Colombia. Doctoral thesis, Universidad Nacional de Colombia - Palmira

Taiwo KA (2006) Utilization potentials of cassava in Nigeria: the domestic and industrial products. Food Rev Intl 22(1):29–42. https://doi.org/10.1080/87559120500379787

Taiwo KA, Fasoyiro S (2015) Women and cassava processing in Nigeria. Int J Dev Res 5(2):3513–3517

Teeken B, Olaosebikan O, Haleegoah J, Oladejo E, Madu T, Bello A, Parkes E, Egesi C, Kulakow P, Kirscht H, Tufan H (2018) Cassava trait preferences of men and women farmers in Nigeria: implications for breeding. Econ Bot 72:263–277

Teeken B, Agbona A, Bello A, Olaosebikan O, Alamu E, Adesokan M, Awoyale W, Madu T, Okoye B, Chijioke U, Owoade D, Okoro M, Bouniol A, Dufour D, Hershey C, Rabbi I, Maziya-Dixon B, Egesi C, Tufan H, Kulakow P (2021) Understanding cassava varietal preferences through pairwise ranking of gari-eba and fufu prepared by local farmer-processors. Int J Food Sci Technol 56(3):1258–1277. https://doi.org/10.1111/ijfs.14862

Thiele G, Dufour D, Vernier P, Mwanga ROM, Parker ML, Schulte Geldermann E, Teeken B, Wossen T, Gotor E, Kikulwe E, Tufan H, Sinelle S, Kouakou AM, Friedmann M, Polar V, Hershey C (2021) A review of varietal change in roots, tubers and bananas: consumer preferences and other drivers of adoption and implications for breeding. Int J Food Sci Technol 56(3):1076–1092. https://doi.org/10.1111/ijfs.14684

Tran T, Da G, Moreno-Santander MA, Velez-Hernandez GA, Giraldo-Toro A, Piyachomkwan K, Sriroth K, Dufour D (2015) A comparison of energy use, water use and carbon footprint of cassava starch production in Thailand, Vietnam and Colombia. Resour Conserv Recy 100:31–40. https://doi.org/10.1016/j.resconrec.2015.04.007

Tran T, Faye M, Hansupalak N, Cantero-Tubilla B (2015a) Technical and economic assessment of cassava starch processing in Thailand. CGIAR Research Program on Roots, Tubers and Bananas (RTB), Lima, Peru. RTB working paper, p 65

Tran T, Faye M, Cantero-Tubilla B (2015b) Technical and economic assessment of cassava starch processing in Northern Vietnam. CGIAR Research Program on Roots, Tubers and Bananas (RTB), Lima, Peru. RTB working paper, p 59

Tran T, Faye M, Cantero-Tubilla B (2015c) Technical and economic assessment of cassava starch processing in Colombia. CGIAR Research Program on Roots, Tubers and Bananas (RTB), Lima, Peru. RTB working paper. 59 pp

UNEP: United Nations Environment Programme (2007) Global environment outlook GEO4: environment for development. UNEP, Nairobi

Yank A, Ngadi M, Kok R (2016) Physical properties of rice husk and bran briquettes under low pressure densification for rural applications. Biomass Bioenergy 84:22–30. https://doi.org/10.1016/j.biombioe.2015.09.015

Zvinavashe E, Elbersen HW, Slingerland M, Kolijn S, Sanders JPM (2011) Cassava for food and energy: exploring potential benefits of processing of cassava into cassava flour and bioenergy at farmstead and community levels in rural Mozambique. Biofuels Bioprod Biorefining Biofpr 5(2):151–164. https://doi.org/10.1002/bbb.272

Chapter 5
Orange-Fleshed Sweetpotato Puree: A Breakthrough Product for the Bakery Sector in Africa

Mukani Moyo (ID)**, Van-Den Truong, Josip Simunovic** (ID)**, Jean Pankuku, George Ooko Abong, Francis Kweku Amagloh** (ID)**, Richard Fuchs, Antonio Magnaghi, Srinivasulu Rajendran** (ID)**, Fredrick Grant** (ID)**, and Tawanda Muzhingi** (ID)

Abstract Replacing some of the wheat flour in breads and pastries with OFSP (orange-fleshed sweetpotato) puree can increase the market demand for these nutritious varieties and would offer economic opportunities for smallholders, including women and youths. The technology to make sweetpotato puree has been well developed in industrialized countries since the 1960s. Techniques fine-tuned by RTB allow OFSP puree to be stored in plastic bags for 6 months, without refrigeration. Private companies in Malawi and Kenya are now manufacturing the puree and selling it to bakeries that substitute OFSP puree for up to 40% of the white wheat flour in bread and other baked goods. Consumers like the bread that is sold in supermarkets and bakeries. Food safety protocols ensure that the puree is part of safe,

M. Moyo (✉)
International Potato Center (CIP), Regional Office for Africa, Nairobi, Kenya
e-mail: Mukani.Moyo@cgiar.org

V.-D. Truong · J. Simunovic · T. Muzhingi
Department of Food, Bioprocessing and Nutritional Sciences, North Carolina State University, Raleigh, NC, USA
e-mail: vtruong@ncsu.edu; simun@ncsu.edu

J. Pankuku
Tehilah Value Chain and Food Processing, Blantyre, Malawi

Root and Tuber Crops Development Trust (RTCDT), Lilongwe, Malawi

G. O. Abong
Department of Food Science, Nutrition and Technology, University of Nairobi, Nairobi, Kenya

F. K. Amagloh
Department of Food Science & Technology, Faculty of Agriculture, Food and Consumer Sciences, University for Development Studies, Tamale, Ghana
e-mail: fkamagloh@uds.edu.gh

145

G. Thiele et al. (eds.), *Root, Tuber and Banana Food System Innovations*,
https://doi.org/10.1007/978-3-030-92022-7_5

healthy products. The OFSP seed is available to smallholder farmers, who are linked with processors who buy the roots. Business models suggest that processing puree is profitable. The Scaling Readiness approach is helping to ensure that more farmers, processors, and consumers benefit from OFSP.

5.1   Introduction

Sweetpotato is a climate-smart root crop that adapts well to low rainfall, different altitudes, and poor soils. It requires minimal labor or other farm inputs. It is a hardy crop with low risk of crop failure (Abidin et al. 2017; Low et al. 2020). These farmer-preferred agronomic traits, combined with relatively high nutritional value, make the sweetpotato an ideal crop for enhancing food security and improving livelihoods in African communities. The biofortified orange-fleshed sweetpotato (OFSP) varieties have high levels of β-carotene, a precursor of vitamin A in the human body. OFSP is a proven, effective, and sustainable source of vitamin A, significantly contributing to the fight against vitamin A deficiency (VAD) in Africa (Low et al. 2007; Girard et al. 2017). OFSP also has vital, life-promoting phytochemicals that enhance protection from peroxides (Abong et al. 2020).

Despite the importance and knowledge of the nutritional benefits of OFSP, it remains generally underutilized, possibly because the roots are highly perishable, which depresses their market value. Low prices can subject OFSP to market losses, discouraging farmers from planting the varieties. Consumers prefer white-fleshed sweetpotato varieties, further reducing demand for OFSP (Moyo et al. 2021). The sweetpotato's seasonality further limits the periods when consumers have access to OFSP. Urban consumers eat less OFSP because they prefer convenient, ready-to-eat foods over roots that need to be cooked. Fresh sweetpotato is eaten boiled, steamed, roasted, or fried in most African communities (Mwanga et al. 2021), so processing OFSP roots into a puree creates more options for consuming them, improving their availability in cities and reducing food waste and losses.

OFSP puree can replace some of the white, wheat flour in baked and fried products. Making puree overcomes the challenge of perishability and seasonality of the roots by processing them into a more shelf-stable product that can be incorporated

R. Fuchs
Natural Resources Institute (NRI), University of Greenwich, Kent, UK
e-mail: R.S.Fuchs@greenwich.ac.uk

A. Magnaghi
Euro Ingredients Limited, Southern By-pass, Kikuyu, Kiambu County, Kenya
e-mail: antonio@euroingredients.net

S. Rajendran · F. Grant
International Potato Center (CIP), Kampala, Uganda
e-mail: srini.rajendran@cgiar.org; F.grant@cgiar.org

into diets even when the fresh roots are off-season (Wanjuu et al. 2018). Bread is a staple food in much of Africa, particularly in cities. Incorporating OFSP puree into bread would significantly increase the number of OFSP consumers and reduce VAD (Awuni et al. 2018; Owade et al. 2018). OFSP puree can replace up to 50% of the wheat flour in bread, while reducing sugar (90%) and fat (50%) and eliminating artificial colorings (egg yellow). The baked bread retains over 50% of the β-carotene, and the OFSP puree improves the texture of wheat products, making them easy to chew and digest (Wanjuu et al. 2018).

OFSP puree in bakery products is new and unique to sub-Saharan Africa (SSA) and shows great potential for expansion. African millers spend millions importing wheat, with East African countries being among the top importers. Kenya's wheat import bills are estimated at $250 million, Tanzania's at $150 million, Uganda's at $53 million, and Rwanda's at $35 million per year. Substituting some of this wheat flour with locally produced OFSP puree in popular baked goods would save African economies much needed foreign currency.

Over the last decade, the International Potato Center (CIP) has gained considerable experience in product development and marketing of bakery products in which 20–45% of the wheat flour is replaced with OFSP puree: successfully commercialized in Kenya, Malawi, Rwanda, and Burkina Faso and piloted in Ethiopia, Uganda, Ghana, Nigeria, Tanzania, The Gambia, South Africa, and Mozambique. The OFSP puree value chain activities have created market opportunities for smallholder sweetpotato farmers, especially women and youths, by increasing the value of the crop. Higher farm-gate prices for roots could encourage greater use of farm inputs, which would help to close the existing yield gap, increasing the availability of sweetpotato in rural areas and the accessibility of nutritious OFSP to urban consumers.

5.2 The Evolution of Sweetpotato Puree Processing

The technology for processing OFSP roots into puree was developed in the United States in the 1960s (Kays 1985), e.g., for baby foods, baked products, and other foods. Processing includes root washing, peeling, cutting, steaming/blanching or precooking, finish cooking (with temperature-time programs suitable for starch conversion by endogenous amylolytic enzymes to obtain the products with targeted maltose levels and viscosities), and grinding into puree. In many countries, processing of sweetpotato paste or puree is traditionally practiced at the household and small-scale levels by boiling or steaming fresh roots and then mashing or grinding them. The puree is an ingredient in many foods, including baby foods, soups, pies, cakes, ice creams, breads, and other products (Collins and Walter Jr 1992; Woolfe 1992; Truong 1989, 1994; Truong and Avula 2010).

The development of OFSP beverages, jam, and ketchup with sensory and nutrient attributes comparable to fruit-based products in the early 1990s raised interest among research institutions in several sweetpotato-producing countries including

India, Japan, Malaysia, the Philippines, and the United States (KNAES 1996; Padmaja 2009; Payton et al. 1992; Sankari et al. 2002; Tan et al. 2004; Truong 1992). Transfer of the processing technologies to farmer organizations, small-scale processors, and food companies for pilot testing and market development was conducted in the Philippines (Truong 1992).

In Japan, the puree from white-, orange-, and purple-fleshed sweetpotato varieties was commercially produced by cooperatives and small manufacturers to make ice creams, jams, juices, confectionary, and breads (Duell 1992; Katayama et al. 2017). However, sweetpotato puree was less common than sweetpotato flour as a substitute for wheat flour in bakeries around the world. In two coastal towns of Peru (Cañete and Chincha), OFSP in the form of grated raw or paste of steamed roots (depending on the availability of processing equipment) was substituted for up to 30% of the wheat flour to improve the β-carotene content of a commercial bread, *pan de camote* (Woolfe 1992). Recently, orange-fleshed sweetpotato puree has replaced 20–50% of wheat flour in cookies, donuts, and breads by some commercial bakeries in Ghana, Kenya, Malawi, Rwanda, and Uganda (Bocher et al. 2017).

Challenges for the puree processing industry include the following: (1) adjusting the process to account for differences in carbohydrate content and amylase activities among variety types, postharvest curing, and storage conditions to produce consistent and high puree quality, and (2) preservation technology that produces a shelf-stable product for convenient use in processed foods. Several techniques were later developed to produce purees with consistent quality, despite the varietal differences in carbohydrate content, starch-degrading enzyme activities, and postharvest handling (Collins and Walter Jr 1992).

5.3 Development and Commercialization of Aseptic Shelf-Stable OFSP Puree by Continuous Flow Microwave Processing

Purees can be preserved by canning, freezing, or refrigerated storage with acidification (with citric acid) or by adding preservatives such as potassium sorbate or sodium benzoate (Bocher et al. 2017; Pérez-Díaz et al. 2008; Musyoka et al. 2018a). However, poor product quality due to excessive heat in conventional cooking, the high cost of investment associated with frozen products, and the use of preservatives are the main hurdles for widespread applications of sweetpotato purees in the food industry. An improved canning process was developed for OFSP puree involving flash sterilization and aseptic filling/packaging to produce a shelf-stable product (Smith et al. 1982). However, validating and scaling up the technology did not proceed to commercial development in the United States.

Continuous flow microwave heating and sterilization technologies were initiated at North Carolina State University (NCSU) in the mid-1990s with support from the

Center for Aseptic Processing and Packaging Studies (CAPPS). The initial push for these developments was to enable industrial processing and commercialization of complex multiphase foods in sterile packaging formats, such as low acid soups and stews, and to retain native, typically heat-sensitive flavors, colors, textures, and nutrients like vitamins and antioxidants. Since then, many studies have been performed on other difficult-to-heat and poorly conductive foods and beverages such as salsas, dairy and plant-based milks, smoothies, baby foods, cheese sauces, other sauces, and cooking blends and pureed, homogenized, and diced vegetables and fruits like strawberries, peaches, mangos, blueberries, bananas, and pineapples. To date, over 70 FDA letters of no objections/no questions (LNO) have been issued for a variety of products processed and commercialized using several configurations of continuous flow microwave processing.

In the early 2000s, a novel process for rapid sterilization and aseptic packaging of OFSP puree using a continuous flow microwave system was successfully developed. The product packaged in flexible plastic containers had similar color, carotene retention, and apparent viscosity as the non-sterilized puree and was shelf-stable for at least 12 months (Coronel et al. 2005; Kumar et al. 2008). With this technology, consistently high-quality puree from sweetpotato roots (30–40% of the crop) could be packaged into virtually unlimited container sizes (up to 200 or 250 liters) as a functional ingredient for the food processing industry (Fig. 5.1). In 2008, the developed technology was transferred to a start-up company, Yamco. This large-scale factory has operated with an output of up to 20,000 kg of aseptic puree per day to supply a high quality, shelf-stable functional ingredient to companies manufacturing baby foods, beverages, soups, chips, baked products, and pet foods, among others. The Yamco factory is the first in the world to use a continuous flow microwave sterilization system for the commercial production of shelf-stable OFSP puree.

This rapid expansion of OFSP as an ingredient in food manufacturing has been influenced by the high retained nutritional values and convenient bulk packaging formats for further processing with easy portioning and conveyance/pumping. Previously available puree bulk packaging formats were in cans, 5-gallon plastic buckets of frozen product, or aseptic packaging using conventional (heat exchanger) methods of sterilization, all of which had significant shortcomings in quality levels, energy usage for storage, and ease of use or stability during preparation or defrosting.

After developing and testing the sterilization and shelf-life protocols, the microwave sterilized puree product was compared to similar products produced by other technologies (Table 5.1).

Advanced thermal (microwave) processing of sweetpotato in the U.S. is now used to make dozens of commercial products, including snacks, pie fillings, ice creams, pastas, bakery products, beverages, distilled alcoholic beverages like sweetpotato vodka and bourbon, soups, sauces, vegetable blends, and baby foods.

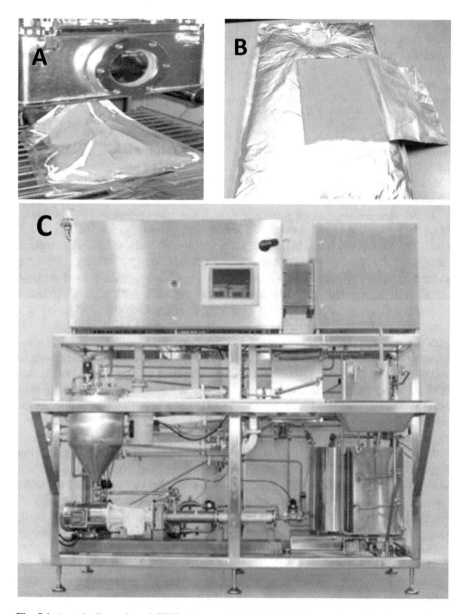

Fig. 5.1 Aseptically packaged OFSP puree produced by the first generation of continuous microwave heating system using a single-head Astepo aseptic filler (**a**) for a pilot scale or LiquiBox StarAsept aseptic filler for commercial production (**b**) in the early 2000s. Photo credits: VD Truong (NCSU). (**c**) Fourth industrial/commercial generation of continuous flow microwave heating/sterilization technology – 24 kW 2450 MHz processing installation at SinnovaTek/FirstWave Innovations in Raleigh, NC. Technology is commercialized by SinnovaTek Inc. under the Nomatic trademark. A factory using this technology has been set up in Kenya since 2019. (Photo courtesy of SinnovaTek Inc.)

Table 5.1 Sweetpotato puree comparisons: continuous flow microwave (first column) aseptic puree vs. other types

Attributes	Puree type				
	Microwave	Fresh	Frozen	Canned	With preservatives
Commercially sterile	Yes	No	No	Yes	Yes
Ambient storage	Yes	No	No	Yes	Yes
Superior natural color	Yes	Yes	No	No	No
Superior flavor	Yes	Yes	No	No	No
Fresh appearance	Yes	Yes	No	No	No
Preferred texture	Yes	No	No	No	No
Ease of use	Yes	No	No	No	Yes
High beta-carotene retention	Yes	Yes	No	No	No
% Preferred (NCSU consumer test)	68	25	1	5	1

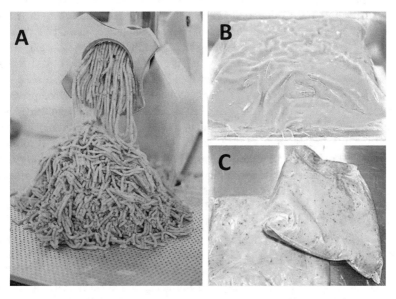

Fig. 5.2 Fresh OFSP puree (**a**) is packed in food-grade plastic bags that are vacuum sealed. (Photo credit: N. Ronoh (CIP)). (**b**) The fresh puree can either be made using peeled roots. (Photo credit: S. Quinn (CIP)). (**c**) A high-fiber puree can be made with unpeeled roots. (Photo credit: CIP)

5.4   The Development of OFSP Puree Inclusive Value Chain in Africa

The rapid growth of African cities is fueling the demand for safe, affordable processed foods, opening up a huge market for OFSP puree as an ingredient in processed foods. Conventionally packaged puree (Fig. 5.2) requires cold chain storage, a major bottleneck to expanding the use of fresh puree in Africa.

Puree processors need to invest in storage facilities to use during the harvest season, to store supplies for the off-season, but this implies a high cost of electricity. Also, the informal food sector is growing in African cities, but few informal traders can afford refrigeration. Therefore, adopting technologies to produce high-quality, shelf-stable OFSP puree in Africa is crucial to allow millions of consumers to access affordable, vitamin A-rich foods throughout the year. This could increase food availability in many food-insecure African countries.

OFSP puree can be treated with preservatives before being packaged and stored for 3–6 months at room temperature. A combination of potassium sorbate, sodium benzoate, and citric acid with vacuum packaging eliminates the need for cold storage for up to 6 months with little loss of β-carotene (Musyoka et al. 2018a). The OFSP puree with preservatives was piloted in Kenya through a partnership with Organi Limited, a sweetpotato processing company based in Homa Bay.

OFSP puree can also be packaged using the aseptic filling technology. This process sterilizes both the product and the packaging material, eliminating the need to add preservatives while prolonging the shelf life of the puree. The OFSP puree can then be stored at room temperature for up to 12 months, enabling informal vendors without cold storage facilities to safely use the product to make cooked foods. A source of affordable packaging pouches was identified, which had been a major bottleneck for scaling this approach. This technology for OFSP puree processing has been piloted in Kenya, Uganda, Ethiopia, and Rwanda through partnerships with private sector entities. It is now being introduced under the BioInnovate project with CIP and partners in Kenya, Rwanda, Ethiopia, and Uganda (BioInnovate 2019). However, further research is needed for this technology to be fully implemented due to the high risk of spore survival and growth of proteolytic *Clostridium botulinum* during storage at room temperatures, with the accompanying high hazard of botulinum toxin formation. Ultimately, a preservative-free OFSP puree is more marketable with more consumers seeking chemical-free food.

As described in Sect. 5.3, OFSP can also be processed into aseptic purees using a continuous flow microwave system to obtain a shelf-stable product. With funding from the UK's DFID, now Foreign, Commonwealth & Development Office (FCDO), under the Development and Delivery of Biofortification at Scale (DDBio) project, CIP partnered with a Kenyan agro-processing company specializing in preserving fruits and other crops, Burton and Bamber (B&B) Limited, to establish the first continuous flow microwave system for sweetpotato purees in Kenya, which will be commissioned in the second half of 2021. CIP worked with researchers from NCSU and SinnovaTek LLC in Raleigh, NC, to provide B&B support with farmer training, certification for Global GAP, and crop management. B&B buys fruits and sweetpotatoes from farmers and sells the produce at supermarkets. B&B encouraged sweetpotato farmers to act as decentralized vine multipliers (DVMs) to supply disease-free sweetpotato planting material (seed) to nearby farmers. Farmers in east and central Kenya were trained in global GAP and to supply OFSP fresh roots for puree processing to B&B.

While investment is increasing in the food sector, small and medium agribusinesses in most developing countries are still capital constrained. New innovative products like biofortified foods require greater visibility and positioning for future commercial investing. CIP's investment in B&B to acquire OFSP puree processing equipment and aseptic packaging is expected to return at least the nominal principal. Positive social and environmental impact is part of the business strategy. CIP also worked with Euro Ingredients Limited (Kenya) to supply the OFSP puree processing technology from Proteo International, Italy. SinnovaTek shipped a microwave processor to Kenya in early 2019. SinnovaTek and Euro Ingredients Limited worked together to install the puree processing and aseptic puree packaging equipment (Fig. 5.1).

CIP and Kenya Agricultural and Livestock Research Organization (KALRO) released new OFSP varieties (Irene and Sumaia) in Kenya that have consumer-preferred appearance and processing traits. Dielectric properties of 12 African genotypes were measured at 915 MHz and 2450 MHz in the temperature range of 22–130 °C. Dielectric properties play a significant role in developing commercial processing systems using microwave energy.

5.5 Scaling and Commercialization of OFSP Puree Processing in Malawi and Kenya

Most wheat flour is imported in Africa: 62% in Kenya, and 99% in Malawi. Scaling and commercializing OFSP puree to make baked and fried foods could replace up to 40% of the wheat flour with a cheaper, nutritious, and locally available product. Bakery products are popular across Africa, where many people eat them for breakfast, lunch, and supper. Bread is widely consumed, with diverse recipes. Its enticing flavor and convenience appeal to consumers, enhancing its popularity (Nwosu et al. 2014).

In Malawi, the Tehilah Bakery and Value Addition Centre has played a major role in scaling OFSP puree. Tehilah is a food processing business, established in 2017, that produces nutritious and healthy foods and adds value to locally grown crops. It was founded by Jean Pankuku, a food technologist, following her practical experience in the food industry, especially in the area of new product development and food processing. Tehilah processes OFSP puree in the conventional form (cold chain) and as baked and fried products. The business currently buys fresh roots from over 1000 OFSP farmers in eight districts (Blantyre, Zomba, Chiradzulu, Mulanje, Thyolo, Balaka, Chikwawa, and Nsanje), increasing economic opportunities not just for sweetpotato growers but also for transporters and workers at the bakery. Tehilah substitutes OFSP puree for 40% of the wheat flour in bread and buns. The bakery in Blantyre sells through over 20 outlets there, including Chichiri Shoprite (Fig. 5.3) as well as to Chikwawa district and Lilongwe, the capital city. The bakery plans to add OFSP cookies and various confectioneries to target new markets such as schools, hotels, offices, and manufacturing companies. Tehilah received recognition from the Ministry of Industry and Agriculture for the successful

Fig. 5.3 Tehilah Bakery substitutes OFSP for 40% of the wheat flour in bread (**a**), sold at Chichiri Shoprite (**b**) and elsewhere in Malawi. (Photo credit: J. Pankuku (Tehilah))

Fig. 5.4 Bread made with OFSP on sale at (**a**) Tuskys Supermarket. (Photo credit: CIP) and (**b**) at Naivas Bakery in Nairobi, Kenya. (Photo credit: C. Bukania (CIP))

commercialization of OFSP puree and notable community impacts by creating jobs for women and youths. Other private companies, such as Mother's Holding Limited, are now also actively involved in OFSP puree processing and product development.

In Kenya, Organi Limited has been actively involved in OFSP puree processing for over 6 years. It is based in Homa Bay County; the cofounder and managing director is Ms. Consolata Bryant. The company has established a network of nearly 200 smallholder farmers of whom at least 50% are women and 20% are youths in the western Kenyan counties of Homa Bay, Migori, Kakamega, Busia, Kericho, and Kisumu. The farmers sell OFSP roots to Organi, which makes about 500 kg of puree daily, with the help of 16 employees. The puree is sold to supermarkets (Fig. 5.4) in Nairobi and Kiambu County. In 2017, the company expanded its product line to include OFSP bread and buns with 40% wheat flour substitution. Organi Ltd is planning to gradually increase the number of fresh root suppliers to at least 1000 by 2025 and to extend the bakery section to industrial capacity by June 2022. This will enable the company to expand the range of OFSP products and to sell to more regions. The Kenyan Ministry of Agriculture and the Ministry of Health have recognized the role of OFSP in improving food security and hidden hunger (e.g., vitamin deficiencies) and have added orange-fleshed sweetpotato to the list of foods that are rich in micronutrients (pro-vitamin A-rich carotenoids). The Ministry of Trade and Industrialization took note of the increasing use of OFSP puree and has now developed specifications to facilitate its use in trade and commercial production. Other private companies, such as Euro Ingredients Limited, are now also actively involved in OFSP puree processing and product development.

The commercialization of OFSP puree in Malawi and Kenya has taken relatively similar paths, with both countries managing to penetrate the bakery industry, with beneficial impacts along the OFSP value chain.

5.6   Consumer Acceptance of OFSP Puree Composite Bread

This section describes consumer acceptability of composite bread from OFSP puree. The focus is on bread because it is widely eaten in Africa (Wanjuu et al. 2019; Atuna et al. 2020). Previous studies show that OFSP puree can be blended with wheat flour to bake consumer-acceptable flat or leavened bread (Table 5.2). The composite bread has a superior aroma, color, and soft texture, which contribute to its acceptability. Other benefits include low production costs and improved vitamin A content (Amagloh 2019). Since favorable sensory assessment scores for OFSP-based bread are one of the key drivers for patronage by consumers (Okello et al. 2021), this product has the potential to be scaled in African communities.

5.7   Food Safety in OFSP Puree Processing in Africa

To maintain consumer trust in OFSP products, the puree must adhere to the highest standard of food safety. Understanding the microflora associated with sweetpotatoes is important if processes are to be designed to control growth of spoilage and pathogenic organisms. While the skin of the root is intact, its flesh is protected from microbial attack. However, once the roots are prepared for puree production, the stages of peeling and cutting disrupt the cell structure, releasing nutrients that can be metabolized by microorganisms (Brackett 1994). Bacteria and fungi make up most of the microflora in low-acid, starchy vegetables (Brackett 1987). How the roots are processed determines the microflora of the puree. First, the OFSP roots are boiled or steamed, killing any vegetative organisms present, but allowing spores to survive. One organism that is associated with many types of foods, including root vegetables, and that produces one of the most potent toxins known is *Clostridium botulinum*, an anaerobic bacterium. Vacuum packaging a puree provides the conditions for the spores to germinate and for this deadly toxin to be produced. Cooking/

Table 5.2 Composite breads including OFSP puree and their consumer assessments

Reference	Flesh color of sweetpotato	Form used	Wheat substitution level (%)	Panelist numbers	Results/deduction
Low and van Jaarsveld (2008)	Orange	Puree Flour	38	98 shoppers in 2004 112 shoppers in 2005	Strong preference for composite bread over wheat-only bread owing to its heavier texture, superior taste, and golden color
Wu et al. (2009)	Not specified	Puree	5, 10, 20, and 30	20 students	Incorporating puree led to a relatively high preference for texture and overall acceptability. The 30% blend was the most preferred
Bonsi et al. (2014)	Orange	Puree	10, 20, and 30	192 consumers	Mean acceptability was 6.25 on a 7-point scale, indicative of high consumer preference for the composite bread
Bonsi et al. (2016)	Orange	Puree	40	50 undergraduates	On a 9-point scale, the overall acceptability for the OFSP-composite bread and wheat-only bread is similar. All other attributes had a score of at least 7, indicating high consumer preference
Awuni et al. (2018)	Orange	Puree	46%	310 consumers	Paired preference assessments indicated that 77% of the respondents preferred the OFSP-composite bread over the existing wheat-only one
Ouro-Gbeleou (2018)	Orange	Puree	Not stated	387 consumers	Scores of respondents showed a preference for OFSP-composite bread based on its sweet taste and soft texture
Wanjuu et al. (2019)	Orange	Puree	30%	1024 consumers	Softness, color, taste, aftertaste, smell, and overall acceptability had scores above 7.4 on a 9-point scale, i.e., "very likable"

steaming is crucial for destroying the vegetative organisms, but post-process contamination can allow microorganisms to reenter the product.

A study in a puree processing plant in Kenya found that steaming roots significantly lowered levels of microorganisms but that levels in OFSP puree increased again due to post-process contamination by organisms such as *Staphylococcus aureus*, a foodborne pathogen that is commonly found on peoples' skin and noses and in infected cuts (Malavi et al. 2018). Fresh puree, which is currently produced in African countries, requires cold storage. However, refrigerated storage (<5 °C) will allow the growth of organisms that thrive at low temperatures, such as *Listeria monocytogenes*, a bacterium that can cause life-threatening illness in vulnerable groups such as pregnant women, young children, the elderly, and people with suppressed immune systems. Preservatives could reduce the chances of contamination. Adding 0.06% (w/v) of the preservative sorbic acid or benzoic acid while reducing the pH of the puree to 4.2 with citric acid inhibited the growth of *L. monocytogenes* at 4 °C (Pérez-Díaz et al. 2008). Acidifying the puree can also control the growth of *C. botulinum* spores that are unable to grow at pH below 4.5 (Lund and Peck 2000). Acidification and preservatives result in an ambient temperature-stable OFSP puree (Musyoka et al. 2018a).

These studies highlight the importance of strict food safety measures in OFSP puree production. In Kenya and Malawi, official quality standards for OFSP puree have been developed and approved by the respective Bureau of Standards, and it will be worthwhile for other African countries to follow suit. Food safety training not only helps change behavior and attitudes of the OFSP puree processors but also contributes to the safety of the products (Malavi et al. 2021). Food safety and hygiene training workshops were conducted in 2017 and 2019 by CIP-Kenya under the RTB scaling project (Musyoka et al. 2018b). Forty-two participants (22 males and 20 females) attended from national and county government, academia, food processing companies, and nonprofit organizations from different counties. The training aimed to equip participants with knowledge and skills on food safety and hygiene. A knowledge-retention evaluation (questionnaire) held in January 2021 found that the trainees had retained 70% of what they learned at the 2019 course. Food safety regulations and trainings are important components for scaling the production and use of OFSP puree (Sect. 5.9).

5.8 Agribusiness Development and Entrepreneurship Opportunities for OFSP Puree in Africa

OFSP varieties with proven agronomic advantages and acceptance by smallholder farmers (Girard et al. 2017) play a great role in the short- and long-term food and nutrition security in sub-Saharan Africa (SSA). From the supply side, improving the productivity of OFSP is hindered mainly by poor access to quality planting material. Farmers usually recycle their seed, so it is often heavily infected with virus diseases and pests (especially weevils). Consequently, yields are typically only 6 tons per hectare, as compared to the potential 21 ton/hectare (Low et al. 2017). To strengthen OFSP seed systems and improve the nutritional status of children under 5 years of age, several new improved biofortified varieties of sweetpotato have been introduced in SSA since 2014 through partnerships between CIP and national agricultural research institutions (NARIs) (Rajendran et al. 2017).

A strong, entrepreneurial sweetpotato seed system has evolved among NARIs to sustain their seed production and provide a consistent supply of quality, disease-free early generation seed (EGS) from community-based decentralized vine multipliers (DVMs). The DVMs are linked to EGS producers (mainly NARIs) who can supply healthy vines (i.e., free of major pests and diseases, such as sweetpotato virus disease (SPVD) and sweetpotato weevil) (Namanda et al. 2011; Rajendran et al. 2017), and they are sold to farmers or supplied through government extension efforts and projects at highly subsidized prices or free of charge (Okello et al. 2015; Ogero et al. 2016; Bentley et al. 2018). Although most sweetpotato farmers still obtain their planting material from their own or neighboring farms, a market for quality vines, particularly for OFSP varieties, has started emerging in African countries such as Kenya, Uganda, and Tanzania (Rajendran et al. 2017). This has created huge opportunities for NARIs, sweetpotato seed and root growers, and also for processors of OFSP products.

5.8.1 Business Models for OFSP Puree Processing

Five potential business models (Table 5.3) for OFSP puree processing have been identified, and the financial viability has been measured. Medium-scale enterprises were the focus of this study because the puree processing equipment is imported and is too expensive for small businesses. The models were identified through a "financial cost-benefit analysis (FCBA)" using real and hypothetical data on technical and financial operations, collected through key informant interviews from industry players in SSA with expertise in this domain (Rajendran et al. 2019). *Model 1* focuses on OFSP puree processors who also make products such as cookies, cupcakes, and the donut-like mandazi. They store the puree briefly without vacuum packaging, and the goods made from OFSP are sold directly to end users. This is an example of the "business to consumers (B2C)" model. *Model 2* focuses on OFSP puree processors who use vacuum packaging and lots of freezers for storage. Puree is then sold to bakeries. This model is categorized as "business to business (B2B)." *Model 3* focuses on bakers and others who buy OFSP puree from processors. *Model 4* focuses on OFSP puree processors who use preservatives and vacuum packaging rather than cold storage. *Model 5* focuses on OFSP puree processors who make use of the hot-fill technology that improves the shelf life of the puree without any need for preservatives.

All five business models are financially viable in the long run, with the level of average investment between $30,000 and $43,000 for medium-scale enterprises. Some key informants said that small enterprises can start with as little as $1000. However, the sensitivity of the business, net present values, internal rate of return, return on investment, and payback period (the period of time required to recoup a capital investment) for the investments differ across the five business models (Table 5.4).

Table 5.3 Types of potential OFSP puree business models and minimum required investment

Business model	Equipment	Final products	Model	Area required (in square meters)	Total annual investment required (USD)	Maximum capacity of puree requirement and annual production (kg)
1	High fiber puree (HFP) machine, steamer, depositor, baking equipment	OFSP puree, OFSP baked and fried products	B2C	36	$43,030	16,000[a]
2	High fiber puree, steamer, vacuum machine with label printer	OFSP puree	B2B	100	$33,370	375,000[b]
3	Depositor (cookie machine), baking equipment	OFSP baked and fried products	B2C	36	$31,390	16,700[a]
4	High fiber puree, steamer, vacuum machine with label printer, ribbon mixer	OFSP puree	B2B	100	$34,170	375,000[b]
5	Hot fill machine, high fiber puree (HFP) machine, steamer, vacuum machine with label printer	OFSP puree	B2B	100	$37,170	375,000[b]

Source: Key informant interviews (KII) with private players in 2018–2019, author's calculation
[a]Investors buy puree from puree producers, which is a requirement for investors per year
[b]Investors produce puree at a maximum capacity per year

Table 5.4 Financial feasibility indicators for potential OFSP puree business models

Business model	NPV	Internal rate of return (IRR)	Discount rate[a]	Payback period (years)	Return on investment (RoI) per year (%)
1	$291,560	85%	18%	1.3	164
2	$247,018	90%	18%	1.4	198
3	$262,342	90%	18%	1.3	203
4	$265,404	77%	18%	2.8	252
5	$345,380	82%	18%	1.0	266

NPV net present value, *IRR* internal rate of return
[a]Discount rate is based on the 18% lending rate in Kenya

The return on investment (RoI) for each business model is more than 100%, and the payback period is a maximum of 3 years (Rajendran et al. 2019). All five models show good financial viability with different intensities probably influenced by different levels of risk factors. It is important for investors to understand the market demand for the products, available funds, and the type of business model that is appropriate for their resource base and setting. There are two significant risks involved in this business: (1) lack of consistent supply of sweetpotato roots and (2) market demand fluctuation for the product. If marketing strategies are well executed, models 1, 3, and 5 can be the most viable ones. If there are fluctuations in the market demand for the puree product, then investors should focus on model 4 where preservatives are used for producing OFSP puree. Though model 2 is financially viable, it comes with the highest risk for investors.

5.9 Scaling Readiness and Strategy for OFSP Puree in Africa

In order to go to scale, the core OFSP puree innovation requires several complementary innovations ranging from availability of planting material and contract farming to cold chain and extension services. With support from RTB, CIP implemented a scaling project between 2019 and 2021 in Kenya, Uganda, and Malawi. The main objective of the project was to increase the use of OFSP puree in popular baked and fried goods. The approach implemented included the following: (1) identifying smallholder OFSP farmers to be trained on good agronomic practices to ensure the production of high-quality roots; (2) product development training for bakers to substitute up to 40% of wheat flour with OFSP puree; (3) creating income and employment opportunities for youths and women along the value chain; and (4) demand creation through increase in awareness of the nutritional and economic benefits of OFSP puree processing and product development. The scaling project worked with small and medium-sized processors to introduce and expand their use of puree and encourage the development of local supply chains providing high-quality OFSP roots.

The scaling project was backstopped by scaling experts from RTB. It used Scaling Readiness, a stepwise approach for analyzing the innovations from a scaling-out perspective, diagnosing the bottlenecks that hinder scaling out, developing scaling strategies, building common agreements with the stakeholders on the strategies, and updating them accordingly (Sartas et al. 2020a). *Scaling Readiness* breaks down the core innovation into components. *Innovation readiness* is the capacity of the innovation to contribute to development outcomes in specific locations. This is presented in nine stages showing progress from an untested idea to a fully mature, proven innovation. *Innovation use* is the level of use of the innovation or innovation package by the project members, partners, and society. Use begins with the team that develops the innovation and broadens to other users who are

Table 5.5 Summary definitions of levels of innovation readiness and use

Stage	Innovation readiness	Innovation use
1	Idea	Intervention team
2	Basic model (testing)	Direct partners (rare)
3	Basic model (proven)	Direct partners (common)
4	Application model (testing)	Secondary partners (rare)
5	Application model (proven)	Secondary partners (common)
6	Application (testing)	Unconnected developers (rare)
7	Application (proven)	Unconnected developers (common)
8	Innovation (testing)	Unconnected users (rare)
9	Innovation (proven)	Unconnected users (common)

Source: Sartas et al. (2020a)

completely unconnected with the team or their partners. Scaling Readiness is a function of innovation readiness and innovation use (Table 5.5).

Readiness and use levels are evaluated using conceptual, applied, and experimental evidence. Scaling Readiness helps to identify the gaps in the design of the innovations and to prioritize the research and engineering work to address these gaps (Sartas et al. 2020b). The Scaling Readiness framework is discussed in more detail in Chap. 3.

Using Scaling Readiness, the team identified ten complementary innovations, their expected outputs, and overall functions (Table 5.6).

5.9.1 Strategic Partners for Scaling

In Kenya during 2020, partnerships with actors along the OFSP value chain came into their own during the Covid pandemic. Interactions were more virtual than physical, but because of the previous engagement, there was a positive response that ensured farmers continued to benefit through trade. Farm Concern International helped to strengthen these partnerships and develop new ones to ensure that farmers benefitted. Engagement with informal markets facilitated linkages to farmers. Processors and wholesale buyers were engaged to source OFSP roots from farmers. As a result, farmers were in a position to sell more than 33 metric tons of OFSP roots to various market outlets, both for fresh sweetpotatoes and for puree processing, with an estimated value of 1,205,270 KSh ($11,095). The market share for OFSP in the profiled markets increased as more traders and consumers became aware of the benefits of OFSP. Organi Limited increased its database from 139 to 239 contracted commercial root producers, almost doubling the OFSP puree supply to Naivas supermarket. In Malawi, Tehilah Bakery linked buyers to over 1000 farmers in Nsanje, Balaka, Blantyre, and Thyolo districts. In 2021, there were individual farmers who were selling their roots to the bakery as well as associations. Perisha

Table 5.6 Complementary innovations for scaling of OFSP puree in Kenya, Uganda, and Malawi

Innovation profile – OFSP puree

Complementary innovation	Output 1	Output 2	Output 3	Value chain level	Innovation function
1. Contract farming to provide disease- and pest-free seed	Large OFSP seed multiplication center for seed production	Quality declared planting material law on commercial vine multiplication	Providing incentives and subsidies to commercial vine producers	Planting material	Accessibility (affordability)
2. Processing-friendly OFSP varieties for production by commercial farmers in Kenya	List of processing-friendly OFSP varieties that can be released quickly in Kenya	List of existing OFSP varieties suitable for different climatic areas		Planting material	Accessibility (availability)
3.Delivery of extension services	Training commercial vine producers on vine multiplication and positive selection	Upgrade extension module on disease- and pest-free OFSP production	Commercial OFSP farmer clusters	Planting material	Accessibility (availability)
4. Cold chain for OFSP puree for sales and shelves for shelf-stable puree	Shelves for storing puree before sales to wholesalers and retailers			OFSP puree	Accessibility (availability)
5. OFSP puree processing technologies, equipment for independent processor	OFSP puree packaging equipment			OFSP puree	Capacity to use (hardware)
6. Demonstrations of OFSP puree processing and packaging equipment	Training workshop on OFSP puree business	Guidelines on procedures of puree making	Temporary holding areas for OFSP fresh roots in the processing facility	OFSP puree	Capacity to use (people)
7. Credit access manual for OFSP producers	Credibility assessment guidelines on OFSP production for banks			OFSP roots	Accessibility (affordability)

(continued)

Table 5.6 (continued)

Innovation profile – OFSP puree					
Complementary innovation	Output 1	Output 2	Output 3	Value chain level	Innovation function
8. Advocacy and awareness campaign on benefits of OFSP	Improved awareness along the sweetpotato value chain			OFSP bread or pastry	Motivation (convince)
9. OFSP puree business development models	Manual for OFSP puree business development			OFSP puree	Capacity to use (people)
10. Climate-controlled storage for OFSP roots	Storage facilities for OFSP to reduce postharvest loses			OFSP roots	Accessibility (availability)

Agro and Packaging Enterprises contracted 140 farmers to supply OFSP roots for puree processing as soon as production starts.

Much attention was given to developing the capacity of small and medium enterprises (SMEs) to run competitive businesses that can access credit from financial institutions. In collaboration with NetBizImpact, the project finalized provision of tailor-made business development services to six SMEs – two from Malawi, three from Uganda, and one from Kenya. The six SMEs were Organi Limited (Kenya), Tehilah Enterprises (Malawi), Perisha Agro and Packaging Enterprises (Malawi), Food Biosciences and Agribusiness Innovation Program (Uganda), EADC (Uganda), and BioFresh Ltd (Uganda). The main goal was to offer business solutions that would facilitate economic potential in the OFSP value chain and make the SMEs involved more competitive in the marketplace and achieve scale. During the coaching and mentorship, the SMEs were helped to identify growth opportunities, key constraints, and support to overcome them. NetBizImpact tailor-made business models aimed at enhancing the SMEs' capacity to take advantage of a promising market potential within the OFSP value chain including the puree. SMEs were encouraged to give their feedback on the relevance of the tools and exercises and effectiveness of the coaching and mentoring sessions included in the business development support. Besides coaching, NetBizImpact helped develop quality profiles for each SME, provided tools and templates, did value chain analysis at firm level, and discussed growth plan development. Discussions were held with NetBizImpact to develop a manuscript on the business development service process.

Naivas Supermarket had a 2-day refresher baking training using OFSP puree to retrain Naivas bakers on how to bake with the puree. The training exercise was held with the food safety training exercise. Some samples were collected for nutritional analysis by CIP's Food and Nutritional Evaluation Laboratory (FANEL), and these showed high levels of pro-vitamin A carotenoids.

A cooking demonstration at the University of Nairobi dietetic kitchen (Department of Food Science, Nutrition and Technology) showed how OFSP puree could be incorporated into foods and contribute to dietary vitamin A. This training for kitchen staff explained the benefits of OFSP puree.

In July, the Cleanshelf Supermarket baking team attended a baking training on using OFSP puree where they were trained to bake with puree to encourage the supermarket to start using OFSP puree.

In partnership with DDBio, BBC Africa, under the Life Clinic segment, broadcasted an episode on OFSP work in Africa, creating awareness of the commercial and nutritional benefits of OFSP for a worldwide audience. Euro Ingredients Limited has continued to provide technology demonstrations, equipment support, and trainings on OFSP product development to different partners in both the formal and informal sectors. Recipes for the incorporation of OFSP puree in commonly consumed Kenyan street foods such chapati, mahamri, and bhajia have been developed. In addition, Euro Ingredients Limited has expanded its supply of the OFSP dough and chapati to also include several informal vendors in middle-income sectors in Nairobi.

In Malawi, Tehilah currently produces a minimum average of 200 OFSP loaves of bread per day to a maximum of 560 loaves a day – depending on the capacity and availability of fresh roots. Production is 6 days a week translating to between 5000 and 13,000 loaves a month. A group of women in Balaka were trained to use OFSP puree to make donuts, fritters, and sweet beer (a locally fermented nonalcoholic drink). Over 10 tons of roots (from farmers associations) were supplied to Mothers Holding Ltd, and 12,480 loaves of OFSP bread were produced by mid-2020. The period also saw the setting up of an OFSP puree processing unit and installation of water and three-phase electricity at Perisha Agro, Malawi.

5.9.2 Core and Complementary Innovations for Scaling

The core and complementary innovations were analyzed using the Scaling Readiness approach. In the three implementation countries, delivery of extension services was the most used complementary innovation, while productive varieties were at the highest level of innovation readiness (Figs. 5.5, 5.6, and 5.7). The next highest level of innovation readiness was for processing-friendly varieties and a credit access manual in Kenya (Fig. 5.5), a business development manual in Uganda (Fig. 5.6), and the advocacy and awareness campaigns in Malawi (Fig. 5.7). Many consumers reject the OFSP varieties because they are low in dry matter, but processors preferred these moister varieties. Processing-friendly varieties were still in development in Uganda and Malawi. Financial constraints limited the implementation of cold chain and climate-controlled root storage. OFSP puree business development

OFSP puree for baked and fried products in Kenya in 2021

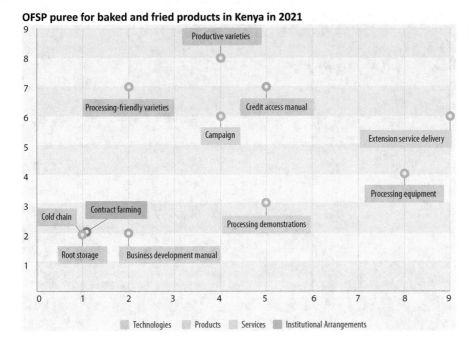

Fig. 5.5 Innovation readiness and use of OFSP puree in Kenya. The *x*-axis represents innovation use, while the *y*-axis represents innovation readiness

OFSP pure for baked and fried products in Uganda in 2021

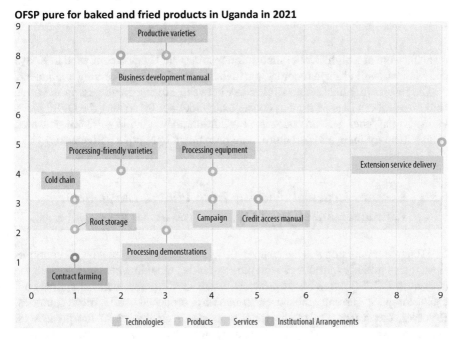

Fig. 5.6 Innovation readiness and use of OFSP puree in Uganda. The *x*-axis represents innovation use, while the *y*-axis represents innovation readiness

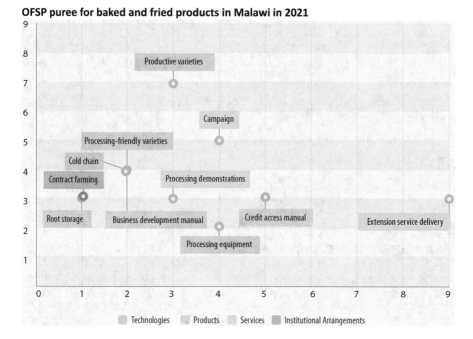

Fig. 5.7 Innovation readiness and use of OFSP puree in Malawi. The *x*-axis represents innovation use, while the *y*-axis represents innovation readiness

manuals were at a high level of innovation readiness in Uganda compared to Kenya and Malawi, while Kenya recorded the highest use of puree processing equipment.

The results highlight some similarities between Scaling Readiness in the three countries and can thus be used as conceptual guidelines for scaling the OFSP puree innovation in other African countries. In addition, they provide guidance on which complementary innovations require more investment to remove bottlenecks.

5.10 Lessons Learned: Moving from Pilot to Large-Scale Commercialization of OFSP Puree in Africa

OFSP puree processing has been successfully piloted in over ten African countries since it was initially introduced with most countries recording steady growth in the demand for OFSP puree-based products. Fresh OFSP puree, which requires cold chain storage, is currently the most common form produced in all African countries. However, the seasonality of the fresh roots poses challenges to the processors. During harvest season, cold storage capacity limits the amount of puree that can be produced and stored in preparation for off-season demand. The need for cold storage also means processors need temperature-controlled transportation to ensure that the puree is supplied to bakeries in good form.

The bakeries typically purchase in small batches to reduce the need for long-term cold storage from their end. This puts pressure on puree processors to continuously increase their cold storage capacity in order to grow their businesses and meet the demand for the nutritious product. To date, the majority of the OFSP processors rely on electricity to power the cold chain storage facilities, which is expensive to maintain in Africa with some regions facing unstable supply, which jeopardizes the quality of the puree. This has been a major bottleneck to scaling the use of OFSP puree in the region. It is this node of the OFSP puree value chain that needs to be addressed in order for large-scale commercialization to become a reality in Africa.

Previous efforts to address the challenges faced by processors included the use of preservatives to eliminate the need for cold storage, which conflicts with consumer preferences. Aseptic shelf-stable puree provides a viable pathway to address this current bottleneck. Yamco LLC in the United States has successfully produced aseptic puree at commercial scale since 2008, supplying the product to various sectors including the bakery industry. Learning from the success story of Yamco, B&B Limited in Kenya has established the first continuous flow microwave system in Africa that is set to commence production of aseptic shelf-stable puree before the end of 2021 (Fig. 5.8). The target market for the product includes large-scale bakeries, restaurants, institutional kitchens, humanitarian organizations for inclusion in nutritional interventions, local governments involved in school feeding programs,

Fig. 5.8 Aseptic puree processing system commercialized by SinnovaTek Inc. under the Nomatic trademark being set up at B&B Limited, Kenya. (Picture credit: Molly Abende, B&B Production Manager)

and informal traders selling street foods, among others. Other countries such as South Africa and Ghana are also in the process of setting up similar systems. Adoption of the continuous flow microwave system in Africa will enable large-scale commercialization of OFSP puree and eliminate the current bottleneck that processors are facing.

5.11 Conclusion

Value addition through processing of agricultural products such as OFSP roots to puree is key to ensuring a stable supply of highly nutritious products to consumers. Puree processing technologies have advanced over the decades from traditional methods involving manual mashing of the cooked roots to highly sophisticated and automated systems such as the fourth generation of continuous flow microwave sterilization systems. To date, steady growth in the commercialization of OFSP puree and the subsequent wheat flour substitution in bakery products has been noted in African countries such as Kenya, Rwanda, Malawi, and Uganda, among others. Positive impacts on revenue generation for small-scale farmers and businesses, employment opportunities for women and youths, and improved nutritional status of target communities are some of the targeted outcomes. These achievements can be attributed to strategic partnerships with stakeholders in agricultural development, public, and private sectors. In addition, financially viable business models have been identified to ensure that OFSP processors make sound financial decisions and maintain a competitive edge in the market. With this rich knowledge base, advanced processing technologies, evidence of Scaling Readiness, and successful piloting of the product, OFSP puree is the breakthrough product for Africa offering the much-needed nutritious products with superior flavor, texture, and appearance accepted by consumers. Further work needs to focus on improving awareness to the multifaceted benefits of OFSP puree along different nodes of the value chain.

Acknowledgments This research was largely undertaken as part of, and funded by, the CGIAR Research Program on Roots, Tubers and Bananas (RTB) and supported by CGIAR Trust Fund contributors. The authors would also like to acknowledge donor agencies including Foreign, Commonwealth & Development Office (FCDO) under the Development and Delivery of Biofortification at Scale (DDBio) project and Swedish International Development Cooperation Agency (Sida) under the BioInnovate Africa Program for the financial support toward the accomplishment of this work. The authors would also like to acknowledge the contributions of private OFSP puree processors in Kenya (Organi Limited, Euro Ingredients Limited, Burton and Bamber Co. Limited), Uganda (Lishe@BIOFRESH), Malawi (Tehilah Bakery and Value Addition Centre), Ethiopia (Duwame Bakery), Rwanda (A Women's Bakery), and Tanzania (AFCO Investment Co. Limited) to the progress reported in this chapter. The authors thank the following individuals who provided input and feedback on earlier drafts of this chapter: Tom Remington, Gordon Prain, and Jeff Bentley. We greatly appreciate Zandra Vasquez's assistance with references and formatting.

References

Abidin PE, Carey E, Mallubhotla S, Sones K (2017) Sweetpotato cropping guide. Africa Soil Health Consortium, Nairobi

Abong GO, Muzhingi T, Okoth MW, Ng'ang'a F, Ochieng' PE, Mbogo DM, Malavi D, Akhwale M, Ghimire S (2020) Phytochemicals in leaves and roots of selected Kenyan Orange fleshed sweet potato (OFSP) varieties. Hindawi-Int J Food Sci 2020:3567972, 11 p

Amagloh FK (2019) Orange-fleshed sweetpotato composite bread: the nutritional and economic potential. In: Aidoo R, Agbenorhevi JK, Wireko-Manu FD, Wangel A (eds) Roots and tubers in Ghana: overview and selected research papers. KNUST Printing Press, Kumasi, pp 268–283

Atuna RA, Sam FE, Ackah S, Amagloh FK (2020) Bread consumption pattern and the potential of orange-fleshed sweetpotato-composite bread in Ghana. Afr J Food Agric Nutr Dev 20(5):16509–16521

Awuni V, Alhassan MW, Amagloh FK (2018) Orange-fleshed sweet potato (Ipomoea batatas) composite bread as a significant source of dietary vitamin A. Food Sci Nutr 6(1):174–179

Bentley JW, Andrade-Piedra J, Demo P, Dzomeku B, Jacobsen K, Kikulwe E, Kromann P, Kumar LP, McEwan M, Mudege N, Ogero K, Okechukwu R, Orrego R, Ospina O, Sperling L, Walsh S, Thiele G (2018) Understanding root, tuber and banana seed systems and coordination breakdown: a multi-stakeholder framework. J Crop Improv 32(5):599–621. https://doi.org/10.1080/15427528.2018.1476998

BioInnovate (2019) Orange-fleshed sweet potato (OFSP) puree for bakery applications in East Africa. https://bioinnovate-africa.org/orange-fleshed-sweet-potato-ofsp-puree-for-bakery-applications-in-east-africa/

Bocher T, Low JW, Muoki P, Magnaghi A, Muzhingi T (2017) From lab to life: making storable orange-fleshed sweetpotato purée a commercial reality. Open Agric 2(1):148–154

Bonsi E, Chibuzo E, Zibawa R (2014) The preliminary study of the acceptability of Ghana bread made with orange sweet potato puree. J Hum Nutr Food Sci 2(5):1045

Bonsi EA, Zabawa R, Mortley D, Bonsi C, Acheremu K, Amagloh FC, Amagloh FK (2016) Nutrient composition and consumer acceptability of bread made with orange sweet potato puree. Acta Hortic 2016(1128):7–14

Brackett RE (1987) Microbiological consequences of minimally processed fruits and vegetables. J Food Qual 10:195–206

Brackett RE (1994) Microbiological spoilage and pathogens in minimally processed refrigerated fruits and vegetables. In: Wiley RC (ed) Minimally processed refrigerated fruits and vegetables. Chapman and Hall, New York, pp 269–313

Collins JL, Walter WM Jr (1992) Processing and processed products. In: Jones A, Bouwkamp JC (eds) Fifty years of cooperative Sweetpotato research 1939–1989, Southern cooperative series. Bulletin no. 369. Louisiana State University Agricultural Center, Baton Rouge, pp 71–87

Coronel P, Truong VD, Simunovic J, Sandeep KP, Cartwright GD (2005) Aseptic processing of sweetpotato purees using a continuous flow microwave system. J Food Sci 70:531–536

Duell BR (1992) Sweetpotato product innovation by small business in Kawagoe, Japan. In: Hill WA, Bonsi CK, Loretan PA (eds) Sweetpotato technology for the 21st century. Tuskeegee University, Tuskeegee, pp 381–389

Girard AW, Grant F, Watkinson M, Okuku HS, Wanjala R, Cole D, Levin C, Low J (2017) Promotion of orange-fleshed sweet potato increased vitamin A intakes and reduced the odds of low retinol-binding protein among postpartum Kenyan women. J Nutr 147(5):955–963. https://doi.org/10.3945/jn.116.236406

Katayama K, Kobayashi A, SakaiT KT, Kai Y (2017) Recent progress in sweetpotato breeding and cultivars for diverse applications in Japan. Breed Sci 67(1):314

Kays SJ (1985) Formulated sweetpotato products. In: Bouwkamp JC (ed) Sweetpotato products: a natural resource for the tropics. CRC Press Inc, Boca Raton, pp 205–218

KNAES (Kyushu National Agricultural Experiment Station) (1996) What's new? Sweetpotato juice. Sweetpotato Breeding. Annual Report for 1996, Number 8. Kyushu National Agricultural Experiment Station, Ministry of Agriculture, Forestry and Foods, Japan

Kumar P, Coronel P, Truong VD, Simunovic J, Swartzel KR, Sandeep KP, Cartwright GD (2008) Overcoming issues associated with the scale-up of a continuous flow microwave system for aseptic processing of vegetable purees. Food Res Int 41(5):454–461

Low JW, Arimond M, Osman N, Cunguara B, Zano F, Tschirley D (2007) A food-based approach introducing orange-fleshed sweet potatoes increased vitamin A intake and serum retinol concentrations in young children in rural Mozambique. J Nutr 137(5):1320–1327. https://doi.org/10.1093/jn/137.5.1320

Low JW, van Jaarsveld PJ (2008) The potential contribution of bread buns fortified with β-carotene-rich sweetpotato in Central Mozambique. Food Nutr Bull 29(2):98–107

Low JW, Mwanga ROM, Andrade M, Carey E, Ball AM (2017) Tackling vitamin A deficiency with biofortified sweetpotato in sub-Saharan Africa. Glob Food Sec 14:23–30. https://doi.org/10.1016/j.gfs.2017.01.004

Low JW, Ortiz R, Vandamme E, Andrade M, Biazin B, Grüneberg WJ (2020) Nutrient-dense orange-fleshed sweetpotato: advances in drought-tolerance breeding and understanding of management practices for sustainable next-generation cropping systems in Sub-Saharan Africa. Front Sustain Food Syst 4:1–22. https://doi.org/10.3389/fsufs.2020.00050

Lund BM, Peck MW (2000) Clostridium botulinum. In: Lund BM, Baird-Parker AC, Gould GW (eds) The microbiological safety and quality of food. Aspen, Gaithersburg, pp 1057–1109

Malavi DN, Muzhingi T, Abong GO (2018) Good manufacturing practices and microbial contamination sources in orange-fleshed sweetpotato purée processing plant in Kenya. Int J Food Sci (2018):11 p. https://doi.org/10.1155/2018/4093161

Malavi DN, Abong GO, Muzhingi T (2021) Effect of food safety training on behavior change of food handlers: a case of orange-fleshed sweetpotato puree processing in Kenya. July 2020. Food Control 119:107500. https://doi.org/10.1016/j.foodcont.2020.107500

Moyo M, Ssali R, Namanda S, Nakitto M, Dery EK, Akansake D, Adjebeng-Danquah J, van Etten J, de Sousa K, Lindqvist-Kreuze H, Carey E, Muzhingi T (2021) Consumer preference testing of boiled sweetpotato using crowdsourced citizen science in Ghana and Uganda. Front Sustain Food Syst. https://doi.org/10.3389/fsufs.2021.620363

Musyoka JN, Abong GO, Mbogo DM, Fuchs R, Low J, Heck S, Muzhingi T (2018a) Effects of acidification and preservatives on microbial growth during storage of orange fleshed sweet potato puree. Int J Food Sci 2018:1–11. https://doi.org/10.1155/2018/8410747

Musyoka J, Njunge F, Muzhingi T (eds) (2018b) Food safety training workshop. Theme: Managing food safety and quality in small-scale food processing for Roots, Tubers and Bananas (RTB) value chains in Sub-Saharan Africa. International Potato Center, Lima, 32 p. isbn:978-92-9060-488-4

Mwanga ROM, Mayanja S, Swanckaert J, Nakitto M, zum Felde T, Grüneberg W, Mudege N, Moyo M, Banda L, Tinyiro SE, Kisakye S, Bamwirire D, Anena B, Bouniol A, Magala DB, Yada B, Carey E, Andrade M, Johanningsmeier SD, Forsythe L, Fliedel G, Muzhingi T (2021) Development of a food product profile for boiled and steamed sweetpotato in Uganda for effective breeding. Int J Food Sci Technol 56:1385–1398. https://doi.org/10.1111/ijfs.14792

Namanda S, Gibson R, Sindi K (2011) Sweetpotato seed systems in Uganda, Tanzania, and Rwanda. J Sustain Agric 35(8):870–884

Nwosu UL, Elochukwu C, Onwurah CO (2014) Physical characteristics and sensory quality of bread produced from wheat/African oil bean flour blends. Afr J Food Sci 8(6):351–355

Ogero K, McEwan M, Pamba N (2016) Chapter 6: Clean vines for smallholder farmers in Tanzania. In: Andrade-Piedra JL, Bentley JW, Almekinders C, Jacobsen K, Walsh S, Thiele G (eds) Case studies of roots, tubers and bananas seed systems, RTB working paper no.2016-3. ISSN 2309-6586., pp 80–97

Okello JJ, Shikuku KM, Sindi K, Low J (2015) Farmers' perceptions of orange-fleshed sweetpotato: do common beliefs about sweetpotato production and consumption really matter? Afr J Food Agric Nutr Dev 15(4):10153–10170

Okello JJ, Shiundu FM, Mwende J, Lagerkvist CJ, Nyikal RA, Muoki P, Mburu J, Low J, Hareau G, Heck S (2021) Quality and psychosocial factors influencing purchase of orange-fleshed sweetpotato bread. Int J Food Sci Technol 56:1432–1446. https://doi.org/10.1111/ijfs.14822

Ouro-Gbeleou T (2018) Boosting demand for biofortified foods: the case of orange fleshed sweet potato bread in Tamale, Ghana. Master's Thesis, University of San Francisco, San Francisco

Owade JO, Abong GO, Okoth MW (2018) Production, utilization and nutritional benefits of orange fleshed sweetpotato (OFSP) puree bread: a review. Curr Res Nutr Food Sci 6(3):644–655

Padmaja G (2009) Uses and nutritional data of sweetpotato. In: Loebenstein G, Thottappilly G (eds) The sweetpotato. Springer Science+Business Media, Germany, pp 189–234

Payton SB, Daniel L, Moore BW, Burnett JA, Stroven JA (1992) Processing method using entire peeled vegetable in a fruit juice/vegetable puree beverage. U. S. Patent 5,248,515

Pérez-Díaz IM, Truong VD, Webber A, McFeeters RF (2008) Effects of preservatives and mild acidification on microbial growth in refrigerated sweetpotato purée. J Food Prot 71:639–642

Rajendran S, Kimenye LN, McEwan M (2017) Strategies for the development of the sweetpotato early generation seed sector in eastern and southern Africa. Open Agric 2(1):236–243. https://doi.org/10.1515/opag-2017-0025

Rajendran S, Magnaghi A, Low J, Muzhingi T (2019) Potential business models and financial feasibility of selected medium-scale business enterprises for orange-fleshed sweetpotato OFSP) value-added products, SASHA brief 22. Sweetpotato Action for Security and Health in Africa Project (SASHA). International Potato Center (CIP), 4 p. https://hdl.handle.net/10568/105951

Sankari A, Thamburaj S, Kannan M (2002) Suitability of certain sweetpotato (*Ipomoea batatas Lam.*) clones for RTS beverage preparation. South Indian Hortic 50:593–597

Sartas M, Schut M, Proietti C, Thiele G, Leeuwis C (2020a) Scaling readiness: science and practice of an approach to enhance impact of research for development. Agric Syst 183:102874. https://doi.org/10.1016/j.agsy.2020.102874

Sartas M, Schut M, van Schagen B, Velasco C, Thiele G, Proietti C, Leeuwis C (2020b) Scaling readiness: concepts, practices, and implementation. CGIAR Research Program on Roots, Tubers and Bananas (RTB). January 2020, pp 217

Smith DA, McCaskey TA, Harris H, Rymal KS (1982) Improved aseptically filled sweetpotato purees. J Food Sci 47:1130–1132, 1142

Tan SL, Omar S, Idris K (2004) Selection of sweetpotato clones with high β- carotene content. In: New directions for a diverse planet: proceedings of the 4th international crop science congress, Brisbane, 26 Sept – 01 Oct 2004

Truong VD (1989) New developments in processing sweetpotato for food. In: Mackay KT, Palomar MK, Sanico RT (eds) Sweetpotato research and development for small farmers SEARCA, Los Baños, pp 213–226

Truong VD (1992) Transfer of sweetpotato processing technologies: some experiences and key factors. In: Scott G, Wiersema S, Ferguson PI (eds) Product development for root and tuber crops. Volume 1 - Asia. Proceedings of the international workshop, April 22–May 1, 1991, VISCA, Philippines, sponsored by CIAT and CIP, Lima, pp 195–206

Truong VD (1994) Development and transfer of processing technologies for fruity food products from sweetpotato. Acta Hortic 380:413–420

Truong VD, Avula RY (2010) Sweetpotato purées and dehydrated forms for functional food ingredients. In: Ray RC, Tomlins KI (eds) Sweetpotatoes: postharvest aspects in food, feed and industry. Nova Science Publishers Inc., New York, pp 117–161

Wanjuu C, Abong G, Mbogo D, Heck S, Low J, Muzhingi T (2018) The physiochemical properties and shelf-life of orange-fleshed sweet potato puree composite bread. Food Sci Nutr 6(6):1555–1563. https://doi.org/10.1002/fsn3.710. PMID: 30258598; PMCID: PMC6145253

Wanjuu C, Bocher T, Low J, Mbogo D, Heck S, Muzhingi T (2019) Consumer knowledge and attitude towards orange-fleshed sweetpotato (OFSP) puree bread in Kenya. Open Agric 4(1):616–622

Woolfe JA (1992) Sweetpotato: an untapped food resource. Cambridge University Press, Cambridge, UK

Wu KL, Sung WC, Yang CH (2009) Characteristics of dough and bread as affected by the incorporation of sweet potato paste in the formulation. J Mar Sci Technol-Taiwan 17(1):13–22

Chapter 6
Turning Waste to Wealth: Harnessing the Potential of Cassava Peels for Nutritious Animal Feed

Iheanacho Okike ⒾⒹ, Seerp Wigboldus ⒾⒹ, Anandan Samireddipalle ⒾⒹ, Diego Naziri ⒾⒹ, Akin O. K. Adesehinwa ⒾⒹ, Victor Attah Adejoh ⒾⒹ, Tunde Amole ⒾⒹ, Sunil Bordoloi ⒾⒹ, and Peter Kulakow ⒾⒹ

Abstract In Nigeria, processing cassava for food and industry yields around 15 million tons of wet peels annually. These peels are usually dumped near processing centres to rot or dry enough to be burned. Rotting heaps release methane into the air and a stinking effluent that pollutes nearby streams and underground water, while burning produces clouds of acrid smoke. However, when properly dried, peels can be an ingredient in animal feed. Previous attempts over two decades to use peels

I. Okike (✉) · P. Kulakow
International Institute of Tropical Agriculture (IITA), Ibadan, Nigeria
e-mail: i.okike@cgiar.org; p.kulakow@cgiar.org

S. Wigboldus
Wageningen University & Research, Wageningen, The Netherlands
e-mail: seerp.wigboldus@wur.nl

A. Samireddipalle
National Institute of Animal Nutrition and Physiology (NIANP), Indian Council of Agricultural Research (ICAR), Bangalore, India

D. Naziri
International Potato Center (CIP), Hanoi, Vietnam
e-mail: d.naziri@cgiar.org

A. O. K. Adesehinwa
Institute of Agricultural Research & Training, Obafemi Awolowo University, Ibadan, Nigeria

V. A. Adejoh
Synergos Nigeria, Maitama Abuja, Nigeria
e-mail: vadejoh@Synergos.org

T. Amole
International Livestock Research Institute, Ibadan, Nigeria
e-mail: t.amole@cgiar.org

S. Bordoloi
Amo Farm Sieberer Hatchery Ltd, Amo Byng Ltd, Awe, Oyo State, Nigeria

G. Thiele et al. (eds.), *Root, Tuber and Banana Food System Innovations*,
https://doi.org/10.1007/978-3-030-92022-7_6

173

in animal feed failed to yield profitable options for drying wet peels at commercial scale, but recent research suggests that cassava peels can be processed into high-quality cassava peel (HQCP) products to be used as nutritious, low-cost animal feed ingredients. The core innovation was to adopt the same steps and equipment used for processing cassava roots into *gari*, the main staple food in the country. When dried, 3 tons of wet peels yield a tonne of healthy and energy-rich animal feed, containing nearly 3,000 kilocalories per kilogram of dry matter (kcal/kgDM). Adopting this innovation at scale in Nigeria's poultry and fish sectors alone has the potential to turn approximately 3.6 million tons of wet peels into 1.2 million tons of feed ingredients capable of replacing approximately 810,000 tons of largely imported maize. The innovation has great potential to increase feed availability and lower its cost while saving cereals for human consumption, reducing the import bill, creating new business opportunities, and protecting the environment. This research was initiated by CGIAR centres and taken up by the CGIAR Research Program on Roots, Tubers and Bananas (RTB) over the past decade with strategic input from the CGIAR Research Program on Livestock to accelerate development of the innovation, and this chapter documents the potential and progress in taking this innovation to scale.

6.1 Introduction

6.1.1 Cassava as an Essential Crop in Nigeria

Cassava (*Manihot esculenta*), also called manioc or tapioca, is a major subsistence and commercial crop in sub-Saharan Africa. Nigeria is the largest producer worldwide, but many of the lessons and opportunities discussed here are applicable to other major cassava-producing countries, particularly in western and central Africa, such as Ghana, Ivory Coast and the Democratic Republic of Congo. Cassava has seen a steady increase in production, averaging 3% per year since 1995, reaching 59 million tons in Nigeria and 178 million tons in Africa (FAOSTAT 2019). Over 90% of the cassava in Africa is consumed as food and very little for industrial processing (Akinpelu et al. 2011). Once considered a subsistence or poor person's crop, cassava is rapidly becoming a commodity with many uses, supporting rural development, poverty alleviation, food security, and value addition with macroeconomic benefits (FAO 2013). Several African governments are promoting cassava to reduce cereal imports through mandatory blending of wheat flour with high-quality cassava flour.

6.1.2 The Growing Demand for Animal Feed and Potential Role of Cassava Peels

Growing population coupled with rising incomes, urbanization and a shift towards protein-rich diets is driving increased demand for meat, dairy and eggs. The demand for animal-derived protein is projected to double by 2050 (FAO 2016), and most of that added demand will come from to low- and middle-income countries (Barry et al. 2012). Between 1999 and 2030, annual meat consumption in these countries will increase from 26 to 37 kg per person, compared with an increase from 88 to 100 kg in industrial countries. Annual per capita consumption of dairy products is expected to rise from 45 to 66 kg in developing countries and from 212 to 221 kg in industrial countries. Egg consumption will grow from 6.5 to 8.9 kg in developing countries and from 13.5 to 13.8 kg in industrial countries (FAO 2011).

The growing demand for animal source food will require more feed (Thornton 2010). Global feed demand is estimated to increase 1.6 times from 1,058 million tons in 2010 to 1,693 million tons by 2050 (OECD-FAO 2013). Growing consensus exists on the tremendous environmental impact of expanding livestock production requiring more land to be brought into production to which the associated increased demand for feed is a major contributor (Willits-Smith et al. 2020). Reducing the pressure on natural resources and ecosystems calls for, among others, a more effective use of available biomass, including recycling (Garnett et al. 2015; Haberl et al. 2014; Karlberg et al. 2015; Smith et al. 2010). Turning crop residues and agricultural waste into feed ingredients is a key strategy to pursue this goal following the circular economy principle.

In Nigeria, the use of cassava roots as animal feed is a traditional backyard practice, but interest from the feed industry for this crop is relatively recent. In 1980, the first international workshop on cassava as livestock feed generated optimism about using cassava roots for replacing cereals, which had become increasingly scarce (Smith 1988). Feed mills in Nigeria were operating at 92% capacity in 1980, but this dropped to 26% in 1997 due to the shortage of local ingredients and the prohibitive cost of imported materials (Azogu et al. 2004), which coincided with an import ban on maize and soybean. Feed mills began searching for substitutes. Cassava chunks, chips and gelatinized grits were the first cassava products to draw attention from feed mills. However, the optimism gradually faded over time as research lagged and cassava was not used in animal feed, largely because yields were stagnant and cassava was growing in demand as food for the growing population.

Livestock can be fed on cassava peels. Farmers and agronomists have known this for years, but the idea had remained largely under-exploited. This situation changed in the 2000s as industry and researchers began to shift their interest from cassava roots to peels and as appropriate technology became available to process wet peels. This development was facilitated by a multi-stakeholder scaling partnership aided by the RTB scaling fund, using the Scaling Readiness approach (Sartas et al. 2020). As a result of this process, high-quality cassava peels (HQCP) have started getting into compound feed for poultry, fish, cattle, sheep and goats in 2018/2019 for some major feed millers, with satisfactory results. We will discuss this process in the following sections.

6.1.3 The Extent of the 'Peel Problem', Its Underlying Causes and Recent Developments

In Africa, processing cassava into food is primarily a manual and labour-intensive activity, typically performed by women, either individually at the household level or collectively in specialized processing groups (Hillocks 2002). More than 90% of cassava processing requires hand peeling. As a result, every year, Nigerians generate 15 million tons of peels, stumps (ends of roots) and undersized and damaged roots from processing cassava into various products (Fauquet 2014).

While cassava peels can be fed to livestock, they are poisonous if not properly processed. Peels have high levels of hydrogen cyanide (HCN), particularly in bitter cassava varieties, which are the most common in Nigeria. Sun drying is an effective way to reduce HCN levels significantly. However, improperly dried peels can be contaminated with mycotoxins, toxic compounds produced by fungi that can have serious health consequences for people who consume meat, milk or eggs.

Sun drying takes about 2–3 days in the dry season, but this is not viable in the rainy season when most cassava is harvested, leaving millions of tons of wet peels to rot. In the rainy season, lactic acid-laden, smelly effluent from rotting heaps contaminates nearby streams, wells and underground water (Fig. 6.1). In the dry season, heaps are often set on fire, and the acrid smoke pollutes the air (Fig. 6.2). Millions of tons of fermenting peels also emit greenhouse gases (Obianwa et al. 2016). During the dry season, about half of the processors in Nigeria try to dry up to 20% of their wet peels to sell or feed their own livestock. In rural Nigeria, peels are often given away to the workers who peel the cassava roots, so the peels can be dried and fed to their animals, especially goats (Azogu et al. 2004). However, these practices absorb only a fraction of the massive amount of peels generated in the country annually.

Manual peeling inevitably removes part of the cassava root's flesh along with the peels. It is estimated that approximately 22% of the fresh root weight (including the peels) is removed when roots are manually peeled. Enhancing hand peeling efficiency can reduce this to approximately 18% and save about 270,000 tons of cassava flesh annually, worth US$[1] 17.5 million, in southwest Nigeria alone (Bennett and Naziri 2013). However, procuring these savings would require the adoption of specially designed knives and considerably increase the time required for peeling.

In spite of recent advances in developing cassava peeling machines in Nigeria, it does not appear that the quantity of peels resulting from manual processing

[1]Exchange rate, USD 1 = NGN 480 at the parallel market where the bulk of foreign currency sourcing is happening

Fig. 6.1 A typical dumping site for cassava peels. (Photo: I. Okike (ILRI))

operations will decline in the foreseeable future. Egbeocha et al. (2016) conducted a review of cassava peeling machines and found their performance to be largely unsatisfactory for small-scale processors, who constitute the backbone of cassava processing in Africa. Kolawole et al. (2016) evaluated seven peeling machines and found that peeling efficiency was 0–96% while flesh loss was 7–96%. The latter compares very poorly with the 4% flesh loss for traditional manual peeling and 2–3% for peeling by women who have been specifically trained by the International Institute of Tropical Agriculture (IITA).

However, in spite of their current low efficiency, mechanical peeling is being increasingly adopted by some medium-to-large-scale processing enterprises that value the higher output capacity and the faster processing times. This is particularly true for industrial starch extraction, which usually requires the rapid processing of large amounts of fresh roots and does not need thorough removal of the external cortex that forms the peel with the periderm. For other cassava-based products, female peelers are still used to apply the finishing touches to clean the root of vestiges of peels after mechanical peeling. Therefore, mechanical peeling does not represent an immediate or foreseeable threat to the commercial exploitation of cassava peels nor to women's job opportunities. Unless the efficiency of mechanical peelers is dramatically increased, their wider adoption would simply generate more peels. These peels, in the meantime, could be processed into HQCP mashes (Fig. 6.3).

Fig. 6.2 Attempts to eliminate cassava peels by burning them. https://www.flickr.com/photos/ilri/43915310274. (Photo: I. Okike, ILRI)

6.2 What Is the Innovation Package?

The core innovation is a processing method for turning wet peels into HQCP mash to use as a feed ingredient. This method circumvents the challenge of drying, which is the most limiting constraint to using cassava peel as feed, reducing drying time from 2–3 days to just 6–8 hours in the dry season by grating, mechanical dewatering, fermenting, pulverizing, sieving and then sun-drying the peels (Fig. 6.4), as detailed by Okike et al. (2015). The steps essentially replicate the age-long processes involved in making cassava storable and safe for human consumption. In the wet season, sun drying is still possible but takes longer due to low intensity of the sun and intermittent periods when the mash spread for drying has to be covered and protected from rain. However, under these conditions, the long time required for reducing the moisture content to the ideal level facilitates fungal growth, which contaminates the product. As such, the sun-drying option is not recommended

Fig. 6.3 An example of peels and wasted roots from a peeling machine, amenable to further processing into HQCP mashes. (Photo: P. Kolawole, IITA)

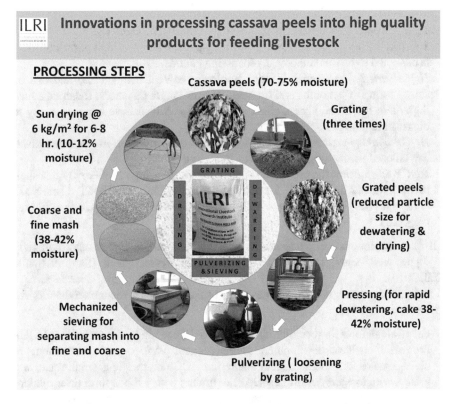

Fig. 6.4 Steps to process fresh cassava peels into high-quality cassava peel products

Table 6.1 Attributes of the four distinct HQCP products

Product type	Attributes	Suitable species to feed
HQCP cake	High moisture (38–42%), low keeping quality (about 7 days). Total energy 2,947 kcal/kgDM and 9.8% fibre	Cattle, goats, sheep and pigs
HQCP whole mash	10–12% moisture and good keeping quality (more than 6 months in storage). Total energy 2,947 kcal/kgDM and 9.8% fibre – Same as for cake	Cattle, goats, sheep and pigs
HQCP fine mash	*Premium product.* 10–12% moisture and good keeping quality (more than 6 months in storage). Total energy 3,039 kcal/kgDM and 8.2% fibre	Poultry, fish and young pigs
HQCP coarse mash	10–12% moisture and good keeping quality (more than 6 months in storage). Total energy 3,039 kcal/kgDM and 15.6% fibre	Cattle, sheep, goats and adult pigs

during the wet season. During wet seasons, additional options for drying include toasting and flash drying.

Regardless of the drying method, four distinct products with different attributes and potential applications result from the process (Table 6.1).

1. *Cassava peel cake* has high moisture content, has a short shelf life of about 1 week and can be fed to cattle, sheep and goats. Due to its high perishability, it is designed to be marketed near processing centres.
2. *HQCP dried whole mash* is drier (moisture content below 12%) and can be stored for up to 6 months and transported over longer distances. It can be a feed ingredient for cattle, sheep, goats and also pigs, which have better capacity to digest fibre than poultry and fish.
3. *HQCP coarse mash* is fibrous and low in energy with a shelf life of 6 months. It can be used for feeding cattle, sheep, goats and adult pigs.
4. *HQCP fine mash* is the premium product, low in fibre, high in energy with a shelf life of 6 months and suitable for poultry, fish and piglets. The HQCP whole mash is mechanically sieved to separate it into HQCP coarse and HQCP fine mashes (Fig. 6.5).

All of these innovative processes reduce the cyanide content to approximately 35 ppm, well below safe levels (90 ppm) for livestock feed (Samireddipalle et al. 2016).

Additional elements of the innovation package include recommendations for feed use, especially:

- FeedCalculator® for formulating least cost balanced rations incorporating HQCP
- Protocols, business development guidelines and plans to guide investments at diverse scales for processing and marketing HQCP mashes (e.g. small scale, two graters with toasters; medium scale, four graters with 4-tonne flash drier related)
- Loan application guidelines for engaging with credit and financial institutions
- District and state-level maps georeferencing cassava processing clusters on Cassava Peel Tracker®, which incorporates information on quantity of peels

Fig. 6.5 A mechanical sieve – coarse mash retained in the wooden box and fine mash captured in the basin after separation https://www.flickr.com/photos/ilri/23003496544/. (Photo: I. Okike (ILRI))

generated daily by the processing clusters by season and contact details of key persons
- Capacity development for processors on best practices in production, equipment maintenance, personnel safety, product and factory hygiene and cassava peel business registration on Cassava Peel Tracker®
- Bookkeeping

Finally, the innovation package includes a WhatsApp™-based community of practice (CoP) platform known as Cassava Peel First Movers for exchanging knowledge and information and promoting HQCP products. Efforts are also ongoing to have HQCP products recognized by the Standards Organisation of Nigeria (SON) and their standard physical and nutritional qualities listed in the SON manuals.

6.2.1 History of CGIAR Research to Develop the Use of Peels as Feed

In the early 2000s, a flurry of government-backed activities brought together researchers and the public and private sectors to herald a cassava revolution in Nigeria with options for developing a livestock feed industry based on cassava and its derivatives. At that time, the focus was still on the use of cassava roots as feed ingredient. In 2007, the International Livestock Research Institute (ILRI) received a

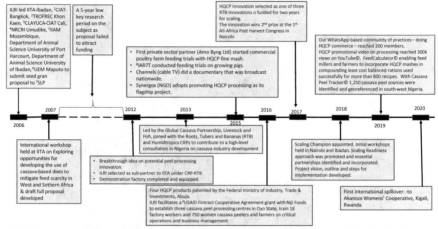

Fig. 6.6 Timeline of CGIAR research to develop the use of cassava peels as feed

seed grant through the CGIAR Systemwide Livestock Programme (SLP) to conduct initial research on cassava peel as livestock feed. A partnership was later established that included IITA and various research institutes from Nigeria, Mozambique, Colombia and Thailand. The initiative marked the beginning of CGIAR research on the use of cassava peels as animal feed. However, it would not be until the early 2010s that the CGIAR would begin exploring the technical viability of processing cassava peel into HQCP products (Fig. 6.6).

6.2.2 Proof of Concept and Feeding Trials Under the Auspices of RTB

After the partnership failed to attract the expected funding, there was a lull in research until 2012, when ILRI engaged in an IITA-led initiative under the CGIAR Research Program on Roots, Tubers and Bananas (RTB). The new initiative aimed at addressing the following research questions:

1. Are HQCP products acceptable to livestock?
2. Are HQCP products economically viable?
3. Are HQCP products have potential for wide adoption and going to scale?

With RTB funding support, a demonstration pilot factory was built, and a multi-disciplinary, multi-institutional team of scientists with expertise in animal science, engineering, food science and pathology came together to investigate options for hygienic and safe processing of wet peels. To complement the RTB effort, the Feed and Forages team of the CGIAR Research Program on Livestock and Fish initiated

Table 6.2 Nutrient profiles and prices of HQCP products compared to maize in March 2021 (dry season)

	HQCP (whole)	HQCP (fine)	HQCP (coarse)	Maize
Starch (%)	66.7	69.0	55.0	68.8
Protein (%)	2.5	2.6	2.8	8.8
Fat (%)	1.4	1.2	1.2	4.1
Crude fibre (%)	9.8	8.2	15.6	2.6
Crude ash (%)	5.8	6.6	3.5	1.4
Total energy (kcal/kgDM)	2947	3039	2495	3840
Price/tonne (naira)	67,000[a]	75,000	50,000	200,000
Price/tonne (US$)	139.56	156.25	104.17	416.67

Source: For HQCP products: www.masterlab.nl. For the HQCP Scaling Project and for maize: https://feedtables.com/content/maize-flour-crude-fibre-2-10

[a]HQCP products are mainly marketed as fine and coarse mashes although produced as HQCP (whole) before separation. HQCP (whole) yields two parts fine for one part coarse, so its price per tonne can be derived as $((2 \times 75,000) + (1 \times 50,000))/3 = NGN\ 67,000$ per tonne

a new research activity to explore related small-scale business development services and technology to capitalize on cassava peel waste by-products to increase animal feed availability within selected livestock and fish value chains.

After producing a few tons of HQCP whole, fine and coarse mashes of consistent nutrient profiles and physical qualities (e.g. texture, particle size, moisture level), laboratory analyses confirmed the products were safe with acceptably low levels of cyanide and aflatoxins. The whole HQCP mash was found to have more than two-thirds the metabolic energy (ME) of maize (2,947 vs. 3,840 kcal/kg of dry matter), and these levels were even higher for HQCP fine mash (Table 6.2).

Feeding trials were then established at the ILRI research station with ruminants, poultry and pigs for assessing the palatability of HQCP. Following satisfactory results of the palatability trials and in an effort to create awareness of the products, samples of HQCP coarse mash were taken to ruminant markets across Oyo State, where cattle, sheep and goats consumed it readily.

Most of those involved were from the CGIAR until 2015, when research began to provide solid evidence about the safety and hygiene of HQCP products for animal feed. CGIAR researchers continued research on product development and assessing the technology, but to prove the concept and move towards scaling, partnerships were expanded to include key private-sector players that had toured IITA labs and feeding trials, including ILRI experiments, and expressed interest in setting up their own trials. These included Durante Fish Industries, Amo Byng Ltd., Nine Stars Fish Farms and Hatchery Ltd., and Niji Foods. Furthermore, the NGO Synergos and the Federal Ministry of Agriculture and Rural Development (FMARD) were also involved.

On-farm and on-station animal feeding trials were conducted in 2016 at the following three locations.

Table 6.3 Results of feeding trials at Amo Byng

	Groups				
HQCP fed (kg/tonne)	Maize based	50	75	100	125
Chicks housed	100	100	100	100	100
Mean age (days)	35	35	35	35	35
Birds harvested	97	98	97	99	97
Mortality (%)	3	2	3	1	3
Feed consumed (kg)	3.092	3.355	3.047	2.815	2.9
Live body weight (kg)	2.23	2.208	2.184	2.135	2.054
Feed conversion ratio (FCR)	1.387	1.519	1.395	1.319	1.412
Total feed consumed/bird (naira)	265	290	266	247	258
Feed cost/kg liveweight (naira)	118.8	131	121.7	115.6	125.6
Sale value/bird (naira)	892	883	874	854	822
Gross margin/bird (naira)	627	593	608	607	564

Source: Amo Byng Ltd., Awe, Oyo State, Nigeria

1. *Amo Farm Sieberer Hatchery Ltd*, a subsidiary of Amo Byng Ltd., Awe, Oyo State, trialled HQCP coarse mash with feedlot cattle and found it could substitute up to 10% of feed resources. They also conducted four on-farm feeding trials with commercial poultry (three with broilers and one with layers) using HQCP fine fraction. They tested different inclusion rates of mash in the total diet (Table 6.3) and found that the 10% inclusion rate (100 kg of HQCP fine mash in 1 tonne of feed) resulted in the best growth rate, lowest feed intake and lowest mortality. The response in broilers was better than the response in layers. These results convinced the company to start incorporating processed cassava peels into poultry and cattle diets, though supply was a major constraint. Furthermore, HQCP fine mash fortified with full fat soy, DL-methionine and soy oil could replace up to 25% of the maize in broiler diets. In spite of fortified HQCP being significantly cheaper than maize, the biofortification option was not further pursued by the project.
2. *Institute of Agricultural Research and Training (IAR&T) of the Obafemi Awolowo University, Moor Plantation, Ibadan, Oyo State*, conducted on-station trials with weaned and growing pigs. They found that HQCP fine mash could replace 75% of maize in the diet of growing pigs, reducing feed costs by 4% with no adverse effect on the growth performance (Adesehinwa et al. 2016).
3. *ILRI farm at IITA, Ibadan, Oyo State*, hosted on-station trials which showed that cassava peels and leaves, fed in the ratio of 70:30 on a dry matter basis, could be the sole feed for cattle, sheep and goats and would reduce feeding costs.

6.2.3 Supporting Private Sector Partners to Take Up the Innovation

In September 2016, ILRI helped Niji Foods acquire a grant under Fintrac's Cooperative Agreement with the US Agency for International Development (USAID). Under this initiative, Niji Foods, with support from ILRI, was to establish three HQCP processing centres in Oyo State, train 18 factory employees, six administrative staff and 750 women cassava peelers and farmers on critical operations and business management. Amo Byng Ltd. committed to buying the HQCP products from Niji Foods, but, ultimately, the purchase agreement with Amo Byng Ltd. did not materialize as Niji Foods was offered better prices from other buyers.

6.2.4 Registration of Products as First Step Towards Developing Product Standards

While conducting feeding trials and broadening the partnership, the HQCP project staff approached the Commercial Law Department, Trade Marks, Patent and Designs Registry of the Federal Ministry of Industry, Trade and Investments in Abuja, who granted an umbrella patent to ILRI that covered four HQCP products under trademarks as follows:

1. Cassa peel mash®, previously referred to as HQCP fine mash
2. Cassa peel bran®, previously referred to as HQCP coarse mash
3. Cassa peel cake®, previously referred to as HQCP cake
4. Cassanules®, granulated HQCP that was designed to replace sorghum for the production of Aflasafe®

6.3 Value and Impacts of Innovation

6.3.1 Societal Value of Innovation (Potential Economic and Environmental Value)

The raw material for HQCP production is freely available as approximately 98% of peels in Nigeria are dumped. As previously mentioned, the drying techniques vary according to the season and scale of operations. With sun drying as the dominant mode of processing, the energy requirement is insignificant. Depending on the drying method (sunshine, toasting or flash drying), production costs of HQCP whole mash vary from US$96 to US$107 per tonne (Table 6.4), while the feed industry

Table 6.4 Production budget for HQCP whole mash in the wet and dry seasons (US$/tonne) and market price projections assuming 25% markup during 2015

Item id.	Item	A Budget for fresh peels and initial processing	B Budget for a tonne of wet cake = A * 2	C Budget for a tonne of whole mash = B * 1.5
a	Fresh peels (dry season)	10.00	20.00	30.00
b	Fresh peels (wet season)	7.00	14.00	21.00
c	Transportation (US$/ tonne/25 km)	1.50	3.00	4.50
d	Loading labour (0.5 h)	0.35	0.70	1.05
e	Off-loading labour (0.5 h)	0.25	0.50	0.75
f	Grating	8.80	17.60	26.40
g	Packing into bags, loading and dewatering by hydraulic press	2.50	5.00	7.50
h	Woven plastic bags (100 reusable bags x 10 kg)	0.50	1.00	1.50
i	**Production of HQCP wet cake and HQCP whole mash (dry season) = (a + c + d + e + f + g + h) of column B**		47.80	71.70
j	**Production of HQCP wet cake and HQCP whole mash (wet season) = (b + c + d + e + f + g + h) of column B**		41.80	62.70
k	**Market price of cake assuming 25% markup (dry season) = i * 1.25**		59.75	74.69
l	**Market price of cake assuming 25% markup (wet season) = j * 1.25**		52.25	65.31
m	Pulverizing and sieving cake			2.40
n	Labour for spreading and stirring wet mash			6.00
o	Labour for toasting wet mash			10.80
p	Fuel (coal) for toasting			16.80
q	Packing finished products into bags and sealing			4.00
r	Woven plastic bags (40 new bags × 25 kg)			10.00
s	**Production of solar dried mash (dry season) = (i + m + n + q + r)**			96.10
t	**Production of toasted mash (wet season) = (j + m + o + p + q + r)**			106.70
u	**Market price for solar dried mash assuming 25% markup (dry season) = s * 1.25**			120.13
v	**Market price for toasted mash assuming 25% markup (wet season) = t * 1.25**			133.38

Source: Analysis based on data from research notes of the cassava peel project

buys HQCP coarse and fine mashes at US$ 100–150 per tonne, respectively, which is less than half the price of maize (Table 6.2). The feed industry is willing to pay for HQCP mashes because of their nutritional content and the satisfactory feedback received from the customers.

Production budgets for both dry and wet seasons indicate that products can reach the market with a markup of 25% on total cost of production and still remain below half the price of maize (weight for weight), which has largely remained the feed industry's benchmark price for HQCP (Table 6.4). Total production cost using sun drying in the dry season is US$97, while the cost of production in the wet season, based on toasting, was US$107 per tonne. A projected market price per tonne of HQCP whole mash of US$120 in the dry season and US$133 in the rainy season would result in a profit margin of 25%, which should keep a business running. However, current market prices (as of Feb 2021) are even higher, approximately US$150 per tonne. It is important to note that while the cost of toasting increases production cost in the rainy season, the price offered for HQCP fine mash is higher during that time because usually, maize prices are highest during the rainy season prior to harvest. Therefore, the additional cost incurred for drying does not necessarily affect profitability adversely.

Cash flow analysis calculated over 5 years of operation for an enterprise producing 2.5 tons of peel cake per day showed that an investment cost of US$9,200 and a working capital of US$100/day would yield a gross revenue of approximately US$39,500 per year, break even in year 4 and result in a net present value (NPV) of US$6,026 (Table 6.5). Other sizes of operations were also analysed with similar results, suggesting a bankable investment (spreadsheets are available on request).

In terms of achievable markets, it is worth noticing that HQCP fine mash is the star product of the core innovation. Because of its lower fibre content and despite

Table 6.5 Five-year budget for an operation processing 2.5 tons of wet cassava peels per day in USD

	Year 0	Year 1	Year 2	Year 3	Year 4	Year 5
Revenue items (USD)						
High-quality cassava peel (wet) cake		37,500	39,375	41,344	43,411	45,581
Gross revenue		37,500	39,375	41,344	43,411	45,581
Cost items (USD)						
Investment (grater, hydraulic press units, well, shed)	9200					
Fresh cassava peels		15,000	15,750	16,538	17,364	18,233
Collation and transportation		1575	1575	1575	1575	1575
Labour and cost of grating		13,200	13,200	13,200	13,200	13,200
Cost of dewatering		5250	5250	5250	5250	5250
Maintenance		200	200	200	200	200
Earnings before taxation		2275	3400	4581	5822	7124
Taxation (1%)		22.75	34	46	58	71
Profit and loss	−9200	2252.25	3366	4535	5763	7053
Present value factor	100%	85%	72%	61%	52%	44%
Present value	−9200	1909	2417	2760	2973	3083
Cumulative present value	−9200	−7291	−4874	−2113	859	3942

Source: Analysis based on data from research notes of the cassava peel project
Internal rate of return = 32%; working capital, US$35,025 in year 1, going up to US$38,258 in year 5

being roughly 50% more expensive than HQCP coarse mash (Table 6.2), the fine mash is preferred by feed millers for inclusion in feed for monogastrics, especially poultry and fish (incidentally, these are the two most industrialized sub-sectors of the animal production industry in Nigeria and capable of absorbing the highest volume of HQCP fine mash). The potential for HQCP inclusion in the poultry sector is particularly remarkable. Nigeria produced 11.5 million tons of maize and imported 0.5 million tons in 2020. About 60%, or 7.2 million, of these 12 million tons went into poultry feed (Agweek 2021). At a 10% inclusion rate (as suggested by the trials hosted by Amo Farm Sieberer Hatchery), the poultry feed industry alone could replace approximately 720,000 tons of maize per year with same amount of HQCP fine mash. Similarly, if 10% of the approximately 900,000 tons of maize currently used in fish feed manufacturing (CORAF 2020) was replaced by HQCP fine mash, this would create an additional market opportunity of approximately 90,000 tons per year. Therefore, the poultry and fish sectors alone have the capacity to absorb up to 810,000 tons of HQCP fine mash. At a selling price of US$100–150/tonne, this would open up an $80–120 million industry annually that could employ an estimated 20,000 people, 80% of whom would be women.

6.3.2 Who Does This Innovation Package Impact?

Business Opportunities

Wide adoption of HQCP can create a new, commercially viable business and generate income opportunities. Cassava producers, workers, entrepreneurs, livestock farmers and consumers of livestock products can benefit from exploiting this upgrade to the value chain:

1. Approximately four million cassava producers would have the opportunity to sell cassava peels, earning new income that could also spur investment in cassava productivity.
2. *Gari* processors, who generate most of the peels (in some 9000 processing centres nationwide) would enjoy new revenue from selling the waste peels while cleaning up their workplaces (usually nearby their homes), thus improving the health and well-being of workers.
3. Feed millers would have access to cheaper raw materials than maize, even when HQCP is fortified with soy products.
4. Incorporating HQCP into feed mills will reduce feed production costs and release maize for human consumption.
5. Livestock and fish producers would have access to cheaper feed of a similar quality to current products, lowering their operating costs.
6. Consumers would benefit from cheaper meat, milk and eggs, contributing to diets richer in protein and essential micronutrients.
7. Due to increased feed availability, cattle could be produced more intensively with less movement, reducing conflicts between pastoralist and farmers.

8. Dairy production would increase, reducing imports of milk.
9. New job opportunities would be created for people collecting, processing and selling HQCP products. See also a YouTube video[2] on 'Transforming cassava peels into animal feed'.

Benefits for Women: An Example from Synergos Project

One of the cassava peel project's partners, Synergos, promoted HQCP to women's groups and cooperatives through the State Partnership for Agriculture Programme (SPA), funded by the Bill & Melinda Gates Foundation (2016–2019). The SPA project sought to support women and nutrition in the shift of agriculture from subsistence to commercial production with the goal of increasing incomes and improving livelihoods. Hence, the project provided a good opportunity to Synergos to create a revenue stream for cassava processors, especially women, by establishing prototype HQCP learning centres where cassava peels were collected, processed into HQCP mashes and sold to herders and feed millers.

The SPA used the prototype learning centres to improve relations with farmer cooperatives in Benue and Kogi. In 2016, Synergos funded the training of two women and two men from each of these states as master trainers at ILRI, Ibadan. Ten satellite clusters were equipped with grating machines to process cassava peels and supply HQCP mashes. Over the years, about 2000 women have benefitted directly or indirectly from this innovation in Benue State.

In Kogi State, 20 women and ten female youths were involved in clusters of cooperatives. Synergos, in collaboration with IITA and ILRI, repaired and converted an existing processing centre in Ojapata into a prototype learning centre and helped procure processing equipment. Here, one of the women, Mrs. Jummai Mohammed, aggregates cassava peels from other processing centres in Ojapata and supplies them to the HQCP centre for making fine mash. She sells the fresh cassava peels at 0.52–0.79 per 100 kg bag, and the new income has allowed her to send her children back to school. With scaling fund support from ILRI and the Technology for African Agricultural Transformation (TAAT) initiative, Synergos set up an additional HQCP processing centre for an all-female cooperative group in Ejule. This work was made possible through a partnership in which the women's group provided land, shelter and security for the centre.

In Ogun State, Valueman International also engaged 32 women in an HQCP processing factory, while IFAD and Synergos trained 50 processors, including 30 women. Currently, over 500 women in Nigeria work in the HQCP value chain, and more are projected to benefit as demand increases nationally for HQCP in animal and fish feed production.

The Synergos project has been successful in promoting the HQCP innovation among women (individually and in groups) and impacted positively on the livelihood of beneficiaries. This benefit can be attributed to the coordinated effort of a

---

[2]The video (http://bit.ly/2j7bRu3) has been viewed more than 336,000 times, sparking 2,000 likes and several questions from young school leavers about where to get training and equipment. Clearly, there is interest from young people to engage in such an enterprise.

wide range of partners, including the Ministry of Agriculture and other national agencies (e.g. Business Innovation Facility), research organizations (ILRI and Kogi State University) and private sector actors (Everest Feed and farmers cooperatives).

Though largely successful, there are still some challenges to be addressed and overcome:

1. Access to processing equipment and maintenance. Skills needed to make the equipment are still very low.
2. Transportation to market. It is expensive for HQCP processors to deliver processed HQCP to final markets.
3. Access to credit. Financial institutions are reluctant to loan funds to HQCP processors to invest in equipment.

6.4 The Approach to Scaling the HQCP Innovation

6.4.1 The Influences of the Scaling Readiness Approach

Initially, there was little emphasis on scaling for HQCP. This changed when the lead innovator was named a 'scaling champion' and HQCP transformation was selected as one of the first of three innovations to be supported through the RTB Scaling Fund Project for 2018–2019. The Scaling Readiness approach (Sartas et al. 2020) makes clear that innovations scale as packages, and packages include core and complementary innovations, which, for HQCP, were unpacked into 'hard' (technical) and 'soft' (culture and socioeconomic) components (Fig. 6.7).

Such unpacking of components helps identify what needs to be addressed to help a core innovation go to scale. This work will inform strategic considerations regarding priorities and the roles needed from potential partners by disciplinary advantage. The scaling project team used the concept of 'Scaling Readiness' to assess the innovation packages. 'Innovation readiness' refers to the demonstrated capacity of an innovation to fulfil its contribution to development outcomes in specific locations. This is presented in nine stages showing progress from an untested idea to a fully mature proven innovation. 'Innovation use' indicates the level of use of the innovation or innovation package by the project members, partners and society. This shows progressively broader levels of use beginning with the intervention team who develop the innovation to its widespread use by users who are completely unconnected with the team or their partners. 'Scaling Readiness' of an innovation is a function of innovation readiness and innovation use. Table 6.6 provides summary definitions for each level of readiness and use adapted from Sartas et al. (2020).

Examining components through the Scaling Readiness approach allows researchers and implementers to critically examine the constrained ones while broadening partnership scoping and reflection.

Fig. 6.7 Unpacking the innovation package to reveal its 'hard' and 'soft' components (brown and blue colours, respectively)

Table 6.6 Summary definition of levels of innovation readiness and use (Sartas et al. 2020)

Stage	Innovation readiness	Innovation use
1	Idea	Intervention team
2	Basic model (testing)	Direct partners (rare)
3	Basic model (proven)	Direct partners (common)
4	Application model (testing)	Secondary partners (rare)
5	Application model (proven)	Secondary partners (common)
6	Application (testing)	Unconnected developers (rare)
7	Application (proven)	Unconnected developers (common)
8	Innovation (testing)	Unconnected users (rare)
9	Innovation (proven)	Unconnected users (common)

For HQCP innovation, unpacking and Scaling Readiness analyses revealed that constraints to scaling existed more within the 'soft' components (e.g. environment and product hygiene, processor/user interfacing, grading and pricing, electricity and labour) rather than around the 'hard' components (e.g. grater, hydraulic press and mechanical sieve). If the 'soft' components are the cultural and social economy of the technology, then more partners with competence in these areas are required on board.

In the diagnosis to assess the innovation package in terms of its innovation readiness and innovation use, its individual components were scored on a scale of 1–9 for component readiness and component use. Owing to the large number of components for the innovation package, the diagnosis was partitioned into hard and soft components (Figs. 6.8 and 6.9).

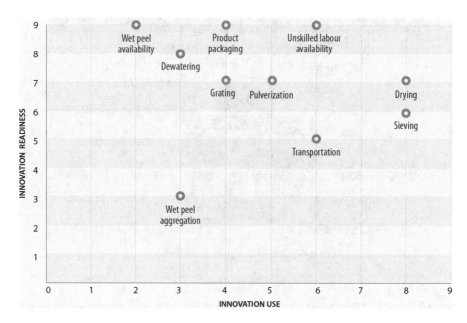

Fig. 6.8 Hard components of the HQCP innovation package. (Source: RTB Scaling Fund Project 2018–2019)

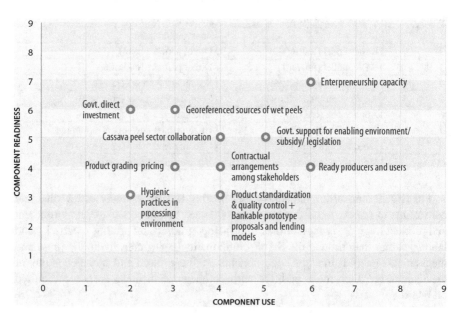

Fig. 6.9 'Soft' components of the HQCP innovation package. (Source: RTB Scaling Fund Project 2018–2019)

From the diagnosis of the hard components, aggregation of wet peels showed the lowest innovation readiness and use, thus constituting an important bottleneck for scaling the HQCP innovation. This triggered the process of developing the scaling strategy considering options to address the identified bottleneck innovations according to Sartas et al. (2020). The initial consideration was to outsource the aggregation of wet peels to a refuse disposal company that would collect and deliver the peels to the processing factories. However, the major problem was the highly dispersed nature of cassava processing centres from which peels have to be collected. Eventually, a web-based platform, called Cassava Peel Tracker®,[3] was developed, and 1,250 locations were surveyed, geo-referenced and mapped in four states in southwest Nigeria to address the bottleneck. The platform provides information on quantity of wet peels generated per location per day (average of 3.3 tons) and the contact details of leaders of the centres. As it turned out, the quantity of peels produced around small enterprises is often sufficient for their operation; however, large enterprises, for whom fresh peels aggregation would be a bottleneck, are yet to emerge.

The diagnosis of the soft components showed that hygiene in the processing environment was at the lowest rung of readiness (Fig. 6.9). The project attempted to tackle this issue at two levels:

(i) At the federal level in the Ministry of Environment, which budgeted to establish six HQCP processing centres in strategic locations to promote hygienic practices. The required funds, however, failed to materialize.
(ii) At the level of individual entrepreneurs through training.

All the options for addressing constraints suggested by the Scaling Readiness approach failed, and as a result, few of the entrepreneurs were able to follow the prescribed hygienic practices (e.g. building underground drainage for effluent). Product standardization and quality control, bankable prototype proposals and lending models were at a similarly low rung of readiness and have remained major constraints to date.

6.4.2 Role of Scaling Workshops in Strengthening Partnerships

The RTB scaling project workshops reiterated the need to focus on the soft components of the HQCP innovation package. This led to the identification of activities needed to address the main bottlenecks for taking the innovation at scale: facilitate access to credit and develop bankable business proposals; establish a cassava peel sub-sector collaboration or community of practice; and ensure product standardization, marketing, promotion, supportive legislation, awareness creation and advocacy.

---

[3] https://seedtracker.org/peeltracker/

Based on this, the Bank of Industry (BoI) was engaged to facilitate access to credit, and the National Office for Technology Acquisition and Promotion (NOTAP) began to help with promotion. Single Spark (makers of FeedCalculator® for feed formulation) and the Raw Materials Research and Development Council (RMRDC) began to support the strengthening of the hard components of the technology. Cassava Peel First Movers groups were established as the community of practice. At this point, the scaling consultant helped the project staff develop a way to track partnerships (Table 6.7).

6.5 Development Outcomes from Making and Using HQCP

This section presents the RTB scaling project's achievements and challenges, which were documented through a short, cross-sectional survey conducted shortly after the end of the intervention. The section is structured around the project's development outcomes.

6.5.1 Development Outcome 1

Two major feed millers in Nigeria are regularly incorporating HQCP mash in their commercial rations.

Among major feed millers, Premium Fish Feeds and Premier Feeds incorporate HQCP fine mash directly into their rations, while Nine Stars uses powdered HQCP (indirectly) as a binder in their fish feed. Premium Fish Feeds, which is one of the top five largest feed mills in Nigeria, uses 7–10 tons of HQCP weekly. Premier Feeds began using 100 tons per week and have now expanded to 200 tons a week. Premier Feeds was initially procuring HQCP from their sister factory, Thai Farms, which eventually could not satisfy their expanding demand. Presently, the additional quantity is purchased from one of the project's trainees, Kofo Agro, at a price of US$ 160 per tonne.

6.5.2 Development Outcome 2

Two hundred female and male processors are organized and up to 60% of them linked to markets and selling HQCP mash to commercial feed millers under contractual arrangements.

Small- and medium-scale enterprises that emerged during scaling activities are taking root around HQCP mash production and sale to feed millers.

The post-intervention survey was conducted on three groups of entrepreneurs: (i) 36 of the more than 200 small-scale entrepreneurs belonging to the Cassava Peel

Table 6.7 List of partners and tracking partners

	What characterizes partner in relation to scaling project						
Prospective partners	Importance for scaling project	Aware of the innovation proposed and its scaling prospects	Supportive and capable of contributing to scaling	Actively participating in helping innovation go to scale	Actively propagating the innovation without support by scaling	Level of effort to help innovation go to scale	Strategy for how to involve them
	Score 1–5	Tick which one applies				Score 1–5	Description
Amo Byng	4	✓	✓	✓		4	Classification, contractual arrangements for commerce, updates through collaborative fora, participation in meetings
Top feeds	4	✓	✓			1	Classification, contractual arrangements for commerce, updates through collaborative fora, participation in meetings
Durante	3	✓	✓	✓			Updates through collaborative fora, participation in meetings
Nine stars	3	✓	✓	✓	✓	4	Updates through collaborative fora, participation in meetings
Niji foods	3	✓	✓			3	Participation in meetings
Synergos	3	✓	✓	✓	✓	4	Factories and with clientele, updates through collaborative fora, participation in meetings

(continued)

Table 6.7 (continued)

What characterizes partner in relation to scaling project							
Prospective partners	Importance for scaling project	Aware of the innovation proposed and its scaling prospects	Supportive and capable of contributing to scaling	Actively participating in helping innovation go to scale	Actively propagating the innovation without support by scaling	Level of effort to help innovation go to scale	Strategy for how to involve them
NOTAP	4	✓	✓			4	Through collaborative fora, participation in meetings
RMRDC	4	✓	✓			3	Through collaborative fora, participation in meetings
FMARD	4	✓	✓			2	Factories with selected cooperative groups, updates through collaborative fora
BoI	5	✓	✓	✓	✓	3	Beneficiaries for credit support, updates through collaborative fora, participation in meetings
Single spark	3	✓	✓			3	Promotion of FeedCalculator, updates
ILRI	3	✓	✓	✓		2	Project, updates through collaborative fora, participation in meetings
CIAT	3	✓	✓	✓		2	Project, updates through collaborative fora, participation in meetings
Indicative summary	3.5	100%	100%	54%	23%	2.8	

Source: RTB Scaling Fund Project 2018–2019

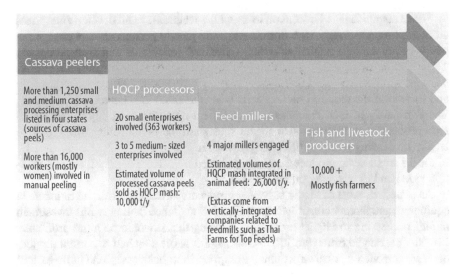

Fig. 6.10 A summary of stakeholders in the HQCP value chain in Nigeria. (Source: End of RTB Scaling Fund project survey, 2020)

First Movers CoP platform, (ii) all three medium-scale entrepreneurs involved with the project and (iii) all four feed millers with whom contractual arrangements were sought. There are 1250 small-scale cassava processors in four states, including more than 200 who belong to the CoP. There are as many as 20 medium-scale entrepreneurs, including those who were not involved with the project (Fig. 6.10).

Small-Scale HQCP Mash Producers

Almost all the 36 small-scale entrepreneurs surveyed had been among the over 200 people trained as a joint RTB-LIVESTOCK CRP activity in the demonstration factory at IITA campus in 2018 and 2019 on best practices in production, equipment maintenance, personnel safety, feed safety, hygiene, cassava peel business registration, bookkeeping, credit application and feed formulation of least-cost balanced rations incorporating HQCP. On a scale of 1 to 5, the training was rated 4.78 for content by respondents. Respondents said the most innovative and useful topics they learned about included how the process mimicked *gari* production, hygiene and feed safety, the use of FeedCalculator® and cash flow analysis.

The respondents already use wet peels as a feed ingredient for their own animals, but now they have identified aggregators for product marketing, good management practices and the promotion of HQCP in feed formulation to reduce costs. For reasons unknown at the time of writing, but probably related to the ease of getting cassava peel, the Cassava Peel Tracker was not popular among small-scale producers and was not used to identify clients or suppliers and/or to buy or sell cassava peels or cassava peel products. Some respondents suggested creating a wider base of processors and buyers to make the website visit worthwhile, while others suggested raising awareness about the existence of the tracker.

Those respondents who went on to adopt the technology and develop enterprises around it (approximately 10% of trainees) were mostly persons already engaged in one or more cassava-related activities – farmers, processors and feed producers. This result is not surprising as the training had a broad waste-to-wealth (w2W) theme around cassava peel that appealed to people already in the cassava business and for whom the required raw material was available free of charge or nearly so. The familiarity of HQCP processing equipment given its similarity to traditional *gari* processing equipment also meant that start-ups or switching from *gari* to HQCP processing was easy for such persons. It is also noteworthy that some 12% of the 36 small-scale producers surveyed started off after watching the demonstration video and/or watching others produce.

As expected, most of the 36 surveyed producers have made fresh investments in equipment and factory construction, while a few switched their previous investment in gari processing to HQCP processing, sensing that it would be more profitable. Investments ranged from US$ 150 to US$ 27,400. More than half of current producers have considered borrowing money to support their businesses, and of these, half of the credit demand is still in preparation. Preferences for credit sources were more for informal and family sources than banks.

A novel industry is developing around the cassava value chain and creating employment in its wake. At the time of the survey, 20 of the 36 respondents were producing HQCP mashes, with enterprises spread across south-east, south-south, southwest and north-central geo-political zones of Nigeria, employing a total of 363 persons (mean 10.8) with weekly production and sales of up to 10 tons of wet peel cake, 14 tons of whole HQCP mash, up to 15 tons of coarse HQCP mash and up to 16 tons of fine HQCP mash (Fig. 6.11).

Medium-Scale HQCP Producers
The three medium-scale HQCP producers who participated in the survey have produced fine mash, coarse mash and cassava peel cake since 2017/2018 for a wide range of clients, including feed millers and fish producers. The medium-scale producers have capacity to process up to 60 tons per week of wet peels and up to 120 tons per week of dried peels. In the past few years, clients have increasingly purchased HQCP products to overcome the problems of traditionally processed cassava peels (dusty and potentially contaminated with HCN and aflatoxins). The customers of the three medium-scale producers rate HQCP as 'most important' on product quality, while availability was the lowest concern. The issue of quality was predicated on nutrient profile (high energy content) and low moisture linked with longer storability and longer shelf life. Longer shelf life helped feed millers even out their HQCP mash supplies and retain better control of their production.

The three medium-scale producers are producing high-quality cassava flour (HQCF), and they source wet peels mainly from their own operations, but they also buy dried peels. The respondents were not sure if they could process more peels than they currently do. An option for expanding operations is to buy wet cake from other producers to process into mashes, but the respondents had no interest in this option being that drying is a major constraint. Nevertheless, these three

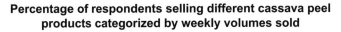

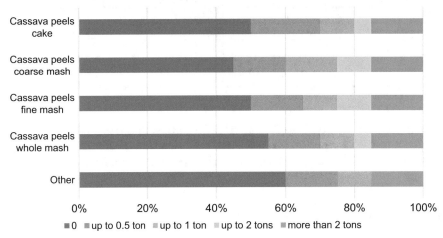

Fig. 6.11 Percentage of the 20 respondents (small-scale processors) selling different cassava peel products categorized by weekly volumes sold. (Source: End of RTB Scaling Fund project survey, 2020)

entrepreneurs plan to expand HQCF production. This development would translate into increased availability of peels, which would expand the HQCP business.

Feed Millers

The four major feed millers who responded to the survey listed several advantages of HQCP, including (1) cheaper than other sources of energy, (2) profitable farmers through reduced feed cost and (3) good source of energy with minimum anti-nutritional factors. They said challenges consisted of (1) low inclusion rate, (2) low crude protein, (3) inconsistent quality and (4) price variation (US$ 125–150). The survey revealed that over 10,000 farmers (mostly fish farmers) have used HQCP products. All of them reported satisfaction with the products, and notably, about one-third of poultry and cattle farmers said they were 'very satisfied'.

Most of feed millers' customers buy poultry and fish feeds. Cattle, small ruminant and other and pig feed clients comprise a small share of the customers (Table 6.8).

Although there was initial reluctance by feed millers, this has been gradually overcome once it became clear that animal production performance remained intact. As such, the incorporation of HQCP in feed has led to increased sales among existing customers and new customers who enjoy lower feed prices. In addition to the 200 tons per week being used by Premier Feeds, up to 520 tons of HQCP is now being incorporated into compound feed by other feed millers every month. This figure is up from 105 tons per month in 2018. Each factory uses different suppliers, but the most reliable supply is through vertical integration (e.g. a cassava flour mill company with a line for HQCP production). All responding feed millers expect to

Table 6.8 Clientele of feed millers that incorporate HQCP mashes by type of feed purchased

Type of feed	No. of clients
Fish	10,132
Poultry	226
Other	21
Cattle	8
Pig	2

Source: End of RTB Scaling Fund project survey, 2020

double their use of HQCP by the end of 2020 barring constraints related to availability, machines for processing, price and consistency in quality and availability. Using more HQCP will lower feed costs and increase factory throughput.

6.5.3 Development Outcome 3

Twenty-five HQCP investors supported to produce successful, bankable proposals and to set up medium to large-scale factories employing youths and women.

The project team developed templates for business plans, including content such as business profiles, vision making and people, business strategy and structure, communication and Information and Communication Technology (ICT) infrastructure, location criteria, operations management strategy, financials and costing, and a marketing and growth program. Different example business plans were prepared for enterprises operating at four different scales. The templates and examples were shared during a 3-day training organized by the project.

Currently, five loan applications are being processed at the Bank of Industry (BoI). BoI staff were part of the trainings and elaborated the requirements for credit. Unfortunately, the gestation period for obtaining credit from BoI, as presented at the training, was overly optimistic. Credit delivery remains encumbered by requirements for securing the loans. However, BoI is not the only financial institution in Nigeria; the business plans are flexible and could be submitted as part of a loan application elsewhere.

6.6 Lessons Learned for Scaling

This section presents the key lessons learned during the implementation of the scaling project. They are categorized in four learning areas.

Capacity/Competencies

The core innovation was developed based on two drying options suitable to small-scale and household enterprises, namely sun-drying and toasting. When the innovation attracted the interest of medium-scale entrepreneurs, additional drying options had to be explored, such as flash drying, rotary drum drying and cabinet drying. The project team lacked the immediate capacities and competencies to respond. Partnership with Niji Foods and collaboration with CIAT experts were critical to overcome these challenges, pointing to the need for careful selection of appropriate partners.

The 'simplicity' of the technology and the decades of processing cassava into *gari* lured us into complacency ('anyone can adopt it') in selecting the initial batches of trainees without sufficient investigation of their motivations. As it turned out, some were interested in the use rather than the production of HQCP, while others were just interested in marketing the products. These are important motivations, but they could not contribute to achieve HQCP production at scale. Based on this experience, interviews were conducted for selecting properly motivated participants to subsequent trainings and to enlist their help to design the training curricula.

HQCP innovation, which builds on age-long processing of cassava into *gari* and other staples, was expected to be a plug-and-play given the similarity of equipment and processes and the huge demand for alternative ingredients in feed production. With an estimated one million households producing and processing cassava into various products and more than 9000 cassava processing centres in Nigeria, we expected immediate massive uptake, but that did not happen. Uptake is never automatic, even when closely related to an existing (technology) way of doing things. However, it is important to note that, when trained in HQCP processing, persons already engaged in some cassava processing had a higher likelihood to become entrepreneurs and adopt the innovation compared to fresh entrants into the cassava processing business.

Strategizing and Scaling Strategy

The high demand for HQCP products could not be met by the small-scale and household producers initially targeted. Meeting this demand will require increasing the numbers of small-scale processors and involve larger-scale investors as well as product aggregators. However, the longer-term aim should be to industrialize the technology so that single factories with capacity to process as much as 50 tons/day become possible and common. There is sufficient demand for HQCP mashes, and fresh peels are available for all scales of producers to the extent that the participation of one group would not adversely affect the other.

Collaboration/Partnership Related

Partnership was clearly biased towards the private sector with a sprinkling of relevant public agencies. The private sector has been involved since the proof-of-concept stage and has stamped a business attitude on the innovation to the extent that it appears that the private sector is leading those big moves.

Early private sector engagement proved helpful to facilitate scaling. For example, with Amo Byng, we performed a technical and economic evaluation of the products and gained their subsequent open endorsement, when bias against HQCP products was still high.

We should do the same and, where necessary, more of reinforcing private sector involvement, engendering transparency and encouraging product standardization to promote commercialization.

Bundling the core innovation with supporting products and services from partners was beneficial to the scaling progress. For example, the collaboration with Single Spark led to the production of FeedCalculator®, which has increased knowledge in feed formulation and promoted the use of HQCP mashes in livestock rations. Practical evidence that inclusion of HQCP mashes into feed did not affect, and sometimes enhanced, livestock performance boosted the confidence of farmers and feed millers and facilitated adoption, especially in poultry and fish production.

The levels of commitment of different partners differed. The partnership with the Bank of Industry, despite being formalized in a contract, did not result in facilitated access to credit by project participants. While formal arrangements can be useful to define roles and responsibilities of partners, these should not be seen as an indicator of commitment to the partnership.

Intervention

Attention to developing the innovation package by adding complementary elements (e.g. hands-on training in production and best business practices, FeedCalculator®, Cassava Peel Tracker® and economic feasibility models) contributed to attract interest to the core innovation – be it for HQCP mash production, feed formulations, pure buying and selling of HQCP products or any combination of the above. These possibilities meant that some of the critical concerns along the value chain regarding new products were addressed within the innovation package.

Overall, it was clear over the course of the entire exercise that scaling had to be a deliberate and proactive effort. *That is, scaling does not just happen, scaling has to be made to happen.* There were several elements (each of them important on their own) that needed to come together, including financial viability and a facilitating business environment. For scaling to happen, each of these elements should be conceptualized not as abstract entities but as having a human face, manifested as interest groups (stakeholders). At this point, one should ask whether these stakeholders desire goals they cannot achieve without certain, mutually beneficial partnerships. If this condition exists, the next step would be to continue to use available, noncostly media to do the groundwork to highlight opportunities created by the technology. Hopefully, this work will arouse different interests and suggest avenues for meeting each other's needs and facilitating agreements that promote commerce. In addition, scaling champions need to follow the positive vibes and remain flexible. There are no blueprints for them to follow.

6.7 Conclusions

Successful scaling requires that the core innovation itself must be easy to use and profitable, like the cassava peel dryers for making HQCP mashes. In this case, the raw materials (cassava peels) were not only free, but they were also an environmental nuisance. At the same time, the rapidly growing livestock industry was demanding more feed. Using HQCP mashes was a chance to turn waste into wealth by using cassava peels to make top quality animal feed. The main challenge was not the hardware (technology), but the software, especially linking small-scale cassava processors (and a few mid-sized ones) with large-scale feed millers. So far, this new value chain is functioning, but it is too soon to tell if the actors will continue to work together.

Doubts remain but these are not insurmountable:

- Will the new product (HQCP mashes) be profitable along the whole value chain?
- Will mutual mistrust remain in spite of the partnerships? Will the lack of formal credit continue to hamper investment?
- Will feed millers continue to buy from small-scale processors (who are mostly women), or will the factories make their own HQCP mashes or buy them from larger processors?
- Will medium-sized producers be able to keep making a quality product, or will they undermine trust in their goods by making them from carelessly dried cassava peels recovered from processor waste piles?

As project funding has ended, it is expected that the work initiated to encourage SON to list HQCP products and set their physical and nutritional standards will provide the guardrails for production of higher-quality products and eliminate poorer quality ones through market forces. Fortunately, the process is still ongoing within SON even in the absence of the project.

The research and scaling efforts have been worthwhile, starting from scratch and gradually nurturing an industry that is in its infancy. Much progress has been made, but support is still needed – especially to buy equipment after training, to standardize products and to promote and market HQCP mashes.

Collaboration with other projects can trigger interest in adopting the HQCP innovation. For instance, the RUNRES project in Rwanda sponsored a training program for a women group in October 2020. During the trip, a skilled fabricator was coached to manufacture the key HQCP equipment. That group now produces about 3 tons weekly based on sun drying even during the rainy season. Another aspect of future work will be to develop HQCP mashes with higher protein content, for instance, through biofortification, which has shown promising preliminary results. This would likely boost the popularity of HQCP mashes among fish and poultry feed millers. The process ferments cassava waste using selected microorganisms, and initial results are positive and indicate the possibility to reach up to 60% conversion rate from waste to feed ingredient on a dry matter basis (Obadina et al. 2006; Aro 2008; Crécy 2012; Tefera et al. 2014).

This innovation has a huge potential to go to scale, since cassava is an important crop in many other countries besides Nigeria. The HQCP project is applicable in all cassava-producing countries. If the importance of cassava in the agricultural sector were to be followed, the Democratic Republic of Congo, Ghana, Cote d'Ivoire, Senegal and Sierra Leone could be possible candidates to scale to innovation. However, the needs of their feed industry should be carefully investigated.

Acknowledgements This research was largely undertaken as part of, and funded by, the CGIAR Research Program on Roots, Tubers and Bananas (RTB) with contributions from the CGIAR Research Program on Livestock, its predecessor, the CGIAR Research Program on Livestock and Fish and the CGIAR Research Program on Integrated Systems for the Humid Tropics and supported by *CGIAR Trust Fund contributors* and initial core funding from the International Livestock Research Institute (ILRI) and the CGIAR Systemwide Livestock Programme.

References

Adesehinwa AOK, Samireddypalle A, Fatufe AA, Ajayi E, Boladuro B, Okike I (2016) High quality cassava peel fine mash as energy source for growing pigs: effect on growth performance, cost of production and blood parameters. Livest Res Rural Dev 28(11)

Agweek (2021) Market output report: Nigeria maize market outlook 2021/22. Agweek 1-1, January 2021

Akinpelu AO, Amamigbo LEF, Olojede AO, Oyekale AS (2011) Health implications of cassava production and consumption. J Agric Social Res (JASR) 11(1):118–125

Aro SO (2008) Improvement in the nutritive quality of cassava and its by-products through microbial fermentation. Afr J Biotechnol 7(25):4789–4797

Azogu I, Tewe O, Ezedinma C, Olomo V (2004) Cassava utilization in domestic feed market in Nigeria. National Centre for Agricultural Mechanization (NCAM), Federal Ministry of Agriculture and Rural Development (FMARD), Abuja, November 2004, 156 pp

Barry MP, Linda SA, Shu WN (2012) Global nutrition transition and the pandemic of obesity in developing countries. Nutr Rev 70(1):3–21. https://doi.org/10.1111/j.1753-4887.2011.00456.x

Bennett B, Naziri D (2013). Market Study for a range of potential cassava and yam waste product solutions in Ghana, Nigeria, Thailand and Vietnam. https://www.researchgate.net/publication/267096062_Market_Study_for_a_range_of_potential_cassava_and_yam_waste_product_solutions_in_Ghana_Nigeria_Thailand_and_Vietnam

Crécy EF (2012) Experimental evolution of a facultative thermophile from a mesophilic ancestor. Appl Environ Microbiol 78(1):144–155

CORAF (2020) Integrated regional strategy for sustainable management of agri-inputs in West Africa and the Sahel. West and Central Africa Council for Agricultural Research Development (CORAF). Dakar. 2020. 84 pp

Egbeocha CC, Asoegwu SN, Okereke NA (2016) A review on performance of cassava peeling machines in Nigeria. Futo Journal Series (FUTOJNLS) 2(1):140–168

FAO (2011) Mapping supply and demand for animal-source foods to 2030, by T.P. Robinson & F. Pozzi. Animal Production and Health Working Paper. No. 2. FAO Rome

FAO (2013) Save and Grow Cassava- A guide to sustainable production intensification. ISBN 978-92-5-107641-5. FAO Rome

FAO (2016) Meat and meat products. FAO's Animal Production and Health Division. http://www.fao.org/ag/againfo/themes/en/meat/home.html

FAOSTAT (2019) Food and Agriculture Organization of the United Nations, 2019. Production: Crops. http://faostat.fao.org.

Fauquet MC (2014) Estimating cassava wastes in Nigeria: a consultancy report submitted to the International Livestock Research Institute (ILRI) Nairobi

Garnett T, Roos E, Little DL (2015) Lean, Green, Mean, obscene...? What is efficiency? and is it sustainable? Animal production and consumption reconsidered Technical Report. Food Climate Research Network, University of Oxford, 48 pp. http://hdl.handle.net/1893/24127

Haberl H, Erb K-H, Krausmann F (2014) Human appropriation of net primary production: patterns, trends, and planetary boundaries. Annu Rev Environ Resour 39:363–391. https://doi.org/10.1146/annurev-environ-121912-094620

Hillocks RJ (2002) Cassava in Africa. In: Hillocks RJ, Thresh JM, Bellotti A (eds) Cassava: biology, production and utilization. CABI Publishing, New York, pp 41–54

Karlberg L, Hoff H, Flores-López F, Goetz A, Matuschke I (2015) Tackling biomass scarcity—from vicious to virtuous cycles in sub-Saharan Africa. Curr Opin Environ Sustain 15:1–8. https://doi.org/10.1016/j.cosust.2015.07.011

Kolawole P, Samuel T, Alenkhe B, Diallo T, Abass A, Kulakow P (2016) Evaluation of commercial cassava peeling machines in Nigeria. Unpublished research manuscript. International Institute of Tropical Agriculture (IITA) Ibadan

Obadina AO, Oyewole OB, Sanni LO, Abiola SS (2006) Fungal enrichment of cassava peels proteins. Afr J Biotechnol 5(3):302–304. http://www.academicjournals.org/AJB

Obianwa C, Uyoh EA, Igile G (2016) Bioethanol production from cassava peels using different microbial inoculants. Afr J Biotechnol 15(30):1608–1612. https://doi.org/10.5897/AJB2016.15391

Okike I, Samireddypalle A, Kaptoge L, Fauquet C, Atehnkeng J, Bandyopadhyay R, Kulakow P, Duncan AJ, Alabi T, Blummel M (2015) Technical innovations for small-scale producers and households to process wet cassava peels into high quality animal feed ingredients and aflasafe™ substrate. Food Chain 5(1–2):71–90. https://doi.org/10.3362/2046-1887.2015.005

Samireddipalle A, Bolakale O, Fabamise A, Okike I, Blümmel M (2016) Performance of West African Dwarf rams fed cassava by products as sole feed. ANACON-2016.X Biennial Conference of Animal Nutrition Association. 09–11 November 2016; College of Veterinary Sciences, SVVU, Tirupati

Sartas M, Schut M, Proietti C, Thiele G, Leeuwis C (2020) Scaling readiness: science and practice of an approach to enhance the impact of research for development. Agric Syst 183(102874). https://doi.org/10.1016/j.agsy.2020.102874

Smith OB (1988) A review of ruminant responses to cassava-based diets. In: Hahn SK, Len R, Egbunike GN (eds) Cassava as livestock feed in Africa: proceedings of the IITA/ILCA/University of Ibadan workshop on the potential utilization of cassava as livestock feed in Africa, 14–18 November 1988, Ibadan, pp 39–53

Smith P, Gregory PJ, Van Vuuren D, Obersteiner M, Havlík P, Rounsevell M, Woods J, Stehfest E, Bellarby J (2010) Competition for land. Philos Trans R Soc B 365(1554):2941–2957. https://doi.org/10.1098/rstb.2010.0127

Tefera T, Ameha K, Biruhtesfa A (2014) Cassava based foods: microbial fermentation by single starter culture towards cyanide reduction, protein enhancement and palatability. Int Food Res J 21(5):1751–1756

Thornton PK (2010) Livestock production: recent trends, future prospects. Philos Trans R Soc. B 365(1554):2853–2867. https://doi.org/10.1098/rstb.2010.0134

Willits-Smith A, Aranda R, Heller MC, Rose D (2020) Addressing the carbon footprint, healthfulness, and costs of self-selected diets in the USA: a population-based cross-sectional study. Lancet Planet Health 4(3):e98–e106. https://doi.org/10.1016/S2542-5196(20)30055-3

Chapter 7
Transferring Cassava Processing Technology from Brazil to Africa

Alfredo Augusto Cunha Alves (ID), **Luciana Alves de Oliveira** (ID), and **Joselito da Silva Motta** (ID)

Abstract Cassava is currently the fourth most important food production crop in tropical and developing countries. Cassava root and its by-products are the main source of calories for the diets of 800 million people in Africa, South America, and Southeast Asia. Over the past 20 years, the Brazilian Agricultural Research Corporation (Embrapa) and collaborators have been developing innovations for the use and postharvest processing of cassava. These technologies have been transferred and disseminated to technicians, entrepreneurs, producers, and processors of cassava from several African countries. This South-South cooperation has been conducted in Brazil through short trainings, workshops, and technical visits requested by national R&D institutions, cooperatives, cassava producers, and processors associations and sponsored by international agencies and foundations. In this chapter, we present an overview of the technology transfer activities of *Embrapa Mandioca e Fruticultura* carried out for Africa, focusing on technological innovations that result in products and by-products of cassava root processing, especially those with great potential for adoption and opening new markets for Africa (e.g., precooked and frozen cassava, cassava chips, among others). The selection of these innovations was based on observations of the trainees' preferences and interests for technologies that they envisioned willingness to apply and share the technology when returning to their countries.

7.1 Introduction

Cassava is one of the oldest cultivated crops in the world; its domestication occurred approximately 9000 years ago in the Amazon region of Brazil (Alves-Pereira et al. 2018). In tropical and developing countries, cassava holds great economic and nutritional importance and feeds around 800 million people. Of the 278 million tons of cassava roots produced by 101 countries in 2018, about 60% is grown on the African

A. A. C. Alves (✉) · L. A. de Oliveira · J. S. da Silva Motta
Embrapa Mandioca e Fruticultura, Cruz das Almas, Brazil
e-mail: alfredo.alves@embrapa.br; luciana.oliveira@embrapa.br; joselito.motta@embrapa.br

G. Thiele et al. (eds.), *Root, Tuber and Banana Food System Innovations*,
https://doi.org/10.1007/978-3-030-92022-7_7

207

continent, 30% in Asia, and 10% in the Americas (FAOSTAT 2019). Cassava is the fourth most important source of calories for the human diet in Africa, South America, and Southeast Asia, behind wheat, rice, and maize. Five African countries are among the ten largest cassava-producing countries, which account for 75% of world production (FAOSTAT 2018).

Cassava cultivars are classified in two types in Brazil: *sweet cassava*, aka *"aipim,"* *"macaxeira,"* or "table cassava," and *bitter cassava*, aka *"mandioca brava"* or "cassava for industry." These types are classified based on levels of cyanide (HCN) release, a highly toxic substance if ingested. Sweet cassava contains less than 100 mg kg^{-1} of cyanogenic compounds (CC) per fresh root, while bitter cassava contains more than 100 mg (Araujo et al. 2019). High levels of CC require a complete detoxification process to reduce compounds to a safe level for consumption (Montagnac et al. 2009). Cassava roots are very perishable and have a short postharvest shelf life, which severely limits their potential in the market and potential benefits to farmers. The roots exhibit visible symptoms of postharvest physiological deterioration (PPD) within only 24–72 hours of harvest (Morante et al. 2010). The darkening caused by physiological deterioration is an important factor that should be considered in root processing. Like other root and tuber crops, cassava has high water content (~65%), which is probably the major limitation to improving the utilization potential of the crop, requiring rapid processing into intermediate products to reduce transport costs, increase shelf life, and improve storage capacity (Falade and Akingbala 2011).

In Brazil, both the cassava root and the starch extracted are processed and used to make a series of products that can be consumed directly or used as ingredients in a wide variety of foods (Fig. 7.1) (Oliveira et al. 2020a; Tomlins and Bennett 2017).

In the last 20 years, the Brazilian Agricultural Research Corporation (Embrapa) has received several demands from African cassava-producing countries and funding agencies for development projects, requesting technical collaboration from Brazil to transfer technological innovations related to the cultivation and postharvest processing of cassava. Cassava cultivation in Brazil and many parts of Africa shares many similarities between Brazil and Africa in both cultural and social aspects. The profile of cassava producers is similar, and most cassava is produced on family farms. In addition, the agroecological similarities and the great diversity of cassava processing products available in Brazil facilitate the transfer and adoption of these technologies to Africa.

Embrapa Mandioca e Fruticultura international technical cooperation program started in 2000 with the Brazil-Japan Partnership Program as a means to offer courses to technicians from Portuguese-speaking African countries (e.g., Angola, Cape Verde, Guinea-Bissau, Mozambique, São Tomé and Príncipe) and Timor East in Southeast Asia. The program was financed by the Japan International Cooperation Agency (JICA).

In this program, courses were offered to technicians, researchers, and representatives of national agricultural research institutions (NARIs), and the content focused

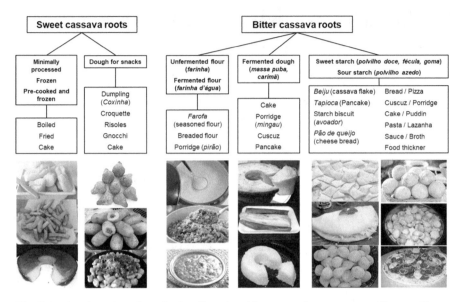

Fig. 7.1 Diversity of products in Brazil made with processed cassava root. (Source: Figure adapted with photos from Alfredo Alves, Joselito Motta, Oliveira et al. (2020a), Podium Alimentos)

on good agronomic practices for growing cassava, including soil preparation, planting, cultivation management, harvesting, and technologies for root postharvest processing.

7.2   Embrapa's Technical Cooperation with Africa: Focus on Cassava

Between 2016 and 2019, *Embrapa Mandioca e Fruticultura*, which also works with tropical fruits (banana, pineapple, papaya, passion fruit, and citrus), received a total of 35 delegations, involving 205 international visitors from 62 institutions and 36 countries from American, African, Asian, and European continents (Fig. 7.2). These visits were aimed at knowledge exchange and technology transfer activities, and most of them (62%) came from the African continent, with greater interest in cassava crop (82%) (Fig. 7.3). For these visits, the trainees explored a variety of topics, including (1) postharvest processing of cassava root and starch, aiming to add value and new products from Brazil that were not yet adopted or adjusted to Africa; (2) good agronomic practices for cassava cultivation, covering technologies from planting to harvest; and (3) production and distribution of cassava seeds with good genetic and phytosanitary quality, focusing on techniques for cleaning planting material via micropropagation, multiplication, and seed distribution. The main

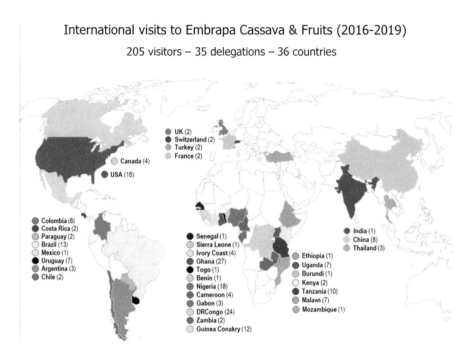

Fig. 7.2 International visits received by *Embrapa Mandioca e Fruticultura* for trainings and technical visits (2016–2019). (Source: NRI-Embrapa Mandioca e Fruticultura (2019))

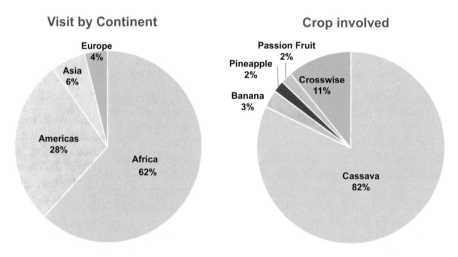

Fig. 7.3 Distribution of international trainings and technical visits by continent and crop of interest. (Source: NRI-Embrapa Mandioca e Fruticultura (2019))

target audience for these trainings were young entrepreneurs (technicians, farmers, and cassava processors) working on their own or in national R&D or private institutions.

7.2.1   Technical Training for Young Africans

In September 2017, the Brazil Africa Institute (IBRAF) launched the Youth Technical Training Program (YTTP) aiming to promote knowledge transfer and develop skills of young Africans in various sectors, including agriculture (AfDB 2017). Financed by the African Development Bank, this program chose the cassava production chain on which to focus the initial training on "propagation, production, and processing of cassava" for young Africans. *Embrapa Mandioca e Fruticultura* was invited to plan and execute this training (Table 7.1).

The training program included theoretical classes, followed by practices in the field and laboratories, and visits to cassava processing units and cassava farmers cooperatives located in the State of Bahia (Fig. 7.4). The main topics covered by trainings included:

1. Postharvest processing of cassava root and starch
2. Good practices for cassava cultivation
3. Production and distribution of disease-free cassava seeds

More details about the training's agenda are available in Table 7.2.

Table 7.1 Trainings for young Africans conducted by *Embrapa Mandioca e Fruticultura* and promoted by the Youth Technical Training Program, 2017–2019

Training	Period (# of days)	# of trainees	# of countries	Country (# of people)
Training on cassava propagation, production, and processing for young Africans	23 Oct to 17 Nov 2017 (27 days)	28 17 women 11 men	14	Benin (1), Burundi (1), Cameroon (1), Congo DR (5), Ghana (1), Ivory Coast (3), Malawi (1), Mozambique (1), Nigeria (4), Senegal (1), Sierra Leone (1), Tanzania (4), Uganda (2), Zambia (2)
Workshop on cassava postharvest processing for young Africans	19 to 21 Nov 2018 (3 days)	8 4 women 4 men	4	Cameroon (3), Ghana (3), Ivory Coast (1), Nigeria (1)
Training in cultivation and cassava processing for young Africans	4 to 8 Nov 2019 (5 days)	26 13 women 13 men	3	Ghana (13), Malawi (6), Nigeria (7)
Total	**2017–2019 (35 days)**	**62 34 women 28 men**	**14**	**Benin (1), Burundi (1), Cameroon (4), Congo DR (5), Ghana (17), Ivory Coast (4), Malawi (7), Mozambique (1), Nigeria (12), Senegal (1), Sierra Leone (1), Tanzania (4), Uganda (2), Zambia (2)**

Source*: NRI-Embrapa Mandioca e Fruticultura (2019)*

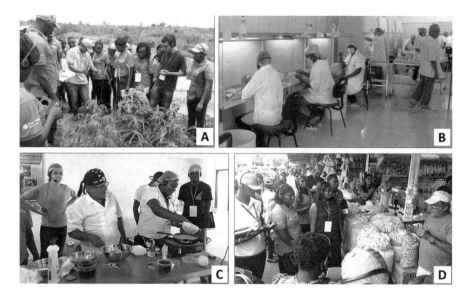

Fig. 7.4 Training in cassava propagation, production, and processing for young Africans at Embrapa. (**a**) Practical class in the field, visiting a collection of cassava varieties and field instructions for cultivation; (**b**) laboratory practices (micropropagation and multiplication techniques for cleaning planting material); (**c**) training in a cassava processing unit for preparing tapioca from starch extracted from the roots; and (**d**) visit to street market (diversity of cassava products for sale). Cruz das Almas-BA, Brazil, 2017. (Photos: Embrapa Cassava & Fruits)

7.2.2   Increasing Performance of the Cassava Industry in West and Central Africa (IPCI Project)

Funded by the International Fund for Agricultural Development (IFAD), this project ran from 2014 to 2017 aiming to provide technical expertise and support to a range of national cassava projects across West and Central Africa (WCA), including Gabon, Ghana, Nigeria, and the Republic of the Congo. As part of this project, IFAD established a partnership for *Embrapa Mandioca e Fruticultura* to engage in IPCI activities and conducted four primary activities in 2017 (NRI 2017):

1. *Promotion of novel and innovative Brazilian cassava products.* In September 2017, Embrapa welcomed a delegation of ten people, consisting of representatives from IFAD and the project coordinating institutions, and chairpersons from cassava processing units, from Gabon, Ghana, and Nigeria for (1) visits to laboratories and experimental areas of *Embrapa Mandioca e Fruticultura*; (2) a visit to cassava production and processing areas in Bahia State; (3) a visit to Embrapa headquarters to discuss opportunities for international technical cooperation between Embrapa and African countries; and (4) technical meetings on cassava processing with the Embrapa cassava technical team and discussion of IPCI project. On this technical visit to Brazil, the IFAD partner researchers explored,

Table 7.2 Content of the training on "Cassava Propagation, Production, and Processing for Young Africans"

Theoretical classes	Field and laboratory practices at Embrapa	Visits to processing/market units and cassava farmers cooperatives
Embrapa's Cassava Breeding Program Cassava agronomic aspects Climate change and cassava water use efficiency Cassava pests and diseases Cassava weed management Reniva project (multiplication and distribution of disease-free cassava seeds) Toxicity and processing cassava Cultivation technologies for small cassava producers	Visit to Embrapa's in vitro and field cassava genebanks Field practices on soil management Visit to physiology and climatology laboratories Visits to entomology, phytopathology, and virology laboratories Root processing practices: farinha production and starch extraction Field practices on cassava cultivation techniques Minimal processing of cassava roots – food technology laboratory Processed product preparation practices	Visits to starch industry in Laje-BA (Bahiamido) and the cassava farmers cooperative in Tancredo Neves-BA (Coopatam) Visits to street markets in Cruz das Almas and Vitória da Conquista-BA (cassava products for sale) Visit to small- and medium-scale *beijus* production units Visits to cassava processing units and Coopasub (Cooperative of Small Farmers in Southwest Bahia) in Vitória da Conquista-BA Visit to Bahia Biofactory (IBB) in Ilhéus-BA: practices on cassava seed production system Practices on culture medium preparation and plant micropropagation at IBB Cassava seed field production: Reniva Project, in Tapera (Cruz das Almas region)

Source: *NRI-Embrapa Mandioca e Fruticultura (2017)*

with Embrapa researchers, Brazilian technologies related to the use and processing of cassava, at different technological levels, from artisanal (small farming) to industrial processing at medium and large scales. The visits occurred in two cassava-producing regions with different profiles regarding the use and processing of cassava roots: (1) the northeast region, where small farmers predominate and the roots are processed mainly for production of flour and starch derivatives; and (2) the southern region, where cassava is grown on a large scale for production and transformation of starch.

2. *Embrapa publications translated.* Embrapa maintains a large stock of publications and gray literature on cassava. With respect to IPCI, literature relating to postharvest and processing aspects of cassava might be especially useful. However, most of these publications are written in Portuguese. Translating these publications to English and French could have a significant impact on the African cassava research environment. Embrapa identified the most appropriate publications for the areas of utilization and postharvest processing of cassava and provided for the translation of 11 publications into French and English. In addition, Embrapa selected multimedia material (videos and documentaries) reedited with English and French subtitles.

3. *Survey of equipment and machinery for cassava processing.* Brazilian manufacturers have developed a wide range of cassava processing machines which might be useful to the cassava processing sector in Africa. Usually, contacts between cassava processors in the target countries and suppliers of equipment are ad hoc. Embrapa made a brief survey of Brazilian suppliers of cassava processing equipment and provided a list of equipment, specifications, and prices, with the key elements translated into English and French (more information available in Table 7.3).

4. *Technical visit of Embrapa experts to target countries.* Three Embrapa cassava experts visited Nigeria and Ghana, where they learned about IFAD's ongoing activities and its projects related to cassava processing through visits to IPCI partner institutions. On the part of farmers and processors, the mission visited cassava production fields, gari processing centers, and family cassava processing units, where visitors gave practical demonstrations on how to prepare *tapioca* from cassava starch (Fig. 7.5a, b). These visits were important to understand the few alternatives for processing cassava roots in those countries, where they concentrate, predominantly, on gari production. The great interest shown by local farmers and processors in the preparation of *tapioca*, for example, highlights the technological gap and potential growth areas for strengthening technology transfer of processing tools from Brazil to Africa.

The trip culminated with "CassavaTech" in Lagos, Nigeria, where the Embrapa team organized an exhibition of cassava products produced in Brazil with practical demonstrations on how to prepare *tapioca* (Fig. 7.5c) and a special session on "Brazilian Technologies for Cassava: What Africa can learn from the Brazilian Cassava Industry."

7.3   Major Innovations for Cassava Root Processing

Cassava roots are processed in various ways to prevent postharvest deterioration, reduce root toxicity, improve the palatability of derived products, and increase the commercial value of cassava.

Based on the training evaluations and the interest shown by the participants, we selected some technological innovations already in use in Brazil and ready for adaptation for Africa, with great potential for adoption and market opening for new products.

The innovations refer to products obtained from the processing of sweet cassava roots and from bitter cassava roots and starch. It is important to note that all products/technologies applied to bitter cassava can also be used in sweet cassava, but the reverse is not always true.

Table 7.3 Information about Brazilian manufacturers and their equipment for processing cassava roots

Manufacturer (website)	Equipment	Web link showing machines operation/catalog
Fankorte – Indústria de Máquinas (www.fankorte.com.br)	Root washer (skin removal) Root peeler Root cutter and sticks maker	https://www.youtube.com/watch?v=NVe9gs5NsPU https://www.youtube.com/watch?v=7zP_2QYWNK4
Jair Pomim (www.youtube.com/channel/UCpEClMBFNwAZoX7i4hnnRHQ)	Mini machine for skin removal and peeling roots Slicing, shredding, and dough maker machine	https://www.youtube.com/watch?v=T23FxjU7v-I https://www.youtube.com/watch?v=rcS70MVGG5g
Maringá Torneadora (https://www.youtube.com/channel/UCrQZ9gbaTEcbXr8ZMIM6EFw)	Mechanized oven for toasted cassava flour	https://www.youtube.com/watch?v=rWfYOnyFdSY
Chapadão Máquinas (www.chapadaomaquinas.com)	All cassava flour processing machines	https://www.youtube.com/watch?v=odXYu900b3w
VMAQ Indústria de Máquinas Especiais (https://vmaq.ind.br/maquinas/)	All the equipment to build a factory for industrial production of starch (polvilho) and cassava flour	https://www.youtube.com/user/vmaqindustria

Fig. 7.5 Practical demonstrations of how to prepare "tapioca" from cassava starch in the District of Baara, Abeokuta (**a** and **b**), and in Lagos, Nigeria, at a CassavaTech event (**c**) in 2017. (Photo credits: Alfredo Alves and Ben Bennet)

7.3.1 Sweet Cassava Processing

For sweet cassava, we highlight cassava minimally processed, frozen cassava, pre-cooked and frozen cassava, cassava chips, and cassava dough for snacks. Oliveira et al. (2020a) describe in detail the preparation steps of these products, which must be carried out following good manufacturing practices.

7.3.1.1 Cassava Minimally Processed

Minimally processed cassava is an alternative to extend the root supply period, providing a product that is easy to prepare and more convenient for consumption (Viana et al. 2010). Minimal processing needs to be carried out in a cold environment (10–15 °C). If this is not possible, use water between 5 °C and 10 °C for root washing and sanitization. Basically, it consists of root reception, selection, washing, sanitizing, cutting, peeling, sanitizing, vacuum sealed packaging, and storage at 5 °C (Fig. 7.6).

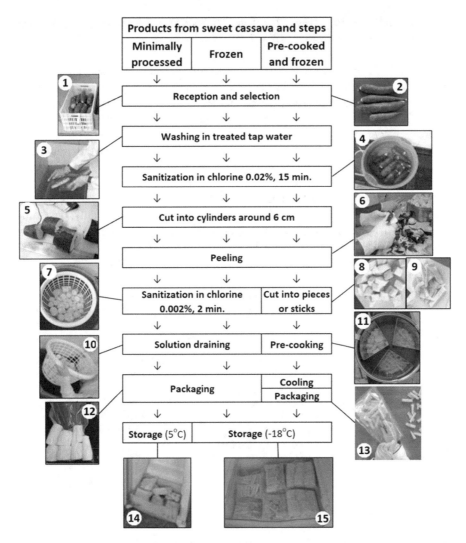

Fig. 7.6 Steps for preparing minimally processed cassava, frozen, and precooked and frozen. (Source: Figure adapted with photos from Oliveira et al. (2020a), Oliveira et al. (2020b), and Viana et al. (2010))

Some steps to prepare minimally processed cassava are also present in the protocol of all products, such as reception and selection, washing, sanitization, peeling, and packaging of the roots, where some good practices must be followed.

Reception and washing of fresh roots To minimize cross contamination, the reception and washing areas for fresh roots must be separated from the internal processing areas. The roots must be harvested on the same day of processing or the day before processing and stored at night. Usually, fresh sweet cassava roots are received in boxes or bags (Fig. 7.6(1)) and washed in treated tap water with a brush to remove soil and dirt residues (Fig. 7.6(3)).

Sanitization Sanitization is performed after washing the roots (with and without peel) and on the materials/tools used in the root processing. This step aims to reduce the number of deteriorating microorganisms and eliminate pathogens (which cause diseases). For each material to be sanitized, different chlorine solutions are used (Table 7.4).

Peeling the roots Peeling the roots must be careful as it can damage the tissues and accelerate the browning process, called physiological and/or microbial deterioration, and increase the water loss. Peeling sweet cassava roots for culinary use is usually done manually with a stainless-steel knife or with suitable peeling equipment. After removing the root tips, if the peeling is carried out manually, the cylinders are cut to approximately 6 cm in length and then peeled. If peeling is done with equipment, cassava is cut into cylinders after peeling. All the peel must be removed to improve the appearance of the product and avoid darkening (Fig. 7.7).

Table 7.4 Sanitization steps performed in the postharvest processing of cassava

Sanitization of	Procedure	How to prepare 1 L of sanitizing solution
Roots with bark	Immersion in *0.02% chlorine solution* for 15 minutes	*0.02% chlorine*: 10 mL of 2% sodium hypochlorite (commercial) in 1 L of water = 200 mg of chlorine/L
Peeled roots	Immersion in *0.002% chlorine solution* for 2 minutes	*0.002% chlorine*: 1 mL of 2% sodium hypochlorite (commercial) in 1 L of water = 20 mg of chlorine/L
Materials used in the processing steps (knives, table/bench surface, boards, sieves, etc.)	Immersion in *0.01% chlorine solution* for 2 minutes or spray with *0.02% chlorine solution*	*0.01% chlorine*: 5 mL of 2% sodium hypochlorite (commercial) in 1 L of water = 100 mg of chlorine/L

Source: Adapted from Oliveira et al. 2020a

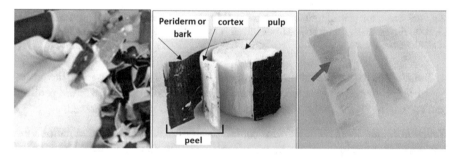

Fig. 7.7 Peeling sweet cassava roots. All peel (periderm and cortex) must be removed. The presence of peel in precooked root impairs the appearance of the product (red arrow on the right). (Photo credits: Eliseth Viana and Luise Sena)

Packaging the roots Packaging protects the product during transportation, distribution, storage, and handling against damage, shock, vibration, and compression that occur during the entire journey. The low-density polyethylene plastic bag is the most commonly used packaging for processed cassava products, as it is resistant to impact and a good water barrier. However, some products, such as flour, may use kraft paper packaging, molded as a box or bag; this material has the advantage of being recyclable and biodegradable. For fried products, such as chips, polypropylene and aluminum plastic film are recommended to prevent oil oxidation and maintain the crispness and flavor of the chips for a longer time. However, it is common to use polypropylene plastic bags for the packaging of fried chips, as they are less expensive, sold in small quantities, and easier to find. The polypropylene packaging is transparent and gives greater visibility to the product, but the chips will have a shorter shelf life due to oil oxidation.

7.3.1.2 Frozen Cassava

Frozen cassava is sold in whole cylinders from 5 to 7 cm in length, in pieces, or crushed to make cakes. The processing steps are the same as for minimally processed cassava with the exception of storage that is carried out at −18 °C in a freezer or cold room (Fig. 7.6).

To obtain frozen cassava, the roots must first be selected, excluding those with hard (fibrous) parts, spots, rot, or other problems (Fig. 7.8). The first selection of the roots usually occurs in the field at harvest time. To evaluate the root quality, a sample of the harvested roots that represents the variety to be processed is cooked. Only the group roots that present desirable quality (e.g., cooking in 30 minutes or less) should be processed. Roots that cook in more time must not be processed, even under pressure. Processing does not improve the quality of the processed root, so it is essential to process good quality roots to obtain a good product (Oliveira et al. 2020b).

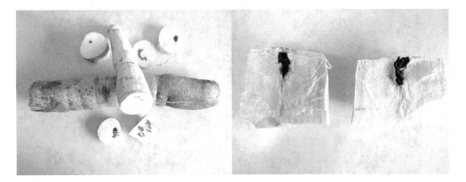

Fig. 7.8 Cassava roots unsuitable for processing due to the presence of dark spots. (Photo credits: Luciana Alves de Oliveira)

After the second sanitization, the sanitizing solution is drained, and the cylinders are packed in more resistant polyethylene packages. Frozen cassava can be sold in portions from 200 g to 2 kg, depending on the consumer market. The packages are sealed and placed at −18°C in a freezer. Vacuum packaging is not recommended as frozen cassava can pierce the packaging.

If the processing is carried out properly, the product will last for at least 4–6 months. The roots must be placed for cooking while still frozen, and this information must appear clearly on the packaging.

Losses in the processing of frozen cassava, such as peel and tips, vary from 25% to 30% of the total weight of the roots. The tips and small pieces of cassava can be used for the preparation of other products, such as dough for snacks.

7.3.1.3 Precooked and Frozen Cassava

Precooked and frozen cassava is sold in the form of sticks or pieces. In this processing, the initial steps are the same as for frozen cassava (root reception, selection, washing in water, sanitation, cut in cylinders, peeling), followed by cutting into sticks or pieces, precooking, cooling, packaging, and storage at −18°C in a freezer or cold room (Fig. 7.6).

The stick is always cut longitudinally to the length of the root in the direction of the fibers. If the stick is cut across the fibers, the percentage of breakage is greater and the product length will be shorter (Fig. 7.9). The stick is cut 1 cm on the side and 5–6 cm long. For cutting the roots into pieces, the cylinder can be cut into 4–8

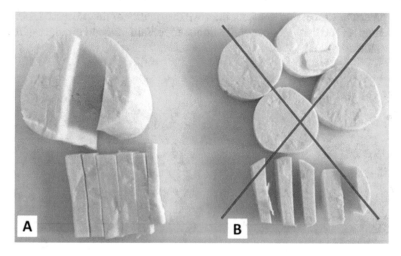

Fig. 7.9 Photo (**a**) shows the proper way to cut cassava lengthwise and with regard to thickness. Photo (**b**) shows improper cutting of cassava for sale, which will diminish its safety and shelf life. (Photo credit: Luise de Oliveira Sena)

Table 7.5 Tests performed to determine the appropriate precooking time for two cassava varieties processed as sticks

| Precooking time | After frying | |
	Variety 1	Variety 2
No precooking	Taste of raw cassava	Taste of raw cassava
2 minutes	Taste of raw cassava	Adequate flavor and texture
4 minutes	Taste of raw cassava	Oil soaked
6 minutes	Adequate flavor and texture	Extremely oil soaked

Source: Oliveira et al. (2020a)

parts. Imperfect sticks, tips, and leftover roots can be used to produce dough for snacks.

After cutting, the precooking starts with 2 L of water for each kilogram of cassava root and 2% salt added to the water (20 g of salt per 1 kg of cassava root) to accentuate the flavor. The cooking water must be changed after three uses. The sticks or pieces of cassava are placed in the water once it is boiling.

The precooking time should be tested for each variety used. For example, precooking can be tested in three time periods: 2, 4, and 6 minutes. For each test, use 250 g of cassava sticks and 0.5 L of water with 5 g of salt. The precooked sticks should be frozen when submerged in water, boiled for the required time, and then fried to verify the quality of the final product. Table 7.5 shows the results of the test performed with two cassava varieties. Each variety was harvested and tested on the same day and at the same age.

Precooking time can vary for the same variety, depending on the root age and environmental conditions (fertilization, rainfall, plant nutrition, soil, drought). The precooking time for a piece of root is longer than that for a root stick. Thus, tests to confirm best precooking time must also be performed.

After precooking, the sticks should be drained and placed in packs made with a resistant (thicker) polyethylene or another resistant material. The sticks tend to adhere to each other. To prevent this, place the sticks in a disorganized manner inside the packaging (Fig. 7.6(13)). The packages should then be sealed and placed in a freezer at −18 °C.

7.3.1.4 Cassava Chips

Another way of adding value to cassava is by producing fried snacks such as chips. The preparation of the chips includes the reception and selection of the roots, washing in water, sanitizing, peeling, sanitizing, slicing, soaking the slices in water, frying, salting, draining surface oil, packaging, and storage (Fig. 7.10).

Peeling the entire root is important to reduce leftover tips. After peeling, the peeled roots should be submerged in treated water to remove any adhering skin. The water used in this stage can be reused for initial root washing.

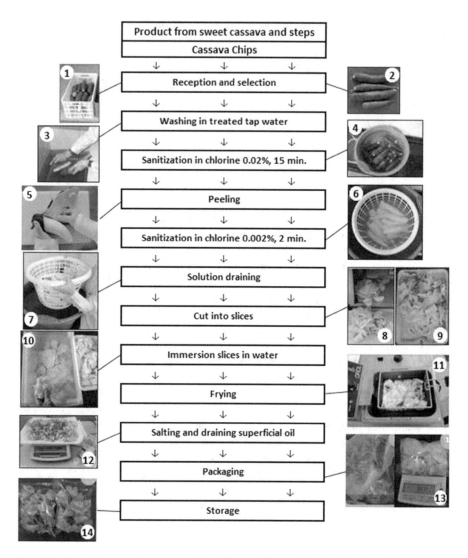

Fig. 7.10 Steps for preparing cassava chips. (Source: Figure adapted with photos from Oliveira et al. (2020a, b)

The root is cut into slices approximately 0.8 mm thick. The thickness of the slice is especially important for the quality of the product (crispness). If the slice is too thick, the chips may become hard. There is cassava slicing equipment available on the market for this purpose. Slicing can occur directly into the fryer; if not available, slices must remain immersed in water to prevent them from sticking. After soaking, the water must be drained and the slices fried. Very thick cassava roots are not suitable for chips as they crack during peeling or slicing (Oliveira and Godoy 2011).

The frying oil temperature should be between 170 °C and 180 °C and not exceed 190 °C to avoid impairing the oil quality. The quality of the oil used for frying will influence the quality and shelf life of the finished product. Oil degradation will be greater the longer the period of its use. To monitor the quality and disposal point of the frying oil, there are quick tests and evaluation equipment on the market. Palm oil has higher oxidative stability (less degradation) than cotton, sunflower, and soybean oils; therefore, it is the most suitable for the production of fried products.

After frying, excess oil is drained and the chips are spread over a sheet of paper to again drain the excess. The product is salted with 1% salt: 10 g for every 1 kg of product. The addition of spices (oregano, chili powder, parsley, onion, among others) can be added at this time also. If the additional seasoning contains salt, the amount of added salt must be decreased.

The slices should be quickly packaged to prevent moisture absorption. The product should be packed in polypropylene bags, in portions of 40–80 g, stored in a dark place, and should be consumed within 15 days.

Chips may vary in quality due to factors such as the quality of the raw material and oil used, inequalities in chip thickness, temperature, and frying time. The color of the final product depends on the color of the cassava variety used, which may be white, cream, yellow, or pink.

7.3.1.5 Cassava Dough for Snacks

For the preparation of dough for snacks, the first steps are the same as for frozen cassava, followed by the steps of cutting the root into four pieces, central fiber removal, cooking, freezing, grinding, salting, shaping, packaging, freezing, and storage at −18 °C in a freezer or cold room (Fig. 7.11). If processed properly, the product will last at least 6 months. Before starting the dough preparation, one batch of roots should be used to evaluate the cooking time of the variety used.

The cooking water is drained (Fig. 7.11(10)), and after the cooked roots return to room temperature, they should be cooled at a temperature between 4 °C (refrigerator) and −18 °C (freezer) (Fig. 7.10(11)) for 24 hours. The cooked root, still frozen, is ground in a meat grinder (Fig. 7.11(12)). Then the crushed dough is mixed with salt, using a ratio of 10 g salt to 1 kg dough (1% salt) in a mixer (Fig. 7.11(13)), adding the salt gradually while mixing the dough.

The dough can be placed in a sausage maker (Fig. 7.11(14)) to make croquettes. The machine's screw system should rotate slowly so that the texture is uniform, and the dough does not acquire air bubbles. When the dough begins to exit the plastic section of the equipment, use hands to press the plastic section outlet (Fig. 7.11(15), red arrow) while the screw system turns. It is important to exert pressure during this step so that all air is removed from the dough. The tray that will receive the dough must be greased with a little oil so the dough does not stick to the tray.

The croquettes must quickly be cut to the same size using a knife (Fig. 7.11(15)) so that the dough does not break. The knife should be rinsed in water to facilitate

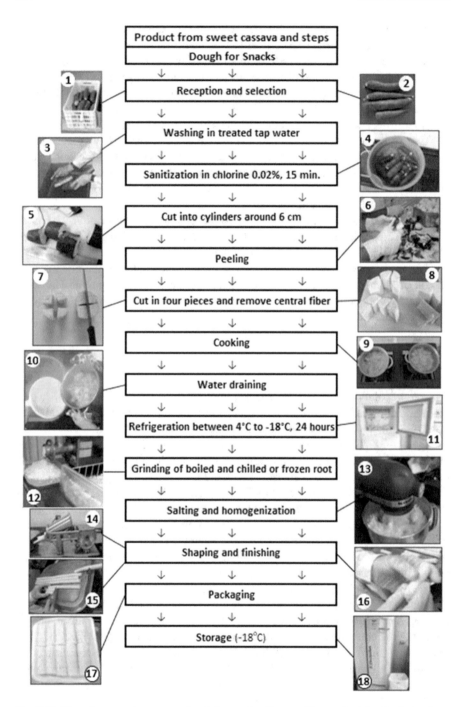

Fig. 7.11 Steps for preparing cassava dough for snacks. (Source: Figure adapted with photos from Oliveira et al. (2020a))

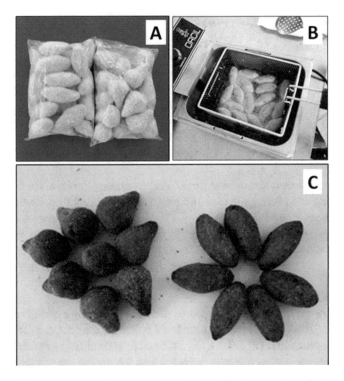

Fig. 7.12 Frozen (**a**), frying (**b**), and fried (**c**) croquettes and "coxinhas" made from stuffed sweet cassava dough. (Photo credits: Luise de Oliveira Sena)

cutting. To finalize the savory pastry, dip fingertips in water and mold the cut ends with moistened fingers (Fig. 7.11(16)) to round the ends of the croquettes.

Croquettes are sold frozen or pre-fried (180 °C for 35 seconds) and frozen. Fry the frozen croquettes in oil at 170 °C for 3–4 minutes in a covered pan, as the croquettes may explode if the dough is too wet.

In addition to salt, other spices can be added to the dough, such as onion, cilantro, basil, oregano, cheese, and parsley. The dough can also be used to make savory foods in different formats ("coxinha," dumpling, rissole, gnocchi) and with desired fillings (bacon, shrimp, ground beef, chicken, cheese, among others) (Fig. 7.12). Savory pastries tend to be quite sticky and need to be breaded. The flour used for breading also removes the sticky quality of the pastries and becomes an oil absorption barrier during frying, which increases the crispness of the product.

7.3.2 Bitter Cassava Processing

In Brazil, bitter cassava roots are processed to produce **flour** (*farinha de mandioca*) and **starch** (known in Brazil as *polvilho*, *fécula*, or *goma*), which can be further processed into a series of products. Below, we will highlight *beiju* (**cassava flake**), *tapioca* (**stuffed beiju**), *avoador* (**starch biscuit**), and **cheese bread**.

7.3.2.1 Flour Processing

Dry flour is consumed throughout Brazil, and its production involves the following steps: washing and peeling the roots, grating, pressing, crumbling, sieving, roasting, sifting, packaging, and storage (Fig. 7.13). Fermented flour, known in Brazil as *farinha d'água* or *farinha puba*, is widely consumed in the Amazon region (North Brazil). Its production involves the same steps as dry flour, except that, before being grated, the roots are macerated (softening of the roots in a fermentative process in water, known as *pubagem*) for 4 days. After *pubagem*, the same steps as dry flour should be followed (Fig. 7.13).

To prepare the flour, it is important to use the same variety and the same age of cassava roots. Root washing prior to manual peeling (or maceration for fermented flour) is recommended (Fig. 7.13(1)). Processing should begin immediately after harvest. Roots must be completely peeled where the bark is not white. Partial peeling can influence the quality of the flour and lead to dark spots and/or browning in the final product.

The peeled root should not be placed directly on the ground but rather in plastic boxes and washed with tap treated water to remove sticky peels, sand, and other adhered dirt.

The next step is root grating, which is done using an electric grater, also called *cevador* or *caititu* (Fig. 7.13(6)). The grater has a cylinder, preferably with serrated stainless-steel blades (Fig. 7.13(4, 5), red arrow) that can be replaced as needed.

Excess water in the grated dough must be removed before roasting using manual (Fig. 7.13(9)) or hydraulic pressing equipment. The grated root dough is placed in raffia bags (Fig. 7.13(7)) or on fine mesh nylon screens and pressed onto pallets (Fig. 7.13(8)) to remove excess water (also called *manipueira*. The pressing time depends on the amount of crushed mass and the equipment used. *Manipueira* and cassava peel can be used in animal feed, fertilization, and irrigation.

The dough becomes compact due to compression and needs to be crumbled. This procedure is performed using the same root grater (Fig. 7.13(6)).

To get a fine dry flour (which is preferred in northeast Brazil), it is recommended to pass the crumbed dough through a hand or mechanical sieve (Fig. 7.13(10)) with a galvanized 26–28 mesh to standardize the grain size of the flour and remove the fibers (central root fiber) before drying and toasting. The higher the mesh number, the finer the flour obtained.

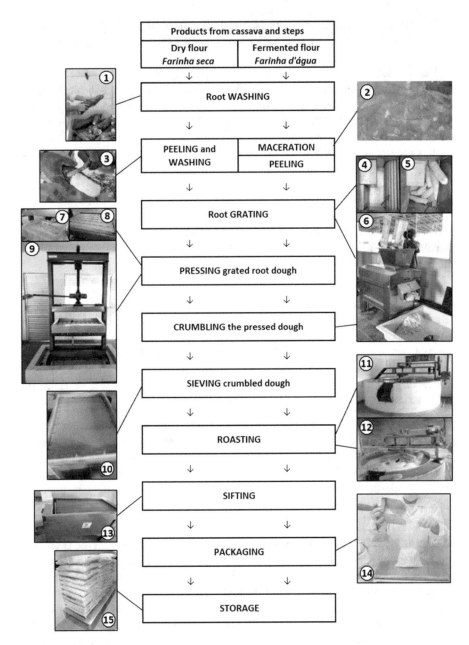

Fig. 7.13 Steps for preparing dry and fermented cassava flours. (Source: Figure adapted with photos from Oliveira et al. (2020a) and SENAR (2018))

The dough is manually placed in the oven using a scraper (Fig. 7.13(11)). The dough should be added slowly into the oven with the stalks moving so as not to tangle.

Initially, the dough should be roasted for 20–30 minutes at 90 °C. After that time, increase the oven temperature to approximately 160 °C for up to 3 hours to reduce moisture (Fig. 7.13(12)). After roasting, the flour must be cooled and sifted to evaporate the remaining moisture (Fig. 7.13(13)). The flour grain size depends on the demand of a particular consumer market.

To keep the flour crispy, packaging for storage and sale is important. When flour is ready and cold, 50 kg plastic bags are recommended. For smaller portions, 1 kg packages of polypropylene are preferred (Fig. 7.13(14)), which preserves the low moisture of the flour. Polyethylene keeps the product crisp for a shorter time.

The flour must be stored in a dry and ventilated place with screened windows. It is recommended that the bags be stacked on pallets or crates, leaving room for air circulation between the packs (Fig. 7.13(15)). According to Brazilian legislation, cassava flour must have a maximum humidity of 13%.

Gari vs farinha d'água The most consumed processed cassava product in Brazil is unfermented cassava flour, which is not made in other regions of the world. Outside Brazil, the main cassava product is **gari**. *Gari* was introduced in Africa in the early nineteenth century after freed slaves returned to the African continent. Today, it is the main cassava product in sub-Saharan Africa, representing 70% of the cassava consumed in Nigeria. Meanwhile, in Brazil, *farinha d'água* is produced, almost exclusively, in the Amazon region (Folegatti et al. 2005; Bechoff et al. 2019; Ndjouenkeu et al. 2021).

Both flours are made from fermented cassava. However, the fermentation of *farinha d'água* occurs in anaerobic conditions by immersing the whole root in water, while gari fermentation is aerobic: the grated root mass is pressed slowly for 2–6 days (Fig. 7.14). Gari also includes a step for "garification," which combines stages of gelatinization (starch cooking) and gari roasting, which affect the physico-chemical and rheological properties of the starch, providing this food its unique sensory and functional characteristics (Folegatti et al. 2005).

Massa puba **(puba dough) or** *carimã* *Massa puba* is produced from the spontaneous fermentation of fresh cassava roots (with or without peel). The roots are placed in water for approximately 5 days until they soften and begin to release the bark. Then, they are crushed in sieves and washed until only the fibers remain. The separated mass must be washed several times and dried in the sun or in dryers until a humidity of 50% is reached for wet puba (or *puba úmida*) and 13% for dry puba (or *puba seca*) (Ferreira Filho et al. 2013). *Carimã* is a fine flour made by crushing dried fermented roots. *Puba* and *carimã* are used in the preparation of cakes, porridge, couscous, and other dishes.

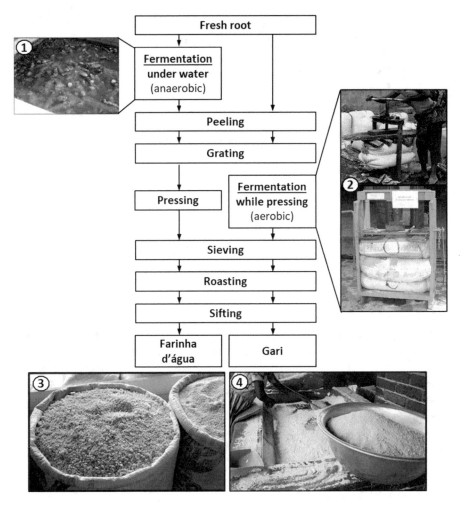

Fig. 7.14 The steps to prepare *farinha d'água* and *gari*, including the processing differences between the two products. (1) With *farinha d'água*, fermentation is anaerobic and the whole roots are immersed in water, (2) while in gari, the fermentation is aerobic and the grated root mass is left to ferment during pressing. (Photos: Alfredo Alves (2-4) and Folegatti et al. 2005 (1))

7.3.2.2   Small-Scale Starch Extraction ("Polvilho Doce" or "fécula")

In the past, the cassava starch was considered the by-product of flour production. Due to the increased demand by the consumer markets for cassava products, the starch has been used to manufacture starch biscuits, tapioca, and cheese bread (SENAR 2018).

The steps for starch extraction (Fig. 7.15) are the same as for dry flour until the grating step. The crushed root dough (Fig. 7.15(6)) should pass through the starch

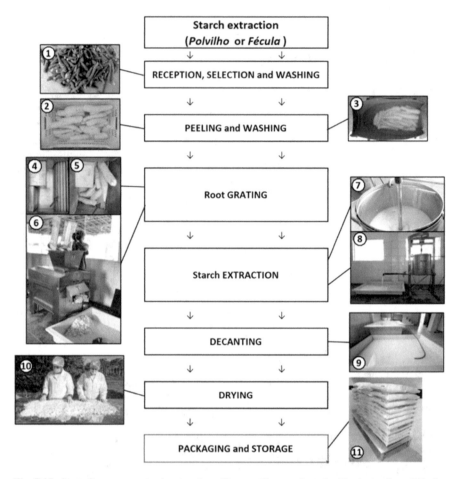

Fig. 7.15 Steps for cassava starch extraction. (Source: Figure adapted with photos from Oliveira et al. (2020a) and SENAR (2018))

extractor (Figs. 15.7 and 15.8) and mix with water while in the starch extractor. As the water drains out, it must be received in a glass fiber trough to separate the starch via decantation (Fig. 7.15(9)). The decanted starch is crumbled and placed to dry in the sun until it reaches a humidity of approximately 13%. The dried starch is then packaged and stored in a cool and airy place. Starch can be stored safely up to 12 months.

Sour starch ("polvilho azedo") The production of sour starch follows the same steps as natural starch, except that in the decantation stage, the starch is left to ferment in the tanks for a period ranging from 15 to 40 days, depending on the ambient temperature, until the product reaches an acidity of approximately 5%. The settling tanks must be under cover to avoid impurities and must be covered with ceramic or fiber tiles. During the fermentation period, the starch must remain covered by a 10 cm layer of water. If not covered, the starch exposed to air acquires a blue or purple color, which reduces its quality. After the fermentation period, the starch is removed from the tanks and taken to dry.

7.3.2.3   "Beijus" (Cassava Flakes) and "Tapiocas" (Stuffed Beiju)

The *beijus*, or tapiocas as they are known in Brazil, are produced from the starch, gum, *fécula*, or *polvilho*. Two types of this product are common: (1) *beijus* roasted on hot plate, crispy, without fillings and (2) soft and folded with various fillings, often called stuffed tapioca, *beiju de tapioca*, or simply tapioca (Motta, 2012).

These products are made with cassava-hydrated starch (*farinha de tapioca* or *polvilho doce úmido*), which is obtained by mixing a part of water with two parts, by weight, of the natural starch (*polvilho doce seco*) and pinches of salt. The mixture is homogenized and sieved to make it thinner.

To make traditional roasted beiju, hydrated starch is spread on a hot plate or heated pan (Fig. 7.16(1)). The drying point of the dough can be identified when the edges come loose and become brittle, like pancakes or a dry crepe. The *beijus* are cut on the hot plate (Fig. 7.15(2)), cooled to room temperature (Fig. 7.16(3)), packed in plastic bags or pots (Fig. 7.16(4)), and sealed, labeled, and stored (Fig. 7.16(4,5,6)).

Colored "Beijus." Traditionally white, *beijus* have been produced in new colors, smells, flavors, and nutrients when the water is replaced by fruit pulps or vegetable extracts. Tasting evaluations found that consumers preferred these products flavored with beets, onion, pineapple, guava, and passion fruit (Fig. 7.17a) (Motta, 2012). To boost cassava sales in schools, producers have also manufactured gum biscuits, cookies, and other products (Fig. 7.17b).

Stuffed tapioca ("beiju de tapioca" or "tapioca recheada")   To make tapioca, start by hydrating the natural starch (*polvilho doce* or *tapioca seca*) and adding a part of water to two parts of starch (Fig. 7.18(1)). Mix well with hands until it crumbles and add a pinch of salt. Sieve the hydrated starch (Fig. 7.18(2)) and spread it over a preheated frying pan (Fig. 7.18(3)). To spread the sieved starch, you can use the sieve or your hands. Let the starch heat up until you notice that the grains have stuck to each other, and the starch pancake becomes loose from the pan (around 1 minute) (Fig. 7.18(4)), and turn it upside down, like a pancake. Add any kind of filling (Fig. 7.18(5)) and fold the tapioca in half (Fig. 7.18(6)). Heat the tapioca a bit more on both sides to complete the process. The frying pan does not need to be greased; it should be clean and without grooves. The type of filling depends on consumer preference, such as butter and cream cheese, condensed milk, ham and cheese, meats, and a range of food mixes (Fig. 7.19).

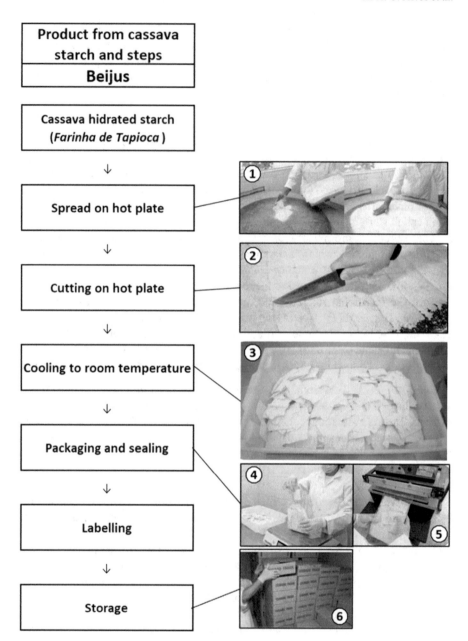

Fig. 7.16 Steps for preparing beijus. (Source: Figure adapted with photos from SENAR (2018))

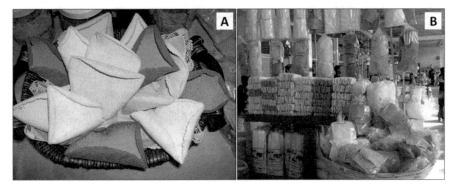

Fig. 7.17 (**a**) Colored *beijus* prepared with fruits and vegetables. (**b**) Diversity of *beijus* for sale at the municipal market in Cruz das Almas, Brazil. (Photo credits: Joselito Motta)

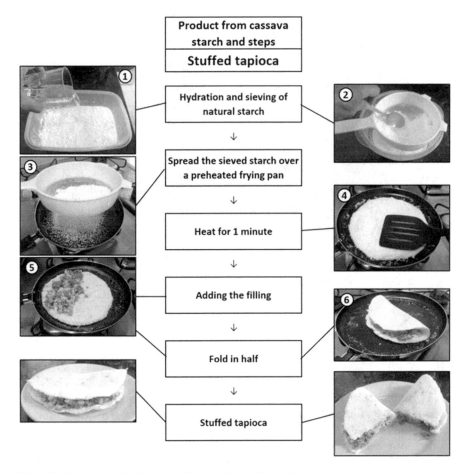

Fig. 7.18 Steps for stuffed tapioca. (Photo credits: Alfredo Alves)

Fig. 7.19 Tapioca with different types of filling: (**a**) plain tapioca with grated coconut filling; (**b**) beet tapioca with scrambled egg filling; (**c**) cabbage tapioca with shredded chicken and mozzarella cheese filling. (Photo credits: Joselito Motta)

Fig. 7.20 Diversity of avoador biscuits and other cassava starch snacks. (Photo credit: Joselito Motta)

7.3.2.4 Avoador (Starch Biscuit)

Avoador, often referred to by its generic name *biscoito de polvilho*, but also known as *peta* or *biscoito avoador*, is a type of snack food in Brazilian cuisine. It is found mainly in the states of Minas Gerais, Bahia, and Rio de Janeiro, but is also consumed in São Paulo and Paraná. Avoador is a popular snack and side dish for breakfast (Fig. 7.20) (Sampaio and Menezes 2020).

Both sour and sweet starch can be used to prepare avoador. For 500 g of starch, mix with 250 mL of milk, 250 mL of water, 250 mL of oil, 1 egg, and 1 teaspoon of salt.

Preheat the oven to low temperature before starting to prepare the recipe. Mix the egg and salt in a bowl. Then add warm oil and milk. Gradually add the starch and stir until you get a consistent dough. Put 250 mL of boiling water into the mix and combine all the ingredients.

Put the dough in a pastry bag (or plastic bag) with a spout. Grease a pan with margarine and make the cookies in the shape you want, but do not let the cookies touch each other while baking. The cookies should be baked at 180°C for 20 minutes.

7.3.2.5 Pão de Queijo (Cheese Bread)

Pão de queijo, or Brazilian cheese bread, is a small, baked cheese roll or cheese bun, and a popular snack and breakfast food in Brazil. It is a traditional Brazilian recipe, originating in the state of Minas Gerais.

Pão de queijo consists of a type of biscuit with sour or sweet *polvilho* plus eggs, salt, vegetable oil, and cheese, of soft and elastic consistency (Fig. 7.21). There are several different recipes that vary the ingredients and type of cheese to be used. Natural and/or sour starches can be used to prepare the bread.

For 500 g of starch, add 300 mL of milk, 150 mL of oil, 3 eggs, 250 g of grated *Minas* cheese, and a teaspoon of salt.

In a bowl, heat the salt, milk, and oil. When boiling, scald the starch with this hot mixture, stir well, and let it cool. Add the eggs one by one, alternating with the cheese and kneading well after each addition. Grease your hands with oil if necessary.

Mold cookies 3–5 cm in diameter and place them on a greased baking sheet. Take to a preheated oven (180 °C) and bake until golden brown (around 15 minutes).

Fig. 7.21 Traditional Brazilian *pão de queijo* (cheese bread). (Photo: Podium Alimentos)

7.4 Closing Remarks

All the technological innovations detailed here are characterized by a high level of technological maturity, which means they are qualified technologies, fully adopted in Brazil, and, potentially, ready to be transferred directly, or with small adaptations, to other international production systems. On the other hand, their positive impacts on society do not seem to have been properly analyzed to create a vision of their success and how these technologies can be framed within the readiness scale recently developed by CGIAR (Sartas et al. 2020).

It is hoped these technology transfer trainings will allow participants to apply the skills in their own contexts. Therefore, it is important to design a training program that is sustainable and replicable in other regions or countries.

Considering this and based on the experiences acquired over the last 15 years, *Embrapa Mandioca e Fruticultura* has been trying to improve the implementation of international training to maximize the returns on investment.

Appropriate selection of participants is crucial. Typically, training candidates are chosen directly by the beneficiary institution, following the view that innovations can be transferred simply by intermediaries and change agents (e.g., extension workers) and then spread throughout the communities of individual beneficiaries. This conception has been widely refuted (Sartas et al. 2020).

In the first training sessions carried out by Embrapa, the African candidates selected were predominantly adult professionals, based in national R&D institutions primarily focused on administrative management activities rather than research. Usually, these individuals held administrative positions; some were not even involved with cassava crop. This lack of proximity to cassava production may have hindered their effectiveness as a technology multiplier inside and outside their institutions.

In the more recent cassava courses carried out by Embrapa, international organizing institutions (usually CGIAR and African NARIs), training sponsors, and funding agencies have selected young candidates with an entrepreneur profile, technical professionals, or recently graduated individuals linked to state institutions, private companies, and/or start-ups. Most young trainees consider training as essential to financial survival, an opportunity for them to introduce innovations in their own business and add value to their products.

Embrapa has specified how its trainings could be adapted or suited to address gender differences; however, we did not see the need to adjust the trainings to meet the preferences of men and women. The instructors – who are women and men – have noticed some differences in participation. Women showed more interest in topics related to techniques for preparing products from cassava processing. Men expressed greater interest in field activities. As the trainings occasionally require overnight stays or long hours away from home, this may create some implications that deserve attention to meet the needs and preferences of both men and women participants.

Participants of the three courses held within the Youth Technical Training Program on Cassava Post-harvest Processing (2017–2019) are already applying what they have learned in Brazil, and many of them provide to IBRAF and Embrapa with great feedback on the results of the Program at the micro level. One of the participants who founded the PaaClee Cassava Processing Company in 2014, for example, started to invest in training sessions with his employees, sharing the knowledge acquired during the 1-week course in Brazil. The YTTP on Post-harvest Processing of Cassava success also resulted in the organization of four YTTP sessions on Smallholder Agriculture with Senegalese learners and one YTTP session on Civic Engagement with Angolan learners, the first ones to be organized in bilateral arrangements, scheduled for 2021 and 2022 (UNDP-IBRAF 2020). During the Fourth International Cassava Conference (GCP21-IV), held in Benin in 2018, the young woman from Benin, *Paula Gnancadja*, a Bachelor in Management of Agricultural and Rural Enterprises, who participated in the YTTP training the previous year, presented the work "A Young Cassava Farmer and Processor Building a New Start-Up in Benin," reporting the success of her start-up (ABB.Sarl), after returning from Brazil, introducing new cassava processing products, which are being very well accepted by Benin consumers. She received the GCP21-IV Best Poster Award in the "entrepreneurship" category.

The technologies that received the greatest interest in the trainings include those products made from starch (*polvilhos*), such as beijus, tapiocas, and biscuits (avoador), and the minimal processing technique as a means for storing roots and using it later as raw material for chips, cakes, and dough for snacks.

One of the great challenges for the implementation of training for Africans is the great cultural and language diversity, coupled with the fact that the majority of Brazilian instructors speak only Portuguese. At the three YTTP courses, most of the African attendees spoke either English or French, while the Brazilian instructors spoke Portuguese, which create some communication difficulties.

The fact that Brazil is the center of origin and domestication of cassava explains the great importance of this crop, as well as the enormous diversity of its use and processing in different regions of Brazil. This time lag and the great genetic diversity of cassava made Brazil a major source of technologies for processing, which would later be mechanized and scaled for industry. Currently, the African cassava-producing countries are investing heavily in both sectors (family and industrial) to develop their own new technologies and/or to adapt or improve Brazilian innovations according to their demands. These efforts are welcome for government development programs, because of the unquestionable importance of cassava to reduce hunger and increase the income of cassava producers and processors. Therefore, we believe that the next cassava technologies transfer initiatives between Brazil and Africa should be extended to technologies, services, and products aimed at large-scale cassava processing and cultivation.

Acknowledgments The authors are grateful for the invitation and support of the CGIAR Research Program on Roots, Tubers and Bananas (RTB) in carrying out this chapter. We also thank the agencies that funded the courses reported here, such as Japan International Cooperation Agency (JICA), International Fund for Agricultural Development (IFAD), and African Development Bank (AfDB), as well as researchers and technicians from Embrapa and collaborators, who acted as training instructors.

References

AfDB (2017) African Development Bank. Brazil to groom African youth in cassava processing. https://www.afdb.org/en/news-and-events/afdb-brazil-to-groom-african-youth-in-cassava-processing-17348. Accessed 5 Jan 2021

Alves-Pereira A, Clement CR, Picanço-Rodrigues D, Veasey EA, Dequigiovanni G, Ramos SLF, Pinheiro JB, Zucchi MI (2018) Patterns of nuclear and chloroplast genetic diversity and structure of manioc along major Brazilian Amazonian rivers. Ann Bot 121:625–639

Araujo FCB, Moura EF, Cunha RL, Farias Neto JT, Silva RS (2019) Chemical root traits differentiate 'bitter' and 'sweet' cassava accessions from the Amazon. Crop Breed Appl Biotechnol 19:77–85

Bechoff A, Tomlins KI, Chijioke U, Ilona P, Bennett B, Westby A, Boy E (2019) Variability in traditional processing of gari: a major food security product from cassava. Food Chain 8:39–57

Falade KO, Akingbala JO (2011) Utilization of cassava for food. Food Rev Intl 27:51–83

FAOSTAT (2018) Food and Agricultural Organization of the United Nations Online Statistics Database. http://www.fao.org/faostat/en/#data/FBS. Accessed 4 Jan 2021

FAOSTAT (2019) Food and Agricultural Organization of the United Nations Online Statistics Database. http://www.fao.org/faostat/en/#data/QC. Accessed 4 Jan 2021

Ferreira Filho JR, Silveira HF, Macedo JJG, Lima MB, Cardoso CEL (2013) Cultivo, processamento e uso da mandioca – Instruções práticas. *Embrapa Mandioca e Fruticultura*, Cruz das Almas-BA, 32p

Folegatti MIS, Matsuura FCAU, Ferreira Filho JR (2005) A indústria da farinha de mandioca. In: Souza LS, Farias ARN, Mattos PLP, Fukuda WMG (eds) Processamento e utilização da mandioca. Embrapa Mandioca e Fruticultura Tropical, Cruz das Almas-BA, pp 61–141

Montagnac JA, Davis CR, Tanumihardjo SA (2009) Processing techniques to reduce toxicity and antinutrients of cassava for use as a staple food. Compr Rev Food Sci Food Saf 8:17–27

Morante N, Sánchez T, Ceballos H, Calle F, Pérez JC, Egesi C, Cuambe CE, Escobar AF, Ortiz D, Chávez AL, Fregene M (2010) Tolerance to postharvest physiological deterioration in cassava roots. Crop Sci 50:1333

Motta JS (2012) Beijus coloridos preparados com frutas e hortaliças. Embrapa Mandioca e Fruticultura, Cruz das Almas-BA, 2 p

Ndjouenkeu R, Kegah FN, Teeken B, Okoye B, Madu T, Olaosebikan OD, Chijioke U, Bello A, Osunbade AO, Owoade D, Takam-Tchuente NH, Njeufa EB, Chomdom ILN, Forsythe L, Maziya-Dixon B, Fliedel G (2021) From cassava to gari: mapping of quality characteristics and end-user preferences in Cameroon and Nigeria. Int J Food Sci Technol 56:1223–1238

NRI (2017) Natural Resources Institute. https://www.nri.org/latest/news/2017/nri-signs-cooperation-agreement-with-embrapa-brazilian-agricultural-research-corporation. Accessed 8 Nov 2020

Oliveira LA, Motta JS, Jesus JL, Sasaki FFC, Viana ES (2020a) Processing of sweet and bitter cassava. Embrapa Mandioca e Fruticultura, Cruz das Almas-BA, 64 p

Oliveira LA, Reis RC, Viana ES, Santos JF, Souza VS, Assis JLJ, Sasaki FFC, Santos VS (2020b) Obtenção de raízes de mandioca de mesa congelada. Embrapa Mandioca e Fruticultura, Cruz das Almas-BA. 8p. (*Embrapa Mandioca e Fruticultura*. Comunicado Técnico, 177)

Oliveira LA, Godoy RCB (2011) Mandioca chips. Embrapa Mandioca e Fruticultura, Cruz das Almas-BA. 6p. (*Embrapa Mandioca e Fruticultura*. Circular Técnica, 101)

Sampaio VS, Menezes SSM (2020) A produção artesanal de biscoitos em Vitória da Conquista-Bahia. Caderno Prudentino de Geografia 42:79–97

Sartas M, Schut M, Proietti C, Thiele G, Leeuwis C (2020) Scaling readiness: science and practice of an approach to enhance impact of research for development. Agric Syst 183:102874, 12 pp

SENAR (2018) Serviço Nacional de Aprendizagem Rural. Agroindústria: produção de derivados da mandioca. Coleção SENAR – 214. Brasília-DF: SENAR, 72 pp

Tomlins K, Bennett B (2017) New uses and processes for cassava. In: Hershey CH (ed) Achieving sustainable cultivation of cassava volume 1: cultivation techniques. Burleigh Dodds Science, Cambridge UK, pp 89–98

UNDP-IBRAF (2020) United Nations Development Programme – Brazil Africa Institute. Good practices on south-south cooperation in the context of food security. UNDP, New York. 90 pp

Viana ES, Oliveira LA, Silva J (2010) Processamento mínimo de mandioca. Embrapa Mandioca e Fruticultura, Cruz das Almas-BA. 4p. (*Embrapa Mandioca e Fruticultura*. Circular Técnica, 95)

Chapter 8
Improving Safety of Cassava Products

Linley Chiwona-Karltun, Leon Brimer, and Jose Jackson

Abstract Cassava was domesticated in the Amazon Basin, where Native Americans selected many bitter varieties, and devised methods for detoxifying them. Cassava reached Africa in the sixteenth century, where rural people soon learned to remove the cyanogenic toxins, e.g., by drying and fermenting the roots. Processing cassava to remove the cyanogenic toxins including the cyanide formed during the processing is time consuming. The work is often done by women, while women and men often prefer bitter cassava varieties for social reasons and superior taste and color. In spite of deep, local knowledge of safe processing, traditional foods made with contaminated water may contain bacterial and fungal pathogens. Improper storage may encourage mycotoxins, such as aflatoxin. Recent advances in industrial processing are developing foods that are free of toxins and microbial contamination. Processing and selling cassava leaves is an emerging but fast-growing sector. Cassava leaves also contain cyanogenic toxins normally in higher concentrations than the cassava roots. In the future, more attention must be paid to the safe processing of cassava leaves and roots, especially as food processing becomes increasingly industrialized worldwide.

L. Chiwona-Karltun (✉)
Department of Urban & Rural Development, Swedish University of Agricultural Sciences, Uppsala, Sweden
e-mail: Linley.chiwona.karltun@slu.se

L. Brimer
Department of Veterinary and Animal Sciences University of Copenhagen, Frederiksberg, Denmark

J. Jackson
Alliance for African Partnership, International Studies and Programs, Michigan State University, East Lansing, MI, USA
e-mail: jacks184@msu.edu

© The Author(s) 2022

G. Thiele et al. (eds.), *Root, Tuber and Banana Food System Innovations*,
https://doi.org/10.1007/978-3-030-92022-7_8

8.1   Improving Safety of Cassava Products in Regional Cassava Production and Processing

The safety of a raw material or a processed food product can be classed into physical, microbiological, and chemical safety. Physical safety, such as sharp particles or contaminants in food, is not discussed here; the focus is on microbiological and chemical threats. The last decade has witnessed a rise in the quality of research and development on the management of pathogenic bacteria and mycotoxin-producing fungi in cassava products. Several research-funding agencies, at times in partnership with the CGIAR programs, have supported much of this development, discussed in this chapter, particularly regarding cassava food safety and human health.

A persistent challenge is how research and education integrate traditional indigenous knowledge on food toxicity, processing, preparation, and food safety into modern research on food technology. Many curricula fail to teach the role of traditional indigenous knowledge, science and technology, and their implications for Africa's food systems. Historically, cassava food safety evolved in the context of practices originating from cassava's region of origin in lowland South America, and elsewhere where the crop was taken since the sixteenth century often linked to slavery and trade. These interactions have laid the foundations for the repertoire of skills in processing, preparation, preferences for taste, cassava products, potential for modification, and consumer preferences.

The chemical safety of a plant-derived food may be affected by the environmental occurrence of toxic anthropogenic compounds. Furthermore, chemical safety is affected by toxic elements concentrated by the plant, e.g., selenium and cadmium, toxic secondary organic metabolites formed by the plant, and toxic organic compounds formed during food processing such as acrylamide.

This chapter focuses on the safety of foods derived from cassava roots and leaves, progress in research and recent consumer legislation worldwide. The chapter also addresses the toxic compounds the plant forms, cyanogenic toxins in the form of cyanogenic glycosides and their products of degradation, as well on aflatoxins that are carcinogenic mycotoxins. The chapter ends with a section on future needs and policy recommendations with emphasis on the opportunities emerging with the African Continental Free Trade Area, particularly for women.

8.2   Introduction

8.2.1   Traditional Indigenous Knowledge in Cassava

According to the World Intellectual Property Organization, "Traditional knowledge (TK) is knowledge, know-how, skills and practices that are developed, sustained and passed on from generation to generation within a community, often forming part of its cultural or spiritual identity" (WIPO 2021). Nowhere is this passing of traditional knowledge so closely interlinked as in the domestication and utilization of cassava as a food in the diet. Cassava is one of about 100 species of the genus *Manihot* that also includes several rubber-producing plants (Schery 1947). This genus is a member of the Euphorbiaceae family, and *Manihot esculenta* is the only plant in that family that produces tuberous roots that have led to its domestication (Schery 1947). Cassava was domesticated in the Amazon region purposively selected for its fleshy roots and wide range in color (white to yellow), starch and taste, particularly based on sugar content (Schaal et al. 2006). Amazonian communities have long interacted with cassava as evidenced by their diverse uses of its roots and leaves (Schwerin 1971). Unlike most other places that have adopted cassava, the whole cassava plant is used, not just the roots (Schaal et al. 2006). During domestication, cassava landraces with high and low levels of cyanogenic glycosides appeared. Toxic types were associated with lower population density and shifting agriculture with more isolated gardens where cyanogens provided protection against damage by animals. The Amazonian peoples developed sophisticated technology to detoxify the bitter types for consumption (McKey et al. 2010). Amazonian peoples have a deep understanding of cassava and of its uses (Carvalho et al. 2004). For example, sugary landraces were used for making natural sweetened drinks; amylose-free starch varieties were fed to young children as this was easier on the stomach; deep yellow colored cassava roots were used for making the traditional drink *tucupi*. Native Amazonian Tukanoan and Jivaroan peoples used cassava leaves as a vegetable, grated and boiled (Dufour 1994). Scientific publications, especially earlier ones, depict cassava as a food that is poor in nutrients, which should be discouraged as food for children. Scientists have learned and still can learn from indigenous communities with deep knowledge of the preparation and consumption of cassava. Such studies demand innovative methods, interdisciplinary teams combining scientists, practitioners, feminist food scholars, and those with a long history of farming and using cassava in Africa as well as worldwide.

As cassava became more and more domesticated and populations moved in search of more land or better opportunities, these agrarian-based families often carried the local and favorite cassava cuttings with them. The Carib- and Arawak-speaking peoples who migrated from South America took cassava with them to the Caribbean (Cousins 1903). For cassava to be readily adopted in Africa, there may have been some existing knowledge of processing toxic plants, such as of processing toxic yams on the west coast of Africa (Ohadike 1981). Because the slave ships often carried cassava as a provision during their travels between the New World and Africa, some processing methods like that of *gari* were brought over to Africa and adapted locally (Northrup and McCammon 1980). In her book on eating wild plants safely, Williams provides ten basic rules about how to enjoy the edible ones without endangering oneself with the poisonous ones. Rule #6 and rule #10 are particularly useful and remain relevant in light of the ensuing section on the chemical food safety of cassava products.

Box 8.1 Basic Rules for Eating Wild Plants Safely
Rule #6
"Know what part of the plant is edible and when it is edible."
Rule #10
"Know the common poisonous plants of the area." A good question to ask when learning any new plant is, "Is there anything poisonous that looks like this?"
Source: Kim Williams 1977. Eating Wild Plants. P 3–5.

8.3 Chemical Safety of Cassava Food Products – The Story About Cyanogens

Of the approximately 374,000 plant species currently described (Christenhusz and Byng 2016), at least 2500 contain cyanogenic glycosides (Zagrobelny et al. 2008). These compounds have important roles in plant defense due to their bitter taste and to their release of the highly toxic compound hydrogen cyanide (HCN) upon plant tissue disruption. Recent research points to additional roles as storage compounds of reduced nitrogen and sugar (Zagrobelny et al. 2008).

Cassava is one of a few plants that contain potentially toxic cyanogenic glycosides, yet are still used as food. Adolf Nahrstedt (1993) reviewed cyanogenesis and food plants, including cassava, lima bean, sorghum, bamboo, flax, bitter almond, apricot, cherry, and laurel cherry. An evaluation of the health risks of cyanogenic glycosides in foods (EFSA 2019) includes the seven mentioned on the Nahrstedt's list and no additional species. The roots and leaves of fresh cassava release hydrogen cyanide (HCN) during chewing and processing and storage. More than 291 million tons of cassava were produced worldwide in 2017, of which Africa accounted for over 60%. In 2017, Nigeria produced 59 million tons, making it the world's largest producer (about 20% of global production) with a 37% increase in the last decade (FAO 2017). The present status concerning production of safe food from cassava worldwide is discussed below.

8.4 Microbial Safety of Cassava Products (Bacteria, Fungi, and Mycotoxins)

Interest in the microflora associated with cassava roots and foods made from them started around the mid-1980s. Many of the root-derived foods are fermented, such as the sun-dried sour starch, which is mainly produced in Colombia and in Brazil.

Table 8.1 Microflora of cassava flour processed by domestic heap fermentation

Microflora	Mean [10]LogN/g (6 samples)
Total aerobic mesophiles	8.45
Lactobacillus spp.	7.50
Micrococcaceae	7.30
Pseudomonas spp.	6.60
Enterobacteriaceae	6.53
Bacterial endospores	6.30
Coliform bacteria	6.28
Escherichia coli	4.00
Bacillus cereus	<2.7
Clostridium perfringens	<1.7
Staphylococcus aureus	<1.7
Salmonella	Absent in 25 g
Yeasts	6.62
Filamentous fungi	5.00

Source: Based on Essers et al. (1995a)

The production of this and other fermented cassava root products involves a lactic acid fermentation and a decrease in pH down to around 3.5. However, the quality varies widely (Morlon-Guyot et al. 1998).

Studies of the fungal flora show that the mucoraceous fungus *Rhizopus oryzae*, associated with the postharvest spoilage of the roots, effectively degrades linamarin (a type of cyanogenic glucoside) (Padmaja and Balagopal 1985). By the end of the 1980s, certain cassava root products, such as flours, were included in Brazilian studies on the occurrence of mycotoxins in foods (Soarez and Rodrigues-Amaya 1989).

During the 1990s, studies began to report on the total microflora, i.e., species of bacteria, yeasts, and filamentous fungi (Essers et al. 1995a, b, c). In the study entitled "Reducing Cassava Toxicity by Heap-Fermentation in Uganda," Essers et al. (1995a) identified and quantified (10LogN/g flour) a number of species of bacteria, some bacterial groups, total yeasts, and total filamentous fungi in flour made from cassava roots that had undergone domestic heap fermentation (Table 8.1).

Fourteen filamentous fungi were identified to species (Table 8.2). Extracts of 25 flours and one sample of dried scraped-off molds were analyzed for mutagenicity and cytotoxicity using the Ames test with the tester strains *Salmonella typhimurium* TA100 and TA98, with and without the addition of S9 liver homogenate, and also for the presence of any aflatoxins. The aflatoxins were analyzed by High Performance Thin Layer Chromatography (HPTLC) with a detection limit of 2 microgram per kilogram (Essers et al. 1995a).

The authors of the above study concluded that except for *Neurospora sitophila* and *Geotrichum candidum*, all other fungi identified (Table 8.2) were able to produce mycotoxins under cultured conditions. The high number of *Escherichia coli* in

Table 8.2 Filamentous fungi occurring in cassava flour processed by domestic heap-fermentation

Fungi	Number of samples in which detected ($n = 6$)
Aspergillus fumigatus	3
Aspergillus niger	1
Aspergillus oryzae	4
Aspergillus parasiticus	1
Fusarium sporotrichioides	1
Geotrichum candidum	6+
Mucor circinelloides	2
Mucor racemosus	3
Neurospora sitophila	6+
Neurospora crassa	1
Penicillium frequentans	1
Penicillium waksmanii	1
Rhizopus oryzae	5+
Rhizopus stolonifer	4

Note: + Numerous in each positive sample
Source: Based on Essers et al. (1995a)

one sample was linked to unhygienic processing, while the spore-forming *Clostridium perfringens* and *Bacillus cereus*, found sporadically, might produce toxins after cooking ugali, a maize dish popular in Tanzania and Kenya, if the grain had been stored for a long time. However, finding no cytotoxicity, mutagenicity, nor any aflatoxins, it was concluded that such mycotoxins were not formed in quantities that are detrimental to public health (Essers et al. 1995a).

The study by Essers et al. (1995a) points to the threat of pathogenic bacteria from contaminated water. A study in Nigeria found that the bacterial and coliform counts of fermented cassava flour (kpor umilin) were reduced considerably if the product was made with potable water and additional washing (Tsav-wua et al. 2004; Inyang et al. 2006).

8.4.1   The Development After Essers et al.

From 2007 to 2019, research on the safety of cassava food products suddenly increased. Thirteen articles are summarized in Table 8.3.

Table 8.3 Summary of articles on microbiological food safety for cassava products (2007–2019)

Year	Country	Cassava product(s)	Potentially hazardous microorganism identified (y/n/NA)	+Mycotoxin(s) identified (y/n/NA)	Recommendations/ conclusions given (y/n)	++ Ref. number
2007	Nigeria	Gari, lafun, ogiri	Y	NA	N	1
2008	Benin	Cassava chips	Y	N	N	2
2008	Nigeria	Fufu	Y	NA	Y (HACCP strategy recommended)	3
2009	Nigeria	Lafun, fufu, gari	N	NA	N	4
2010	Uganda	Cassava chips	Y	Y (aflatoxins)	Y (drying methods recommended)	5
2010	Nigeria	Wet fufu	Y	NA	Y (GMP recommended)	6
2012	Nigeria	Gari	NA (only total counts analyzed)	NA	Y (type of packaging material recommended)	7
2012	Benin	Cassava chips	Y	N	Y (maize and groundnuts are higher in aflatoxins)	8
2013	Nigeria	Fermented cassava flour "lafun"	Y	NA	Y (improved water quality and health inspections during production)	9
2014	Nigeria	Gari	NA (only total counts analyzed)	NA	Y (keep moisture low by proper packaging)	10
2016	Kenya	Cassava chips, flour	Y (heavy contamination with coliforms and for one sample with *E. coli*)	NA	Y (proper training of producers)	11
2017	Nigeria	9 products ++	NA	Y (aflatoxins and fumonisins)	Y (mycotoxin levels are low)	12
2019	Nigeria	Cassava root (white and yellow) fermented 4 days	Y/N (*E. coli* found after 48 hours of fermentation, but not after 72 and 96 hours)	NA	N	13

(continued)

Table 8.3 (continued)

+NA (not analyzed for)

++References by numbers: 1, Ljabadeniyi (2007); 2, Gnonlonfin et al. (2008); 3, Obadina et al. (2008); 4, Obadina et al. (2009); 5, Kaaya and Eboku (2010); 6, Obabina et al. (2010); 7, Adejumo and Raji (2012); 8, Gnonlonfin et al. (2012a, b); 9, Adebayo-Oyetoro et al. (2013); 10, Olopade et al. (2014); 11, Gacheru et al. (2016); 12, Abass et al. (2017); 13, Obi and Ugwu (2019); 14, Odu and Maduka (2019)

+++Cassava starch, HQCF, lafun, fufu flour, tapioca, fine yellow gari, fine white gari, yellow kpo-kpo gari, white kpo-kpo gari

Note: This microbiological research dealt with the most analyzed commercialized products bought at selected markets or supermarkets, while a few of the studies included products produced by the scientists themselves in the laboratory. The analytical methods may have different limits of detection or of quantification for mycotoxins

8.4.2 The Developments Made Possible

One pleasant surprise from this research was the absence or exceptionally low level of aflatoxins in some of the studies, some of which even failed to find the fungus *Aspergillus flavus*, which causes aflatoxins (Abass et al. 2017; Kaaya and Eboku 2010; Gnonlonfin et al. 2012a, b), especially considering that these studies did use advanced analytical methods. During this period, several manuals were published, such as *Potential for Commercial Production and Marketing of Cassava: Experiences from the Small-Scale Cassava Processing Project in East and Southern Africa* (Abass et al. 2013), *Quality Management Manual for Production of High Quality Cassava Flour* (Dziedzoave et al. 2006), and *Quality Management Manual for the Production of Gari* (Abass et al. 2012), suggesting improving conditions for developing commercial cassava food products on a larger scale. Such more advanced business developments even can find further support by the fact that an expert group with representatives from (1) University of Natural Resources and Life Sciences in Vienna, Austria, (2) Kwara State University in Nigeria, (3) International Fund for Agricultural Development, and (4) IITA in 2017 published the extremely important paper "Assessment of the Potential Industrial Applications of Commercial Dried Cassava Products in Nigeria" (Awoyale et al. 2017). In this paper, they summarized seven functional properties and seven pasting properties for 13 cassava product groups, based on samples collected from cassava processors and marketers.

8.4.3 The Cyanogenesis of Cassava

The major cyanogenic compound in cassava is linamarin. As it degrades, it releases the toxic HCN, through a two step process. That is the reactions with a glucosidase (linamarase) followed by a reaction with a second enzyme (hydroxynitrile lyase). Further in-depth details are shown in the risk assessment by the European Food Safety Authority (EFSA 2019, p. 11). The reactions start when the cyanogenic glucoside and the enzymes are brought in contact through rupture of the plant tissues, in which the three components have been stored separately (Behera and Ray 2016). Intact cassava roots may release less than 10 mg HCN/kg fresh tissue or as much as 1100 (EFSA 2019).

In 2003, Codex Alimentarius Commission released the Codex Standard for Sweet Cassava (CODEX STAN 238-2003), which has been amended three times, i.e., 2005, 2011, and 2013. This standard among others regulates the maximum content of cyanogenic glucosides.

> **Box 8.2 Codex Standard for Sweet Cassava (CODEX STAN 238-2003)**
> "Sweet varieties of cassava are those that contain less than 50 mg/kg hydrogen cyanide (fresh weight basis). In any case, cassava must be peeled and fully cooked before being consumed."

In 2010, the 238-standard was followed by another standard, this time for bitter cassava root.

> **Box 8.3 CODEX STANDARD FOR BITTER CASSAVA1 (CODEX STAN 300-2010)**
> "This Standard applies to commercial bitter varieties of cassava roots grown from *Manihot esculenta* Crantz, of the Euphorbiaceae family, to be supplied fresh to the consumer, after preparation and packaging. Cassava for industrial processing is excluded. Bitter varieties of cassava are those containing more than 50 mg/kg of cyanides expressed as hydrogen cyanide (fresh weight basis). In addition to the requirements of the General Standard for the Labelling of Prepackaged Foods (CODEX STAN 1-1985), the following specific provisions apply: Each package shall be labelled as to the name of the produce and type (bitter) and may be labelled as to the name of the variety. A statement indicating the following is required: cassava must not be eaten raw; cassava shall be peeled, de-pithed, cut into pieces, rinsed and fully cooked before consumption; and cooking or rinsing water must not be consumed or used for other food preparation purposes."

While the commercial marketing of bitter cassava according to the CAC standard (CODEX STAN 300-2010) is limited, studies in Denmark (Kolind-Hansen and Brimer 2010), Ireland (O'Brien et al. 2013), and Australia (Burns et al. 2012) demonstrated many cassava roots sold as sweet were in fact bitter. Cassava safety becomes important as African migrants, 22 million in Europe alone, seek to consume foods from their home continent.

The toxicity of a cyanogenic food plant such as cassava has to do with the amount of HCN and cyanohydrins formed before or while eating. This is because the intact glycosides (e.g., linamarin) are stable in the digestive tract, but the cyanohydrins will split into HCN and its constituent aldehyde/ketone. The non-hydrolyzed part of the linamarin is absorbed and excreted unchanged in the urine (Brimer and Rosling 1993). The primary mode of action for acute toxicity of HCN is the inhibition of the cell's oxidative phosphorylation. This leads to anaerobic energy production. Due to their high oxygen and energy demand, the brain and heart are among the most sensitive organs. Thus, cyanide can result in hypoxia, metabolic acidosis, and impairment of several vital functions. The CONTAM Panel of EFSA in 2019 concluded that there are no data indicating what the acute reference dose (ARfD) for cyanide of 20 µg/kg bw, established in 2016, should be (EFSA 2019). Certain neurological disorders, e.g., spastic paraparesis (konzo) and tropical ataxic neuropathy, have been associated with chronic dietary exposure to cyanide. However, this has always been in populations where cassava was the main food source. In areas with low iodine intake, chronic cyanide exposure from cassava has also been associated with hypothyroidism and goiter. In 2012, the JECFA concluded that the epidemiological association between cassava consumption and konzo was consistent, even though the etiological mechanism of konzo is still unknown (cited in EFSA 2019).

Clearly, the content of cyanogenic glycosides in raw cassava roots and the effects of food processing on these toxins are of major importance. Below, we will take a further look.

8.4.4 Traditional Processing: Reasons and the Products

According to Olsen and Schaal (1999), wild populations of *M. esculenta* occur primarily in west central Brazil and eastern Peru. This region is also the home for Native Americans, including the Tukanoans, where Dufour studied 14 cassava varieties and their cyanogenic potentials (Dufour 1988). The mean potentials varied from around 310–561 mg HCN equivalents/kg f.w. for the *Kii* (= toxic cultivars) and down to around 170 for *Makasera* (nontoxic/safe), the only nontoxic cultivar grown. The toxic varieties were preferred for growing (Dufour 1988). These values can be compared with the accepted highest content of cyanogenic glycosides in sweet cassava roots as accepted by the CAC standard, i.e., 50.

Ten years later after Dufour's publication, a study by Chiwona-Karltun et al. (1998) found similar results in Malawi, where cassava had been brought by the Portuguese starting in the 1500s (Leestma 2015). The study found that bitter cultivars were often preferred, because of experience and traditional indigenous knowledge in processing, managing theft, and most of all ensuring food security (Chiwona-Karltun 2001; Chiwona-Karltun et al. 2000). Toxicity was not seen as a problem, although such roots needed processing for detoxification (Chiwona-Karltun et al. 2000) and the women were extremely adept at correlating bitter taste with toxicity (Chiwona-Karltun et al. 2004). In 1986, Fresco wrote: "throughout Africa, bitter varieties are much more common than sweet ones, notwithstanding the fact that sweet varieties have been promoted by some colonial administrations out of concern with the toxicity of bitter varieties. There is no satisfactory explanation in the literature for the predominance of bitter varieties" (Fresco 1986, p. 153). From the work in Malawi and studies elsewhere, it is clear that bitter varieties where they constitute a major part of the diet are socially preferred, notwithstanding the amount of work that is required and mostly undertaken by women, to process and render them safe for consumption.

8.5 Gender-Preferred Traits and Traditional Indigenous Foods

Now we understand that a considerable part of cassava roots grown worldwide needs to be processed in order to remove a major part of the cyanogenic glycosides present, thereby making a safe product, and in addition obtain one of several kinds of foods. Long processing times leave women with little time for other activities (Nweke 1999). This could have profound effects on the dietary diversity of a household as observed in the Amazon Basin (Dufour 1994) where women spent much time processing cassava. Processing is essential to reduce the water content and preserve the very perishable roots in a way they can be more easily transported.

Processing is also related to gendered variety and trait preference. In the works published by Chiwona et al. in (1998), the reasons for the preferences for bitter varieties were linked with experience and possession of traditional indigenous knowledge in processing, managing social changes in the society such as theft, and most of all ensuring food security. The studies also revealed that bitter cassava cultivars had certain traits that provided products that were deemed to be of superior quality, e.g., processed cassava flour is perceived to be finer and whiter in the making of the staple dish *kondowole* (Chiwona-Karltun et al. 1998). Cassava leaf foods from bitter varieties are regarded as tastier, also verified in Zambia (Chiwona-Karltun et al. 2015). Most importantly, bitter cassava reduced the fear of being without food, especially for women that were "vulnerable in society." These women were usually in female-headed households, and they needed to protect their food sources based on the deep knowledge of food processing and preparation. Incorporating gender-preferred traits continues to be a challenge in breeding programs (Weltzien et al. 2019) especially when these priorities compete with urban demand for food, changing diets and preferences for processed foods (Raheem & Chukwuma 2001).

It is worthwhile reflecting on how to preserve traditional food knowledge while understanding these preferred cassava traits from a gendered perspective. For example, in Jamaica, the popularity of *bammy bread* (Fig. 8.1), a preferred traditional food has been preserved largely through the women to women transfer of knowledge and expertise. By way of the kitchen and cuisine identity, bammy bread has successfully been incorporated in modern bakeries and extended to beyond export markets (Canevari-Luzardo et al. 2020). As climate change dampened the production and availability of cassava, Jamaica stopped exporting bammy and only served the national market. This calls for paying closer attention to the roles that women and men play in cassava value chains (Andersson et al. 2016), including how specific products impact women and men and their well-being. As traditional foods are

Fig. 8.1 Traditional fried
bammy bread. Photo credit
Sam Biehl

Fig. 8.2 Mortar used for pounding cassava leaves (left) (Photo credit Brian LeniKanjeri Msukwa, Malawi). Preferred locally pounded cassava leaves being cooked (right). (Photo credit D. Dufour (CIRAD))

increasingly appreciated for their contribution to health and well-being, demand for locally made and processed foods will gain more traction.

Cassava leaves also play important roles. Fresh or dried cassava leaves are rarely sold on the market, largely due to the focus on cassava roots as an energy provider. But now cassava leaf production and marketing are well established in the Democratic Republic of the Congo, Rwanda, Angola, and Malawi to mention a few examples and have even reached, e.g., European markets in frozen form as sold in special shops. The technology for processing, packaging, and marketing remains a challenge. To render cassava leaves safe to eat, they must be disintegrated in order for the endogenous enzyme to initiate the breakdown of the cyanogenic glucosides (Bokanga 1994). Pounding for at least 15–20 minutes is crucial. Few affordable leaf-processing machines are available (Fig. 8.2).

8.5.1 The Traditional Processing Reasons and the Products

Several cassava root products were developed in the center or origin of the crop. The manual *Cassava Products of Brazil* lists starch, sour starch, *puba* flour (similar to African *fufu*), fermented roasted flour (similar to gari), *tucupi* (fermented savory syrup), tiquira, unfermented gari (*farinha de mandioca*), and cassava crisps (Crenn 2019). Various scientific articles and book chapters published after 2010 have described the different unit processes and their impact on the composition of cassava root-based food products (Gnonlonfin and Brimer 2013; Gnonlonfin et al. 2012a, b). Important processes include drying, soaking, cooking (baking, frying, steaming), size reduction (grating), fermentation (dry in heap or wet), and smoking (to make the smoked balls called *kumkum*).

Fermentation can be used to make cassava products that are free of cyanogenic toxins, as discussed in the following section.

8.5.2 The Progress in Knowledge and Technology

Gari is the most well-known fermented cassava food product found throughout much of West Africa. Today, it comes in several professional and internationally marketed brands. Yet in many places, gari is still produced locally to eat at home or to sell. The CGIAR organizations have been involved in training and research on gari.

Starting with the manual *Quality Management Manual for the Production of Gari* (Abbas 2012) and following up with workshops like "Nigeria's Gari Revolution: Improving Efficiency and Equity of a Staple Food" (organized by Global Cassava Partnership for the 21st Century in 2016), this important fermented product is becoming professionalized. However, depending on the naturally occurring microorganisms for its fermentation, gari and other fermented cassava products still have room for improvement. The possibilities include (1) the selection and characterization of microorganisms found in safe products and (2) microorganisms which upon testing may be candidates for more complete degradation of the cyanogenic glycosides in cassava roots.

Penido et al. (2018) selected and characterized microorganisms (lactic acid bacteria and yeasts) originating from the production of Brazilian sour cassava starch, while Kostinek et al. (2007) isolated 375 predominantly lactic acid bacteria from fermenting cassava in different African countries. *Lactobacillus plantarum* isolated from fermenting cassava can hydrolyze (degrade) cyanogenic glycosides (e.g., linamarin), as can different *Bacillus subtilis* isolates collected from low pH starters used to produce *gergoush* (a Sudanese fermented snack) (Lei et al. 1999; Abban et al. 2013). See also the chapter by Behera and Ray (2016) on "Microbial Linamarase in Cassava Fermentation."

Fig. 8.3 Grating sweet cassava by hand (left), (Photo credit Joyce NaMtawali Mwagomba, Malawi) and sun drying the grated cassava on mats to produce high-quality cassava flour (HQCF, right). (Photo credit D. Dufour (CIRAD))

8.6   Conclusions and Looking Forward

Traditional indigenous cassava-based food products have been developing greater food safety. Worldwide, more and more popular drinks like *bubble* or *boba* tea with tapioca "pearls" have been developed with very low levels of cyanogens (Bulathgama et al. 2020). New food products based on cassava must focus on these toxins to develop safe products.

As more women venture into value addition especially at local scale and hoping to capitalize from the African Continental Free Trade Area and the opportunity it may accord women, we need to enable women's participation. However, there are some immediate challenges for women and men, such as lack of access and skills to resources to scale-up production. For example, large-scale processing of high-quality cassava flour or cassava flour and cassava leaves is still rudimentary. Packaging and meeting standards for export requires training and investments that often are beyond the reach of women or small-scale farmers (Fig. 8.3).

References

Abass AB, Dziedzoave NT, Alenkhe BE, James BD (2012) Quality management manual for the production of gari. IITA, Ibadan, Nigeria.

Abass A, Mlingi N, Ranaivoson R, Zulu M, Mukuka I, Abele S, Bachwenkizi B, Cromme N (2013) Potential for commercial production and marketing of cassava: experiences from the small-scale cassava processing project in East and Southern Africa. IITA, Ibadan, pp 13–88

Abass AB, Awoyale W, Sulyok M, Alamu EO (2017) Occurrence of regulated mycotoxins and other microbial metabolites in dried cassava products from Nigeria. Toxins 9(7):207. https://doi.org/10.3390/toxins9070207

Abban S, Brimer L, Abdelgadir WS, Jacobsen M, Thorsen L (2013) Screening for Bacillus subtilis group isolates that degrade cyanogens at pH 4.5–5.0. Int J Food Microbiol 161:31–35

Adebayo-Oyetoro AO, Oyewole OB, Obadina AO, Omemu MA (2013) Microbiological safety assessment of fermented cassava flour "Lafun" available in Ogun and Oyo states of Nigeria. Int I Food Sci 213: Article ID 845324, 5 pages. https://doi.org/10.1155/2013/845324

Adejumo BA, Raji AO (2012) Microbial safety and sensory attributes of gari in selected packaging materials. Acad Res Int 3(3):153–162

Andersson K, Lodin JB, Chiwona-Karltun L (2016) Gender dynamics in cassava leaves value chains: the case of Tanzania. J Gender Agric Food Security 1(2):84–109

Awoyale W, Abass AB, Ndavi M, Maziya-Dixon B, Sulyok M (2017) Assessment of the potential industrial applications of commercial dried cassava products in Nigeria. Food Measure 11:598–609

Behera SS, Ray RC (2016) Microbial linamarase in cassava fermentation. In: Ray RC, Rossel CM (eds) Microbial enzyme technology in food applications. Food biology series. CRC Press/Taylor & Frances Group, Boca Raton

Bokanga M (1994) Processing of cassava leaves for human consumption. Acta Hortic 375:203–208

Brimer L, Rosling H (1993) Microdiffusion method with solid state detection of cyanogenic glycosides from cassava in human urine. Food Chem Toxicol 31(8):599–603

Bulathgama BEAU, Gunasekara GDM, Wickramasinghe I, Somendrika MAD (2020) Development of commercial tapioca pearls used in bubble tea by microwave heat–moisture treatment in cassava starch modification. EJERS (Eur J Eng Res Sci) 5(1):103–106. https://doi.org/10.24018/ejers.2020.5.1.1455

Burns AE, Bradbury JH, Cavagnaro TR, Gleadow RM (2012) Total cyanide content of cassava food products in Australia. J Food Compos Anal 25(1):79–82

Canevari-Luzardo LM, Berkhout F, Pelling M (2020) A relational view of climate adaptation in the private sector: how do value chain interactions shape business perceptions of climate risk and adaptive behaviours? Bus Strateg Environ 29(2):432–444

Carvalho LJCB, de Souza CRB, de Mattos Cascardo JC, Junior CB, Campos L (2004) Identification and characterization of a novel cassava (Manihot esculenta Crantz) clone with high free sugar content and novel starch. Plant Mol Biol 56:643–659

Chiwona-Karltun L (2001) A reason to be bitter : cassava classification from the farmers' perspective. Institutionen för folkhälsovetenskap/Department of Public Health Sciences, Dissertation. Karolinska Institutet. Sweden.

Chiwona-Karltun L, Mkumbira JS, Saka J, Bovin M, Mahungu NM, Rosling H (1998) The importance of being bitter—a qualitative study on cassava cultivar preference in Malawi. Ecol Food Nutr 37(3):219–245

Chiwona-Karltun L, Tylleskar T, Mkumbira J, Gebre-Medhin M, Rosling H (2000) Low dietary cyanogen exposure from frequent consumption of potentially toxic cassava in Malawi. Int J Food Sci Nutr 51(1):33–43

Chiwona-Karltun L, Brimer L, Kalenga Saka JD, Mhone AR, Mkunbira J, Johansson L, Bokanga M, Mahungu NM, Rosling H (2004) Bitter taste in cassava roots correlates with cyanogenic glucoside levels. J Sci Food Agric 84:581–590

Chiwona-Karltun L, Nyirenda D, Mwansa CN, Kongor JE, Brimer L, Haggblade S, Afoakwa EO (2015) Farmer preference, utilization, and biochemical composition of improved cassava (Manihot esculenta Crantz) varieties in southeastern Africa. Econ Bot 69(1):1–15

Christenhusz MJM, Byng JW (2016) The number of known plants species in the world and its annual increase. Phytotaxa 261(3):201–217. https://doi.org/10.17660/ActaHortic.1994.375.18

Cousins HH (1903) Analysis of 17 varieties introduced from Colombia and grown by Mr. Robert Thompson at Half-Way-Tree. Jamaica Bull Dept Agric Jam 1(2):35

Crenn (2019) Cassava products of Brazil, IFAD and Natural Resources Institute. http://projects.nri.org/cassava-ipci/images/documents/Cassava_products_of_Brazil_a_production_guide_for_Africa.pdf

Dufour DL (1988) Cyanide content of cassava (Manihot esculenta, Euphorbiaceae) cultivars used by Tukanoan Indians in Northwest Amazonia. Econ Bot 42:255–266

Dufour DL (1994) Cassava in Amazonia: lessons in utilization and safety from native peoples. International Workshop on Cassava Safety. Acta Hortic 375:175–182. https://doi.org/10.17660/ActaHortic.1994.375.15

Dziedzoave NT, Abass AB, Amoa-Awua WK, Sablah M, Adegoke GO (2006) Quality management manual for production of high quality cassava flour. International Institute of Tropical Agriculture (IITA), Ibadan, p 68

EFSA (2019) Evaluation of the health risks related to the presence of cyanogenic glycosides in foods other than raw apricot kernels. EFSA J 17(4):5662–5740

Essers AJA, Ebong C, van der Grift RM, Nout MJR, Otim-Napa W, Rosling H (1995a) Reducing cassava toxicity by heap-fermentation in Uganda. Int J Food Sci Nutr 46(2):125–136

Essers AJA, Jurgens CMGA, Nout MJR (1995b) Contribution of selected fungi to the reduction of cyanogen levels during solid substrate fermentation of cassava. Int J Food Microbiol 26(2):251–257

Essers AJA, Bennik MHJ, Nout MJR (1995c) Mechanisms of increased linamarin degradation during solid-substrate fermentation of cassava. World J Microbiol Biotechnol 11:266–270

FAO (2017) Food outlook – Food and Agriculture Organization of the United Nations. Biannual report on global food markets November 2017

Fresco LO (1986) Cassava in shifting cultivation: a systems approach to agricultural technology development in Africa. Royal Tropical Institute, Amsterdam

Gacheru PK, Abong GO, Okoth MW, Lamuka PO, Shibairo SA, Katama CKM (2016) Microbiological safety and quality of dried cassava chips and flour sold in the Nairobi and coastal regions of Kenya. Afr Crop Sci J 24(suppl sI):137–143

Gnonlonfin GJB, Brimer L (2013) Cassava (Manihot esculenta Crantz) as a source of chemically safe food: a critical review. In: Chloe M, Jones DB (eds) Processed foods. Quality, safety characteristics and health implications. Nova Publishers, New York, pp 59–81

Gnonlonfin GJB, Hell K, Fandohan P, Siame AB (2008) Mycoflora and natural occurrence of aflatoxins and fumonisin B1 in cassava and yam chips from Benin. West Afr Int J Food Microbiol 122:140–147

Gnonlonfin GJ, Sanni A, Brimer L (2012a) Preservation of cassava (Manihot esculenta Crantz): a major crop to nourish people worldwide. In: Bhat R, Alias AK, Paliyath G (eds) Progress in food preservation. Wiley-Blackwell, Chichester, pp 331–342

Gnonlonfin GJB, Adjovi CSY, Katerere DR, Shephard GS, Sanni A, Brimer L (2012b) Mycoflora and absence of aflatoxin contamination of commercialized cassava chips in Benin, West Africa. Food Control 23:333–337

Inyang CU, Tsav-Wua JA, Akpapunam MA (2006) Impact of traditional processing methods on some physico chemical and sensory qualities of fermented cassava flour "Kpor umilin". Afr J Biotechnol 5(20):1985–1988

Kaaya A, Eboku D (2010) Mould and aflatoxin contamination of dried cassava chips in eastern Uganda: association with traditional processing and storage practices. J Biol Sci. https://doi.org/10.3923/jbs.2010.718.729

Kolind-Hansen L, Brimer L (2010) The retail market for fresh cassava root tubers in the European Union (EU): the case of Copenhagen, Denmark — a chemical food safety issue? J Sci Food Agric 90(2):252–256

Kostinek L, Sprecht VA, Edward VA, Pinto C, Egounlety M, Sossa C, Mbugua S, Dortu C, Thonart P, Taljaard M, Mengu C, Franz CMAP, Holzaplel WH (2007) Characterisation and biochemical properties of predominant lactic acid bacteria from fermenting cassava for selection as starter cultures. Int J Food Microbiol 114(3):342–351

Leestma J (2015) A tropical flour: Manioc in the Afro-Brazilian world, 1500–1800. https://digitalrepository.unm.edu/

Lei V, Amoa WKAA, Brimer L (1999) Degradation of cyanogenic glycosides by lactobacillus plantarum strains from spontaneous cassava fermentation and other microorganisms. Int J Food Microbiol 53(2–3):169–184

Ljabadeniyi AO (2007) Microbiological safety of gari, lafun and ogiri in Akure metropolis. Nigeria Afr J Biotechnol 6(22):2633–2635

McKey D, Cavagnaro TR, Cliff J, Gleadow R (2010) Chemical ecology in coupled human and natural systems: people, manioc, multitrophic interactions and global change. Chemoecology 20:109–133. https://doi.org/10.1007/s00049-010-0047-1

Morlon-Guyot J, Guyot JP, Pot B, de Haut IJ, Raimbault M (1998) Lactobacillus manihotivorans sp., a new starch-hydrolysing lactic acid bacterium isolated during cassava sour starch fermentation. Int J Syst Bacteriol 48:1101–1109

Nahrstedt A (1993) Cyanogenesis and foodplants. In: van Beek TA, Bretelser H (eds) Proceedings of the phytochemical society of Europe, Phytochemistry and Agriculture (Chapter 7). Clarendon Press, Oxford, pp 106–129

Northrup SH, McCammon JA (1980) Simulation methods for protein structure fluctuations. Biopolymers 19(5):1001–1016. https://doi.org/10.1002/bip.1980.360190506

Nweke FI (1999) Gender surprises in food production, processing, and marketing with emphasis on cassava in Africa (No. 19). IITA

O'Brien BGM, Weir RR, Moody K, Liu PWS (2013) Cyanogenic potential of fresh and frozen cassava on retail sale in three Irish cities: a snapshot survey. Int J Food Sci Technol 48(9):1815–1831

Obadina AO, Oyewole OB, Sanni LO, Tomlins KI, Westby A (2008) Identification of hazards and critical control points (CCP) for cassava fufu processing in South-West Nigeria. Food Control 19(1):22–26

Obadina AO, Oyewole OB, Odusami AO (2009) Microbial safety and quality assessment of some fermented cassava products (lafun, fufu, gari). Sci Res Essay 4(5):432–425

Obadina AO, Oyewole OB, Sanni LO, Tomlins KI, Westby A (2010) Improvement of the hygienic quality of wet "fufu" produced in South West Nigeria. Food Control 21:639–643

Obi CN, Ugwu CJ (2019) Effects of microbial fermentation on cyanide contents and proximate composition of cassava tubers. Nigerian J Biol 33(2):4493–4502

Odu NN, Maduka N (2019) Microbial analysis and molecular characterization of bacterial and fungal isolates present in exposed and packaged cassava, plantain and yam flour sold in selected markets in Port Harcourt, Rivers State, Nigeria. Am J Microbiol Res 7(2):63–72

Ohadike DC (1981) The influenza pandemic of 1918-19 and the spread of cassava cultivation on the lower Niger: a study in historical linkages. J Afr Hist 22(3):380–391

Olopade BK, Oranusi S, Ajala R, Olorunsola SJ (2014) Microbiological quality of fermented cassava (gari) sold in Ota Ogun state Nigeria. Int J Curr Microbiol App Sci 3(3):888–895

Olsen KM, Schaal BA (1999) Evidence on the origin of cassava: Phylogeography of Manihot esculenta. Proc Natl Acad Sci U S A 96(10):5586–5591

Padmaja G, Balagopal C (1985) Cyanide degradation by Rhizopus oryzae. Can J Microbiol 31:663–669

Penido FCL, Piló FB, Sandes SHDC, Nunes ÁC, Colen G, Oliveira, EDS, ... & Lacerda, ICA (2018) Selection of starter cultures for the production of sour cassava starch in a pilot-scale fermentation process. Braz J Microbiol 49:823–831

Raheem D, Chukwuma C (2001) Foods from cassava and their relevance to Nigeria and other African countries. Agric Hum Values 18:383–390

Schaal BA, Olsen KM, Carvalho LJCB (2006) Evolution, domestication, and agrobiodiversity in the tropical crop cassava. In: Darwin's harvest: new approaches to the origins, evolution, and conservation of crops, pp 269–284

Schery R (1947) Manioc-a tropical staff of life. Econ Bot 1(1):20–25. https://www.jstor.org/stable/4251837

Schwerin KH (1971) The bitter and the sweet, some implications of techniques for preparing manioc. In: Unpublished paper presented at the 1971 Annual Meeting of the American Anthropological Association

Soarez LMV, Rodrigues-Amaya DB (1989) Survey of aflatoxins, ochratoxin A, zearalenone, and sterigmatocystin in some Brazilian foods by using multi-toxin thin-layer chromatographic method. J Assoc Off Anal Chem 72(1):22–26

Tsav-wua JA, Inyang CU, Akpapunam MA (2004) Microbiological quality of fermented cassava flour "kpor umilin". Int J Food Sci Nutr 55(4):317–324

Weltzien E, Rattunde F, Christinck A, Isaacs K, Ashby J (2019) Gender and farmer preferences for varietal traits: evidence and issues for crop improvement. Plant Breed Rev 43:243–278

Williams K (1977) Eating wild plants. Mountain Press Pub

WIPO. Accessed 2021-05-24. https://www.wipo.int/tk/en/tk/

Zagrobelny M, Bak S, Møller BL (2008) Cyanogenesis in plants and arthropods. Phytochemistry 69:1457–1468

Part III
Enhancing Productivity

Chapter 9
Innovative Digital Technologies to Monitor and Control Pest and Disease Threats in Root, Tuber, and Banana (RT&B) Cropping Systems: Progress and Prospects

Jan Kreuze ⓘ, Julius Adewopo ⓘ, Michael Selvaraj ⓘ,
Leroy Mwanzia ⓘ, P. Lava Kumar ⓘ, Wilmer J. Cuellar ⓘ, James P. Legg ⓘ,
David P. Hughes ⓘ, and Guy Blomme ⓘ

Abstract This chapter provides the first comprehensive review of digital tools and technologies available for the identification, monitoring, and control of pests and diseases, with an emphasis on root, tuber, and banana (RT&B) crops. These tools include systems based on identification keys, human and artificial intelligence-based identification based on smart applications, web interfaces, short messages

J. Kreuze (✉)
International Potato Center (CIP), Lima, Peru
e-mail: j.kreuze@cgiar.org

J. Adewopo
International Institute of Tropical Agriculture, Kigali, Rwanda
e-mail: j.adewopo@cgiar.org

M. Selvaraj · L. Mwanzia · W. J. Cuellar
The Alliance of Bioversity International and the International Center for Tropical Agriculture
(CIAT), Cali, Colombia
e-mail: m.selvaraj@cgiar.org; l.mwanzia@cgiar.org; w.cuellar@cgiar.org

P. L. Kumar
International Institute of Tropical Agriculture (IITA), Ibadan, Nigeria
e-mail: l.kumar@cgiar.org

J. P. Legg
International Institute of Tropical Agriculture (IITA), Dar es Salaam, Tanzania
e-mail: j.legg@cgiar.org

D. P. Hughes
Pennsylvania State University, Department of Biology, University Park, PA, USA
e-mail: dph14@psu.edu

G. Blomme
The Alliance of Bioversity International and the International Center for Tropical Agriculture
(CIAT), Addis Ababa, Ethiopia
e-mail: g.blomme@cgiar.org

261

G. Thiele et al. (eds.), *Root, Tuber and Banana Food System Innovations*,
https://doi.org/10.1007/978-3-030-92022-7_9

services (SMS), or combinations thereof. We also present ideas on the use of image recognition from smartphones or unmanned aerial vehicles (UAVs) for pest and disease monitoring and data processing for modeling, predictions, and forecasting regarding climate change. These topics will be presented in the context of their current development and future potential but also the challenges, limitations, and innovative approaches taken to reach end users, particularly smallholder farmers, and achieve impacts at scale. Finally, the scope and limitation of private sector involvement demonstrates the need of publicly funded initiatives to maximize sharing of data and resources to ensure sustainability of unbiased advice to farmers through information and communication technology (ICT) systems.

9.1 Introduction

Globalization and climate change are exacerbating condition under which we lose 20–40% of global crop production to pests and diseases annually, especially in food-deficit regions with fast-growing populations. Weak existing pest and disease surveillance systems in developing countries have resulted in slow responses to large-scale outbreaks and epidemics in Africa and Asia, e.g.:

- Various cassava (*Manihot esculenta* Crantz) virus diseases.
- Desert locust (*Schistocerca gregaria* Forskål).
- Fall armyworm (*Spodoptera frugiperda* J.E. Smith) in maize (*Zea mays* L.) and many other crops.
- *Fusarium* wilt (*Fusarium oxysporum* f. sp. *cubense* (E.F. Sm.) W.C. Snyder & H.N. Hansen) in banana (*Musa* sp.)
- Maize lethal necrosis disease (MLND; coinfection with maize chlorotic mottle virus (MCMV) and one of several viruses from the Potyviridae group).
- Wheat (*Triticum aestivum* L.) rust (*Puccinia triticina* Erikss).

All of these have had devastating impacts on food security. Pests and diseases continue to devastate agricultural production and are expected to intensify with ever-increasing movement of people and planting material, disruption of natural habitats by encroachment through human activities, and climate change. As we write, the COVID-19 pandemic reminds us that the keys to controlling these threats are early detection, rapid and reliable diagnosis, and efficient tracking of spread – all of which can be enhanced with evolving digital technologies.

9.2 Digital Disease Identification Tools

Crop diseases are responsible for significant economic losses in agriculture worldwide. Monitoring of crop health and early detection of new diseases are essential to reduce disease spread and facilitate effective field management practices. Crop disease detection, generally carried out through scouting or field inspections, is often supplemented by diagnostic tools based on serological methods and polymerase chain reaction (PCR) tests. Field inspection for early disease detection is assessed as prevalence (present or not) at the field, farm, village, and/or landscape levels. This process is large scale, challenging, and time-consuming (Johansen et al. 2014). These limitations on direct field inspection methods have led scientists to investigate advanced and novel techniques that could rapidly and economically obtain crop health information (Heim et al. 2019; Steward et al. 2019). Several novel and noninvasive methods have been developed in the last decade, which are sensitive, reliable, standardized, high throughput, rapid, and cost-effective (Golhani et al. 2018). Frontline remote sensing (RS) methods coupled with machine learning (ML) is one of the emerging approaches to provide reliable and precise technical support for real-time and large-scale crop disease detection and monitoring. Remote sensing permits the noninvasive measurement of crops' biophysical and biochemical parameters and thus allows for nondestructive monitoring of crop health status (Ramcharan et al. 2017; Lu et al. 2015). Various imaging sensors – visible, thermal, multispectral, and hyperspectral – have been studied for crop disease detection (Mishra et al. 2020). The applications of these techniques have been gradually developed from novel sensor development, high-throughput image acquisition, processing, and computing, leading to image segmentation and disease classification with algorithm development. In the following sections, examples of these techniques will be described with a focus on root, tuber, and banana (RT&B) crops.

9.2.1 Smartphone Image-Based Disease Detection and Classification

Deep learning is an innovative method for image processing and object detection providing high accuracy in the classification of various crop diseases (Kamilaris and Prenafeta-Boldú 2018). Smartphone-based AI-powered apps could alert farmers and expedite disease diagnosis, potentially preventing or limiting pest and disease outbreaks. Even though many developing countries' farmers do not have access to these advanced tools, increased Internet infiltration, smartphone penetration, and offline models offer new outfits for infield crop disease detection.

Cassava provides food for more than 500 million African people every single day. Several diseases affect cassava, causing larger yield losses. Farmers (and even extension workers) struggle to correctly identify the various diseases. To that end, scientists at Penn State University have created an innovative solution that uses AI through Google's open-source TensorFlow technology. In collaboration with the International Institute of Tropical Agriculture (IITA), and working through the CGIAR Research Program on Roots, Tubers and Bananas (RTB) and the CGIAR Platform for Big Data in Agriculture, the team developed PlantVillage Nuru, an application built by annotating over 200,000 images of diseased cassava plants to train a machine to recognize various diseases and make predictions about a farmer's crop's health in less than one second (Fig. 9.1). It can identify symptoms of the cassava mosaic disease (CMD; caused by different species of plant pathogenic viruses), cassava brown streak virus disease (CBSD), and the feeding damage of mites (e.g., the green mite, *Mononychellus tanajoa* (Bondar)). With PlantVillage Nuru, an extension worker or farmer can point their phone over a specific cassava leaf, and a box will pop around any areas with symptoms with the diagnosis. In some cases, it will guide the farmer to scan several leaves from different parts of the plant to arrive to a more reliable result (Mrisho et al. 2020; Ramcharan et al. 2017; Ramcharan et al. 2019). Once a disease is diagnosed, farmers can simply push a button to request advice on how to respond.

Collaborations with other research centers and organizations have helped extend the power of PlantVillage Nuru to other crops:

- International Potato Center (CIP) for potato: (*Solanum tuberosum* L.) diseases (late blight, *Phytophthora infestans* (Mont.) de Bary; early blight, *Alternaria solani* Sorauer) and various viral diseases.
- United Nations Food and Agricultural Organization (FAO) and the International Maize and Wheat Improvement Center (CIMMYT) for maize: fall armyworm infestations.

The app is available for android mobile phones through the Play Store and for Apple phones through its App Store. The same AI technology is also being used extensively in support of the FAO's surveillance of desert locust using the PlantVillage-developed app eLocust3m and FAMEWS (Fall Armyworm Monitoring and Early Warning System).

For many years, banana disease surveillance and mapping has relied on in-person field surveys by knowledgeable scientists and trained field staff. This approach is restricted by high cost, limited human capacity, and inability to access more remote locations. In addition, the lack of capacity for farmers and some local extension workers to identify and/or differentiate between the different biotic diseases and constraints often hampers timely intervention efforts. Scientists from the Alliance of

Fig. 9.1 Using PlantVillage Nuru to identify plant diseases with a cellphone

Bioversity International and CIAT developed an AI-powered smartphone app called Tumaini ("hope" in the Swahili language) that is capable of identifying and differentiating symptoms of six banana diseases:

- *Xanthomonas* wilt of banana (BXW) caused by the bacterium *Xanthomonas vasicola* pv. *musacearum.*
- Banana bunchy top disease (BBTD) caused by the banana bunchy top virus (BBTV).
- The soilborne fungal disease *Fusarium* wilt caused by *Fusarium oxysporum* f. sp. *cubense* (E.F. Sm.)
- Banana blood disease caused by the blood disease bacterium (*Pseudomonas celebensis*) of banana.
- Black leaf streak disease (BLSD) or black sigatoka caused by the fungus *Pseudocercospora fijiensis*, formerly known as *Mycosphaerella fijiensis*
- Banana weevil (*Cosmopolites sordidus* Germar).

The Tumaini AI app was developed using a dataset of over 18,000 field images collected by banana experts. The app is freely available on Google Play Store.

The integration of these AI-powered apps in a multilevel sensing system for disease surveillance (e.g., comprising satellites, UAVs, AI apps, and ground truthing; covered below) is necessary to monitor crop health at different scales. Data collected through the smartphone app and obtained from drone and satellite image analysis is expected to be fed into the PestDisPlace (Sect. 9.2.5) surveillance and mapping platform, to create an early warning system, and advise research, extensions and advisory services, and National Plant Protection Organizations (NPPOs) of ongoing banana disease spread, linked risks, and priority zones for surveillance. See, for example, the PlantVillage warning maps for locust[1] or dedicated ArcGIS boards where the data from smartphone-derived observations is integrated with NASA soil moisture data and NOAA HYSPLIT wind models and soil maps from ISRIC.[2]

9.2.2 Aerial (UAV or Satellite) Image-Based Disease Detection and Classification Tools

Rapid technology development in unmanned aerial vehicles (UAVs) and the availability of low-cost UAVs carrying sensors provide the opening to capture high spatial and spectral resolution data, especially for disease detection across fields and landscapes. The ability to capture the crop phenotypic differences in this complex multidimensional system is necessary to better understand host-pathogen interactions (Steward et al. 2019). The combination of aerial image information and

---

[1] https://plantvillage.psu.edu/panel/analytics/locust_surveys

[2] https://arcg.is/0aHGHi

AI-based approaches can provide an accurate, high-throughput method for crop disease detection under real-world conditions (Boulent et al. 2019; Selvaraj et al. 2020). Working Benin and the Democratic Republic of the Congo, Selvaraj et al. (2020) demonstrated that banana fields and production zones can be mapped through both drone and satellite image analysis. In addition, these authors developed algorithms to detect BXW- and BBTD-infected plants or mats on drone images. These outputs will contribute to health status mapping in banana production zones (Kamilaris and Prenafeta-Boldú 2018; Selvaraj et al. 2019; Ramcharan et al. 2017; Selvaraj et al. 2020).

In potato research, detection of diseases through remote sensing has been advanced for late blight (Gold et al. 2020; Ray et al. 2011; Duarte-Carvajalino et al. 2018) and viruses (Chávez et al. 2009; Chávez et al. 2010; Griffel et al. 2018; Polder et al. 2019). Whereas there are clear challenges to implementing such technologies routinely at a landscape or regional level, obvious direct application could be available through automated screening for resistance in breeding trials and rapid screening for diseases in seed multiplication plots. Indeed, CIP is currently validating the use of airborne multispectral and high-resolution RGB (red, green, and blue) multitemporal imagery acquired by UAV to monitor late blight infection in breeding plots, potentially reducing disease scoring time from several days to hours.

To our knowledge, current disease detection systems focus on single sensorbased solutions and often lack the integration of multiple information sources. Moreover, using UAV to monitor larger landscapes is often challenging; thus, at larger geographic scales, satellite-based machine learning (ML) models could help classify overall crop health. Such general crop health information could then be further assessed using UAVs and AI-powered smartphone-based sensors to detect the specific reason(s) of an observed poor plant health status. Therefore, it is essential to combine high-resolution imagery data with advanced ML algorithms to acquire ground truths through mobile apps. UAVs and high-resolution satellites can capture a large number of high-quality spectral-temporal aerial images, becoming the ultimate technology for classifying crop yield for purposes of monitoring plant health and determining economic value (Burke and Lobell 2017; Ji et al. 2018). The real-time tracking of crop disease spread and impact at the regional, national, and global scales can be realized if large-scale data integration analysis is achieved. With the development of multisource remote sensing data, the fusion of multisource data may be a future development trend.

9.2.3   Tools for Modeling, Including Backcasting and Forecasting of Pest and Disease Incidence to Deliver Decision Support Intelligence to Smallholder Farmers

The application of models can be a powerful tool to understand pest and disease epidemiology. Simulations and forecasts with these models can provide advice to farmers and other actors in the agricultural sector to manage crop protection.

Different agroclimatic modeling approaches have been developed that range from simple analytical to complex simulation models and use actual and/or predicted climate data such as temperature, precipitation, humidity, and, in the case of atmospheric transport models, wind currents (Orlandini et al. 2020). Their applications can be roughly divided into two categories: risk assessment and forecasting.

Risk assessment tools, typically including maps and longer time scales (seasonal/annual), can be used to predict the potential distribution and damage of nonindigenous or indigenous pests due to climate change. They can also support farmers and decision-makers in preseason planning for pest and disease management and/or advise on timing and location for monitoring and surveillance. Forecasting tools, on the other hand, generate predictions within the growing season to provide farmers with information for daily or weekly decision-making.

Under RTB, the Insect Life Cycle Modeling (ILCYM) software was further developed as a generic open-source software platform that enables the development of phenology models and linked to geographic information systems (GIS) to generate risk maps at different geographic scales (Tonnang et al. 2013). The ILCYM software facilitates the selection and compilation of temperature functions for development, survival, and reproduction of different life stages of insects of interest, including parameters for variability in these processes. This information is used to generate life table parameters for a given range of constant and fluctuating temperatures as input data. The software can then simulate pest population development with their specific age-stage structure that allows temporal and special simulations of population parameters according to real or interpolated temperature data. ILCYM has been applied to understand pest potential spread and impacts under current and future climates for a wide range of crops including potato and cassava (Kroschel et al. 2013; Sporleder et al. 2008; Mwalusepo et al. 2015; Aregbesola et al. 2020; Khadioli et al. 2014; Rao and Prasad 2020; Mujica et al. 2017; Azrag et al. 2017; Soh et al. 2018; Fand et al. 2014; Fand et al. 2015; Rebaudo et al. 2016). It was also used to compile a Pest Distribution and Risk Atlas for Africa[3] for several crops to support the preparedness of policymakers and farmers to implement timely adaptation strategies (Kroschel et al. 2016). Models have also been developed with ILCYM for biocontrol agents (such as parasitoids) of invasive pests to provide advice on the probable success of establishment in pest-invaded regions and control efficiency under varying climatic conditions (Kroschel et al. 2016). Researchers have also used these models to study the effect of different biocontrol application intervals and rates (e.g., for *Phthorimaea operculella* granulovirus) for controlling the potato tuber moth, *Phthorimaea operculella* (Zeller) (Sporleder and Kroschel 2008). Recently, the approach was expanded to include risk assessment and guide surveillance for insect-transmitted viruses as demonstrated for the potato yellow vein virus (PYVV) (Gamarra et al. 2020a; Gamarra et al. 2020b).

The use of risk maps to guide surveillance can be further refined by combining establishment risks as modeled (e.g., by ILCYM) with geographic cropland patterns

---

[3] http://cipotato.org/riskatlasforafrica/

which represent an important factor contributing to risk of pest entry. The cropland connectivity risk patterns were modeled for RT&B crops globally by Xing et al. (2020), and the approach of combining these maps with establishment, activity, and generation risk indices modeled by ILCYM is currently being tested to monitor the possible spread of the North American potato psyllid (*Bactericera cockerelli* (Sulc)) into Colombia and Peru, which moved through Ecuador in recent years, causing significant crop losses and environmental impacts (Douthwaite 2020). Local in-country spreading pathways can be further analyzed by studying seed networks, as was done for potato in Ecuador (Buddenhagen et al. 2017). Although ILCYM is currently geared for risk assessment through generating complete life cycle models, these models can be also used to simulate in-season population development based on prevailing weather data – thus appropriate for forecasting as well (Kroschel et al. 2013). Initial steps to implement and validate such forecasts have been initiated in the central Andes by CIP and its partners for a number of important potato pests in the region.

Models for plant diseases have been developed primarily for airborne diseases. Among RT&B crops, late blight of potato has received much attention with at least 16 different models[4] developed to predict onset and progress of the disease. These models have been implemented with relative success in Europe and North America, providing farmers with alerts and recommendations about when to start and how frequently to use control measures based on prevailing weather conditions and cultivars planted.

The LATEBLIGHT model, originally developed to simulate late blight epidemics in the Andes (Andrade-Piedra et al. 2005b; Andrade-Piedra et al. 2005c), was later shown to perform quite well over a range of tropical environments (Andrade-Piedra et al. 2005a; Blandón-Díaz et al. 2011). More recently, BLIGHTSIM, a mechanistic model that accounts for diurnal oscillating temperatures, was developed to predict changes in severity that might be expected due to climate change (Narouei-Khandan et al. 2020).

Temperature, humidity, and rainfall (and leaf wetness) are the main drivers of these models, and they work well when reliable weather data are available in the relevant geographies. Unfortunately, this is rarely the case in most low-income countries. Although this limitation may be alleviated soon by using downscaled weather information modeled from satellite-acquired remote sensing data, simple models based on farmers' weather observations may provide a low-tech solution that still can provide meaningful advice for control. With that in mind, a simple handheld tool for late blight management was developed consisting of three concentric circles: the two outer circles can be moved to select options for answers to two questions and the inner third circle provides the recommendation (Fig. 9.2) (Pérez et al. 2020). The answers to each question provide a number. The individual numbers should be added, and if the sum is above a certain threshold, the advice is to apply control measures. The three different circles are used depending on the

---

[4] http://ipm.ucanr.edu/DISEASE/DATABASE/potatolateblight.html

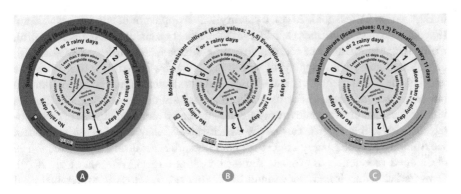

Fig. 9.2 Handheld decision support tool discs help determine pest and disease control measures for (**a**) susceptible, (**b**) moderately resistant, and (**c**) resistant potato varieties (Pérez et al. 2020)

susceptibility and resistance of the potato cultivars planted, which are classified as low, medium, and high. Cultivars will have to be categorized for each country and geographic region depending on what varieties are grown. During validation experiments, handheld tool for late blight management gave relatively good disease control in all cases, but was more effective among highly susceptible cultivars compared to traditional local spraying regimes in Ecuador and Peru. The application of the tool resulted in equal or lower numbers of fungicide sprayings per season and reduced the environmental impact (Pérez et al. 2020).

9.2.4 *Nucleic Acid (NA) Sequence-Based Digital Surveillance Systems for Pathogens and Pests*

There is high potential for the use of genetic data of pests and pathogens to strengthen current conventional approaches for surveillance. High-throughput sequencing (HTS) technologies applied to field diagnostics are increasing the amounts of nucleic acid sequences that become available as data. The availability of genetic data will turn out to be most relevant for early warning detection of the introduction and potential spread of different strains of pests and pathogens. These technologies were adapted early for plant virus research and discovery with sweetpotato and a few other crops (Kreuze et al. 2009; Adams et al. 2009).

Since its inception, many new viruses and variants in RT&B crops have been discovered and characterized, contributing to the rapid identification of novel diseases or pathogen variants and subsequent to the design of specific diagnostic assays (De Souza et al. 2013; Wang et al. 2013; Abad et al. 2013; Fuentes et al. 2012; Cuellar et al. 2011a; Cuellar et al. 2011b; Souza Richards et al. 2014; Kreuze et al. 2013; Kreuze et al. 2009; Monger et al. 2010; Carvajal-Yepes et al. 2014; Hanafi et al. 2020; Kreuze et al. 2020; Kutnjak et al. 2014; Leiva et al. 2020). HTS technologies have also been used to evaluate, e.g., the diversity and evolution of CBSD in East Africa (Alicai et al. 2016), a countrywide mapping of potato viruses in Peru[5] (Fuentes et al. 2020; Fuentes et al. 2019a; Silvestre et al. 2020; Fuentes et al. 2019b), regional characterization of cassava viruses in Southeast Asia (Siriwan et al. 2020), and a continent-wide mapping of sweetpotato viruses in Africa[6]. This work has enabled the identification of new viruses, but also indicated prevalence of known but understudied viruses. For example, the analysis of the African sweetpotato virome revealed that begomoviruses are the third most common viruses of sweetpotato in Africa, meriting further study, which showed they could cause significant yield impacts despite being almost symptomless, thus requiring clear attention in clean seed production as well as setting breeding targets (Wanjala et al. 2020). Conversely, badnaviruses were found to be almost ubiquitous in sweetpotato but lacked any detectable effect on sweetpotato, and thus were ruled out as a matter of concern for production (Kreuze et al. 2020).

The recent Ebola virus and SARS-CoV-2 epidemics have revealed the potential of HTS technologies to track the evolution of virus epidemics in near real time and have provided the scientific community with novel tools and platforms that can be applied to any other pathogen. One can foresee that massive sequence information will continue accumulating and be made available through different public databases. One excellent example is the International Nucleotide Sequence Database Collaboration, which comprises the DNA DataBank of Japan (DDBJ), the European Nucleotide Archive (ENA), and GenBank (Arita et al. 2021). These three organizations curate and exchange DNA sequence data daily and, at present, host more than 70 million sequences corresponding to bacteria, fungi, and viruses. Even so, affected countries still hesitate to share data because these are public domain databases, through which data can be accessed and used anonymously, which can create policy and/or intellectual property issues. However, there is hope as evidenced by the GISAID-Initiative,[7] developed for near-real-time reporting of viruses related to the bird flu virus (H5N1) and SARS-CoV-2.

The availability of large amounts of DNA sequences requires robust bioinformatic pipelines that can analyze the data in real time (preferably open source). With that in mind, the Nextstrain system was developed (Hadfield et al. 2018), which is currently heavily used to track the evolution of SARS-CoV-2, but can be applied to any virus of interest. So far, only a few examples of the use of such tools in surveillance of crop pathogens have been published (Siriwan et al. 2020; van de Vossenberg et al. 2020; Leiva et al. 2020). As more novel, portable, and affordable next generation sequencing (NGS) technologies take over classic time-consuming diagnostic tools based on biological isolation, PCR-based approaches, and phenotypic evaluation of disease response in specific host genotypes (Boonham et al. 2013), their use and applications will expand to the surveillance of other major crop pests and pathogens. Recent efforts to integrate such information with field data (incidence of

---

[5] http://potpathodiv.org/index.html

[6] http://bioinfo.bti.cornell.edu/virome/

[7] www.gisaid.org

symptoms and pests) are currently implemented in the platform PestDisPlace (Cuellar et al. 2018).

9.2.5 Data Management and Open Data

Efforts have been made to aggregate data and monitor the spread of crop disease by integrating data into one platform that combines different surveillance methods, such as field inspections, PCR test results, published literature, and ML diagnosis using mobile applications. One such effort is PestDisPlace[8] (Cuellar et al. 2018), an online database, surveillance, and visualization platform, used to standardize the collection, analysis, and sharing of surveillance data acquired by multiple research teams. The platform uses these disparate data sources to create user-friendly geolocated visualizations of longitudinal pest and disease data from different spatial locations. The data visualization has enabled scientists working in the field to use PestDisPlace to understand trends and use the platform as an easy-to-understand communication tool for engaging with decision-makers and scientists in disease-affected countries. PestDisPlace is currently crowdsourcing disease monitoring data from CGIAR researchers and their partners, including national plant protection institutions mandated with crop protection.

PlantVillage has a similar database that visualizes pest and disease reports collected by its AI-powered apps (Nuru, FAMEWS, eLocust3m).

The tools described in the previous sections all generate huge amounts of data that can be used further as input for more extensive analysis and other kinds of modeling. Good data management and open access are crucial components of knowledge discovery and innovation (Wilkinson et al. 2016). However, in academia and research, only peer-reviewed publications have traditionally been considered outputs of science. But that tendency is changing a bit now as data are earning recognition as a legitimate research product that can be validated, preserved, cited, and credited (Kratz and Strasser 2015a; Kratz and Strasser 2015b; Data Citation Synthesis Group: Joint Declaration of Data Citation Principles. 2014).

Even though agriculture trails most sectors in terms of digitalization (Manyika et al. 2015), more and more data are being produced by research organizations, academia, governments, and farmers through mobile phones and social media. The management, dissemination, and reuse of data can contribute to building resilience to food system shocks. The CGIAR Big Data Platform has shown that organizing and disseminating open data on agriculture, applying analytics on this data, and working with agricultural stakeholders, such as farmer organizations, can help building resilience to pest and disease outbreaks, climate impacts, and land degradation (Jimenez and Ramirez-Villegas 2018).

---

[8] https://pestdisplace.org/

9.3 Decision Support Systems (DSS)

The fourth global evolution of agriculture will be causally linked to the emergence and use of technologies including smartphones and smart applications, Internet of Things (IoTs), artificial intelligence, cloud computing, remote sensing, and others (Zhai et al. 2020). These technologies are progressively evolving, increasingly used under various agricultural production contexts, and generating unprecedented volume of data. However, stakeholders and farmers often find it challenging to access, process, and digest these data into practical knowledge that can guide decision-making (Taechatanasat and Armstrong 2014).

In many smallholder farming systems where RT&B crops are cultivated, extension and advisory services are often overwhelmed or nonfunctional (Fabregas et al. 2019). Thus, to manage and mitigate threats of pests and diseases, farmers require reliable information to guide timely action. Digital DSS platforms that support and complement existing extension systems hold promise for sustainable RT&B crop production. These digital DSS vary in their overall function, content, and sophistication, but they can meet user needs in ways that transcend the capabilities of traditional extension systems and advisory services.

It should be noted that DSS tools are not exclusively Internet dependent and often built with the intent of democratizing information access across age, education, gender, and socioeconomic classes. Experience with smallholder RT&B farming systems suggests that male and female farmers face similar constraints in accessing DSS tools that are smartphone based due to limited or lacking Internet access and capability to use such devices. Yet, recent evidence suggests that disparity exists in access to (and ownership of) basic phones relative to gender and age class among farmers (Adewopo et al. 2021). Equitable capacity building of target tool users can accelerate adoption as these various tools are deployed across diverse geographies.

9.3.1 Short Messaging Service Systems

Short messaging service (SMS) systems have been adapted for rapid delivery of agricultural advisory services, including specific recommendations on best practices to tackle biotic threats on farm, information on specific precautions to mitigate incidence, or timely alerts for risks of pest outbreaks or disease infection and spread. SMS systems have proven to be an effective entry point to empower smallholder farmers because they are easy to deploy on basic cellphones and do not cost additional investment to receive information. Depending on use contexts and goals, farmers can informally exchange messages with each other, or respond to messages promptly to engage the sender, within a DSS setup. Several SMS-based DSS have been successfully tested and deployed to support data collection and information exchange on RT&B pests and diseases. In coastal Tanzania, the Commonwealth

Agricultural Bureau International (CABI) and other extension and advisory partners have facilitated SMS messaging among farmers to share information about cassava whiteflies (*Bemisia* spp.), a major pest in the region, and how to identify and remove infected cassava plants[9]. Project-based pilots of SMS for decision support, social enterprises, and private businesses – like Esoko[10] and EcoFarmer[11] – are advancing the frontiers of SMS application for other purposes, such as exchange of information on prices and market information on potato in Zimbabwe (Ifeoma and Mthitwa 2015). While there are limited examples of SMS-based DSS that are exclusively focused on RT&B pest or disease surveillance, vendors or service providers have the flexibility to configure their systems to meet emerging demands for basic surveillance information at village or farm level. Therefore, RT&B-focused SMS systems can be enriched with periodic information on pest and disease threats, enhanced as a bidirectional information exchange tool, or integrated with more robust systems to offer the recipients direct access to further information or resources.

Despite its potential advantages, SMS deployment for surveillance and control of RT&B pests and diseases can be constrained by some factors. Although SMS is a low-cost pathway to reach a vast number of farmers and enables quick dissemination of information, user response to the message can be passive because the SMS nudge does not suffice as an incentive for immediate action. Similarly, the literacy levels of recipient can limit the interpretation of the message; therefore, smallholder RT&B farmers who have little or no formal education are often at a disadvantage. Furthermore, it is easy for recipients to ignore SMS-based DSS (e.g., incident alerts, control measures, preventive practices) when the risk or threat is not considered imminent.

Recently, the Bill & Melinda Gates Foundation has invested funds in PlantVillage to test the hypothesis that data on important diseases can be collected from SMS. This experiment will take advantage of a collaboration between PlantVillage, iShamba (which sends messages to farmers), and the TV show *Shamba Shape Up*, a popular show watched by over nine million farmers in Kenya. Using USSD (Unstructured Supplementary Service Data), the project will ask farmers what they grow and what pests they observe among their crops.

9.3.2   Interactive Voice Response (IVR) Advisory

Interactive voice response (IVR) technology allows users to interact with a pre-programmed host computer system through a telephone keypad or by speech recognition. Generally, the IVR-based DSS for farmers has a major advantage over

---

[9] https://cabi-uptake.netlify.app/

[10] www.esoko.com

[11] www.ecofarmer.co.zw

SMS-based systems because it does not require basic literacy for effective use – only the ability to listen, select numbers on phone, or verbally respond to voice prompts. However, developing and operationalizing IVR systems is more complex because it requires voice coding and algorithms that route user selection through the system and return relevant content. In addition to the complexity of the back-end system, higher costs of connectivity are associated with IVR-based DSS because users often need to dial into the system to initiate information exchange, which often accrue mobile connection fees by the service provider.

Despite the promising aspects of IVR systems and their compatibility with the contexts of smallholder farmers, it is rare to find real-world application of IVR for reporting or monitoring of RT&B pests and diseases. One of the earliest assessments of IVR to support agricultural decision-making (Patel et al. 2010) reported a significant improvement in information sharing among smallholder farmers in rural India. It is noteworthy that prior to the testing of IVR within agricultural systems, this technology has been successfully used to curate timely information on chronic health problems (Piette 2000), and this success could be adapted for agriculture contexts, especially for pest and disease surveillance. Recently, digital agriculture enthusiasts have begun exploring entry points to introduce DSS directly to farmers or to support adoption and integration into extension and advisory services.[12] As these initial efforts mature, it is likely that vendors and service providers will have compelling information to design IVR systems as versatile DSS that offer broad information services and products to the target end users.

9.3.3   Smart Applications (Smart Apps)

The growing use of smart devices (primarily tablets and phones) in smallholder farming systems is creating unprecedented opportunity to engage farmers with robust and content-rich DSS for RT&B pest and disease monitoring and control. Smart devices have unique functions that allow users to access and exchange digital contents in various formats, either separately or complementarily to achieve desired user engagement and experience. In addition to the multi-format functions, smart devices are generally built with capacity to record actual locations (geocoordinates) of user inputs and observations, thereby facilitating rapid acquisition of georeferenced datasets, a critical input for spatially explicit near-real-time assessment of status and risks of pests and diseases at local, regional, or national scales.

---

[12] For instance, in 2020, a project funded by Bill & Melinda Gates Foundation to develop an IVR system for plant pest and disease surveillance in Ghana, where farmers are engaged to ensure timely reporting and control (https://www.ausvet.com.au/ivr-for-plant-health/). Similarly, in 2017, the CGIAR supported Viamo and Voto Mobile to develop an IVR system that connects farmers to market intelligence among farmers in Nepal (https://bigdata.cgiar.org/inspire/inspire-chal-lenge-2017/using-ivr-to-connect-farmers-to-market/)

In most smallholder farming systems, the use of smart devices remains relatively low, but the high interest in smart digital tools constitutes a viable entry point for innovation among stakeholders, including farmers (McCampbell et al. 2018). Therefore, various smart apps are emerging for the diagnosis, surveillance, and control of pests and diseases in RT&B crops. These include:

- The Plantix app,[13] developed by PEAT GmbH (Germany), which supports rapid diagnosis for detection of pests, diseases, and nutritional deficiencies in banana, cassava, potato, sweetpotato (*Ipomoea batatas* (L.) Lam.), and other crops.
- The BXW app[14] (Fig. 9.3), developed by a consortium of partners (led by IITA), supports monitoring of banana *Xanthomonas* wilt (BXW) disease with combined functionality of awareness messaging, diagnosis, control, and agronomic recommendations.
- The Crop Disease Surveillance (CDS) app, developed by IITA, is designed for cost-effective surveillance for cassava virus diseases in Nigeria; it facilitates rapid diagnosis through digital image-based analysis, communication among the quarantine officials, notification of pest risk, and facilitation of emergency response. This app was later expanded for use against banana bunchy top virus (BBTV) surveillance in Nigeria.

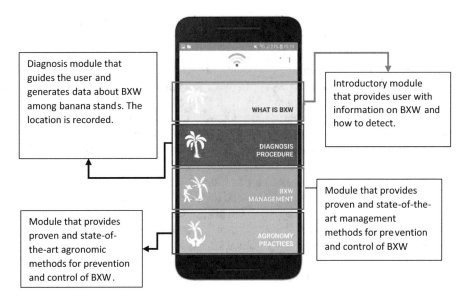

Fig. 9.3 The homepage interface of the smartphone-based BXW app, an example of a smart decision support system (DSS) for the surveillance and control of banana *Xanthomonas* wilt (BXW) disease

---

[13] www.plantix.net

[14] www.ict4bxw.com

- The PlantVillage Nuru app (Fig. 9.1) was collaboratively developed by several researchers from Penn State University, FAO, IITA, CIP, and CIMMYT for the diagnosis of plant diseases, including cassava and potato (Ramcharan et al. 2017; Ramcharan et al. 2019; Mrisho et al. 2020).
- Others (including Tumaini and PestDisPlace apps) that are highlighted in the next section.

Most of these smart apps have been tested in the field, are available through the Android Play Store, and are being evaluated or tweaked to optimize user experience.

The sustainability of smart applications depends somewhat on various technical and contextual factors. For example, scaling up the use of apps across geographies depends on the proliferation of smart devices among smallholder farmers or last-mile extension delivery agents. Similarly, although some of the apps work in offline mode, the availability of Internet coverage is critical for initial downloads, data exchange with back-end servers, and synchronization of the app and content updates. The commitment of vendors and/or institutional hosts is indispensable to progressively iterate over the core and ancillary functionalities of these smart apps in response to user demands.

Several systems are already having impact in improving production of RT&B crops at the community level:

- PlantVillage Nuru has been available for free download since June 2018, and since that time, more than 15,000 reports have been generated in 32 cassava-growing countries of the tropics, using the cassava pest and disease identification component. An important feature of this work has been the promotion of the role of "lead farmers" who are equipped with the basic smartphones required to download and use the app and work with their communities to help farmers to learn about the symptoms and damages caused by the main pests and diseases of cassava, as well as guiding them in the application of appropriate control practices. Users in Tanzania and Nigeria are also invited to check on the availability of certified planting material through using the SeedTracker app[15], which is a system for the registration and certification of farmers producing high-quality planting material. The widespread use of the PlantVillage Nuru app has made an important contribution to improving the health of cassava production systems in western Kenya, and farmers have given testimony to the value that these changes are making to their livelihoods[16,17] PlantVillage Nuru is also being expanded for application to other RT&B crops, notably potato and sweetpotato, and this development will extend the benefits being realized by producers to other parts of the world.
- The Tumaini AI-powered app, which detects various diseases and a pest of banana, is free available for download on Google Play Store since June 2019;

---

[15] www.seedtracker.org

[16] https://bigdata.cgiar.org/blog-post/success-story-from-cassava-farmers-in-busia-county-kenya/

[17] https://plantvillage.psu.edu/blogposts

since that time, the app has been downloaded over 2000 times, mostly by users in Asia. For example, in India, the Tumaini app was discussed and demonstrated on two famous radio and TV channels of Tamil Nadu where banana cultivation is omnipresent. In addition, the Tumaini app utility and functions are being promoted at farmer's exhibitions in Tamil Nadu, while in collaboration with Indian state agricultural universities, farmers and growers are being trained in the app use through agricultural college students in the framework of village stay programs.

- The BXW app is being co-validated in the field with a network of 65 village-level extension agents (known as farmer promoters) who have successfully reached over 4200 farmers (Fig. 9.4), and independently completed over 2500 diagnosis within banana farms, across 8 districts in Rwanda. Based on the data flow, robust georeferenced data on BXW incidence is being generated, through farm-level usage of the app, to assess spread and potential factors that impact the dynamics of the disease. Further, the agronomic information and BXW control methods were disseminated through an IVR-based system, under the auspices of Viamo[18], to circumvent the constraints associated with low smartphone usage among smallholder farmers. The information on banana was accessed by over 10,000 unique callers (mainly farmers) over a period of 8 weeks, with 87% of randomly surveyed users reporting their knowledge of agronomic and control practices improved after accessing the digital platform.

Fig. 9.4 Enthusiastic Rwandan banana farmers who are interacting with beta version of ICT4BXW app during field testing. (Photo credits: IITA)

---

[18] https://viamo.io

Despite what is described in the examples above, assumptions about the impact of the innovations have to be tested over time and relevant M&E frameworks can help to build up evidence while identifying the limitations of assessing direct impact (e.g., lagged effect and indirect impacts).

9.4 Future Perspectives

9.4.1 Development of Integrated and Interoperable Systems and Major Challenges to Overcome

Many different databases and apps have been developed to track and identify diseases globally, but are generally used only for a single crop or disease or promoted by a single research center, thus limiting the use of the data to a specific community of practice. Once developed, these communities become vested in their systems and unlikely to change to another platform, because their home platforms have been tailored to their needs. However, if the data captured by these different platforms could be shared more broadly, it would benefit all involved and could be reused to support many specific use cases, including the tracking of epidemics, modeling, providing farmers with advice, and supporting seed systems.

Within RTB collaborating organizations and the broader CGIAR collaborating partners, several such databases and apps have been developed. To enable the sharing of data between these platforms, the concept of an AgDx alliance was proposed between platforms of the CGIAR and other publicly funded, nonprofit organizations dedicated to research for development and education in agriculture with the aim to improve plant health globally through ICT tools. As an alliance of individual platforms, tools and databases are aimed at supporting agricultural health in different aspects AgDx commits to address former shortcomings through the development of application program interfaces (APIs) to facilitate data sharing between platforms and/or by providing links between complementary ICT tools where appropriate.

The AgDx alliance is taking steps to create data interoperability among platforms, starting with data harmonization. Through a collaborative and community-based process, AgDx standardized a core set of data elements that form the first data interoperability step. This standardization process ensures that data elements will have the same semantic meaning by agreeing on definitions and scale and enable effective exchange of data.[19]

---

[19] The AgDx community created a cross-reference of all common data fields and the standardization process divided the data fields into two groups: one for pests and one for disease data. Examples of basic shared data fields include the affected crop, the location, severity, incidence, and scale data. Disease data fields include the disease name, pathogen type, and pathogen name. In comparison, pest name, pest origin, and pest type are pest-specific examples. The dataset includes

An application program interface (API) will facilitate data exchange and interoperability between various data systems and platforms of the AgDx alliance. An API is a set of protocols and definitions that work as an intermediary that allows computing units to communicate, exchange, or retrieve information or perform a function. Specifically, the AgDxAPI will be a Web Service API. Web services allow computing systems to communicate with each other over a network. The AgDx database or platform exposing a web service will define resources in the form of computer files such as images and audio files or outputs of a computing function as standard computer formats such as JSON (Ecma International 2017) and XML (W3C 2009). Using the REST API architecture style, a de facto standard for creating Web Service APIs (Fielding et al. 2017), AgDxAPI will allow implementing systems to communicate which services and resources are available and how to request these resources. Using APIs has been proven successful in creating interoperability across agricultural research, such as in the plant breeding community (Selby et al. 2019).

All participating databases and platforms will implement the AgDxAPI specification into their systems. Each database will customize the API so that service calls will retrieve data in the agreed format and scale. Harmonizing the core exchange dataset with existing ontologies such as the Crop Ontology (Shrestha et al. 2010) and the in-development Plant Crop Stress Ontology[20] will be the next step to enhance data standardization. To this end, AgDx is engaging with the Ontology Community of Practice of the CGIAR Big Data Platform (Arnaud et al. 2020). This community has experience in creating and integrating ontologies with platforms and APIs in agricultural research. Another next step for the AgDxAPI specification will be to include an expanded list of data elements for specific domains. For example, AgDxAPI imaging could be an extension to allow for more detailed retrieval of images stored in different databases. This feature is critical for machine learning.

9.4.2 Approaches to Scaling (Opportunities, Packaging, Strategies, Public and Private Partnerships, Expected End Users, and Impacts)

Digital tools for pest and disease identification and surveillance have become a busy "space" in recent years, with rapid developments in several component technology areas, such as high-throughput sequencing, remote sensing, artificial intelligence, and the development of web platforms and phone apps. These changes are being built on rapidly expanding communications architecture that is making it increasingly easy to share digital technologies and associated information with rural

---

both the common names and scientific names for crops, pathogens, pests, and linkages to other related computer systems such as the GenBank (Sayers et al. 2021).

[20] https://github.com/Planteome/plant-stress-ontology

farmers. According to the GSM Association[21], some of the most rapid gains are being made in sub-Saharan Africa, where mobile Internet users are expected to increase from 272 million (26%) in 2019 to 475 million (39%) by 2025, and smartphone connections are expected to increase from 44% to 65% over the same period. Meanwhile, the interoperability of tools will likely be a major factor for effectively scaling applications among smallholder RT&B farmers. This has been an important feature of the work of the CGIAR's RTB research program, where significant efforts have been undertaken to foster links and information exchange among a range of digital surveillance tools.

Efforts are underway to incorporate this approach into plans for the new OneCGIAR, which would aim to bring together the majority of the digital surveillance tools being used throughout the CGIAR at present. Promoting the use of this anticipated platform of applications at farm level will require a diverse set of innovative partnerships with the public and private sectors. Recent experience with the rollout of test and trace digital tools for SARS-CoV-2 has demonstrated that specific solutions may need to be tailored for different countries and geographies and that applications will achieve greatest impact where there are strong partnerships between private technology providers and public institutions. Ultimately, the likelihood of any individual farmer using a specific digital tool will depend on the perceived benefit that the tool delivers. This highlights the importance of ensuring that disease surveillance tools are packaged with business development tools for the same crop. Many commercial digital platforms have been set up in recent years to provide a range of e-extension services, including a growing number established by phone providers as well as others set up by dedicated electronic agro-support services, such as Esoko in Africa and ImpactTerra in Southeast Asia. Building coalitions with these types of providers and national extension systems appears to be the best opportunity for scaling digital surveillance solutions and meeting the needs of farmers down to the last mile.

9.4.3 911 For Planet Earth

In 1937, England established the world's first emergency number: 999. By calling this number, any person could access the emergency services. All countries now have such numbers, allowing persons to make phone calls for help without coins or credits.

African farmers, and smallholder farmers around the world, need extension advisory support to cope and adapt to climate change. There are a lot of advices available such as planting drought-tolerant crops, promoting soil moisture conservation, engaging in water harvesting, tree planting, and other activities that are known to increase resiliency to climate shocks. As more pest data becomes available, we can

---

[21] https://www.gsma.com/mobileeconomy/sub-saharan-africa/

integrate them with weather forecasts and satellite observations as demonstrated by the response to desert locust crisis of 2020/2021 (discussed above, https://arcg. is/0aHGHi). Available advices from various public sources (including scientific inputs from CGIAR, NASA, FAO, NOAA) are considered as public good, and they are often synthesized by organizations like the United Nations, who readily draw from the global community of scientists (CGIAR, universities, federal agencies). But because of very high data costs in places like Africa, smallholder farmers, or the communities to which they belong, cannot afford to access this "free advice."

Millions of smallholder farmers are unable to access digital information as public goods because they cannot afford airtime for basic or smartphone usage. Considering extant global-scale emergency and crises, it will be relevant to deploy an emergency response system, similar to 911. Such a system can be developed to function in a way that it provides science-based advice and coping strategies to farmers through free and accessible platforms. Such platforms offer contents in various forms and formats, including well-illustrated videos on proven farming practices, advice on climate-resilient crops, and strategies to combat pests.

9.5 Conclusions and Ways Forward

Digital tools and systems for monitoring and controlling pest and disease threats in smallholder farming systems are crucial for ensuring food and nutrition security around the world and into the future. However, achieving and sustaining this goal requires cohesive engagement between digital tool developers, researchers, extension and advisory services, and farmers to define specific problems and opportunities and develop suitable digital solutions. Despite the array of digital DSS tools and platforms that have been developed and deployed for other use cases within and beyond agriculture, their adaption or use is still in its infancy with RT&B cropping systems. This suggests wide opportunity to innovate for impacts in RT&B farming systems, leveraging existing knowledge and experiences, and rapidly integrating tools and methods for efficient mitigation of pest and disease risks at scale.

Smartphone penetration and usage is expected to increase globally, especially in developing countries, while Internet coverage will improve, reaching even the most remote regions through new satellite systems (GSMA Intelligence, 2020). Consumer smartphone capabilities, including image capture and integration with other sensors, are also expected to evolve, thus creating new opportunities for innovative applications to advance plant health, including major RT&B crops. Machine learning algorithms are likely to improve simultaneously with new generation satellite-derived remote sensing data on various agrometeorological variables (including precipitation, soil moisture, temperature), which can be accessed and used for decision support in near real time. Combining these assets with improved analytical methods for image processing, downscaling data, and crowdsourcing of information, the prospects are good for accurate forecasting and prediction of pest and disease risks at global and local scales to support timely decision-making and action.

Despite the gains in coverage and use of ICT tools across developing countries described in this chapter, there is a considerable gender gap, especially in regard to Internet use. This gender gap is significantly larger in most developing countries and even more so in rural settings. To date, we lack studies analyzing the digital gender divide in the agricultural sector or how ICT tools may influence gender disparities in agricultural settings. Such studies are urgently needed in this era when digital agriculture is rising; we must take care not to inadvertently fuel more gender inequality.

The vision of a dynamic and robust pest and disease surveillance system that allows for reflexive learning and inclusive iteration of tools can be achieved with proper consideration of user needs, contextual realities, and aspirational thinking for technology development and deployment. For instance, efficient information exchange from different data sources is indispensable for scalable analytics on pest and disease dynamics in RT&B cropping systems, and this aspect requires common data definition standards – an effort which has been initiated by CGIAR researchers through the AgDx alliance initiative. An initial version of AgDxAPI has been developed based on a similar approach implemented by crop breeders (named BrAPI; Selby et al. 2019). The AgDxAPI is currently in its pilot stage, leveraging PestDisPlace and PlantVillage tools under field conditions. As an agricultural research for development organization, the OneCGIAR may have an important role to play as a trusted broker for harnessing and disseminating data streams across diverse sources and institutions.

In conclusion, the impact of monitoring and controlling RT&B pests and diseases can be quite nuanced, especially when accounting for direct net impacts on yield. However, the cost of inaction would be perilous. Assessing returns on investment in DSS for RTBs should include various ancillary benefits, including gender-sensitive equitable access to information, youth empowerment for last-mile service delivery, and ease of resource demand on national extension advisory systems.

Acknowledgments This research was largely undertaken as part of, and funded by, the CGIAR Research Program on Roots, Tubers and Bananas (RTB) and supported by CGIAR Trust Fund contributors. The authors would also like to acknowledge donor agencies including United States Agency for International Development (USAID), the Bill & Melinda Gates Foundation, German Association for International Cooperation (GIZ), UK Biotechnology and Biological Sciences Research Council (BBSRC), UK Department for International Development (DFID), National Program for Agricultural Innovation in Peru (PNIA), Spanish Agency for International Development Cooperation (AECID), Belgian Directorate General for Development Cooperation and Humanitarian Aid (DGDC) through the Consortium for Improving Agriculture-based Livelihoods in Central Africa (CIALCA – www.cialca.org), the Australian Centre for International Agricultural Research (ACIAR), and the CGIAR Platform for Big Data in Agriculture for financial support towards the accomplishment of this work. The authors thank the following individuals who provided input and feedback on earlier drafts of this chapter: Jill Lenne and Christopher Butler. We greatly appreciate Zandra Vasquez's assistance with references and formatting.

References

Abad JA, Li R, Fuentes S, Kreuze JF, Loschinkohl C, Bandla P (2013) Interception and identification by deep sequencing of a "caulimo-like" virus in a potato germplasm accession imported from South America. Paper presented at the APS-MSA Joint Meeting, Austin, Texas, U.S.A., August 10–14, 2013

Adams IP, Glover RH, Monger WA, Mumford R, Jackeviciene E, Navalinskiene M, Samuitiene M, Boonham N (2009) Next-generation sequencing and metagenomic analysis: a universal diagnostic tool in plant virology. Mol Plant Pathol 10(4):537–545. https://doi.org/10.1111/j.1364-3703.2009.00545.x

Adewopo JA, McCampbell M, Mwizerwa C, Schut M (2021) A reality check for digital agricultural extension tool development and use. Int J Rural Dev 55(1):23–25

Alicai T, Ndunguru J, Sseruwagi P, Tairo F, Okao-Okuja G, Nanvubya R, Kiiza L, Kubatko L, Kehoe MA, Boykin LM (2016) Cassava brown streak virus has a rapidly evolving genome: implications for virus speciation, variability, diagnosis and host resistance. Sci Rep 6:36164

Andrade-Piedra JL, Forbes GA, Shtienberg D, Grünwald NJ, Chacón MG, Taipe MV, Hijmans RJ, Fry WE (2005a) Qualification of a plant disease simulation model: performance of the LATEBLIGHT model across a broad range of environments. Phytopathology 95(12):1412–1422

Andrade-Piedra JL, Hijmans RJ, Forbes GA, Fry WE, Nelson RJ (2005b) Simulation of potato late blight in the Andes. I: modification and parameterization of the LATEBLIGHT model. Phytopathology 95(10):1191–1199

Andrade-Piedra JL, Hijmans RJ, Juárez HS, Forbes GA, Shtienberg D, Fry WE (2005c) Simulation of potato late blight in the Andes. II: validation of the LATEBLIGHT model. Phytopathology 95(10):1200–1208

Aregbesola O, Legg J, Lund O, Sigsgaard L, Sporleder M, Carhuapoma P, Rapisarda C (2020) Life history and temperature-dependence of cassava-colonising populations of Bemisia tabaci. J Pest Sci 93(4):1225–1241

Arita M, Karsch-Mizrachi I, Cochrane G (2021) The international nucleotide sequence database collaboration. Nucleic Acids Res 49(D1):D121–D124

Arnaud E, Laporte M-A, Kim S, Aubert C, Leonelli S, Cooper L, Jaiswal P, Kruseman G, Shrestha R, Buttigieg PL (2020) The ontologies community of practice: an initiative by the cgiar platform for big data in agriculture. https://doi.org/10.2139/ssrn.3565982

Azrag AG, Murungi LK, Tonnang HE, Mwenda D, Babin R (2017) Temperature-dependent models of development and survival of an insect pest of African tropical highlands, the coffee antestia bug Antestiopsis thunbergii (Hemiptera: Pentatomidae). J Therm Biol 70:27–36

Blandón-Díaz JU, Forbes GA, Andrade-Piedra JL, Yuen JE (2011) Assessing the adequacy of the simulation model LATEBLIGHT under Nicaraguan conditions. Plant Dis 95(7):839–846

Boonham N, Kreuze J, Winter S, van der Vlugt R, Bergervoet J, Tomlinson J, Mumford R (2013) Methods in virus diagnostics: From ELISA to next generation sequencing. Virus Res. Available on Line (0). https://doi.org/10.1016/j.virusres.2013.12.007

Boulent J, Beaulieu M, St-Charles P, Théau J, Foucher S (2019) Deep learning for in-field image-based grapevine downy mildew identification. In: Proceedings of the 12th European conference on precision agriculture (ECPA). Montpellier, France, pp 8–11

Buddenhagen C, Hernandez Nopsa J-F, Andersen KF, Andrade-Piedra J, Forbes G-A, Kromann P, Thomas-Sharma S, Useche P, Garrett K (2017) Epidemic network analysis for mitigation of invasive pathogens in seed systems: potato in Ecuador. Phytopathology 107(10):1209–1218

Burke M, Lobell DB (2017) Satellite-based assessment of yield variation and its determinants in smallholder African systems. Proc Natl Acad Sci 114(9):2189–2194. https://doi.org/10.1073/pnas.1616919114

Carvajal-Yepes M, Olaya C, Lozano I, Cuervo M, Castano M, Cuellar WJ (2014) Unraveling complex viral infections in cassava (Manihot esculenta Crantz) from Colombia. Virus Res 186:76–86

Chávez P, Zorogastúa P, Chuquillanqui C, Salazar L, Mares V, Quiroz R (2009) Assessing potato yellow vein virus (PYVV) infection using remotely sensed data. Int J Pest Manage 55(3):251–256

Chávez P, Yarlequé C, Piro O, Posadas A, Mares V, Loayza H, Chuquillanqui C, Zorogastúa P, Flexas J, Quiroz R (2010) Applying multifractal analysis to remotely sensed data for assessing PYVV infection in potato (Solanum tuberosum L.) crops. Remote Sens 2(5):1197–1216

Cuellar WJ, Cruzado RK, Fuentes S, Untiveros M, Soto M, Kreuze JF (2011a) Sequence characterization of a Peruvian isolate of sweet potato chlorotic stunt virus: further variability and a model for p22 acquisition. Virus Res 157(1):111–115. https://doi.org/10.1016/j.virusres.2011.01.010

Cuellar WJ, De Souza J, Barrantes I, Fuentes S, Kreuze JF (2011b) Distinct cavemoviruses interact synergistically with sweet potato chlorotic stunt virus (genus Crinivirus) in cultivated sweet potato. J Gen Virol 92(5):1233–1243. https://doi.org/10.1099/vir.0.029975-0

Cuellar W, Mwanzia L, Lourido D, Garcia C, Martínez A, Cruz P, Pino L, Tohme J (2018) PestDisPlace: Monitoring the distribution of pests and diseases. Version 2.0 International Center for Tropical Agriculture (CIAT). Available at: https://pestdisplace.org

Data Citation Synthesis Group: Joint Declaration of Data Citation Principles (2014) In: Martone M (ed). https://doi.org/10.25490/a97f-egyk

De Souza J, Fuentes S, Savenkov S, Cuellar W, Kreuze J (2013) The complete nucleotide sequence of sweet potato C6 virus: a carlavirus lacking a cysteine-rich protein. Arch Virol 158(6):1393–1396. https://doi.org/10.1007/s00705-013-1614-x

Douthwaite B (2020) Control of potato purple top in Ecuador: evaluation of CGIAR contributions to a policy outcome trajectory. International Potato Center (CIP), Lima, Peru. https://doi.org/10.4160/9789290605553

Duarte-Carvajalino JM, Alzate DF, Ramirez AA, Santa-Sepulveda JD, Fajardo-Rojas AE, Soto-Suárez M (2018) Evaluating late blight severity in potato crops using unmanned aerial vehicles and machine learning algorithms. Remote Sens 10(10):1513

Ecma International (2017) The JSON data interchange syntax. Ecma International. https://www.ecma-international.org/publications-and-standards/standards/ecma-404/

Fabregas R, Kremer M, Schilbach F (2019) Realizing the potential of digital development: the case of agricultural advice. Science 366(6471):eaay3038. https://doi.org/10.1126/science.aay3038

Fand BB, Tonnang HE, Kumar M, Kamble AL, Bal SK (2014) A temperature-based phenology model for predicting development, survival and population growth potential of the mealybug, Phenacoccus solenopsis Tinsley (Hemiptera: Pseudococcidae). Crop Prot 55:98–108

Fand BB, Sul NT, Bal SK, Minhas P (2015) Temperature impacts the development and survival of common cutworm (Spodoptera litura): simulation and visualization of potential population growth in India under warmer temperatures through life cycle modelling and spatial mapping. PLoS One 10(4):e0124682

Fielding RT, Taylor RN, Erenkrantz JR, Gorlick MM, Whitehead J, Khare R, Oreizy P (2017) Reflections on the REST architectural style and" principled design of the modern web architecture"(impact paper award). In: Proceedings of the 2017 11th Joint Meeting on Foundations of Software Engineering, pp 4–14

Fuentes S, Heider B, Tasso RC, Romero E, Zum Felde T, Kreuze JF (2012) Complete genome sequence of a potyvirus infecting yam beans (Pachyrhizus spp.) in Peru. Arch Virol 157(4):773–776. https://doi.org/10.1007/s00705-011-1214-6

Fuentes S, Jones RA, Matsuoka H, Ohshima K, Kreuze J, Gibbs AJ (2019a) Potato virus Y; the Andean connection. Virus Evol 5(2):vez037

Fuentes S, Perez A, Kreuze J (2019b) Dataset for: The Peruvian potato virome. https://doi.org/10.21223/P3/YFHLQU

Fuentes S, Gibbs AJ, Adams IP, Wilson C, Botermans M, Fox A, Kreuze J, Boonham N, Kehoe MA, Jones RA (2020) Potato virus a isolates from three continents: their biological properties, phylogenetics, and prehistory. Phytopathology®:PHYTO-08-20-0354-FI

Gamarra H, Carhuapoma P, Cumapa L, González G, Muñoz J, Sporleder M, Kreuze J (2020a) A temperature-driven model for potato yellow vein virus transmission efficacy by Trialeurodes vaporariorum (Hemiptera: Aleyrodidae). Virus Res 289:198109

Gamarra H, Sporleder M, Carhuapoma P, Kroschel J, Kreuze J (2020b) A temperature-dependent phenology model for the greenhouse whitefly Trialeurodes vaporariorum (Hemiptera: Aleyrodidae). Virus Res 289:198107

Gold KM, Townsend PA, Herrmann I, Gevens AJ (2020) Investigating potato late blight physiological differences across potato cultivars with spectroscopy and machine learning. Plant Sci 295:110316

Golhani K, Balasundram SK, Vadamalai G, Pradhan B (2018) A review of neural networks in plant disease detection using hyperspectral data. Inf Process Agric 5(3):354–371

Griffel L, Delparte D, Edwards J (2018) Using support vector machines classification to differentiate spectral signatures of potato plants infected with potato virus Y. Comput Electron Agric 153:318–324

GSMA Intelligence (2020) The state of mobile internet connectivity 2020. Accessed 05/12/2021; Available at https://www.gsma.com/r/wp-content/uploads/2020/09/GSMA-State-of-Mobile-Internet-Connectivity-Report-2020.pdfnstraints often hampers 61p

Hadfield J, Megill C, Bell SM, Huddleston J, Potter B, Callender C, Sagulenko P, Bedford T, Neher RA (2018) Nextstrain: real-time tracking of pathogen evolution. Bioinformatics 34(23):4121–4123

Hanafi M, Tahzima R, Kaab SB, Tamisier L, Roux N, Massart S (2020) Identification of divergent isolates of banana mild mosaic virus and development of a new diagnostic primer to improve detection. Pathogens 9(12):1045

Heim R, Wright I, Allen A, Geedicke I, Oldeland J (2019) Developing a spectral disease index for myrtle rust (Austropuccinia psidii). Plant Pathol 68(4):738–745

Ifeoma OD, Mthitwa HT (2015) An analysis of the impact of the use of mobile communication technologies by farmers in Zimbabwe. A case study of Esoko and EcoFarmers platforms. In: Proceedings of SIG GlobDev Pre-ECIS Workshop. SIG GlobDev Munster, Germany

Ji S, Zhang C, Xu A, Shi Y, Duan Y (2018) 3D convolutional neural networks for crop classification with multi-temporal remote sensing images. Remote Sens 10(1):75

Jimenez D, Ramirez-Villegas J (2018) Unlocking big data's potential to strengthen farmers' resilience: the platform for big data in agriculture. Ospina, AV big data for resilience storybook: experiences integrating big data into resilience programming. International for Sustainable Development, Winnipeg, pp 97–108. https://www.iisd.org/library/big-data-resilience-storybook

Johansen K, Sohlbach M, Sullivan B, Stringer S, Peasley D, Phinn S (2014) Mapping banana plants from high spatial resolution orthophotos to facilitate plant health assessment. Remote Sens 6(9):8261–8286

Kamilaris A, Prenafeta-Boldú FX (2018) Deep learning in agriculture: a survey. Comput Electron Agric 147:70–90

Khadioli N, Tonnang Z, Muchugu E, Ong'amo G, Achia T, Kipchirchir I, Kroschel J, Le Ru B (2014) Effect of temperature on the phenology of Chilo partellus (Swinhoe) (Lepidoptera, Crambidae); simulation and visualization of the potential future distribution of C. partellus in Africa under warmer temperatures through the development of life-table parameters. Bull Entomol Res 104(6):809

Kratz JE, Strasser C (2015a) Making data count. Sci Data 2(1):1–5

Kratz JE, Strasser C (2015b) Researcher perspectives on publication and peer review of data. PLoS One 10(2):e0117619

Kreuze JF, Perez A, Untiveros M, Quispe D, Fuentes S, Barker I, Simon R (2009) Complete viral genome sequence and discovery of novel viruses by deep sequencing of small RNAs: a generic method for diagnosis, discovery and sequencing of viruses. Virology 388(1):1–7. https://doi.org/10.1016/j.virol.2009.03.024

Kreuze J, Koenig R, De Souza J, Vetten HJ, Muller G, Flores B, Ziebell H, Cuellar W (2013) The complete genome sequences of a Peruvian and a Colombian isolate of Andean potato latent virus and partial sequences of further isolates suggest the existence of two distinct potato-infecting tymovirus species. Virus Res 173(2):431–435

Kreuze JF, Perez A, Gargurevich MG, Cuellar WJ (2020) Badnaviruses of sweet potato: symptomless coinhabitants on a global scale. Front Plant Sci 11:313

Kroschel J, Sporleder M, Tonnang H, Juarez H, Carhuapoma P, Gonzales J, Simon R (2013) Predicting climate-change-caused changes in global temperature on potato tuber moth Phthorimaea operculella (Zeller) distribution and abundance using phenology modeling and GIS mapping. Agric For Meteorol 170:228–241

Kroschel J, Mujica N, Carhuapoma P, Sporleder M (2016) Pest distribution and risk atlas for Africa. Potential global and regional distribution and abundance of agricultural and horticultural pests and associated biocontrol agents under current and future climates. International Potato Center (CIP), Lima, Peru

Kutnjak D, Silvestre R, Cuellar W, Perez W, Müller G, Ravnikar M, Kreuze J (2014) Complete genome sequences of new divergent potato virus X isolates and discrimination between strains in a mixed infection using small RNAs sequencing approach. Virus Res 191:45–50

Leiva AM, Siriwan W, Lopez-Alvarez D, Barrantes I, Hemniam N, Saokham K, Cuellar WJ (2020) Nanopore-based complete genome sequence of a Sri Lankan cassava mosaic virus (Geminivirus) strain from Thailand. Microbiol Res Announcements 9(6)

Lu J, Miao Y, Huang Y, Shi W, Hu X, Wang X, Wan J (2015) Evaluating an unmanned aerial vehicle-based remote sensing system for estimation of rice nitrogen status. In: 2015 fourth international conference on agro-geoinformatics (Agro-geoinformatics). IEEE, pp. 198–203

Manyika J, Ramaswamy S, Khanna S, Sarrazin H, Pinkus G, Sethupathy G, Yaffe A (2015) Digital America: a tale of the haves and have-mores

McCampbell M, Schut M, Van den Bergh I, van Schagen B, Vanlauwe B, Blomme G, Gaidashova S, Njukwe E, Leeuwis C (2018) Xanthomonas Wilt of Banana (BXW) in Central Africa: opportunities, challenges, and pathways for citizen science and ICT-based control and prevention strategies. NJAS-Wageningen J Life Sci 86:89–100

Mishra P, Polder G, Vilfan N (2020) Close range spectral imaging for disease detection in plants using autonomous platforms: a review on recent studies. Curr Robot Rep 1(2):43–48

Monger WA, Alicai T, Ndunguru J, Kinyua Z, Potts M, Reeder R, Miano D, Adams I, Boonham N, Glover R (2010) The complete genome sequence of the Tanzanian strain of cassava brown streak virus and comparison with the Ugandan strain sequence. Arch Virol 155(3):429–433

Mrisho LM, Mbilinyi NA, Ndalahwa M, Ramcharan AM, Kehs AK, McCloskey PC, Murithi H, Hughes DP, Legg JP (2020) Accuracy of a smartphone-based object detection model, PlantVillage Nuru, in identifying the foliar symptoms of the viral diseases of cassava–CMD and CBSD. Front Plant Sci 11:1964

Mujica N, Sporleder M, Carhuapoma P, Kroschel J (2017) A temperature-dependent phenology model for *Liriomyza huidobrensis* (Diptera: Agromyzidae). J Econ Entomol 110(3):1333–1344

Mwalusepo S, Tonnang HE, Massawe ES, Okuku GO, Khadioli N, Johansson T, Calatayud P-A, Le Ru BP (2015) Predicting the impact of temperature change on the future distribution of maize stem borers and their natural enemies along East African mountain gradients using phenology models. PLoS One 10(6):e0130427

Narouei-Khandan HA, Shakya SK, Garrett KA, Goss EM, Dufault NS, Andrade-Piedra JL, Asseng S, Wallach D, van Bruggen AH (2020) BLIGHTSIM: a new potato late blight model simulating the response of Phytophthora infestans to diurnal temperature and humidity fluctuations in relation to climate change. Pathogens 9(8):659

Orlandini S, Magarey RD, Park EW, Sporleder M, Kroschel J (2020) Methods of agroclimatology: modeling approaches for pests and diseases. Agroclimatol Link Agric Clim 60:453–488

Patel N, Chittamuru D, Jain A, Dave P, Parikh TS (2010) Avaaj Otalo: a field study of an interactive voice forum for small farmers in rural India. In: Proceedings of the SIGCHI conference on human factors in computing systems, pp. 733–742

Pérez W, Arias R, Taipe A, Ortiz O, Forbes GA, Andrade-Piedra J, Kromann P (2020) A simple, hand-held decision support designed tool to help resource-poor farmers improve potato late blight management. Crop Protection. 105186

Piette JD (2000) Interactive voice response systems in the diagnosis and management of chronic disease. Am J Manag Care 6(7):817–827

Polder G, Blok PM, de Villiers HA, van der Wolf JM, Kamp J (2019) Potato virus Y detection in seed potatoes using deep learning on hyperspectral images. Front Plant Sci 10:209

Ramcharan A, Baranowski K, McCloskey P, Ahmed B, Legg J, Hughes DP (2017) Deep learning for image-based cassava disease detection. Front Plant Sci 8:1852

Ramcharan A, McCloskey P, Baranowski K, Mbilinyi N, Mrisho L, Ndalahwa M, Legg J, Hughes DP (2019) A mobile-based deep learning model for cassava disease diagnosis. Front Plant Sci 10:272

Rao MS, Prasad T (2020) Temperature based phenology model for predicting establishment and survival of Spodoptera litura (Fab.) on groundnut during climate change scenario in India. J Agrometeorol 22(1):24–32

Ray SS, Jain N, Arora R, Chavan S, Panigrahy S (2011) Utility of hyperspectral data for potato late blight disease detection. J Ind Soc Remote Sensing 39(2):161–169

Rebaudo F, Faye E, Dangles O (2016) Microclimate data improve predictions of insect abundance models based on calibrated spatiotemporal temperatures. Front Physiol 7:139

Sayers EW, Beck J, Bolton EE, Bourexis D, Brister JR, Canese K, Comeau DC, Funk K, Kim S, Klimke W (2021) Database resources of the national center for biotechnology information. Nucleic Acids Res 49(D1):D10

Selby P, Abbeloos R, Backlund JE, Basterrechea Salido M, Bauchet G, Benites-Alfaro OE, Birkett C, Calaminos VC, Carceller P, Cornut G (2019) BrAPI—an application programming interface for plant breeding applications. Bioinformatics 35(20):4147–4155

Selvaraj MG, Vergara A, Ruiz H, Safari N, Elayabalan S, Ocimati W, Blomme G (2019) AI-powered banana diseases and pest detection. Plant Methods 15(1):92

Selvaraj MG, Vergara A, Montenegro F, Alonso Ruiz H, Safari N, Raymaekers D, Ocimati W, Ntamwira J, Tits L, Omondi AB, Blomme G (2020) Detection of banana plants and their major diseases through aerial images and machine learning methods: a case study in DR Congo and Republic of Benin. ISPRS J Photogramm Remote Sens 169:110–124. https://doi.org/10.1016/j.isprsjprs.2020.08.025

Shrestha R, Arnaud E, Mauleon R, Senger M, Davenport GF, Hancock D, Morrison N, Bruskiewich R, McLaren G (2010) Multifunctional crop trait ontology for breeders' data: field book, annotation, data discovery and semantic enrichment of the literature. AoB plants 2010

Silvestre R, Fuentes S, Risco R, Berrocal A, Adams I, Fox A, Cuellar WJ, Kreuze J (2020) Characterization of distinct strains of an aphid-transmitted ilarvirus (Fam. Bromoviridae) infecting different hosts from South America. Virus Res 282:197944

Siriwan W, Jimenez J, Hemniam N, Saokham K, Lopez-Alvarez D, Leiva AM, Martinez A, Mwanzia L, Becerra LA, Cuellar WJ (2020) Surveillance and diagnostics of the emergent Sri Lankan cassava mosaic virus (Fam. Geminiviridae) in Southeast Asia. Virus Res. 197959

Soh BSB, Kekeunou S, Nanga Nanga S, Dongmo M, Rachid H (2018) Effect of temperature on the biological parameters of the cabbage aphid Brevicoryne brassicae. Ecol Evol 8(23):11819–11832

Souza Richards R, Adams IP, Kreuze JF, De Souza J, Cuellar W, Dulleman AM, Van Der Vlugt RAA, Glover R, Hany U, Dickinson M, Boonham N (2014) The complete genome sequence of two isolates of potato black ringspot virus and their relationship to other isolates and nepoviruses. Arch Virol 159(4):811–815. https://doi.org/10.1007/s00705-013-1871-8

Sporleder M, Kroschel J (2008) The potato tuber moth granulovirus (PoGV): use, limitations and possibilities for field applications. Integrated Pest Management for the Potato Tuber Moth-a Potato Pest of Global Importance Tropical Agriculture 20:49–71

Sporleder M, Simon R, Juarez H, Kroschel J (2008) Regional and seasonal forecasting of the potato tuber moth using a temperature-driven phenology model linked with geographic information systems. In: Integrated Pest Management for the Potato Tuber Moth Phthorimaea Operculella Zeller—A Potato Pest of Global Importance Weikersheim. Margraf Publishers, Germany, pp 15–30

Steward BL, Gai J, Tang L (2019) The use of agricultural robots in weed management and control. In: Billingsley J (ed) Robotics and automation for improving agriculture, vol 44. Agricultural and Biosystems Engineering Publications. Burleigh Dodds Science Publishing, Cambridge. https://doi.org/10.19103/AS.2019.0056.13

Taechatanasat P, Armstrong L (2014) Decision support system data for farmer decision making. In: Perth WA (ed) Proceedings of Asian federation for information technology in agriculture. Society of Information and Communication Technologies in Agriculture. https://ro.ecu.edu.au/ecuworkspost2013/855/., Australian, pp 472–486

Tonnang E, Juarez H, Carhuapoma P, Gonzales J, Mendoza D, Sporleder M, Simon R, Kroschel J (2013) ILCYM-insect life cycle modeling. A software package for developing temperature-based insect phenology models with applications for local, regional and global analysis of insect population and mapping. International Potato Center, Lima, Peru, p 193

van de Vossenberg BT, Visser M, Bruinsma M, Koenraadt HM, Westenberg M, Botermans M (2020) Real-time tracking of tomato brown rugose fruit virus (ToBRFV) outbreaks in the Netherlands using Nextstrain. PLoS One 15(10). https://doi.org/10.1371/journal.pone.0234671

W3C (2009) Extensible Markup Language (XML) 1.0 (Fifth Edition). W3C. https://www.w3.org/TR/xml/

Wang M, Abad J, Fuentes S, Li R (2013) Complete genome sequence of the original Taiwanese isolate of sweet potato latent virus and its relationship to other potyviruses infecting sweet potato. Arch Virol 158(10):2189–2192. https://doi.org/10.1007/s00705-013-1705-8

Wanjala BW, Ateka EM, Miano DW, Low JW, Kreuze JF (2020) Storage root yield of Sweetpotato as influenced by Sweetpotato leaf curl virus and its interaction with Sweetpotato feathery mottle virus and Sweetpotato chlorotic stunt virus in Kenya. Plant Dis 104(5):1477–1486

Wilkinson MD, Dumontier M, Aalbersberg IJ, Appleton G, Axton M, Baak A, Blomberg N, Boiten J-W, da Silva Santos LB, Bourne PE (2016) The FAIR guiding principles for scientific data management and stewardship. Sci Data 3(1):1–9

Xing Y, Hernandez Nopsa JF, Andersen KF, Andrade-Piedra JL, Beed FD, Blomme G, Carvajal-Yepes M, Coyne DL, Cuellar WJ, Forbes GA, Kreuze JF, Kroschel J, Kumar PL, Legg JP, Parker M, Schulte-Geldermann E, Sharma K, Garrett KA (2020) Global cropland connectivity: a risk factor for invasion and saturation by emerging pathogens and pests. Bioscience 70(9):744–758. https://doi.org/10.1093/biosci/biaa067

Zhai Z, Martínez JF, Beltran V, Martínez NL (2020) Decision support systems for agriculture 4.0: survey and challenges. Comput Electron Agric 170:105256

Chapter 10
Scaling Banana Bacterial Wilt Management Through Single Diseased Stem Removal in the Great Lakes Region

Enoch Kikulwe (ID)**, Marsy Asindu** (ID)**, Walter Ocimati** (ID)**, Susan Ajambo** (ID)**, William Tinzaara** (ID)**, Francois Iradukunda** (ID)**, and Guy Blomme** (ID)

Abstract *Xanthomonas* wilt (XW) of banana caused by *Xanthomonas vasicola* pv. *musacearum* (Xvm) is an important emerging and non-curable infectious disease which can cause up to 100% yield loss. At the start of the XW epidemic, complete uprooting of diseased mats (CMU) was recommended. There was little adoption of CMU, especially by women farmers, because it was labor-intensive and it sacrificed banana production for up to 2 years. CMU assumed that infection on a single plant would systemically spread to all plants in a mat. However, field experiments showed that Xvm did not spread systemically in a mat and that latent infections occurred. As a result, not all shoots on an infected plant show symptoms. This led to the idea of removing only the visibly infected banana plants, referred to as single diseased stem removal (SDSR). The SDSR package comprises three innovations: (1) regularly cutting symptomatic stems at ground level, (2) sterilizing cutting tools with fire, and (3) early male bud removal using a forked stick. The SDSR package was promoted jointly with a set of complementary practices: (i) avoiding infections by browsing animals, (ii) using clean planting materials, (iii) bending leaves at the petiole level when intercropping in infected fields, (iv) training on disease recognition and epidemiology, and (v) demand-specific extension and knowledge sharing. Several

E. Kikulwe (✉) · W. Ocimati · S. Ajambo · W. Tinzaara
Alliance of Bioversity International & CIAT, Kampala, Uganda
e-mail: e.kikulwe@cgiar.org; w.ocimati@cgiar.org; s.ajambo@cgiar.org; w.tinzaara@cgiar.org

M. Asindu
International Livestock Research Institute, Kampala, Uganda
e-mail: m.asindu@cgiar.org

F. Iradukunda
Alliance of Bioversity International & CIAT, Bujumbura, Burundi
e-mail: f.iradukunda@cgiar.org

G. Blomme
The Alliance of Bioversity International and the International Center for Tropical Agriculture (CIAT), Addis Ababa, Ethiopia
e-mail: g.blomme@cgiar.org

© The Author(s) 2022
G. Thiele et al. (eds.), *Root, Tuber and Banana Food System Innovations*,
https://doi.org/10.1007/978-3-030-92022-7_10

approaches that have been used for scaling out XW management technologies are documented in this chapter. This review looks at the process, practices, challenges, lessons learned, and future policy implications associated with scaling of XW management practices.

10.1 Introduction

10.1.1 Importance of the Banana Crop

Bananas and plantains (*Musa* spp.), hereafter called banana, are an economically important food and income security crop (FAO 2017). The crop is a principal source of carbohydrates for millions of people worldwide and fetches a large revenue share in domestic and international markets (Siddhesh and Thumballi 2017). It is currently cultivated in over 130 countries, on over 11 million hectares with a global production of about 155 million tons (FAO 2019). About one-third of the global banana production comes from Africa of which more than 50% is produced in the Great Lakes Region (GLR) including Burundi, Rwanda, the Democratic Republic of the Congo (DR Congo), Uganda, Kenya, and Tanzania (FAO 2019). Bananas of four use types are grown in the region, including cooking, brewing, dessert, and roasting bananas. Choice of a specific banana type varies by location, farmers' preferences, market demand, and climate. Cooking bananas dominate in most areas, followed by dessert, brewing, and roasting types (Lusty and Smale 2003). The GLR is the highest banana-consuming region of the world with annual per capita consumption almost 15 times the world's average and triple that of Africa (FAO 2019) (Fig. 10.1).

Bananas are also processed into puree, used for making banana drinks (Adeniji et al. 2010). Ripe bananas with a sweet taste and fine flavor and texture are also made into jam (Aurore et al. 2009). Leaves are used for thatching, for weaving baskets and mats, as a food wrapper for marketing and cooking, coverings over food, tablecloths, and plates, while extracted fibers are used as raw material for making specialty papers (Muraleedharan and Perumal 2010). The banana plant also holds medicinal potential and offers feed for animals. Starch extracted from the fruit has industrial uses, while the banana field confers high social status on the owners (Lusty and Smale 2003) who are predominantly men (Ajambo et al. 2018). The crop also controls soil erosion, sequesters carbon, and recycles nutrients (Lufafa et al. 2003; Ocimati et al. 2018; Kamusingize et al. 2017). Despite the great contribution of bananas to the GLR, average farm productivity has been declining since 2014 as opposed to African and global trends over the same period (FAO 2019) (Fig. 10.2).

This progressive decline in banana productivity in the GLR is attributed to suboptimal management, declining soil fertility, increased moisture stress, postharvest losses, gender inequalities in access to resources, and biotic constraints. The burden of pests and diseases has especially been on the rise in the past three decades (Jones 2000; Tushemereirwe et al. 2004; Nyombi et al. 2013; Tinzaara et al. 2018). In the

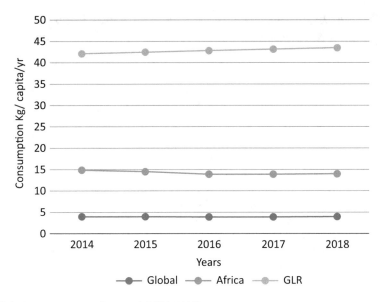

Fig. 10.1 Banana consumption trend (2014–2018)

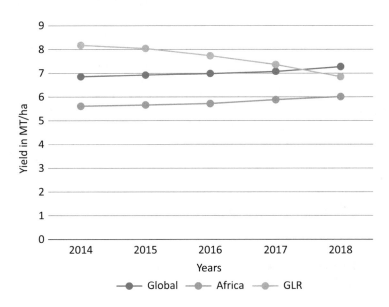

Fig. 10.2 Banana productivity trends (2014–2018)

late 1990s, the fungal diseases black leaf streak or black sigatoka and *Fusarium* wilt had been considered the most important banana diseases in the region (Blomme et al. 2017a). However, the introduction and spread of a bacterial disease called *Xanthomonas* wilt, with significant impacts on banana yields and productivity, has become a force to reckon with in the past two decades.

10.1.2   *Xanthomonas Wilt of Banana and Enset Emerges as a Serious Problem*

Xanthomonas wilt (XW) of banana and enset, caused by the bacterium *Xanthomonas vasicola* pv. *musacearum* (Xvm), is currently one of the biggest threats to banana production in the GLR (Tripathi et al. 2009; Nkuba et al. 2015). First reported in Ethiopia (Castellani 1939; Yirgou and Bradbury 1968, 1974), XW was later observed in Uganda and the Eastern DR Congo in 2001 (Tushemereirwe et al. 2004; Ndungo et al. 2006). Since then, the disease has spread to the entire GLR, compromising production and food and income security (Reeder et al. 2007; Ndungo et al., 2008; FAO 2012). Xvm is mainly spread by insect vectors, contaminated cutting tools, and infected planting materials (Eden-Green, 2004; Tinzaara et al. 2006; Shimwela et al. 2016a, b, 2017). XW also spreads through long-distance trade (bunches, leaves used as wrapping material); other vectors such as birds, bats, and browsing animals are also common (Smith et al. 2008; Buregyeya et al. 2008; Blomme et al. 2014; Kikulwe and Asindu 2020).

Symptoms during the vegetative stage of a plant are yellowing and wilting of leaves, while floral infections start with wilting of the male buds and rachis, followed by premature ripening and rotting of the fruits, and finally the death of the plant (Ocimati et al. 2013, 2015; Nakato et al. 2014; Blomme et al. 2017a; Kikulwe and Asindu 2020). XW causes acute infections that can lead to the complete loss of a banana garden. It infects all edible *Musa* cultivars in the GLR and causes up to 100% yield loss, severely compromising livelihoods and food security for banana farming households (Kalyebara et al. 2006; Abele and Pillay 2007). Farmers respond to XW by abandoning bananas in favor of other crops (Desire et al. 2016; Ocimati et al. 2018, 2020). There are currently no chemicals, biocontrol agents, or resistant varieties available to control XW (Tripathi et al. 2009). XW management has relied on cultural controls that have evolved or been improved through iterations of co-innovation with farmers over the past two decades. For instance, it was the farmers who first pointed out that SDSR was effective (Andrade-Piedra et al., 2016). Some of these cultural techniques, their evolution, and the approaches adopted for their wider scaling are discussed in this chapter.

10.1.3   *Technology Development for Xanthomonas Wilt Management*

Promoting complete disease mat uprooting was only partially successful. During the initial years of the XW epidemic in the GLR, management recommendations were based on practices used in Asia for controlling Moko and blood disease (caused by the bacteria *Ralstonia solanacearum* and *Ralstonia syzygii* subsp. *celebesensis*, respectively) that have similar spread mechanisms (Blomme et al. 2014). These practices included (1) rogueing entire mats or fields when a single plant was infected within a mat, followed by disposal of rogued plants (by burying or burning), referred to here as the complete uprooting of diseased mats (CMU), (2) sterilizing garden tools between farm operations using fire or a 3.5% sodium hypochlorite solution, (3) timely removal of male buds using a forked stick to prevent transmission of bacteria by insects, and (4) the use of clean planting materials (Karamura et al. 2006; Blomme et al. 2014, 2017b, 2019).

These control strategies were enforced through massive awareness creation campaigns and bylaws forcing the use of the practices. However, XW control failed and the disease continued to spread to new areas. It was later realized that smallholders greatly resisted this impractical, costly strategy (Tushemereirwe et al. 2006; Blomme et al. 2014; Kikulwe et al. 2018). Uprooting and destroying whole banana mats (i.e., the mother plant and attached lateral shoots) was time-consuming, labor-intensive, and expensive, especially in fields with many diseased mats. This made it impractical, especially for elderly and female farmers who were not up to this back-breaking task or could not afford to hire labor. Besides CMU, debudding was also labor-intensive and unattractive for women who reported they got tired and their chests hurt after debudding. Women also voiced concerns of being stung by bees and having male buds fall on their heads (FGD with female farmers in central Uganda). Farmers did not see any benefits from using the practices, and the disease continued to spread across farms over the years (Mwangi and Nakato 2009).

10.1.3.1 Learning Leads to Single Diseased Stem Removal

The drawbacks of CMU stimulated further research to gain insights into the epidemiology of the disease and varietal responses in order to develop farmer-friendly management options (Blomme et al. 2014). These studies found that Xvm bacteria do not colonize all lateral shoots in a physically interconnected mat (Ocimati et al. 2013, 2015). Healthy-looking plants were still observed and banana bunches were still harvested even in heavily infected fields. These observations formed the basis of a new XW control package that comprised the cutting of only the diseased banana plants in a mat, a practice referred to as the single diseased stem removal (SDSR) technique (Blomme et al. 2017b).

The SDSR package comprises (1) regularly (at least weekly at onset) cutting, at soil level, of all symptomatic stems; (2) removing the apical meristems of cut plants in the vegetative stage to prevent regrowth; (3) adding soil onto the cut surface to minimize vector contact with bacterial ooze; (4) disinfecting the cutting tools with fire after cutting all visibly diseased stems; and (5) disposing of cut stems at the edge of the garden or in a compost heap (leaving the removed stems intact, as additional cuts would enhance the oozing out of bacteria). Farm tool use for de-leafing and de-suckering excess asymptomatic plants is avoided or minimized during the first 3–6 months of SDSR application while using a forked wooden stick to remove male buds after the formation of the last hand of the bunch (Blomme et al. 2017b). Where necessary, e.g., to let in more light for intercrops, excess leaves are only bent at the petiole level and aligned along the pseudostem (Blomme et al. 2017b). The regular (weekly) use of SDSR in controlled trials and on farm reduced and maintained XW incidence to below 1% within 10 months (Blomme et al. 2017b, 2019).

Single diseased stem removal is based on the idea that the continued removal of only the diseased plants in a mat will reduce the amount of Xvm inoculum, reduce the risk of Xvm spread to other plants in a mat, and ultimately bring down disease incidence to an acceptable level. The method also demands less time and labor than removing a complete mat, so it is more acceptable to farmers (Blomme et al. 2021). The total time for SDSR application on a single plant has been observed to be 88% less, averaging 4.3 minutes compared with 36.5 minutes for CMU (Blomme et al. 2021). Consequently, the cost of SDSR was 96% lower than that for CMU (Blomme et al. 2021). Moreover, farmers who use CMU lose several rounds of harvests from a mat, whereas with SDSR they continue harvesting healthy bunches as the field gradually recovers from the disease. Compared to CMU, the labor-saving SDSR increased the potential of women to use management practices in Burundi (Iradukunda et al. 2019). In Uganda, female-headed households were more likely to adopt SDSR, with an average of 15% non-adopters compared to 28% low adopters and 24% full adopters (Gotor et al. 2020).

10.1.3.2 Outcome of SDSR Adoption

SDSR was observed to reduce and maintain XW incidence on farm to below 1% within 10 months, resulting in positive and significant impacts on banana production and sales (Blomme et al. 2017b, 2019). An on-farm study in Rwanda showed that SDSR reduced disease as effectively as CMU, but offered a 96% recovery in control costs (Blomme et al. 2021). Success on farm has also been shown to vary with the level of adoption (Ocimati et al. 2019; Kikulwe et al. 2019). For instance, maximum benefits have been shown to occur when farmers adopt the full SDSR package, with the value of banana production increasing by US$462/ha/year (Kikulwe et al. 2019).

10.1.3.3 Potential Hindrances to the Success of SDSR

Despite the huge ray of hope for XW management through SDSR, latent infections have been shown to occur in infected lateral shoots which are attached to an infected mother plant, and it can take up to 24 months for symptoms to resurface in some mats, even though most latently infected plants continue to grow vigorously and produce edible bunches (Ocimati et al. 2013, 2015). Thus, follow-up over several years is recommended when practicing SDSR, as symptoms may still appear later.

Tool sterilization, a core component of SDSR and CMU, was infrequently used due to the expense and low availability of household bleach (3.5% sodium hypochlorite solution) in remote rural areas and because tools were damaged by repeated heating on fire (Blomme et al. 2014, 2019; Kikulwe et al. 2019). A later study showed that washing tools with cold water and soap or detergent (more readily available and cheaper) and immersing tools in boiling water for a minute are as effective as bleach (Ocimati et al. 2021). Using soap or detergent and boiling water will increase the options for tool sterilization and encourage more farmers to sterilize their tools.

Control of XW with clean planting materials (such as tissue culture plantlets) was also poorly adopted because of its high cost. There is little private-sector investment in banana seed production, because the crop is perennial and farmers rely on naturally regenerated planting materials (suckers) that are readily available in villages. However, informally sourced suckers are associated with a higher risk of disease spread. And farmers do not have the technical means to verify whether healthy-looking planting materials are XW-free or not (Jogo et al. 2013; Rutikanga et al. 2013; Kikulwe and Asindu 2020), especially given latent infections and incubation periods as long as 24 months (Ocimati et al. 2013, 2015). So, there is still a high risk of transmitting the disease through infected planting materials, even when farmers apply the other control practices effectively. The realization that XW causes less damage to varieties with persistent male bracts and flowers has led to the identification of varieties like KK (*Musa* ABB genome) as alternatives to the farmer preferred but highly susceptible Pisang Awak (*Musa* ABB genome) variety, which is called Kayinja in Uganda.

XW control may be uneven if farmers and traders fail to fully implement cultural practices. Women farmers often face constraints such as limited access to information, resources, and limited decision-making power (Kikulwe et al., 2018; Ajambo et al. 2020). The need to understand gender relations in terms of norms and who does what in banana production is crucial for informed targeting of SDSR. Men and women may be responsible for different production activities and fields; for instance, in Burundi, men do most of the production work for the commercial fields, while the women manage plantains grown for home consumption. Men are also more likely to attend trainings on bananas (Iradukunda et al. 2019). SDSR scaling approaches ought to integrate these dynamics and ensure access to information by those who do the actual work.

10.2 General Conceptual Framework Adopted for Scaling

For successful scaling to occur, stakeholders must understand the development problem at hand before exploring the appropriate innovations that can address the problem (IDIA 2017). The next steps are to foster dynamic relationships among multiple actors who enable scaling, facilitate learning interactions, and invest in local capacity and leadership to support sustainability (Hartmann and Linn 2008; Wigboldus et al. 2016; IDIA 2017). Scaling XW management in the GLR used this framework highlighting the disease as the baseline scenario (problem); its incidence was assessed and full impact on banana production was estimated (Fig. 10.3). The base scenario (XW) then informs on the need to innovate and generate technologies to combat XW considering the broader socioeconomic needs and benefits.

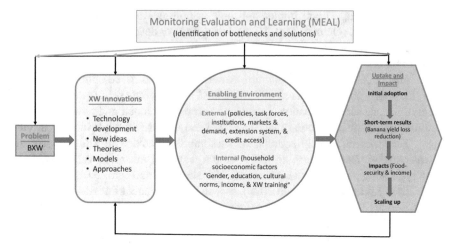

Fig. 10.3 Conceptual framework for XW scaling in East and Central Africa

Developed technologies and innovations are then piloted among target groups, also known as the diagnostic phase of scaling innovation (commonly done through trials, demonstration plots, farmer field schools, exchange visits, and lead farmers).

After testing the innovations, extra effort is allocated to creating an enabling environment for wider scaling, by identifying external and internal factors deemed influential to scaling the technologies. For instance, institutions are identified and examined for their capacity, resource availability, possible roles in scaling, and other elements necessary for collaborative scaling. Existing technologies and scaling channels are also critically examined for their readiness to be used in the fight against XW. This is possible through scaling readiness assessments in which scores are allocated to different technologies and scaling channels with the final choice based on overall scores for each component (Sartas et al. 2020).

Once the enabling environment is laid out and rules are defined (policies, legislative frameworks, etc.), strategic plans to take innovations to scale are drawn up, and the implementations are rolled out. The plans are drafted by a team comprising the developers of the innovation, farmer representatives, scaling partners, and scaling champions. Once the innovations are implemented, a monitoring and evaluation (M&E) system is set in place to document outcomes and impacts. This is also called the navigation stage (Sartas et al. 2020). The M&E is facilitated by the scaling champions and includes the primary scaling partners, farmer representatives, and the secondary scaling partners who also double as extension agents. Based on lessons learned from continuous M&E, systems are set in place to ensure that the scaling strategy is sustained. Depending on performance, innovations are improved or replaced with others that perform better. The overall process is generally overseen by the M&E framework, which draws lessons at every stage to inform the next steps.

10.2.1 XW Scaling Interventions in GLR

In the last two decades since the emergence of XW, several management interventions have been designed and scaled through collaborations of national and international institutions including National Agricultural Research Organization (NARO) and the Ministry of Agriculture, Animal Industry and Fisheries (MAAIF) in Uganda, Institut des Sciences Agronomiques du Burundi (ISABU) and Bureau Provinciale de L'Environnement, de l'Agriculture et de l'Élevage (BPEAE) in Burundi, Institut National pour l'Étude et la Recherche Agronomique (INERA) and Inspection Provinciale de l'Agriculture, Pêche et Élevage (IPAPEL) in DR Congo, Rwanda Agriculture Board (RAB), universities, Bioversity International, the International Institute of Tropical Agriculture (IITA), Catholic Relief Services (CRS), Réseau Burundi 2000Plus (RBU), and various universities and community-based farmer organizations (CBFOs). Some of these interventions have always been crosscutting, while others are country specific (Table 10.1). However, the level of scaling of these technologies differed in each country. To bridge this gap and make the XW management innovations accessible, several interventions were implemented over the past decade, mainly by NARIs and CGIAR centers. The scaling approaches evolved over time, as later efforts tried to address the limitations identified in the first ones.

The eradication and containment strategy of XW spread was deployed in the early years of the epidemic (2001–2007) by NARIs and their partners in Uganda, Rwanda, and DR Congo where localized outbreaks would be identified; then teams would be dispatched and paid to cut and bury the diseased mats (Mwangi and Nakato 2009). This costly approach was later replaced by the community sensitization for action strategy, practiced mainly in Uganda, involving printing and distribution of information leaflets, factsheets, brochures, and posters about XW. It was assumed that this information would trigger action by farmers, but it was later observed that although over 85% of farmers were aware, fewer than 30% were practicing (Bagamba et al. 2006; Tushemereirwe et al. 2006). As a countermeasure, the participatory development communication (PDC) approach was developed, tested, and promoted by NARO. This approach mobilized community stakeholders to

Table 10.1 XW disease management approaches used in SDSR by country

Approaches	Country			
	Uganda	Burundi	DR Congo	Rwanda
Eradication and containment strategy	✓		✓	✓
Community sensitization for action	✓			
Participatory development communication (PDC)	✓			
Community action intervention	✓			
Farmer field schools (FFS)	✓			
Learning and Experimentation Approaches For Farmers (LEAFF)	✓			
Scaling readiness assessment approach	✓	✓	✓	✓

explore solutions to XW, which they constituted into an action plan that detailed what needed to be done, when, where, and how to do it, who would take what responsibility and mechanisms for M&E. The PDC led to a drop in XW prevalence by 70% (Kubiriba and Tushemereirwe 2014). Unfortunately, this approach was limited by its dependence on an external PDC resource person and thus was unsustainable.

The community action intervention was later introduced by NARO in Uganda. The community action intervention was an improvement on the PDC that integrated mechanisms for enforcing the actions generated through the PDC. Its implementation led to the development and enforcement of bylaws at community level that led to a drop in XW prevalence by 68% and a banana yield recovery of 22% (Kubiriba et al. 2012).

The farmer field school (FFS) was introduced to reinforce adoption of XW strategies. The FFS mobilized community members for participatory discovery, decision-making, problem-solving, and stimulating local innovation while using the field as their classroom under the guidance of a technical facilitator (Ochola et al. 2015). The implementation of FFS in Uganda in 2006–2008 was accompanied by a drop in XW prevalence of 43% (Kubiriba et al. 2012; Kubiriba and Tushemereirwe 2014). A shortcoming of the FFS was its tendency to encourage the one-way flow of information from facilitators to farmers.

To address the limitations associated with the earlier approaches, Learning and Experimentation Approaches For Farmers (LEAFF) was developed by Bioversity International in collaboration with NARO in Uganda, with support from the McKnight Foundation. LEAFF aimed to foster sustainable and inclusive community-based management of XW (Tinzaara et al. 2019). LEAFF is similar to FFS: both are group-based, adult learning approaches that teach farmers how to experiment and solve problems independently. LEAFF, however, is more inclusive and involves farmers in the development of control measures through farmer action research networks which support experimentation and learning, value chain strengthening, enterprise development, and knowledge sharing. The community-level networks act like knowledge-sharing channels through which adoption challenges are explored.

LEAFF farmers were connected through mobile phones and the Internet to increase interactions within and between the groups, which are often lacking in FFS. This increased farmer-to-farmer interactions to enhance the sharing of experiences and skills in the quest to increase the effectiveness of the control measures. LEAFF exposed farmers to experimentation skills, including constraint identification, prioritization, hypothesis setting and testing, data collection and analysis, participatory M&E and reporting with emphasis on XW management, while tapping into the farmers' own experiences. LEAFF was found to be a powerful approach against XW; when applied in the study areas of Bushenyi and Kiboga districts, it contributed to a reduction in disease prevalence by 27% and then to its eradication in 82% of the farms vs 64% where the approach was not applied (Tinzaara et al. 2019). LEAFF worked directly with 220 farmers, but through trainings, meetings, and radio talk shows, the program reached an estimated 22,000 farmers.

10.2.2 Scaling Readiness Across Three Countries in GLR

Supported by the RTB scaling project, scaling readiness (see Chap. 3) was implemented in four countries in GLR: Uganda, Rwanda, DR Congo, and Burundi. The 2-year, RTB-funded scaling readiness project (January 2018 to December 2019) aimed to come up with a strategy to promote the implementation of previously identified XW management practices. The project kicked off with a series of workshops, including one in Kampala, Uganda, in April 2018, where scaling champions and primary scaling partners from the four project countries met to conduct scaling readiness assessment exercises. During this participatory effort, three teams from Burundi, Uganda, and DR Congo devised country-specific "rich pictures" and "visions of success." The need for gender-responsive scaling using action research and social learning approaches was highlighted during the workshop (Rietveld 2021).

The rich picture from the April 2018 workshop created a shared vision of how different institutions would work together to see that SDSR went to scale in each country. Figure 10.4 depicts the rich picture for Uganda showing how NARO would conduct research and disseminate the results through multiple channels to men and women farmers. The channels included local government institutions and secondary scaling partners who would help to train and pass on the information to other banana farmers through their extension staffs. Information on SDSR was also expected to reach the farmer through radio and TV shows, schools, churches, and local NGOs that promoted banana production. It was thought that religious and community leaders could be equipped with the SDSR messages and empowered to teach other farmers and that the traders in banana markets could share SDSR messages with farmers as they brought their fruit to sell.

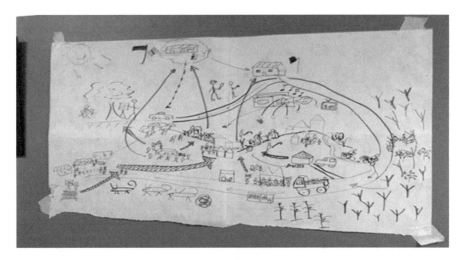

Fig. 10.4 A picture drawn by community members showing actions to scale SDSR in Uganda

The vision of success (also created during the March workshop) depicted the elements that needed to mesh together to enable SDSR interventions to go to full scale. It was drawn on two joined flipcharts, pointing out the institutions to be engaged and the activities they were to carry out to take SDSR to scale. The vision map also showed the situation before the SDSR interventions and the desired situation that would result from them. During this exercise, several public- and private-sector actors were identified to take part in SDSR scaling. There were calls for training and building the capacity of partners to design and implement gender actions to improve the impact of scaling (Fig. 10.5). The vision of success was later used to design the strategic plan for implementing XW management in each country. An assessment was also done of the institutions that were identified as scaling partners; only those with a shared vision to scale XW management practices and with strong community ties and skilled staff were picked to help implement the practices. The scaling champions from the CGIAR centers also used the vision map to develop their scaling action plans and select scaling channels.

After the workshop in Kampala, teams from DR Congo, Uganda, and Burundi devised country-specific scaling strategies. The three core innovations received above-average scores in Uganda possibly because these techniques had been refined over time and promoted in prior projects. However, these projects were not on a large scale, with full packages like it was planned for the SDSR scaling project. As such, they were deemed ready for scaling. Basing on the scoring of complementary components for Uganda (Table 10.2), the strategy focused on developing radio messages and factsheets with gender-sensitive illustrations of the XW management practices and on mobilizing agents to scale out the innovations to banana farmers. During the workshop in Uganda, tool sterilization was ranked the lowest of the core components, because fire damages metal tools. Farmers did not know in advance how long the tools should be kept in the fire, but this was later addressed by scaling

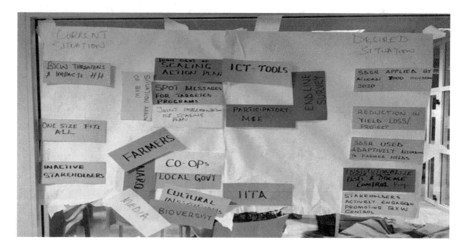

Fig. 10.5 Shared vision for SDSR scaling in Uganda

Table 10.2 Scaling readiness score for SDSR innovations (Uganda, May 2018)

May 2018			Readiness	Use	Critical component	Scaling readiness	Substitute	Outsource	Improve
Core components	Regularly remove diseased stems		8	4.5	N	12.5	0	0	
	Sterilize tools with fire		8	3.5	N	11.5	0	0	
	Remove male buds		8	4	N	12	0	0	
Complementary components									
	Factsheet		7.7	3.7	N	11.4	0	0	
	Posters		7.3	3.3	N	10.6	0	0	
	Radio messages		8	4.7	N	12.7	0	0	
	Short videos		4.3	2.3	Y	6.6	0	0	
	Written phone-based messages		6	1.3	Y	7.3	0	0	
	Scaling agents (ToT)		7.7	4.7	N	12.4	0	0	
	Community leaders' training on XW	Cultural leaders	6	1	Y	7	0	0	
		Political leaders	6.7	3	Y	9.7	0	0	
		Cooperative leaders	5.7	2	Y	7.7	0	0	
	Task force + bylaws		4.5	2.5	Y	7	0	0	
					Scale used:		Readiness (1 to 8)		
							Use (1 to 5)		

agents who promoted sodium hypochlorite (bleach) for sterilizing knives and machetes.

In a similar exercise held in DR Congo, tool sterilization with fire also received a low readiness score, for the same reasons, resulting in the recommendation to use cutting brigades to inspect and cut infected banana plants together with the community members instead of relying only on individual farmers. It was hoped that the cutting brigades would ensure regular monitoring and ensure the sustainability of the XW management activities.

Scaling readiness assessments were repeated twice for each country at an interval of 6 months under the supervision of the scaling project leader (principal investigator) in collaboration with the scaling champions of the respective countries, the primary scaling partners, farmer representatives, and secondary scaling agents (extension agents). During the reviews, core components were retained because their scores did not drop. But those complementary components with declining scores were replaced with others that earned low scores during the first review but gained by the second assessment. Bottlenecks were identified during implementation and alternatives were proposed. For instance, in Uganda, following the second phase of scaling readiness scoring, short videos were added, using drama centered on XW management. For the overall scaling readiness score for Uganda at the first readiness assessment, see Table 10.2.

In Burundi, the scaling readiness assessment exercise was used to refine and validate (1) the SDSR core components discussed during the RTB Scaling Fund kickoff workshop in March 2018 and (2) the additional complementary component shared during the scaling readiness meeting held in Uganda during May 2018. Both core and complementary components of SDSR were scored high and ready to go to scale. Although video and radio messages scored highly, participants in the scaling readiness assessment realized that these approaches would be too expensive given the lack of community-based communication infrastructure in Burundi. Therefore, media were abandoned in favor of a comprehensive factsheet (Fig. 10.6). To increase the innovation readiness of the factsheet, it was validated for comprehension and efficacy in the field (Table 10.3).

10.2.3   Enabling Policy and Institutional Factors for Success of XW Management Scaling

The institutional and policy environment was critical for the success of scaling SDSR innovations in all four countries. At the start of the project, Rwanda had no policy framework which would allow implementing the proposed innovations, so the 2-year project was used for trials in farmers' fields comparing the effectiveness, labor cost, and time demands of CMU and SDSR to influence policy decisions and enable the innovations to sail through to the implementation stage (Blomme et al. 2021). There was also an initiative termed information and communication

Fig. 10.6 Factsheet used in Burundi to scale SDSR

technologies for banana *Xanthomonas* wilt (ICT4BXW) to collect digital data from extension providers and farmers for developing an early warning system to curb the spread of XW and provide decision support on cost-effective XW control.

In Uganda, there was already a government buy-in by MAAIF for the SDSR package, since it replaced the mat uprooting technique originally promoted by NARO, its partners, and the government. More importantly, there was an established development strategy and investment plan (by MAAIF), policies (bylaws), and a partnership capable of taking XW innovations to scale. The success of the scaling project in Uganda was enhanced by integrating it into the Bill & Melinda Gates Foundation (BMGF)-funded Banana Agronomy project, which had a component for linking banana farmers to markets by selling collectively through the banana cooperative union. The linkage with markets was a motivation for adopting XW control, as farmers envisioned higher yields and a surplus they could sell.

In Burundi, farmers embraced the SDSR innovations as cost-effective and feasible. This played a pivotal role toward renewed trust in banana as a preferred food and cash crop. BPEAE (a government agricultural extension agency) staff quickly acknowledged and supported SDSR innovations for XW control. Even though SDSR had not been officially integrated into national policy, the many staff trained in the technique resulted in wider scaling. The SDSR team also received requests for advice from colleagues in other provinces facing XW. The head of the banana program of ISABU, the Burundi NARI, also supported the SDSR innovations. Much thought went into the scaling activities in the DR Congo, since design of interventions had to take into account the post-conflict situation, poor coordination of field activities, and the limited involvement of youth and women in managing

Table 10.3 Scaling readiness assessment score for SDSR innovations (Burundi, May 2018)

Component	Readiness	Use	Critical component	Scaling readiness	Substitute	Outsource	Improve
Core components							
Regular removal of diseased plants	9	7	No	16	No		
Tool sterilization by fire	8	7	No	15	No		
Male bud removal	9	7	No	16	No		
Complementary components							
Video +	3	0	Yes	3	Yes		
Screening videos for farmers	4	4	Yes	8	Yes		
Bending leaf	8	1	No	9	No	No	Yes
Field day	7	6	No	13	No		
Coordination meetings with DPAE	4	6	No	10	No		
Factsheet/brochure[a]	5	6	No	11	No		
Cascade training model	7	6	No	13	No		

[a]Investments required to increase readiness: real-world testing and validation of the factsheet

banana plantations. The lack of an enabling environment was exacerbated by limited government commitment and the poor skills of government extension services, even though there was an existing legal framework for XW detection and control. This left most of the coordination to be done by international NGOs, with some assistance from INERA.

10.2.4   Case Studies of Scaling Outcomes

During the implementation of the SDSR package, case stories on XW management were collected from farmers and documented for wider scaling. Some of the inspiring stories documented from the field in Uganda are highlighted below.

10.2.4.1   Case Story 1: SDSR Package for XW Management

Federesi Kenema, an amiable, 38-year-old mother of five, says her banana farming fortunes have gotten better since she was trained by NARO. Before the training, Kenema was in the dark about how to deal with XW, which had plagued her garden. Her crops in Kakinga village in Rwimi sub-county in Bunyangabu district were always devastated by the wilt, and as a control measure, she would ill-advisedly uproot most of her bananas. "I would notice that many of my bananas had gotten ripe whilst still young. Their leaves would turn yellow and wilt and when I would cut the pseudostem, it would ooze a yellowish liquid. It seemed as if the wilt was spreading from one sick plant to another, but much as I racked my brain, I could not come up with any plausible clues on how it was happening."

"The experts who were conducting the training came to my farm and saw firsthand how it had been damaged by XW. Then, I was trained in SDSR and other techniques like sterilizing cutting tools and bending dried leaves in heavily infected banana gardens," Kenema explains. Before, a farmer friend had talked her into cutting all of her diseased banana plants, rationalizing that it was the only cure. Unknowingly, however, the practice damaged the soil structure and soil fertility and it was a lot of work to boot. After seeing how counterproductive the practice of complete mat removal had been, Kenema adopted SDSR.

"I wanted to control the spread of XW and I was told that I would have to regularly cut each infected stem at ground level, instead of uprooting the whole mat," Kenema recounts. "Because bees frequently spread the wilt by moving back and forth between the male buds of banana plants, I was advised to remove male buds from my healthy banana plants with a forked stick, instead of using knives or pangas, which easily spread the infection." Kenema adds, "The experts also advised me to sterilize farm cutting tools like knives and pangas with fire or bleach whenever I used them to prune. There was less disease after I started using those practices. My banana garden is flourishing. I now harvest bigger banana bunches, meaning I can now sell at more competitive prices to the ever increasing banana buyers from across the country and from Rwanda."

In the past, at the peak of XW, Kenema could only harvest 7–12 small bunches every fortnight and each bunch went for the paltry price of 7000 Uganda shillings ($1.86). Now, thanks to the new practices she employs, she harvests 20–40 bunches every fortnight. She sells each bunch, depending on the size and season, between 10,000 and 15,000 Uganda shillings ($2.65 to $3.97). Kenema has been able to improve her family's standards of living, with the better monthly profits, worth 650,000 Uganda shillings ($172).

10.2.4.2 Case Story 2: Rising from the Ashes of XW

Mzee Stanley Rwabuchye, a banana, bean, and coffee farmer in Ntaza II, Ishaka municipality, regained his normal life after an intervention of NARO, the district, and Bioversity International. He grew up in a banana-growing family, and in his own words, "bananas have always been our source of food and income. The disease started in 2008 when the banana leaves started yellowing, but we continued with our work. Alarm bells went off when the banana bunches started yellowing and getting rotten while immature. During this time, a team from Bioversity International, the district and NARO came to our village. They told us to cut and bury the infected plants. The more we buried the infected plants, the more the disease spread. The team gave us posters to guide in controlling the disease."

He still has the banana bacterial wilt (XW) poster he was given in 2008 at the height of the outbreak (Fig. 10.7). The paper is brown with age, but the way he holds onto it says something about the importance of the information on it. "The team came back and advised us to stop burying the infected plants, but instead to chop them up and leave them on the ground to dry. We were also told to stop pruning the dry banana leaves and to use a forked stick to remove the male bud. By this time, my bananas were almost wiped out so I decided to cut down everything. After all the few remaining plants were also going to get infected. Shortly, NARO gave me new, disease-free suckers, which I planted and started my banana garden afresh. We were encouraged to disinfect our tools using bleach or fire."

Rwabuchye says, "I want to tell you about the time when I cut down my banana plants. Life was difficult. We lost our source of food and income, and started depending on donations. After our bananas were wiped out, we started growing maize, cassava and sweetpotatoes. These helped with food, but not for money. It was a desperate time, I tell you."

"Life is now back to normal. The plantations are thriving and trucks have come back to our village to buy bananas. Before the XW outbreak I used to harvest 80 bunches every month, but the price was as low as 2000 shillings ($0.53), even 500 per bunch and I'd earn 250,000 shillings per month ($66). Today, I harvest half of that, 40 bunches, but the price is better, 15,000 to 20,000 shillings per bunch, so I earn about 500,000 shillings per month.

Fig. 10.7 Poster describing the spread and control of XW

"I thank Bioversity International and NARO for teaching us this technology and saving our livelihood. If they had not come to our rescue, we would have lost everything and become destitute. Matooke (plantain) is our food and livelihood; my parents used to make beer from bananas and pay school fees for us. I have also educated my children with income from bananas."

10.2.5   Project Learning Outcomes

The rate of spread and the severity of bacterial wilt disease caught farmers and research scientists off guard. Nobody quite knew how to control this disease. Some said diseased banana fields should be destroyed. Others said that you only had to remove the male bud on an infected plant, while others insisted that sterilizing farm tools was the key. One thing was certain, the disease disrupted many countries' food security and created social unrest among the farmers. Farmers and scientists alike were alarmed by its virulence, for it could destroy the entire bunch as well as up to 100% of the plants in a field within a matter of weeks. Among the most affected varieties was Kayinja (Pisang Awak), which is used to produce beer in Uganda. As XW spread, researchers realized that bees spread the disease from plant to plant as they collected nectar, pollen, and sap from the male floral bud. Farmers have always removed male buds from young bunches of cooking varieties to allow the bunches to fill up and grow bigger. Beer bananas would retain their buds as farmers believed that their removal would affect the quality of juice and beer (Bagamba et al. 2006). This explains why the disease was rife in areas where Kayinja beer bananas predominated, which hampered efforts of scaling XW control packages. It was also realized that the disease caused less damage to varieties with persistent male and female floral bracts and neuter flowers. Although farmers were taught and adopted practices for managing the disease, developing resistant varieties offered the only effective long-term solution in terms of both time and cost.

Farmers who grow bananas as a business are more responsive to adopting XW management technologies in Uganda. Through the scaling readiness scoring approach, communication materials for scaling XW management measures were harmonized for all stakeholders. At the end, implementing teams were in a much better position to transfer timely, coherent, and effective messages on XW management to banana farmers. To improve efficiency of scaling through radios, fan clubs were created. This led to closer interaction between the radio DJs and fans. The closer ties fostered a favorable environment for hands-on practice sessions on some of the SDSR practices that had previously been aired on radio talk shows. The continued DJ-fan interaction saw XW management interventions move quickly beyond the initial project sites, i.e., wider scaling of the SDSR interventions.

Working with partners already implementing agricultural projects enhanced the sustainability of the project as they did not request additional funding to add XW management components to their existing activities. This is one reason why the implementation of the scaling project in Uganda was integrated into the BMGF-funded Banana Agronomy project led by NARO in collaboration with IITA, Bioversity International, Makerere University, and CABI. Lastly, a quick and sustainable model for wider scaling was realized through the use of "Star performing model farmers." These farmers were identified at the start of project implementation, trained on SDSR through demonstration plots in the banana farming communities. Each demo host was tasked to train 15 other banana farmers on a voluntary basis. This not only made it easier to communicate SDSR messages to other banana farmers but also quickly increased the number of beneficiaries of the scaling project.

We learned from experience that XW control requires commitment of all stakeholders along the banana value chain as there is a need for continued sensitization. In sites with land shortages, intercropping affected adoption of some of the XW technologies. Suspension of sterilized and unsterilized cutting tool use in heavily infected fields is not appropriate for farmers and for increased banana productivity; XW management should include all other recommended banana agronomic packages which do require the use of these cutting tools.

In Rwanda, approval by the government to scale the innovations to the farmers was still pending. This induced the scaling team to undertake a unique approach centered on doing trials to convince the government that SDSR controlled XW much better than CMU (see Blomme et al. 2021).

In DR Congo, the SDSR scaling project brought together many institutions to complement each other and to share the same message on XW management. These included farmers and their groups, community leaders, Bioversity International, IITA, INERA Mulungu (the National Agriculture Research body of DRC), IPAPEL (the DRC extension body), and local, national, and international NGOs. The NGOs that participated included Food for the Hungry International, World Vision, and Projet Agricole de Buhengere (PABU). Different communication methods were used, including music, field demonstrations, radio programs, training workshops, and posters. Despite this, women's participation and access to extension services remain low. For instance, only 34% of women participated in the trainings on SDSR. Where bananas are intercropped, the banana mats are predominantly managed by male farmers, while women manage annual crops (Blomme et al. 2017b). Conflicts may arise if men try to prune bananas without taking into account the women's intercropped bean plants, so it is important to integrate gender when designing XW scaling projects to ensure that both genders benefit equally.

Interactions with farmers revealed that they rarely sterilized tools with household bleach or fire in any of the scaling sites. This led to the exploration of cheaper, readily accessible, and easy-to-use alternatives. Washing metals in cold water with laundry soap or detergent or immersing tools in boiling water for a minute were found to be as effective as household bleach (Ocimati et al. 2021). Soap and detergents are readily available in shops even in remote communities. Washing tools to sterilize them is easy to use even where other techniques are not applicable, e.g., it may be hard to light a fire on rainy days or fire be a hazard, especially where farms are far from the homestead. Washing tools for a minute in soap and water was long enough to sterilize them (Ocimati et al. 2021). Repeated heating in fire will certainly weaken steel tools, and farmers are loathe to buying new ones any more frequently than is necessary.

Failure to take gender into account can hamper uptake of management recommendations. In Burundi, for instance, women were not motivated to use XW strategies because men sell the bananas and keep the money. In addition, access to information was unequal, with men participating in more trainings while women had limited access to radios and other sources of new ideas (Iradukunda et al. 2019).

Ugandan women had limited access to extension, the main channel used to share information (Kikulwe et al. 2018). The labor requirements of the XW control strategies also influenced uptake (Gotor et al. 2020). Women were less likely than men to adopt backbreaking techniques like CMU. However, most of the approaches developed to scale out XW strategies integrated gender issues.

10.2.6   Impact of XW Scaling Interventions

In all SDSR scaling efforts, the focus has been on recovering banana productivity, improving food security, income, and well-being of banana farming households in the GLR. Quantification of such impacts saw various impact assessment studies, focus group discussions, key informant interviews, field visits, and observations in the project countries. SDSR has been extended to over 100,000 farmers in Burundi, DR Congo, and Uganda, with a recovery in banana production area of about 50,000 ha and an estimated annual value of total production ranging from US$10–30 million or US$100 to 300 per farmer/annum was attributed to these efforts (Rietveld 2021). Additional and detailed country-specific impacts are documented as follows.

Burundi: In the province of Muyinga, the household survey conducted in November 2019 showed that the number of households having XW on-farm fell from 57% to 10% and the number of households with an XW incidence of 10% declined from medium (level 3) to low (level 2) on a four-point scale: 1 = very low (one or two plants), 2 = low (only a few plants), 3 = medium (many plants), and 4 = high (more than one-quarter of plants are symptomatic).

About 65% of banana-growing households have been reached by the scaling project and about 70% of those are already applying the SDSR innovations. During a field visit by senior project scientists in the later stages of the project, no XW diseased plants nor male buds beyond the expected cutting stage were seen in the banana fields in the province of Muyinga. Previously abandoned banana fields were being quickly restored, showing that farmers were regaining their trust in bananas as a preferred food and cash crop. The BPEAE has become a committed supporter of the SDSR innovations, even though this has not yet been officially integrated into Burundi's national policy. With this warm enthusiasm and high number of staff trained on SDSR, it is expected that future posting of such staff to new locations will help scale out the innovations to other provinces. There are already informal requests for advice from other provinces facing XW, a move likely to spark wider scaling in Burundi.

DR Congo: In one of the earlier projects in which single diseased stem removal was deployed, the results showed that 1 year after starting field trials in the North and South Kivu provinces of DR Congo, the proportion of diseased plants had fallen to less than 0.5% (InfoMus@'s News and Analysis 2014). For the latest interventions by scaling project (2018/2019), DR Congo took strides to further contain the spread of the disease by cutting sick plants and removing male buds, even though the impact of these interventions has not yet been assessed. However, indications are that most (75%) banana farmers now apply SDSR and some (10%) use both SDSR and CMU.

Uganda: Bioversity International conducted a quantitative survey with 1224 banana farmers to assess impact of adopting SDSR. The study found that adoption is constrained by limited knowledge of SDSR and influenced by the importance of the banana crop to the household (Kikulwe et al. 2019). The study further showed that training women farmers increased the chances of households adopting all three core practices under the SDSR package. Adoption of the full package (all three SDSR core practices) was more profitable than using one or two of them. Adopters obtained considerable increased earnings from their banana production compared to non-adopter, as adopting XW control measures reduced the disease incidence and increased banana yields and sales. Extrapolations from this survey suggested that of an estimated 800,000 banana-growing households in Uganda, about 600,000 had adopted some of the BXW control strategies.

Focus group discussions (FGDs) were held in six districts in the three major banana-producing regions, including Central, Midwestern, and Southwestern Uganda. The focus group discussions examined the contribution of the SDSR interventions to household food security and banana yields. The results showed that farmers could identify XW diseased plants, and they understood its spread mechanisms and how to use SDSR. Both men and women, young and old, had fully embraced male bud removal. Men and women were routinely sterilizing their tools, some on a daily basis.

Banana farming communities had started growing much less of the Kayinja variety, the most susceptible to XW. Those who had initially left banana production (about 25–50% in central and western Uganda) or who had reduced the area of land devoted to bananas were quickly replanting their fields after learning the SDSR package. Food security had improved and households that had adopted SDSR were eating more meals per day. The number of banana-based meals increased, and this was more appreciated in the central region where banana is expected to be on the table for every meal.

10.2.7 Future Interventions

Despite the lack of currently known genetic resistance against Xvm in cultivated banana in East and Central Africa, resistance has been reported in nonedible species such as *Musa balbisiana*, while varieties of the *Musa acuminata* subsp. *zebrina* (AA) set were also identified as potentially useful sources of Xvm resistance (Ssekiwoko et al. 2006, 2015; Nakato et al. 2019; Tripathi et al. 2019). It would take many years to incorporate these genes into edible varieties through conventional breeding. The difficulties of conventional breeding with this sterile crop call for the need to innovate transgenic approaches for XW control.

Genetically modified (GM) XW resistant bananas have already been developed and were field-tested in Uganda (Namukwaya et al. 2012; Tripathi et al. 2017).

However, policies allowing the release of these materials are lacking. The public is skeptical of GM and consumer resistance is a major constraint to GM crop production (Panzarini et al. 2015). Given consumer rejection of GM crops, clustered regularly interspaced short palindromic repeats (CRISPR) may offer an alternative. Because CRISPR edits genes out instead of introducing new ones, it has brought about rapid positive changes in the attitude of many producers toward genetically engineered crops (Waltz 2018). Banana breeding using CRISPR is still in the early stages, but shows promise (Tripathi et al. 2019, 2021). Future scaling approaches need to address farmer and consumer attitudes about crops bred with CRISPR.

10.3   Lessons Learned and Policy Implications

SDSR effectively manages XW. Farmers using it report reduced banana yield losses. In much of East Africa, SDSR has contributed to a tremendous improvement in food security and incomes.

Men and women farmers accept SDSR because it saves money while effectively controlling XW without all the work of CMU. A single stem is cut while healthy bunches are still harvested, making SDSR more farmer-friendly. Nevertheless, it is important to further reduce the labor requirements of the SDSR package. Action plans for SDSR training and information dissemination have to consider gender from the beginning in order to guarantee ideal participation for women and youth. Barriers to overcome include work burdens, unequal allocation of cash and other benefits from the banana crop, and conflicts arising as men trample women's beans while pruning bananas.

Tool sterilization with fire is still being promoted for XW management, even though fire weakens steel tools. To make sterilization cheaper, without ruining farmers' tools, knives and pangas can be washed in cold water with soap or detergents.

The Rwanda case suggests the importance of engaging policymakers at all stages of technology innovation to ensure their buy-in and subsequent scaling to end users. Platforms are also essential for bringing stakeholders together and scaling innovations.

Producing banana as a business is key to the continued use of XW management practices. A sustainable banana market may enhance scaling of XW recommendations. In the GLR countries where banana is still considered a man's crop, allowing women to take decisions and make money from bananas will boost the uptake of the SDSR package. XW management practices must continue to be used even when disease incidence drops.

Scaling agricultural technologies such as the XW control strategies require a clear linkage with food systems, e.g., by integrating with a systems-based project. In Uganda, for instance, linking with the ongoing Banana Agronomy project enabled contextualization of the strategies and improved the results.

Acknowledgments This research was largely undertaken as part of, and funded by, the CGIAR Research Program on Roots, Tubers and Bananas (RTB) and supported by CGIAR Trust Fund contributors. The authors with great gratitude would also like to acknowledge donor agencies including the McKnight Foundation for the financial support toward the accomplishment of this scaling work.

References

Abele S, Pillay M (2007) Bacterial wilt and drought stresses in banana production and their impact on economic welfare in Uganda: implications for banana research in east African highlands. J Crop Improv 19:173–191

Adeniji TA, Tenkouano A, Ezurike JN, Ariyo CO, Vroh-Bi I (2010) Value-adding post-harvest processing of cooking bananas (Musa spp. AAB and ABB genome groups). Afr J Biotechnol 9(54):9135–9141

Ajambo S, Rietveld A, Nkengla LW, Niyongere C, Dhedha DB, Olaosebikan DO, Nitunga E, Toengaho J, Lava KP, Hanna R, Kankeu R, Kankeu S, Omondi A (2018) Recovering banana production in bunchy top-affected areas in Sub-Saharan Africa: developing gender-responsive approaches. Acta Hortic 1196:219–228. https://doi.org/10.17660/ActaHortic.2018.1196.27

Ajambo S, Mbabazi EG, Nalunga A, Kikulwe EM (2020) Gender roles and constraints in the green cooking banana value chain: evidence from southwestern Uganda. Acta Hortic 1272:135–144. https://doi.org/10.17660/ActaHortic.2020.1272.17

Andrade-Piedra J, Bentley JW, Almekinders C, Jacobsen K, Walsh S, Thiele G (eds) (2016) Case studies of roots, tubers and bananas seed systems. CGIAR research program on roots, tubers and bananas (RTB), Lima: RTB Working Paper No. 2016–3. ISSN 2309-6586. 244 p

Aurore G, Parfait B, Fahrasmane L (2009) Bananas, raw materials for making processed food products. Trends Food Sci Technol 20(2):78–91

Bagamba F, Kikulwe E, Tushemereirwe WK, Ngambeki D, Muhangi J, Kagezi GH, Green S (2006) Awareness of banana bacterial wilt control in Uganda: farmers' perspective. Afr Crop Sci J 14(2):157–164

Blomme G, Jacobsen K, Ocimati W, Beed F, Ntamwira J, Sivirihauma C, Ssekiwoko F, Nakato V, Kubiriba J, Tripathi L, Tinzaara W, Mbolela F, Lutete L, Karamura E (2014) Fine-tuning banana Xanthomonas wilt control options over the past decade in east and Central Africa. Eur J Plant Pathol 139:271–287

Blomme G, Dita M, Jacobsen KS, Pérez VL, Molina A, Ocimati W, Poussier S, Prior P (2017a) Bacterial diseases of bananas and Enset: current state of knowledge and integrated approaches toward sustainable management. Front Plant Sci 8:1290. https://doi.org/10.3389/fpls.2017.01290

Blomme G, Ocimati W, Sivirihauma C, Lusenge V, Bumba M, Kamira M, van Schagen B, Ekboir J, Ntamwira J (2017b) A control package revolving around the removal of single diseased banana stems is effective for the restoration of Xanthomonas wilt infected fields. Eur J Plant Pathol 149:385–400

Blomme G, Ocimati W, Sivirihauma C, Lusenge V, Bumba M, Ntamwira J (2019) Controlling Xanthomonas wilt of banana: influence of collective application, frequency of application, and social factors on the effectiveness of the single diseased stem removal technique in eastern Democratic Republic of Congo. Crop Prot 118:79–88

Blomme G, Dusingizimana P, Ntamwira J, Kearsley E, Gaidashova S, Rietveld A, Van Schagen B, Ocimati W (2021) Comparing effectiveness, cost-and time-efficiency of control options for Xanthomonas wilt of banana under Rwandan agro-ecological conditions. Eur J Plant Pathol:1–15

Buregyeya HG, Tusiime J, Kubiriba J, Tushemereirwe WK (2008) Evaluation of distant transmission of banana bacterial wilt in Uganda. Paper presented at the conference on Banana and Plantain in Africa: Harnessing International Partnerships to Increase Research Impact, October 5–9, 2008, Mombasa, Kenya

Castellani E (1939) Su un marciume dell' Ensete. L'Agricoltura Coloniale Firenze 33:297–300

Desire RM, Bahananga JBM, Romain L, Barhahakana C, Amato S (2016) Analyse de l'impact socioéconomique du flétrissement bactérien du bananier et réponses paysannes dans la région du Bushi Kivu à l'Est de la République Démocratique du Congo. Int J Innov Appl Stud 18:66

Eden-Green S (2004) How can the advance of banana xanthomonas wilt be halted? InfoMusa 13:38–41

FAO (2012) Banana production systems at risk: effectively responding to banana wilt disease in the Great Lakes Region, FAO Banana Factsheet. 2012. Available online: http://www.fao.org/fileadmin/user_upload/emergencies/docs/FAO%20BANANA%20factsheet%20ENGLISH.pdf

FAO (2017) The future of food and agriculture: trends and challenges. FAO, Rome. http://www.fao.org/publications/card/en/c/d24d2507-41d9-4ec2-a3f888a489bfe1ad/

FAO (2019) Production statistics. FAO Rome, Italy. http://www.fao.org/faostat/en/#data/QC

Gotor E, Di Cori V, Pagnani T, Kikulwe E, Kozicka M, Caracciolo F (2020). Public and private investments for banana Xanthomonas Wilt control in Uganda: the economic feasibility for smallholder farmers. Afr J Sci Technol Innov Dev. 1-12

Hartmann A, Linn J (2008) Scaling up: a framework and lessons for development effectiveness from literature and practice. Wolfensohn Center for Development at Brookings Institute (Working Paper 5)

International Development Innovation Alliance (IDIA) (2017) Insights on scaling innovation. Available at: https://static.globalinnovationexchange.org/s3fs-public/asset/document/Scaling

Iradukunda F, Bullock R, Rietveld A, van Schagen B (2019) Understanding gender roles and practices in the household and on the farm: implications for banana disease management innovation processes in Burundi. Outlook Agric 48:37–47

Jogo W, Karamura E, Tinzaara W, Kubiriba J, Rietveld A (2013) Determinants of farm-level adoption of cultural practices for Banana Xanthomonas wilt control in Uganda. J Agric Sci 5(7):70–82

Jones DR (2000) Diseases of banana, abaca and enset. CABI, London

Kalyebara MR, Ragama PE, Kikulwe E, Bagamba F, Nankinga KC, Tushemereirwe WK (2006) Economic importance of the banana bacterial wilt in Uganda. Afr Crop Sci J 14:93–103

Kamusingize D, Majaliwa JM, Komutunga E, Tumwebaze S, Nowakunda K, Namanya P, Kubiriba J (2017) Carbon sequestration potential of East African Highland Banana cultivars (*Musa* spp. AAA-EAHB) cv. Kibuzi, Nakitembe, Enyeru and Nakinyika in Uganda. J Soil Sci Environ Manag 8(3):44–51. https://doi.org/10.5897/JSSEM2016.0608

Karamura E, Kayobyo G, Blomme G, Benin S, Eden-Green JS, Markham R (2006) Impacts of BXW epidemic on the livelihoods of rural communities in Uganda. P. 57. In: Proceedings of the 4th international bacterial wilt symposium, 17–20 July 2006. Central Science Laboratory, New York

Kikulwe ME, Asindu M (2020) A contingent valuation analysis for assessing the market for genetically modified planting materials among banana producing households in Uganda. GM Crops & Food 11(2):113–124

Kikulwe EM, Okurut S, Ajambo S, Gotor E, Ssali TR, Kubiriba J, Karamura E (2018) Does gender matter in effective Management of Plant Disease Epidemics? Insights from a survey among rural Banana farming households in Uganda. J Dev Agric Econ 10(3):87–98

Kikulwe E, Kyanjo J, Kato E, Ssali R, Erima R, Mpiira S, Ocimati W, Tinzaara W, Kubiriba J, Gotor E, Stoian D, Karamura E (2019) Management of Banana Xanthomonas Wilt: evidence from impact of adoption of cultural control practices in Uganda. Sustainability 11(9):1–18

Kubiriba J, Tushemereirwe WK (2014) Approaches for the control of banana Xanthomonas wilt in east and Central Africa. Afr J Plant Sci 8(8):398–404

Kubiriba J, Karamura EB, Jogo W, Tushemereirwe WK, Tinzaara W (2012) Community mobiliza-
    tion: a key to effective control of banana Xanthomonas wilt. J Dev Agric Econ 4(5):125–123
Lufafa A, Tenywa MM, Isabirye M, Majaliwa MJG, Woomer PL (2003) Prediction of soil erosion
    in a Lake Victoria basin catchment using a GIS-based universal soil loss model. Agric Syst
    76(3):883–894. https://doi.org/10.1016/S0308-521X(02)00012-4
Lusty C, Smale M (2003) Assessing the social and economic impact of improved banana variet-
    ies in East Africa. Proceedings of an Interdisciplinary Research Design Workshop, held at
    Kampala (UGA) from 07 to 11/11/2003. Montpellier (FRA): INIBAP, 18 p
Muraleedharan H, Perumal K (2010) Ecofriendly handmade paper making. Booklet published
    from Shri AMM Murugappa Chettiar Research Centre, Taramani, Chennai
Mwangi M, Nakato G (2009) Key factors responsible for the Xanthomonas wilt epidemic on
    banana in East and Central Africa. Acta Horticulture 828:395–404
Nakato GV, Ocimati W, Blomme G, Fiaboe KKM, Beed F (2014) Comparative importance of infec-
    tion routes for banana Xanthomonas wilt and implications on disease epidemiology and man-
    agement. Can J Plant Pathol 36(4):418–427. https://doi.org/10.1080/07060661.2014.959059
Nakato GV, Christelova P, Were E, Nyine M, Coutinho TA, Dolezel J, Uwimana B, Swennen R,
    Mahuku G (2019) Sources of resistance in Musa to *Xanthomonas campestris* pv. *musacearum*,
    the causal agent of banana Xanthomonas wilt. Plant Pathol 68:49–59. https://doi.org/10.1111/
    ppa.12945
Namukwaya B, Tripathi L, Tripathi NJ, Arinaitwe J, Mukasa SB, Tushemereirwe WK (2012)
    Transgenic banana expressing Pflp gene confers enhanced resistance to Xanthomonas wilt dis-
    ease. Transgenic Res 21:855–865
Ndungo V, Eden-Green S, Blomme G, Crozier J, Smith J (2006) Presence of banana xanthomonas
    wilt (*Xanthomonas campestris* pv. *musacearum*) in the Democratic Republic of Congo (DRC).
    Plant Pathol 55(294). https://doi.org/10.1111/j.1365-3059.2005.01258.x
Ndungo V, Komi F, Mwangi M (2008) Banana Xanthomonas wilt in the DR Congo: impact, spread
    and management. J Appl Biosci 1:1–7
Nkuba J, Tinzaara W, Night G, Niko N, Jogo W, Ndyetabula I, Mukandala L, Ndayihazamaso P,
    Niyongere C, Gaidashova S, Rwomushana I, Opio F, Karamura E (2015) Adverse impact of
    Banana Xanthomonas wilt on farmers livelihoods in eastern and Central Africa. Afr J Plant Sci
    9(7):279–286
Nyombi K (2013) Towards sustainable highland banana production in Uganda: opportunities and
    challenges. Afr J Food Agric Nutr Dev 13:7544–7561
Ochola D, Jogo W, Tinzaara W, Odongo M, Onyango M, Karamura BE (2015) Farmer field school
    and Banana Xanthomonas wilt management: a study of Banana farmers in four villages in
    Siaya County, Kenya. J Agric Ext Rural Dev 7(12):311–321
Ocimati W, Ssekiwoko F, Karamura E, Tinzaara W, Eden-Green S, Blomme G (2013) Systemicity
    of *Xanthomonas campestris* pv. *musacearum* and time to disease expression after inflorescence
    infection in East African highland and Pisang Awak bananas in Uganda. Plant Path 62:777–785
Ocimati W, Nakato GV, Fiaboe KM, Beed F, Blomme G (2015) Incomplete systemic move-
    ment of *Xanthomonas campestris* pv. *Musacearum* and the occurrence of latent infections in
    Xanthomonas wilt-infected banana mats. Plant Pathol 64(1):81–90. https://doi.org/10.1111/
    ppa.12233
Ocimati W, Groot JCJ, Tittonell P, Taulya G, Blomme G (2018) Effects of Xanthomonas wilt
    and other banana diseases on ecosystem services in banana-based agroecosystems. Acta
    Horticulturae 1196:19–32. https://doi.org/10.17660/ActaHortic.2018.1196.3. https://www.
    actahort.org/books/1196/1196_3.htm
Ocimati W, Bouwmeester H, Groot JCJ, Brown D, Blomme G (2019) The risk posed
    by Xanthomonas wilt disease of banana: mapping of disease hotspots, fronts and vulnerable
    landscapes. PLoS One 14:e0213691. https://doi.org/10.1371/journal.pone.0213691
Ocimati W, Groot JJ, Tittonell P, Taulya G, Ntamwira J, Amato S, Blomme G (2020) Xanthomonas
    wilt of banana drives changes in land-use and ecosystem services across infected landscapes.
    Sustainability 12(8):3178

Ocimati W, Tazuba AF, Blomme G (2021) Farmer friendly options for sterilizing farm tools for the control of Xanthomonas Wilt disease of banana. Front Agron 3:655824. ISSN:2673–3218

Panzarini HN, Bittencourt MVJ, De Aville Mantos SAE, Wosaick AP (2015) Biotechnology in agriculture: the perceptions of farmers on the inclusion of Genetically Modified Organisms (GMOs) in agricultural production. Afr J Agric Res 10(7):631–636

Reeder RH, Muhinyuza JB, Opolot O, Aritua V, Crozier J, Smith J (2007) Presence of banana bacterial wilt (*Xanthomonas campestris* pv. *musacearum*) in Rwanda. Plant Pathol 56(6):1038–1038

Rietveld AM (2021) End-of-Project report 'Broadening the scaling of Banana Xanthomonas Wilt (BXW) management in East and Central Africa'. 12 p. https://hdl.handle.net/10568/113202

Rutikanga A, Sivirihauma C, Murekezi C, Anuarite U, Ocimati W, Lepoint P, Ndungo V (2013) Banana Xanthomonas wilt management: effectiveness of selective mat uprooting coupled with control options for preventing disease transmission. Case study in Rwanda and eastern Democratic Republic of Congo. In: Banana systems in the humid highlands of sub-Saharan Africa. CABI, London, pp 116–124

Sartas M, Schut M, Proietti C, Thiele G, Leeuwis C (2020) Scaling readiness: science and practice of an approach to enhance impact of research for development. Agric Syst 183(102874):1–12

Shimwela MM, Ploetz CR, Beed DF, Jones BJ, Blackburn KJ, Mkulila IS, van Bruggen A (2016a) Banana Xanthomonas wilt continues to spread in Tanzania despite an intensive symptomatic plant removal campaign: an impending socio-economic and ecological disaster. Food Secur 8:939–951

Shimwela M, Ploetz R, Beed F, Jones J, Blackburn J, Mkulila S, van Bruggen A (2016b) Banana xanthomonas wilt continues to spread in Tanzania despite an intensive symptomatic plant removal campaign: an impending socio-economic and ecological disaster. Food Secur 8. https://doi.org/10.1007/s12571-016-0609-3

Shimwela MM, Blackburn JK, Jones JB et al (2017) Local and regional spread of banana Xanthomonas wilt (BXW) in space and time in Kagera, Tanzania. Plant Pathol 66:1003–1014

Siddhesh G, Thumballi G (2017) Genetically modified bananas: to mitigate food security concerns. Sci Hortic 214:91–98

Smith JJ, Jones DR, Karamura E, Blomme G, Turyagyenda FL (2008) An analysis of the risk from *Xanthomonas campestris* pv. *musacearum* to banana cultivation in Eastern. Central and Southern Africa. Bioversity International, Montpellier, France. Musalit. Org

Ssekiwoko F, Tushemereirwe WK, Batte M, Ragama EP, Kumakech A (2006) Reaction of banana germplasm to inoculation with *Xanthomonas campestris* pv. *musacearum*. Afr Crop Sci J 14:151–155

Ssekiwoko F, Kiggundu A, Tushemereirwe WK, Karamura E, Kunert K (2015) *Xanthomonas vasicola* pv. *musacearum* down-regulates selected defense genes during its interaction with both resistant and susceptible banana. Physiol Mol Plant Pathol 90:21–26

Tinzaara W, Gold CS, Ssekiwoko F, Tushemereirwe W, Bandyopadhyay R, Abera A, Eden-Green SJ (2006) Role of insects in the transmission of banana bacterial wilt. Afr Crop Sci J 14:105–110

Tinzaara W, Stoian D, Ocimati W, Kikulwe E, Otieno G, Blomme G (2018) Challenges and opportunities for smallholders in banana value chains. In: Kema GHJ, Drenth A (eds) Achieving sustainable cultivation of bananas. Burleigh Dodds Science, pp 65–90. https://doi.org/10.19103/AS.2017.0020.10

Tinzaara W, Ssekiwoko F, Kikulwe E, Karamura E (2019) Effectiveness of learning and experimentation approaches for farmers as a community-based strategy for banana Xanthomonas wilt management. J Agric Ext Rural Dev 11(7):128–138

Tripathi V, Mwangi M, Abele S, Aritua V, Tushemereirwe WK, Bandyopadhyay R (2009) Xanthomonas wilt. A threat for banana production in East and Central Africa. The American Phytopathological Society. Plant Dis 93(5):440–451

Tripathi L, Atkinson H, Roderick H, Kubiriba J, Tripathi JN (2017) Genetically engineered bananas resistant to Xanthomonas wilt disease and nematodes. Food Energy Secur 6(2):37–47. https://doi.org/10.1002/fes3.101

Tripathi L, Tripathi JN, Shah T, Samwel Muiruri K, Katari M (2019) Molecular basis of disease resistance in banana progenitor *Musa balbisiana* against *Xanthomonas campestris* pv. *musacearum*. Sci Rep 9:7007. https://doi.org/10.1038/s41598-019-43421-1

Tripathi L, Ntui VO, Tripathi JN, Kumar PL (2021) Application of CRISPR/Cas for diagnosis and Management of Viral Diseases of Banana. Front Microbiol 11(609784):1–13

Tushemereirwe WK, Kangire A, Kubiriba J, Nakyanzi M, Gold CS (2004) Diseases threatening banana biodiversity in Uganda. Afr Crop Sci J 12(1):19–26

Tushemereirwe WK, Okaasai O, Kubiriba J, Nanakinga C, Muhangi J, Odoi N, Opiuo F (2006) Status of banana bacterial wilt in Uganda. Afr Crop Sci J 14:73–82

Waltz E (2018) With a free pass, CRISPR-edited plants reach market in record time. Nat Biotechnol 36(1):6–7

Wigboldus S, Klerkx L, Leeuwis C, Schut M, Muilerman S, Jochemsen H (2016) Systemic perspectives on scaling agricultural innovations. A review. Agron Sustain Dev 36:46. https://doi.org/10.1007/s13593-016-0380-z

Yirgou D, Bradbury JF (1968) Bacterial wilt of enset (*Ensete ventricosum*) incited by *Xanthomonas musacearum* sp. n. Phytopathology 58:11–112

Yirgou D, Bradbury JF (1974) A note on wilt of banana caused by the enset wilt organism *Xanthomonas musacearum*. East Afr Agric Forestry J 40:111

Chapter 11
Toolbox for Working with Root, Tuber, and Banana Seed Systems

Jorge L. Andrade-Piedra ⓘ, Karen A. Garrett ⓘ, Erik Delaquis ⓘ,
Conny J. M. Almekinders ⓘ, Margaret A. McEwan ⓘ,
Fleur B. M. Kilwinger ⓘ, Sarah Mayanja ⓘ, Lucy Mulugo ⓘ,
Israel Navarrete ⓘ, Aman Bonaventure Omondi ⓘ, Srinivasulu Rajendran,
and P. Lava Kumar ⓘ

Abstract Root, tuber, and banana (RT&B) crops are critical for global food security. They are vegetatively propagated crops (VPCs) sharing common features: low reproductive rates, bulky planting materials, and vulnerability to accumulating and

J. L. Andrade-Piedra (✉)
International Potato Center, Lima, Peru
e-mail: j.andrade@cgiar.org

K. A. Garrett
Plant Pathology Department and Food Systems Institute, University of Florida,
Gainesville, FL, USA
e-mail: karengarrett@ufl.edu

E. Delaquis
Alliance Bioversity-CIAT, Rome, Italy
e-mail: e.delaquis@cgiar.org

C. J. M. Almekinders · F. B. M. Kilwinger · I. Navarrete
Wageningen University, Wageningen, The Netherlands
e-mail: conny.almekinders@wur.nl; fleur.kilwinger@wur.nl

M. A. McEwan
International Potato Center, Regional Office for Africa, Nairobi, Kenya
e-mail: M.McEwan@cgiar.org

S. Mayanja · S. Rajendran
International Potato Center (CIP), Kampala, Uganda
e-mail: S.Mayanja@cgiar.org; srini.rajendran@cgiar.org

L. Mulugo
Makerere University, Kampala, Uganda

A. B. Omondi
Alliance Bioversity-CIAT, Cotonou, Benin
e-mail: b.a.omondi@cgiar.org

P. L. Kumar
International Institute of Tropical Agriculture (IITA), Ibadan, Nigeria
e-mail: L.Kumar@cgiar.org

© The Author(s) 2022

319

G. Thiele et al. (eds.), *Root, Tuber and Banana Food System Innovations*,
https://doi.org/10.1007/978-3-030-92022-7_11

spreading pathogens and pests through seed. These crops are difficult to breed, so new varieties may be released slowly relative to new emerging threats. VPC seed systems are complex and face several challenges: poor-quality seed of existing varieties, low adoption rates of improved varieties, and slow varietal turnover, limiting yield increases and farmers' ability to adapt to new threats and opportunities. Addressing these challenges requires first identifying key knowledge gaps on seed systems to guide research for development in a holistic and coherent way. Working together across 10 crops and 26 countries in Africa, Asia, and Central and South America, the CGIAR seed systems research community has developed a "Toolbox for Working with Root, Tuber, and Banana Seed Systems," which introduces 11 tools and a glossary to address four major gaps: (1) capturing the demand characteristics of different types of farmers; (2) identifying effective seed delivery pathways; (3) ensuring seed health and stopping the spread of disease; and (4) designing effective policies and regulations. We describe the toolbox and its creation and validation across 76 crop-and-country use cases, and illustrate how the tools, applied individually or in combination, are addressing the key knowledge gaps in RT&B seed systems. The tool developers are actively working to scale the toolbox, including identifying new partners and models for collaboration, developing new tools, and supporting new applications in VPCs, as well as for fruit, vegetable, grain, and pulse seed systems.

11.1 Introduction

Quality seed, i.e., healthy, at the right physiological stage, in good physical condition, and of an appropriate variety (McGuire and Sperling 2011; Bentley et al. 2018), is the basis of all agricultural productivity. Roots, tubers, and bananas (RT&B) are backbones of food security in tropical and subtropical regions, which overlap considerably with the so-called least developed countries (LDCs). Bananas (including plantains), cassava, potato, sweetpotato, and yam are vegetatively propagated crops (VPCs) and, as such, their planting material or *seed* (suckers, stems, tubers, vines, roots) is bulky, perishable, prone to seed degeneration, and has low multiplication rates (Bentley et al. 2018). Farmers overwhelmingly use seed from previous harvests for the next season, exchange seed with family and friends, or buy seed from informal seed traders, with little or no access to certified seed and improved varieties (Sperling et al. 2020). This places the farmers at the center of VPC seed systems and at the heart of research to support the development of these systems. Research and policy on seed systems has traditionally focused on sexually propagated crops (SPCs), leaving major knowledge gaps for VPCs. In this chapter, we present the "Toolbox for Working with Root, Tuber, and Banana Seed Systems" (referred to as "the toolbox"), a collection of biophysical and socioeconomic tools which together offer a systematic approach to diagnose, evaluate, and improve the seed systems of banana, cassava, potato, sweetpotato, and yam. The chapter is structured around four knowledge gaps, discussed below, that prevent seed systems from functioning effectively and limit dissemination of improved varieties (McEwan et al. 2021a). These gaps were identified while conducting the many case studies that applied the toolbox across crops and contexts worldwide.

We start the chapter by describing the four main knowledge gaps and the challenges facing seed systems for RT&B crops. Then we describe how the toolbox was developed, its context among similar initiatives, the entry points for using the toolbox, combinations of tools, and the level of use. We then describe the tools and discuss how they have been used and the results achieved. We finish the chapter with lessons learned and perspectives.

11.1.1 Key Challenges and Knowledge Gaps

Seed systems for RT&B crops in the tropics and subtropics face three main challenges. First, poor-quality seed limits the yields of existing varieties. In the case of potato in sub-Saharan Africa (SSA), for example, Harahagazwe et al. (2018) estimated that the absolute yield gap was 58.3 t/ha, and they identified poor-quality seed as the top driver of the yield gap. The second and third challenges are related to improved varieties: low adoption and slow varietal turnover. In SSA, Walker et al. (2015) found that the adoption ceiling of modern varieties (improved varieties released after 1970) was less than 40% of the total area, with cassava having the highest adoption rate (39.7%) and banana the lowest (6.2%). Similarly, the average varietal turnover (expressed as varietal age) was 15.2 years (Thiele et al. 2020), with potatoes having the slowest (19.4 years) and bananas the fastest turnover (10.2 years) (Walker et al. 2015). The combination of poor-quality seed, low adoption rates, and slow varietal turnover affects farmers' livelihoods, because yield capacity is reduced and farmers are unable to seize genetic gains obtained by crop breeders (Rutkoski 2019; CGIAR Excellence in Breeding Platform 2011), as they do not benefit from improved varieties with traits such as better yields, high nutritional value, resistance or tolerance to biotic and abiotic stresses, and market-preferred characteristics.

Addressing these challenges is a complex task and requires first identifying key knowledge gaps on seed systems to guide research for development (R4D) in a holistic and coherent way. McEwan et al. (2021a) described four gaps: (1) capturing the demand characteristics of different types of farmers; (2) identifying effective seed delivery pathways; (3) ensuring seed health and stopping the spread of disease; and (4) designing effective policies and regulations.

11.1.2 Toolbox

Filling in these knowledge gaps requires "tools," i.e., methods, models, approaches, or information and communication technologies (ICT), that can be applied systematically and repeatedly in different contexts to study, diagnose, evaluate, and ultimately generate evidence to improve RT&B seed systems. The toolbox includes 11 biophysical and socioeconomic tools and a glossary, each designed to address specific knowledge gaps (Table 11.1). The general goal of the toolbox depends on the user: researchers with the goal of studying seed systems; policymakers with the goal of developing, strengthening, and supporting seed systems; practitioners with the goal of designing, implementing, and evaluating seed system projects; and plant breeders who define product/client profiles (Andrade-Piedra et al. 2020).

The toolbox was created by the *Seed Systems Community of Practice of the CGIAR Research Program on Roots, Tubers and Bananas (RTB)*, a group of biophysical and social scientists. Since 2012, this group has designed and tested the tools in East, West, and Southern Africa; Central, South, and Southeast Asia; and Latin America (Table 11.2). Most of the tools were adapted from other crops or from other fields of study and were usually developed for one crop and then adapted for the others. The tools integrate gender-responsive strategies as much as possible to enable users to explore different interests, preferences, opportunities, and constraints for different categories of users and social groups. The tools vary in the level of expertise and time required to implement them. For example, the multistakeholder framework (MSF) requires basic expertise on seed systems and workshop facilitation, and it is applied in a period of up to 2 months (Bentley et al. 2020), while new applications of the seedHealth model in R currently require collaborators with experience using the R programming environment and, if analyses are intended for publication in scientific journals, might take up to a year for completion (Garrett and Xing 2021). Documentation for each tool includes a peer-reviewed journal article discussing how the tool was created or adapted and applied, a user guide, a description sheet, communication materials, and software (if applicable), all available at https://tools4seedsystems.org/ (Andrade-Piedra et al. 2020).

11.1.3 Other Initiatives

The challenge of improving seed systems has inspired the development of various initiatives. For example, Seed System (https://seedsystem.org/) provides guidance to improve seed security in high stress and vulnerable areas using tools such as the seed system security assessment (SSSA) (Sperling 2008). The Seed Commercial, Legal, and Institutional Reform (SeedCLIR) diagnostic tool addresses the role of legal and institutional components (USAID 2013). Bioversity International's

Table 11.1 Tools in the RTB toolbox, their purpose, and the knowledge gap each addresses

Tool	Purpose	Seed system knowledge gaps[a]			
		Gap 1: seed demand	Gap 2: seed delivery	Gap 3: seed health	Gap 4: policies and regulations
1. Multi-stakeholder framework (MSF)	Identify stakeholders, coordination breakdowns, bottlenecks; rapid assessment of seed availability, access, and quality	✓	✓	✓	✓
2. Impact network analysis (INA)	Evaluate the likely outcomes for the current seed system, and for potential interventions in it, in scenario analyses		✓	✓	✓
3. Seed Tracker (ST)	Organize information to enable quality seed production, certification, and market linkages and to integrate the seed value chain		✓	✓	✓
4. Integrated seed health (ISH) approaches and models	Evaluate how a scenario for the potential use of formal seed, disease resistance, and on-farm management are likely to affect crop health			✓	✓
5. Seed tracing (STg)	Map parts of the seed system such as volume of seed distributed, transaction types, or types of varieties		✓		
6. Small N exploratory case stud y(SN)	Understand farmers' use of seed	✓	✓		
7. Four-square method (FSM)	Characterize seed and variety diversity and use	✓	✓		
8. Means-end chain analysis (MEC)	Understand farmers' motivations for preferring particular seed types and sources and the expected benefits	✓	✓		
9. Experimental auctions (EA)	Elicit individual's willingness to pay (WTP) and willingness to accept (WTA) seed traded in the market	✓			✓

(continued)

Table 11.1 (continued)

		Seed system knowledge gaps[a]			
Tool	Purpose	Gap 1: seed demand	Gap 2: seed delivery	Gap 3: seed health	Gap 4: policies and regulations
10. Seed regulatory framework analysis (SRFA)	Analyze seed regulatory frameworks and implications for vegetatively propagated crops from different stakeholder perspectives			✓	✓
11. Sustainable early generation seed business analysis tool (SEGSBAT)	Prepare a business plan and analyze the financial sustainability of a seed enterprise	✓	✓		
12. Glossary of root, tuber, and banana seed systems	Cites published literature to define and explain important terms in seed systems research and development	✓	✓	✓	✓

[a]Gap 1, capturing demand for varieties and seed; Gap 2, identifying effective seed delivery pathways; Gap 3, ensuring seed health; Gap 4, effective policies and regulation

Resource Box for Resilient Seed Systems (http://www.seedsresourcebox.org/) emphasizes participatory plant breeding approaches and linking seed producers to local and international gene banks, while their Seeds for Needs initiative (https://www.bioversityinternational.org/seeds-for-needs/) uses access to genetic resources to minimize risks from climate change. Other relevant initiatives include the Integrated Seed Sector Development approach (https://issdseed.org/) and SeedSAT (https://seedsat.org/), among others.

The toolbox introduced in this chapter builds on these initiatives and is designed to be complementary and to fill a neglected niche. For example, the MSF presented below was adapted from the SSSA for targeted application with RT&B crops. The toolbox also includes tools that more explicitly address crop health in seed systems and strategies for protecting seed health which are particularly important in VPCs. We also provide scenario analysis to inform decision-making among potential strategies for deploying new varieties, sampling for disease, and managing disease.

Although several other seed system-oriented educational products and research or development toolkits exist, the toolbox is unique in its breadth of coverage of topics and disciplines, its focus on the specific needs of major VPCs, and its depth of peer review and scientific validation in real-world contexts before the tools were released.

Table 11.2 Use of the tools across countries and crops

Country	1. Multi-stakeholder framework (MSF)	2. Impact network analysis (INA)	3. Seed Tracker (ST)	4. Integrated seed health (ISH)	5. Seed tracing (Stg)
Burundi	Banana (6)[a]				
Cambodia		Cassava (4, 10)			Cassava (4, 10)
Democratic Republic of the Congo	Cassava (6)				
Ecuador	Potato (6, 22)	Potato (7)		Potato (8)	
Ethiopia	Teff, wheat (20)				Potato (33)
Georgia	Potato (5)	Potato (5)		Potato (5)	
Ghana	Banana, cassava (6)				
Haiti		Banana, mango (9, 12)			
Kenya	Banana, potato (6, 19)	Potato (13)		Potato (8)	
Lao People's Democratic Republic		Cassava (4, 10)			Cassava (4, 10)
Malawi	Cassava, potato (6)				
Mozambique	Cassava (6)				
Nigeria	Cassava, yam (6, 29, 34)		Cassava, yam (28, 35)		
Nicaragua	Cassava (6)				
Peru	Potato (6)				
Philippines					Forages (18)
Rwanda	Sweetpotato (6)				Cassava (15)
Sierra Leone	Cassava (6)				
Thailand		Cassava (4, 10)			Cassava (4, 10)
Uganda	Banana, sweetpotato (2, 6)	Sweetpotato (3)			
United Republic of Tanzania	Cassava, sweetpotato (6, 24)		Cassava (35)	Sweetpotato (25)	
Vietnam	Cassava, potato (14)	Cassava (4, 10)			Cassava, forages (4, 10, 18)

(continued)

Table 11.2 (continued)

Country	6. Small N exploratory case study (SN)	7. Four-square method (FSM)	8. Means-end chain analysis (MEC)	9. Experimental auctions (EA)	10. Seed regulatory framework analysis (SRFA)	11. Sustainable early generation seed business analysis tool (SEGSBAT)
Burkina Faso						Sweetpotato (30)
Cameroon	Banana (23)					
Ethiopia	Potato (32)		Sweetpotato (26)			Sweetpotato (30)
Ghana						Sweetpotato (30)
Kenya			Sweetpotato (26)		Potato (19)	Sweetpotato (30)
Lao People's Democratic Republic				Cassava (11)		
Malawi						Sweetpotato (30)
Malaysia		Forest species (1)				
Mozambique						Sweetpotato (30)
Nigeria	Cassava (29)				Cassava (31)	Sweetpotato (30)
Rwanda						Sweetpotato (30)
Uganda	Banana (16)	Banana (16, 21)	Banana (17)			Sweetpotato (30)
United Republic of Tanzania			Potato (27)			Sweetpotato (30)
Vietnam					Cassava, potato (14)	
Zambia						Sweetpotato (30)

[a]References 1, Aini et al. 2017; 2, Ajambo et al. in preparation; 3, Andersen et al. 2019; 4, Andersen et al. 2020; 5, Andersen Onofre et al. 2021; 6, Bentley et al. 2018; 7, Buddenhagen et al. 2017; 8, Buddenhagen et al. 2022; 9, Dantes et al. 2020; 10, Delaquis et al. 2018; 11, Delaquis et al. unpublished data; 12, Fayette et al. 2020; 13, Gachamba et al. 2022; 14, Gatto et al. 2021; 15, Kilwinger et al. 2021b; 16, Kilwinger et al. 2019; 17, Kilwinger et al. 2020; 18, Leyte et al. 2021; 19, McEwan et al. 2021c; 20, Mulesa et al. 2021; 21, Mulugo et al. unpublished data; 22, Navarrete et al. 2019; 23, Nkengla-Asi et al. 2020; 24, Ogero et al. 2015; 25, Ogero et al. 2019; 26, Okello et al. 2018a; 27, Okello et al. 2018b; 28, Ouma et al. 2019; 29, Pircher et al. 2019; 30, Rajendran et al. 2017; 31, Spielman et al. 2021; 32, Tadesse et al. 2017a; 33, Tadesse et al. 2017b; 34, Wossen et al. 2020; 35, https://seedtracker.org/

11.1.4   Entry Points

Questions about seed systems (for research or for development projects) are the basis for selecting a tool. The following examples of questions are described in more detail elsewhere (Andrade-Piedra et al. 2020):

- For MSF: Who are the specific stakeholders of a seed system?
- For INA: What types of interventions are likely to lead to wider adoption of a new variety?
- For FSM: What local and improved varieties do farmers grow?
- For EA: What is the real market value for seed?
- For SRFA: What types of public policies and regulations are in place for the subject crops in a country?

These questions can be broadly grouped into the four knowledge gaps described above (seed demand, seed delivery, seed health, and policies and regulations), with each tool addressing at least one knowledge gap (Table 11.1). While some tools (e.g., the multi-stakeholder framework) address all knowledge gaps in an exploratory manner, others address one or two knowledge gaps at a deeper level, such as the four-square method and the means-end chain analysis that focus on understanding seed demand and seed delivery.

In addition to using questions about seed systems as the basis for selecting appropriate tools for a given case, tools can also be selected using two other entry points: the seed value chain (Fig. 11.1) and the project implementation life cycle (Fig. 11.2). A seed value chain includes components from plant genetic resources through breeding, early generation seed (EGS), decentralized multiplication, farmer and trader use, and markets and consumers (Fig. 11.1).

Tools from this toolbox may be selected and combined for implementation considering what tools might be important at each stage of a seed value chain (Fig. 11.1). First, a quick and comprehensive overview of the bottlenecks and gaps of the seed value chain can be conducted using the multi-stakeholder framework. Then, for planning optimal use of plant genetic resources, the seed regulatory framework analysis can provide input on relevant policies for genetic resource management, while the four-square method can provide input on what genetic resources would support variety development for the needs farmers express.

To inform crop breeding, the same tools can be used for evaluating plant genetic resources, along with means-end chain analysis and experimental auctions to provide information about the basis for farmer selection of variety traits and farmer willingness to pay for specific types of varieties, respectively. For planning EGS production, INA can evaluate how effectively seed movement is linked to the next stages of the system; ISH approaches and models can be used to evaluate how much seed would be needed for optimal farmer purchase of quality-declared seed to manage seed-borne disease; SEGSBAT can evaluate seed production needs for business efficacy downstream; and the Seed Tracker can be used to keep track of where seed is available and how it moves downstream, and it can be used in regulatory oversight.

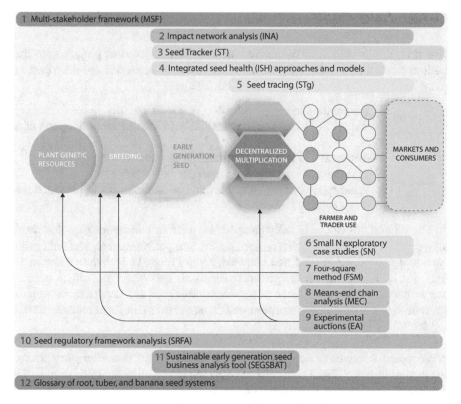

Fig. 11.1 The components of a seed value chain and the range of components for which each of the tools can be used. Tools corresponding to each number are listed in Table 11.1

For planning decentralized multiplication of seed, the same tools can be used as for EGS production, along with seed tracing to characterize seed movement based on surveys and group discussions. To evaluate options for improvements at the stage of farmer and trader use, small N exploratory case studies, the four-square method, means-end chain analyses, and experimental auctions all offer different perspectives on how farmers choose varieties, cultivation methods, and whether to purchase seed or save their own seed.

To understand the influence of markets and consumers on the seed value chain, the multi-stakeholder framework can be used to create an overview of all actors in the system. The seed regulatory framework sheds light on policy effects and policy options on seed markets. INA can evaluate how seed and crop products move through from EGS to the market. Seed Tracker and seed tracing can continue to track seed and product movement, and small N exploratory case studies can help to understand market decisions.

A project designed to improve seed systems typically has the stages illustrated in Fig. 11.2. The tools may be used at various stages in the project cycle to support regular evaluation and improvements. Most of the tools can be used in defining the

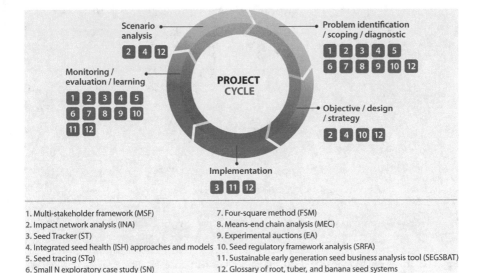

Fig. 11.2 A seed system project cycle and the tools that can be used at each stage. Tools corresponding to each number are listed in Table 11.1

problem the project will address and the strategy for addressing it. Some of the tools can be used during project implementation. All can be used as part of project monitoring and evaluation (M&E). Some tools are useful during scenario analysis to understand the results of M&E, to help the project succeed (Fig. 11.2).

11.1.5   Combining Tools

The tools gain strength when used together. For example, data from several tools can inform the scenario analyses provided by INA. And the results from INA can inform other tool applications (Fig. 11.3). In the Republic of Georgia, the MSF, INA, and ISH approaches were used together (Andersen Onofre et al. 2021).

11.1.6   Use of the Toolbox

The 11 tools in the toolbox have been applied on 10 crops in 26 countries for a total of 76 applications (Table 11.2, Fig. 11.4). Of the crops, cassava has the most applications (24), followed by sweetpotato (18), potato (17), banana (9), and yam (2). Besides RT&Bs, the tools have been applied on forage crops (twice), forest species (once), mango (once), and on teff and wheat (once each). The tools have been applied in 15 countries in Africa, 7 in Asia, and 4 in Central and South America

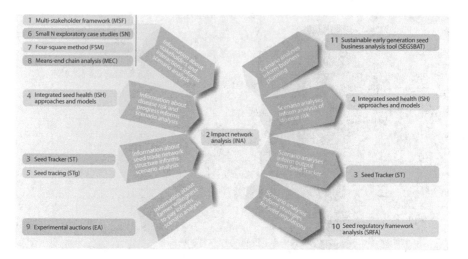

Fig. 11.3 Tools can be used together to inform seed system strategies. For example, INA can use data from other tools as input and can provide output data for use by other tools

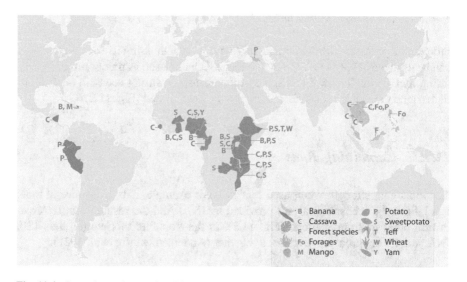

Fig. 11.4 Countries and crops in which projects have applied the tools in this toolbox

(Fig. 11.4). The tools have been used most frequently in Kenya and Uganda (six times each), Tanzania and Ethiopia (five each), and Nigeria and Vietnam (four each). The tools most often applied are MSF (25 applications), SEGSBAT (11), INA (10), and seed tracing (7).

11.2 Tools

11.2.1 Multi-stakeholder Framework (MSF)

Grasping the complexity of seed systems is a challenge for those who are working in a new location or crop, whether to understand the existing seed systems or to conduct projects to improve them. The multi-stakeholder framework (MSF) addresses this challenge by providing a snapshot of a seed system in a specific crop, location, and time.

The MSF is an adaptation of the seed system security assessment (SSSA) (Remington et al. 2002; Sperling 2008; FAO 2015) built around concepts derived from food security: access, availability, and utilization (quality). The MSF considers seed regulations and policies, sustainability, and gender as crosscutting themes. The MSF was tested in 13 case studies, finding that gender roles are important in seed systems and that ignoring the differences between women and men can lead to coordination breakdowns that can threaten seed security (Bentley et al. 2018). The MSF has been applied in 17 countries and 7 crops (Table 11.2).

The MSF has been used to:

- Understand the seed sourcing behavior of cassava farmers and to identify entry points for decentralized stem multipliers (DSMs) in Nigeria (Pircher et al. 2019).
- Identify stakeholders in the potato production system in Georgia (Andersen Onofre et al. 2021).
- Review a sweetpotato project promoting a systematic reflection from different stakeholder perspectives in Tanzania (Ogero et al. 2015).
- Identify participants and design key informant interviews and focus group discussions to explore regulations for potato and cassava seed in Vietnam, Nigeria, and Kenya (Wossen et al. 2020; Gatto et al. 2021; McEwan et al. 2021c; Spielman et al. 2021).
- Identify stakeholders and the main features and bottlenecks of potato seed systems to refine research questions and design a household survey in Ecuador (Navarrete et al. 2019).
- Estimate seed security of teff and wheat in Ethiopia (Mulesa et al. 2021).

The MSF can be used as the starting point for a comprehensive analysis of bottlenecks in a seed system, to monitor an intervention (McEwan et al. 2021a), and for cross crop/region comparison among interventions (Bentley et al. 2018). Users of the MSF usually gain an understanding of the complexity in structure and interactions between stakeholders, including tensions which are not obvious prior to using the tool. The MSF is multidisciplinary and transdisciplinary, which may be a challenge during workshops or field visits, but also a benefit, encouraging the users to take a more holistic view of seed systems.

11.2.2 Impact Network Analysis (INA)

Impact network analysis (INA) is a tool for anticipating the outcomes of a seed system project that is underway or in planning. It is a new tool based on modeling a system as a combination of (1) a network of people or institutions who may influence each other and have transactions and (2) a network of the movement between farms of seed, varieties, and potentially of pathogens or pests (Garrett et al. 2018; Garrett 2021a, 2021b). Results from scenario analyses support decision-making by researchers, policymakers, and practitioners.

INA includes an R package that simulates outcomes for scenarios defined by the user (Garrett 2021a, 2021b, updates at garrettlab.com/ina/). It provides scenario analyses in stochastic simulations to evaluate questions such as the following: (1) How likely is a new variety to spread through a seed system, and how could changes in the system make it spread further? (2) What will be the most effective sampling strategy for monitoring disease spread through a seed system? (3) What strategy for managing disease spread is likely to be most effective? (4) Do men and women (and other social groups) receive equitable benefits from the current seed system, or what changes would be necessary?

INA has been applied in combination with the MSF and the ISH approach to help design a new potato seed system in the Republic of Georgia, taking into account risks from diseases such as potato wart (caused by *Synchytrium endobioticum*), and identifying key locations to monitor the disease to prevent losses (Andersen Onofre et al. 2021). INA is currently being applied with seed tracing to develop strategies for deploying clean seed to slow the spread of cassava mosaic disease in SE Asia (Delaquis et al. 2018; Andersen et al. 2020). The INA framework was applied to understand potato seed systems in Ecuador (Buddenhagen et al. 2017) and sweetpotato seed systems in Uganda (Andersen et al. 2019), where these two studies provided groundwork for applications in new systems. INA is also currently being applied in collaboration with the Kenya Plant Health Inspectorate Service (KEPHIS) to evaluate strategies for managing disease in Kenyan potato seed systems (Gachamba et al. 2022) and for banana and mango disease and pest risk assessment in Haiti and the Caribbean (Dantes et al. 2020; Fayette et al. 2020).

11.2.3 Seed Tracker (ST)

The Seed Tracker (ST) is an ICT tool that digitally links seed value chain actors, tracks seed production, and organizes seed information for stakeholders. The ST's digital data collection tools are usable on any Internet-enabled device with an Android operating system. It offers secure individual and group accounts and a database with analytics and geographic information system (GIS) tools. The ST covers all stages of the seed value chain and the needs of stakeholders: researchers, extensionists, regulators, seed producers, traders, service providers, and farmers. It supports seed production planning, seed traceability, seed inventory management, and quality assurance. The Seed Tracker allows regulatory authorities to monitor the production of certified seed and allows real-time information exchange between seed producers and regulators. It is also a business tool that helps to link seed producers with customers. It offers real-time information on seed production by seed class, variety, volume, and location. The ST can be customized to fit different crops, national seed regulations, and user-defined needs. ST can potentially map gender-disaggregated information which policymakers and extension services can use to inform seed delivery strategies.

11.2.4   Integrated Seed Health (ISH) Approaches and Models

Seed systems that spread diseases or pests can do more harm than good. Understanding seed degeneration, and how to manage it, is important for supporting better seed systems. The "integrated seed health approach" combines three management components to help farmers decide how to manage seed health: periodic purchase of healthy seed, disease resistance, and on-farm disease management (Thomas-Sharma et al. 2016). A model called "seedHealth" identifies the combinations of these three components most likely to be successful, to support training and decision-making by researchers, policymakers, and practitioners (Thomas-Sharma et al. 2017; using an online dashboard link at garrettlab.com/seedhealth/; Garrett and Xing 2021).

At regional scales, a fourth component of ISH approaches is phytosanitary management to prevent the introduction of new pathogens and pests. The seedHealth model can be used to answer questions such as: (1) How frequently would it benefit farmers to buy certified/quality-declared seed, and/or to access a new variety? (2) What would the effect be of strengthening particular types of on-farm management, e.g., training to support positive selection? (3) Do differences in men's and women's access to, and use of, the components of seed health management lead to different levels of success?

ISH approaches and the seedHealth model are being used to study seed health in systems such as sweetpotato in Tanzania (Ogero et al. 2019) and potato in Kenya (Gachamba et al. 2022). The ISH approach has been applied in combination with the MSF and INA to help design a new potato seed system in the Republic of Georgia that protects against the spread of diseases such as potato wart, balancing on-farm management, resistant varieties, and new seed certification standards (Andersen Onofre et al. 2021). Seed health in a potato system in Ecuador was studied to support "management performance mapping," identifying locations in the Andes and Kenya where support for positive selection of farmer-saved seed is likely to have the greatest benefit (Buddenhagen et al. 2022). The seedHealth model can also be applied to evaluate better phytosanitary standards, to address the trade-off between higher availability of seed and poorer seed health (Choudhury et al. 2017). In banana bunchy top management, the model has been used to visualize and predict the strategies for managing disease spread and seed degeneration and to compare the performance of specific control options under different field conditions (I. Nduwimana, pers. comm.).

11.2.5   Seed Tracing (STg)

An important issue in the seed systems of RT&B is how new varieties diffuse. These systems are mostly informal, so the exchange between farmers is the main avenue for distributing new varieties. Seed tracing can be used to understand the diffusion of seed from formal to informal networks, and within farmer networks, thus providing strategic information for seed interventions and for policymakers.

In Ethiopia, a seed tracing study found that an NGO distributed seed potato of new varieties to wealthier farmers, who shared seed tubers frequently, including with poor farmers who rarely shared seed (Tadesse et al. 2017b). The wealthier farmers were key in variety diffusion, but also potentially in spreading pests and diseases.

Among Rwandan cassava farmers, seed tracing was used to inform the design of commercial seed businesses (Fig. 11.5). As in the Ethiopian case, better-off growers were more likely to obtain a new variety from formal sources, while poor farmers accessed new varieties from fellow farmers. Most (60%) seed transactions were for

Fig. 11.5 Moving cassava stakes and banana suckers between two farms. Seed tracing can map such exchange networks. (Photo: F. Kilwinger/WUR)

cash, suggesting that there are market opportunities. Yet, they were all one-time acquisitions. Once they obtained the variety, all farmers multiplied their own material, and 80% shared this with fellow farmers (Kilwinger et al. 2021b). This is common with the introduction of new VPC varieties: after a first spike of demand, the new variety gets absorbed into the informal seed system (Barker et al. 2021), discouraging commercial seed businesses.

A study of legume seed (Almekinders et al. 2020) found that men most often shared with men, and women with women, but men shared with women more often than the other way around. Such patterns of gendered seed flow could have implications for introducing new varieties. Yet, there was little effect on who the seed eventually spread to. The study allowed the project to report to donors that it had reached an estimated 2.5 million farmers in Africa through direct distribution and spontaneous diffusion of seed over the course of the project (Almekinders et al. 2020; Sikkema 2020).

11.2.6 Small N Exploratory Case Study (SN)

A small N exploratory case study collects data on formal and informal seed systems of a crop at the level of the farmers: what varieties do they grow and what are their patterns of seed saving, replacement, and sourcing? This is useful when diagnosing a seed system and identifying the challenges in improving local availability and access to quality seed: a first step that leads to deeper reconnaissance and seed system intervention. Typically, the core of the data collection is a survey with well-targeted questions for 35–50 farmers in a few communities. Because seed use practices, variety preferences, and needs of better-off and poor farmers often differ, it is worthwhile to collect data on both types of farmers and on male and female farmers.

A small team can collect the data relatively quickly. These inexpensive studies can be designed and carried out by staff members of an NGO or an agency that is active in the area and may later support seed activities. In contrast to surveys with many farmers and hired enumerators (i.e., large N surveys of 400 farmers or more), small N surveys can be enlightening for the data collectors, who may later help to implement the seed project. Our survey experience in Nigeria showed that joint analysis and discussion of the data were important learning opportunities for the local staff to arrive at a joint understanding of the cassava seed systems of the farmers they worked with (see Pircher et al. 2019).

This type of study belongs to a family of small N approaches (White and Phillips 2012) and has proven to be publishable, especially when gathering the first information about a seed system. For example, in the RTB case of banana seed systems, there was limited understanding of how management of the mat and suckers influenced variety choice and how it related to the farmer's age and gender (Kilwinger et al. 2019). In these situations, the additional information was acquired through semiformal interviewing or focus group discussions and use of the four-square method (see below).

11.2.7 Four-Square Method (FSM)

The four-square method (FSM) originally meant identifying a community's common, unique, and endangered crop varieties for genetic conservation (Grum et al. 2008). It comprises four squares that are drawn on the ground or on a chart. Each of the four squares holds the names of varieties of interest based on their abundance, i.e., if a variety is grown by many or few households, on a large or a small area:

1. Many households on large area.
2. Many households on small area.
3. Few households on large area.
4. Few households on small area.

The FSM has been adapted to assess crop diversity and popularity within a community and to create discussions around seed systems. The method can generate an inventory of varieties grown in a particular place and discuss their importance with farmers. Such information helps to identify seed interventions needed to conserve crop varieties and to highlight desirable traits in new varieties. The classification can also be a quick way of assessing the penetration of new varieties or changes in the popularity over time in response to seed systems or environmental stressors (Simbare et al. 2020). The FSM is often used in a focus group discussion with men or women to capture gender-related differences in appreciation of varieties and their traits (Mulugo et al. 2021).

The FSM has been used to study:

- The changes in varietal diversity of East African highland bananas in banana bunchy top disease outbreak areas of Burundi (Simbare et al. 2020).
- Farmers' production objectives regarding banana diversity in central Uganda (Kilwinger et al. 2019).
- Cassava diversity, loss of landraces, and farmers' preference criteria in southern Benin (Agre et al. 2016).
- Yam diversity and production in Southern Ghana (Nyadanu and Opoku-Agyeman 2015).
- Varietal diversity and genetic erosion of cultivated yams in Togo (Dansi et al. 2013).
- Seed interventions and cultivar diversity in pigeon pea in Eastern Kenya (Audi et al. 2008).
- Farmers' limited uptake of tissue culture banana seed in central Uganda (Mulugo et al. unpublished data).

In all these cases, the FSM was complemented with other methods such as literature review, key informant interviews, Venn diagrams, participatory value chain mapping, participatory rapid market appraisal, household surveys, and other tools of the toolbox, e.g., small N exploratory case study.

The FSM has been used to assess different seed system contexts including (1) before an intervention to understand the existing seed systems and to identify key issues for the project and (2) during interventions to monitor or evaluate them. The FSM creates a versatile overview and has been adapted elsewhere in dietary diversity studies (Aboagye et al. 2015) and gender studies in banana seed systems (Nkengla-Asi et al. 2020). The results can help to identify entry points for further research, for example, identifying varieties to study in greater depth using INA (I. Nduwimana, pers. comm.). Nkengla-Asi et al. (2020) adapted the method to classify household seed decision-making based on the level of responsibility and consultation between men and women, to reveal areas of common understanding and potential conflict. Other uses of the method could be developed.

11.2.8 Means-End Chains (MEC)

Means-end chain (MEC) analysis is an approach from the field of consumer studies developed in the 1980s (Reynolds and Gutman 1988). Since its development, the method has been applied in diverse fields such as tourism, food quality and preference, and sustainable behavior. Recently, it has also been used to understand how farmers evaluate, and why they value, different agricultural products, practices, and innovations (e.g., Okello et al. 2019; Urrea-Hernandez et al. 2016). The method is based on several psychological theories and takes into consideration differences among individuals' experiences. The means-end chain analysis identifies such differences as respondents are invited to select and verbalize their own personally relevant attributes to evaluate a product, service, or practice and relate those to their personal values (Walker and Olson 1991).

The method is promising in cross-cultural and exploratory research as it avoids forcing respondents into predetermined categories (Watkins 2010). And the psychological theories on which the method is based are similar to those underlying new approaches to understand adoption (Kilwinger and van Dam 2021a). For example, the framework to understand technological change developed by Glover et al. (2019) is based on the theory of affordances (Gibson 1977), which has considerable overlap with the personal construct theory (Kelly 1955). Also, Tricot trials (van Etten et al. 2019) make use of the principle of asking farmers to differentiate between three choices.

One MEC study with Andean potato farmers found that farmers and experts understand seed quality differently (Urrea-Hernandez et al. 2016), suggesting that understanding farmers' perceptions of quality seed is important for developing effective seed interventions. A MEC study of farmers' perceptions of formal and informal sources of banana planting material in Uganda showed that all farmers (large and small, male and female) had similar goals, but considered different variety traits and the benefits derived from them to achieve those goals (Kilwinger et al.

2020). Some of these Ugandan farmers expected to find the planting material of these varieties in nurseries, while others planned to get it from fellow farmers. These farmers care about variety traits, but also about the source of their planting material. It is important to understand which variety traits farmers prefer, as well as how they like seed to be delivered.

11.2.9   Experimental Auctions (EA)

A key challenge in developing VPC seed systems is understanding and predicting demand for different types and quality of planting material. The viability of the seed system depends on whether farmers perceive the seed as a quality planting material and whether they are willing to pay a premium for that quality. To get those insights, various types of experimental auctions can elicit "true willingness to pay." This tool allows comparing the premium value given to seeds, varieties, or variety traits by different groups, e.g., men and women farmers. This tool can also map out seed market size and segments for various types of customers. Outcomes can support more competitive pricing policies, attract different types of seed producers and customers, and nicely complement other tools (e.g., SEGSBAT described below).

Experimental auctions have recently been conducted among bean farmers in Tanzania and cowpea farmers in Ghana, to evaluate their willingness to pay for certified, quality-declared, and recycled seed, concluding that farmers were willing to pay a slightly higher price for what they perceived as better seed (Maredia et al. 2019).

There has been less use of experimental auction approaches with vegetatively propagated crops. Due to the economic differences between VPC planting material and grain seeds, the method needs to be further evaluated and adapted (several studies to do this are underway led by members of the toolbox group). However, the method has yielded useful preliminary results which are already shaping seed system strategies. In Rwanda, this tool was applied in 29 villages in six leading sweetpotato production areas to estimate willingness to pay for high-quality sweetpotato planting materials and drivers of the demand for these vines (Fig. 11.6). The study also estimated willingness to pay a premium for biofortified varieties (rich in provitamin A) as opposed to the non-biofortified local ones. The preliminary results showed that true willingness to pay a price premium for quality attributes is significantly higher than the current price for the sweetpotato seed.

In Lao PDR, 21 experimental auctions in cassava areas around the country unearthed large differences in stem prices linked to villages' historical experiences with commercial cassava production (Delaquis et al. unpublished data; Fig. 11.7). In all sites, bids were higher for phytosanitary-tested seed and elite varieties than for farmer seed. The auctions also elicited how many bundles of seed were desired, which varied widely, demonstrating different seed purchase strategies (purchasing a few bundles for testing vs. going all-in and buying enough to replace the farmer's whole supply). Demand curves generated from the results are also informing early

Fig. 11.6 A Rwandan farmer, highest bidder for the sweetpotato vines in a second price experimental auction. (Photo: S. Rajendran/CIP)

stage, clean seed multiplication initiatives in the country with price points and preferences that can shape areas of intervention and outgrower strategies. This example demonstrates the tool's use at several stages in the project cycle.

11.2.10 Seed Regulatory Framework Analysis (SRFA)

More than 95% of the seed of RT&B crops flows through informal channels (e.g., farmer saved, purchased from neighbors and local markets). Yet current seed regulatory frameworks do not recognize this and may act as a constraint to improving the quality of vegetatively propagated seed. Most national seed policies and regulations were developed using the experiences from grain seed, especially hybrids. The characteristics of RT&B crops, such as clonal reproduction and specific plant health constraints that contribute to seed degeneration, have not been fully recognized. This means that regulatory processes need to be revised to remain relevant

Multidisciplinary teams of researchers, together with seed regulators, have used the Seed Regulatory Framework Analysis Tool (McEwan et al. 2021b; Spielman et al. 2021) to assess the implications of current seed regulatory frameworks in Kenya, Nigeria, and Vietnam. The teams have asked if implementing regulations

Fig. 11.7 Lao farmers participate in an experimental auction for cassava planting stems in Lao PDR. (Photo: E. Delaquis/Alliance of Bioversity International and CIAT, WUR)

increases the availability and access to quality seed potato, for whom, and with what consequences. In Kenya, stakeholders are gathering around two key narratives. The first narrative, "quality at any cost," ties potato (and other VPCs) to national food security objectives, arguing that yields will only increase within a regulatory framework that provides certified seed at scale, minimizes the risk of pests and diseases, and protects the reputation of seed producers and the hard-earned credibility of the country's regulator. The second narrative, "local quality assurance," introduces "clean" (healthy) seed production models that build off the entrepreneurial spirit of smallholder farmers and their organizations and allows for more relaxed quality standards and informal trade (McEwan et al. 2021c).

In Kenya, the increased understanding that VPC seed faces different challenges than grain seed has led to separate regulations for vegetatively propagated crops, perhaps the first instance in sub-Saharan Africa. In Vietnam, despite strict regulations on the production and trade of VPC seed, the rules are weakly enforced. Instead, seed producers and traders signal quality to farmers through trust, reputation, and long-term relationships. This may be effective at a localized scale, but these informal systems are unlikely to accommodate expansion of the cassava and potato sectors and unlikely to effectively manage increases in pest and disease pressures that result from cross border trade or climate change (Gatto et al. 2021). In Nigeria, findings have led to decentralized policy and regulatory approaches to managing the cassava seed system, prioritizing investment in innovative capacity at the community and enterprise levels (Wossen et al. 2020).

11.2.11 Sustainable Early Generation Seed Business Analysis Tool (SEGSBAT)

The transition from breeder seed to pre-basic (i.e., first generation) seed production is a major bottleneck in the smooth functioning of a formal seed system. An early generation seed (EGS) company requires predictable revenues based on competitive and affordable prices for market-preferred varieties. Using the sustainable early generation seed business analysis tool (SEGSBAT) (Rajendran and McEwan 2021a, 2021b), public and private sector institutions in 11 countries in sub-Saharan Africa analyzed the financial sustainability of their sweetpotato EGS businesses (International Potato Center 2017). Multidisciplinary teams first determined accurate costs of EGS production which were then used to calculate the appropriate price of EGS products and formulate a pricing strategy to attract more customers, increase revenue, and create a positive net cash flow (Fig. 11.8). Partners in six countries improved continuity of funding and met at least 90% of their recurrent seed production costs from season to season. Most institutions reduced the gap between production and sales, which increased marketed surplus. By having a detailed cost structure, users identified and addressed production inefficiencies to reduce the cost of goods sold, e.g., by reducing the number of tissue culture

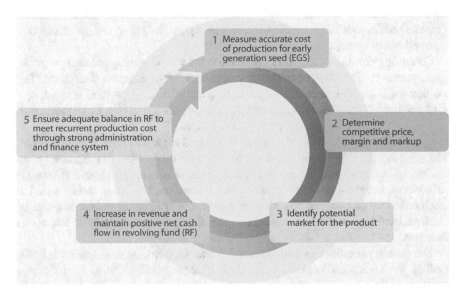

Fig. 11.8 Interconnection of financial performance and sustainability of sweetpotato EGS business. (Source: Rajendran and McEwan (2018))

plantlets required, and optimizing screenhouse production (Rajendran et al. 2017). Partners used SEGSBAT to develop business plans to guide sustainable sweetpotato EGS production, a first for RT&B crops in Africa (Gurmu et al. 2019).

Applying this tool revealed the specific challenges of determining the production costs of VPC seed, including (1) varying multiplication rates due to varietal characteristics, changes in temperature, growing conditions, and ratooning practices; (2) multiple stages in seed production, which may take place in different locations, i.e., pathogen tested tissue culture micro-propagation, hardening tissue culture plantlets for screenhouse multiplication before producing commercial seed in open fields; and (3) because seed is alive, wastage can be high and this must be factored into production costs.

Use of this tool highlighted that current methods for estimating seed requirements for production planning are inadequate (International Potato Center 2017). There are clear opportunities to continue working with public and private EGS producers and their networks of seed entrepreneurs to match SEGSBAT and other tools from the toolbox to the different stages in the product life cycle as part of the handover for commercialization from breeding outputs to seed value chain actors.

11.2.12   Glossary

Discipline-specific jargon can be a big obstacle to reaching a wider audience (Bullock et al. 2019). Seed system initiatives often provide a glossary of terms to help readers to understand key concepts. However, definitions may be generated by the authors themselves, to apply only within the context of their particular initiative. This can lead to confusion as many different interpretations arise and are misused or repeated out of context.

Common definitions are especially important for an emerging research area like seed systems, which brings together concepts from many different disciplines. The toolbox itself contains technical content from economics, behavioral science, network analysis, botany, agronomy, plant pathology, policy analysis, and gender studies, so most readers will encounter unfamiliar terms.

The glossary of RT&B seed systems developed for the toolbox (Delaquis et al. 2020) lends clarity to this issue by compiling definitions cited in literature across disciplines and providing the context of each term, references and links to the original sources, and the date of last modification. Over time, new terms can be added to the interactive glossary on the toolbox website, and existing definitions can be updated. Having definitions in one public place facilitates disambiguation and opens dialogue.

The glossary provides a stand-alone reference, supports the use of all tools in the toolbox by a wide audience, and can track changing definitions as seed systems research evolves and new concepts emerge, serving as a resource for anyone working on seed systems.

11.3   Conclusions

The development and use of the toolbox over a wide range of cases and contexts has led to several higher-level findings and lessons. Implementation has validated the great interest of public and private sector actors in diagnosing and improving VPC seed systems and those of other crops. The four knowledge gaps which formed the basis of this chapter emerged from reflection about these diverse experiences, a direct outcome of structured interactions between the toolbox development community of practice. Assembling the tools in a toolbox made them more accessible, provided an intuitive structure for new users, and helped to clarify which tools and combinations of tools are most useful for addressing different types of challenges. The modular structure with validated tools also inspired confidence and increased the value of lessons learned across crops and locations.

11.3.1   Outcomes

Applying the tools individually or in combination is generating outcomes by addressing key knowledge gaps in seed systems (Sect. 11.1.1; McEwan et al. 2021a) and facing the main challenges of RT&Bs: poor-quality seed of existing varieties, low adoption rates of improved varieties, and slow varietal turnover. Seed Tracker (ST) is helping seed growers and regulators to track yam and cassava seed production and marketing in Nigeria, Tanzania, and Brazil (Ouma et al. 2019; Kumar 2021; www.seedtracker.org/cassava). Tracking the seed improves the delivery of quality seed of improved varieties (knowledge gap 2) and the implementation of policies and regulations (knowledge gap 4). Four-square method (FSM) has been used to facilitate farmer understanding of optimal banana variety use and has helped stakeholders to appreciate the need for banana variety conservation in Uganda (Kilwinger et al. 2019; Mulugo et al. unpublished data), improving the capture of seed demand characteristics (knowledge gap 1).

Although the tools were designed to function as stand-alones, in several cases, they were used in combination to better address knowledge gaps. For example, in the Republic of Georgia, combining the multi-stakeholder framework (MSF), impact network analysis (INA), and integrated seed health (ISH) models provided direction for establishing a new potato seed system (Andersen Onofre et al. 2021). In Southeast Asia, combining INA, ISH, and seed tracing (STg) generated new understanding of cassava seed trade networks and the implications of their structure, guiding deployment of clean seed to manage an emerging cassava mosaic disease epidemic (Delaquis et al. 2018; Andersen et al. 2020). In Tanzania, combining MSF and ISH provided a rapid view of challenges to the sweetpotato seed system and the potential for new disease management strategies to provide economic benefits to growers (Ogero et al. 2015, 2019). These three examples show how the tools contribute to improve seed health and stop the spread of diseases (knowledge gap 3).

There are also many opportunities to link these tools with other methods or initiatives. Other RTB programs address disease testing and disease diagnosis in the field, because effective seed systems often depend on accurate diagnostic testing to protect seed health. For example, ST is being linked to *PlantVillage Nuru* (Mrisho et al. 2020), a smartphone-based artificial intelligence system for infield diagnosis of cassava mosaic disease (CMD) and cassava brown streak disease (CBSD). *PlantVillage Nuru* helps farmers to diagnose the problem and ST tells them where to get healthy seed. Potato seed systems in East Africa must address the spread of the pathogen *Ralstonia solanacearum*, causing bacterial wilt (Gachamba et al. 2022). The analysis of cropland connectivity, i.e., the importance of locations for the potential spread of crop-specific pathogens, has been used to evaluate crop risks for RT&Bs (Xing et al. 2020) and can provide networks for input in INA. In Rwanda, STg was combined with the rural household multiple indicator survey (RHoMIS) (van Wijk et al. 2020) and typology analysis (Hammond et al. 2020) to understand the cassava seed sourcing practices of different farm typologies (Kilwinger et al. 2021b).

While most of the tools strive to incorporate gender, there is a need for greater gender integration in existing tools and for stand-alone gender tools. For example, a study to validate the gendered MSF in Uganda revealed more nuanced gender dimensions in banana seed systems which will require the attention of extension services to address women farmers' limited access to banana varieties and to improve their knowledge of variety performance and the control of pests and diseases (Ajambo et al. in preparation). The tools could also be improved by putting them in the hands of multidisciplinary research teams at study design and implementation to allow for greater integration of social and gendered perspectives. Since the toolbox is a living resource, including stand-alone gender tools in the future will enhance its applicability for a wider range of research propositions by targeted R&D users.

Toolbox development has focused on building its research foundations through the portfolio of applications discussed above. The current portfolio gives examples of applying each tool and some combinations of them. With each use, lessons are learned, and the tools are adapted. The tools are designed to be flexible for new questions and systems. While the tools have initially been applied to RT&B crops, with their particular challenges for seed system development, the tools are also ready for wider application to grains, pulses, fruits and vegetables, and new VPCs. New tools are also in the pipeline, including one to co-develop seed delivery profiles with breeders and seed system scientists to design effective seed delivery (McEwan et al. 2021a).

11.3.2 Scaling

Following the validation of the tools in different crop and country combinations, we have turned our attention to understand how we can use scaling readiness concepts (Sartas et al. 2020) to promote wider use of the toolbox to realize our vision of different types of farmers accessing quality seed and improved varieties. Many seed innovations struggle to survive beyond project support. This may reflect a linear approach to technology adoption and the failure to consider the wider enabling factors around scaling.

Quality seed and improved varieties are technical innovations. But they are used within specific social and natural contexts. The tools provide diagnostic, methodological, and decision support. By promoting the findings and outcomes from using the tools, we seek to encourage a change in mindsets and new ways of thinking about how formal and informal seed systems function, the challenges and how to address them, and the required changes in infrastructure and investments. These elements come together as the toolbox innovation package. We have identified bottlenecks to scaling the toolbox, including lack of awareness by potential users and the need to provide training and mentoring opportunities for national seed system actors (e.g., national research institutions, extension, and NGOs).

To characterize the institutions and better target stakeholders who might use the toolbox, we conducted a landscape analysis of seed systems for sub-Saharan Africa (Cox et al. 2021). A communication strategy targeting different audiences, with video, infographic, and social media toolkits, has been implemented to support the launch of the toolbox. This process is helping us to identify new partners, networks, and types of collaborations to support scaling.

Our vision for the future is a global initiative that will foster collaboration around the tools to improve seed systems research on three levels: First, by continuing to refine the tools themselves. As tool use increases, feedback from more practitioners will promote improvement and synergies with other approaches. Second, through linkages to other initiatives and research programs, the toolbox can be a converging point for a greater community of practice, sharing wider experiences and novel approaches which may help to adapt the tools and expand the toolbox. Third, and most importantly, through continued documentation of impacts and improvements in seed system outcomes, findings can be integrated into higher-order evaluations and cross-case lessons for implementers and policymakers, deepening the scientific understanding and policy relevance of toolbox outputs. We envision the toolbox as a game changer to improve seed systems in the coming years.

Acknowledgments This research was undertaken as part of, and funded by, the CGIAR Research Program on Roots, Tubers and Bananas (RTB), with support from CGIAR Fund Donors, Netherlands Organization for Scientific Research (NWO)—WOTRO Science for Global Development, Wageningen University and Research (WUR), USDA NIFA grant 2015-51181-24257. Support is also appreciated from the CGIAR Research Program on Climate Change, Agriculture and Food Security (CCAFS). Bill & Melinda Gates Foundation grants OPP1019987 (SASHA), OPP121332 (SweetGAINS), and OPP1080975, USDA APHIS grant

11-8453-1483-CA, the USAID Feed the Future Haiti Appui à la Recherche et au Développement Agricole (AREA) project grant AID-OAA-A-15-00039, US NSF Grant EF-0525712 as part of the joint NSF-NIH Ecology of Infectious Disease program, US NSF Grant DEB-0516046, McKnight Foundation grant 16-275, and the University of Florida.

References

Aboagye LM, Nyadanu D, Opoku-Agyeman MO, Owusu SK, Asiedu-Darko E (2015) Survey of diversity and production of yams in four communities in Southern Ghana. Afr J Agric Res 10(24):2453–2459. https://doi.org/10.5897/AJAR2014.9468

Agre AP, Bhattacharjee R, Dansi A, Lopez-Lavalle LAB, Dansi M, Sanni A (2016) Assessment of cassava (*Manihot esculenta* Crantz) diversity, loss of landraces and farmers preference criteria in southern Benin using farmers' participatory approach. Genet Resour Crop Evol 62(8):307–320. https://doi.org/10.1007/s10722-015-0352-1

Aini MF, Elias M, Lamers H, Shariah U, Brooke P, Hafizul HM (2017) Evaluating the usefulness and ease of use of participatory tools for forestry and livelihoods research in Sarawak, Malaysia. Forests Trees Livelihoods 26(1):29–46. https://doi.org/10.1080/14728028.2016.1246213

Ajambo S, Mayanga S, Kikulwe E (in preparation) Validating the gender responsive multi-stakeholder framework (G+MSF) tool: a review of banana and sweet potato seed systems in northern Uganda. https://hdl.handle.net/20.500.11766/12999

Almekinders CJM, Ronner E, van Heerwaarden J (2020) Tracing legume seed diffusion beyond demonstration trials: an exploration of sharing mechanisms. Outlook Agric 49(1):29–38. https://doi.org/10.1177/0030727020907646

Andersen KF, Buddenhagen CE, Rachkara P, Gibson R, Kalule S, Phillips D, Garrett KA (2019) Modeling epidemics in seed systems and landscapes to guide management strategies: the case of sweetpotato in Northern Uganda. Phytopathology 109:1519–1532. https://doi.org/10.1094/PHYTO-03-18-0072-R

Andersen KF, Delaquis E, Newby J, de Haan S, Le Cu TT, Minato N, Legg JP, Cuellar WJ, Alcalá-Briseño RI, Garrett KA (2020) A decision support model for landscape-level management of cassava mosaic disease in Southeast Asia. Plant Health 2020, American Phytopathological Society Annual Meeting abstract. https://apsnet.confex.com/apsnet/2020/meetingapp.cgi/Paper/17189

Andersen Onofre KF, Forbes GA, Andrade-Piedra JL, Buddenhagen CE, Fulton J, Gatto M, Khidesheli Z, Mdivani R, Xing Y, Garrett KA (2021) An integrated seed health strategy and phytosanitary risk assessment: potato in the Republic of Georgia. Agric Syst 191:103144. https://doi.org/10.1016/j.agsy.2021.103144

Andrade-Piedra JL, Almekinders CJM, McEwan MA, Kilwinger FBM, Mayanja S, Mulugo L, Delaquis E, Garrett KA, Omondi AB, Rajendran S, Kumar PL, Thiele G (2020) User guide to the toolbox for working with root, tuber and banana seed systems. Lima, Peru. CGIAR Research Program on Roots, Tubers and Bananas (RTB). RTB user guide. No. 2020-1. https://doi.org/10.4160/9789290605577

Audi P, Nagarajan L, Jones RB (2008) Seed interventions and cultivar diversity in pigeon pea: a farmer based assessment in eastern Kenya. J New Seeds 9(2):111–127. https://doi.org/10.1080/15228860802073016

Barker I, Jones R, Klauser D (2021) Smallholder seed systems for sustainability. In: Klauser D, Robinson M (eds.) The sustainable intensification of smallholder farming systems. Burleigh Dodds Science Publishing, Cambridge, UK, 2021, ISBN: 978 1 78676 430 0 https://doi.org/10.19103/AS.2020.0080.05

Bentley JW, Andrade-Piedra JL, Demo P, Dzomeku B, Jacobsen K, Kikulwe E, Kromann P, Kumar PL, McEwan MA, Mudege N, Ogero K, Okechukwu R, Orrego R, Ospina B, Sperling L,

Walsh S, Thiele G (2018) Understanding root, tuber, and banana seed systems and coordination breakdown: a multi-stakeholder framework. J Crop Improv 32:599–621.

Bentley JW, Mudege N, Andrade-Piedra JL (2020) User guide to the multi-stakeholder framework for intervening in root, tuber and banana seed systems. Lima: International Potato Center on behalf of CGIAR Research Program on Roots, Tubers and Bananas. RTB User Guide 2020–2022. https://doi.org/10.4160/9789290605591

Buddenhagen CE, Hernandez Nopsa JF, Andersen KF, Andrade-Piedra JL, Forbes GA, Kromann P, Thomas-Sharma S, Useche P, Garrett KA (2017) Epidemic network analysis for mitigation of invasive pathogens in seed systems: potato in Ecuador. Phytopathology 107:1209–1218. https://doi.org/10.1094/PHYTO-03-17-0108-FI

Buddenhagen CE, Xing Y, Andrade Piedra JL, Forbes GA, Kromann P, Navarrete I, Thomas-Sharma S, Choudhury RA, Andersen Onofre KF, Schulte-Geldermann E, Etherton BA, Plex Sulá AI, Garrett KA (2022) Where to invest project efforts for greater benefit: A framework for management performance mapping with examples for potato seed health. Phytopathology 112. https://doi.org/10.1094/PHYTO-05-20-0202-R

Bullock OM, Amill DC, Shulman HC, Dixon GN (2019) Jargon as a barrier to effective science communication: evidence from metacognition. Public Underst Sci 28, 845–853. https://doi.org/10.1177/0963662519865687

CGIAR Excellence in Breeding Platform (2011). https://excellenceinbreeding.org/. Accessed 21 May 2021

Choudhury RA, Garrett KA, Klosterman SJ, Subbarao KV, McRoberts N (2017) A framework for optimizing phytosanitary thresholds in seed systems. Phytopathology 107:1219–1228. https://doi.org/10.1094/PHYTO-04-17-0131-FI

Cox CM, Weinrich ES, Xing Y, Andrade-Piedra JL, McEwan MA, Almekinders CJM, Garrett KA (2021) Tools for evaluating and improving seed systems and seed health in roots, tubers and bananas. American Phytopathological Society, Southern Division Annual Meeting abstract. https://apsnet.confex.com/apsnet/SOUTH21/meetingapp.cgi/Paper/18150

Dansi A, Dantsey-Barry H, Dossou-Aminon I, N'Kpenu EK, Agré AP, Sunu YD, Kombaté K, Loko YL, Dansi M, Assogba P, Vodouhè R (2013) Varietal diversity and genetic erosion of cultivated yams (*Dioscorea cayenensis* Poir – *D. rotundata* Lam complex and *D. alata* L.) in Togo. Int J Biodiver Conserv 5(4):223–239. https://doi.org/10.5897/IJBC12.131

Dantes W, Xing Y, Choudhury RA, Fayette J, Andersen KF, Blomme G, Dita Rodriguez MA, Ocimati W, Ophny NC, Hernandez Nopsa JF, Garrett KA (2020) Getting a jump on new banana diseases: risk assessment for Haiti and the Caribbean. Plant Health 2020 American Phytopathological Society Annual Meeting abstract. https://apsnet.confex.com/apsnet/2020/meetingapp.cgi/Paper/17194

Delaquis E, Andersen KF, Minato N, Cu TTL, Karssenberg ME, Sok S, Wyckhuys KAG, Newby JC, Burra DD, Srean P, Phirun I, Le ND, Pham NT, Garrett KA, Almekinders CJM, Struik PC, de Haan S (2018) Raising the stakes: cassava seed networks at multiple scales in Cambodia and Vietnam. Front Sustain Food Syst 2:73. https://doi.org/10.3389/fsufs.2018.00073

Delaquis E, Almekinders CJM, Andrade-Piedra JL (2020) Glossary of root, tuber and banana seed systems. RTB user guide. International Potato Center on behalf of CGIAR Research Program on Roots, Tubers and Bananas. RTB User Guide 2020-5, Lima, Peru. https://doi.org/10.4160/9789290605621

Delaquis E, Slavchevska V, Newby JC, Almekinders CJM, Struik PC (unpublished data) Employing experimental seed auctions to understand cassava seed demand in Lao PDR

Fayette J, Xing Y, Dantes W, Andersen KF, Ophny NC, da Silva Galdino TV, Kumar S, Penca CJ, Szyniszewska AM, Garrett KA (2020) Integrating climate, cropland connectivity, and trade data layers to understand phytosanitary risk for mango in Haiti and the Caribbean. Plant Health 2020, American Phytopathological Society Annual Meeting abstract. https://apsnet.confex.com/apsnet/2020/meetingapp.cgi/Paper/17233

FAO (2015) Household seed security concepts and indicators: discussion paper. Building capacity for seed security assessments. FAO Expert Consultation on Seed System Security,

12–13 December 2013. http://www.fao.org/fileadmin/user_upload/food-security-capacity-building/docs/Seeds/SSCF/Seed_security_concepts_and_indicators_FINAL.pdf. Accessed 23 May 2021

Gachamba S, Xing Y, Andersen Onofre KF, Garrett KA, Miano DW, Mwangombe A, Sharma K (2022) Epidemic networks and potential sources of bacterial wilt infection in a potato seed network in Kenya. BIORXIV/2022/475701

Garrett KA (2021a) Impact network analysis and the INA R package: decision support for regional management interventions. Methods Ecol Evol 12:1634–1647. https://doi.org/10.1111/2041-210X.13655

Garrett KA (2021b) User guide to impact network analysis (INA). CGIAR Research Program on Roots, Tubers and Bananas (RTB). RTB user guide. No. 2021-4. Lima, Peru. https://cgspace.cgiar.org/handle/10568/111326

Garrett KA, Xing Y (2021) User guide to the seedHealth model as part of the integrated seed health approach. CGIAR Research Program on Roots, Tubers and Bananas (RTB). RTB User Guide. No. 2021-5. Lima, Peru. https://cgspace.cgiar.org/handle/10568/111327

Garrett KA, Alcalá-Briseño RI, Andersen KF, Buddenhagen CE, Choudhury RA, Fulton JC, Hernandez Nopsa JF, Poudel R, Xing Y (2018) Network analysis: a systems framework to address grand challenges in plant pathology. Annu Rev Phytopathol 56:559–580. https://doi.org/10.1146/annurev-phyto-080516-035326

Gatto M, Dung Le P, Pacillo G, Maredia M, Labarta R, Hareau G, Spielman DJ (2021) Policy options for advancing seed systems for vegetatively propagated crops in Vietnam. J Crop Improv. https://doi.org/10.1080/15427528.2021.1881011

Gibson JJ (1977) The theory of affordances. Hilldale 1(2):67–82

Glover D, Sumberg J, Ton G, Andersson J, Badstue L (2019) Rethinking technological change in smallholder agriculture. Outlook Agric 48(3):169–180. https://doi.org/10.1177/0030727019864978

Grum M, Gyasi EA, Osei C, Kranjac-Berisavljevic G (2008) Evaluation of best practices for landrace conservation: farmer evaluation. Bioversity International, Rome. https://hdl.handle.net/10568/104791

Gurmu F, Abele W, Tsegaye G, Gezahen G (2019) Sweetpotato seed business model: the case of the South Agricultural Research Institute. Ethiopia J Agric Crop Res 7(8):127–136. https://doi.org/10.33495/jacr_v7i8.19.126

Harahagazwe D, Condori B, Barreda C, Bararyenya A, Byarugaba AA, Kude DA, Lung'aho C, Martinho C, Mbiri D, Nasona B, Ochieng B, Onditi J, Randrianaivoarivony JM, Tankou CM, Worku A, Schulte-Geldermann E, Mares V, de Mendiburu F, Quiroz RQ (2018) How big is the potato (*Solanum tuberosum* L.) yield gap in Sub-Saharan Africa and why? A participatory approach. Open Agric 3:180–189. https://doi.org/10.1515/opag-2018-0019

Hammond J, Rosenblum N, Breseman D, Gorman L, Manners R, Wijk MT van, Sibomana M, Remans R, Vanlauwe B, Schut M (2020) Towards actionable farm typologies: scaling adoption of agricultural inputs in Rwanda. Agric Syst 183:102857. https://doi.org/10.1016/j.agsy.2020.102857

International Potato Center (2017) SASHA Brief 7. Analysis of cost structure and pricing strategy for sweetpotato early generation seed – examining experiences from Kenya and Ghana. https://hdl.handle.net/10568/105961

Kelly GA (1955) The psychology of personal constructs. Norton, New York

Kilwinger FBM, Rietveld AM, Groot JCJ, Almekinders CJM (2019) Culturally embedded practices of managing banana diversity and planting material in central Uganda. J Crop Improv 33(4):456–477. https://www.tandfonline.com/doi/full/10.1080/15427528.2019.1610822

Kilwinger FBM, Marimo P, Rietveld AM, Almekinders CJM, van Dam YK (2020) Not only the seed matters: farmers' perceptions of sources for banana planting materials in Uganda. Outlook Agric 49(2):119–132. https://doi.org/10.1177/0030727020930731

Kilwinger FBM, van Dam YK (2021a) Methodological considerations on the means-end chain analysis revisited. Psychol Mark 38:1513. https://doi.org/10.1002/mar.21521

Kilwinger FBM, Mugambi S, Manners R., Schut, M, Tumwegamire S, Nduwumuremyi A, Bambara S, Paauwe M, Almekinders CJM (2021b) Characterizing cassava farmer typologies and their seed sourcing practices to explore opportunities for economically sustainable seed business models in Rwanda. Outlook Agric 50(4):441–454. https://doi.org/10.1177/00307270211045408

Kumar L (2021) Seed Tracker used by national seed regulators to support national certification schemes in Nigeria, Tanzania and Brazil. https://mel.cgiar.org/projects/yam-improvement-for-incomes-and-food-security-in-west-africa-phase-ii/364/seed-tracker-used-by-national-seed-regulators-to-support-national-certification-schemes-in-nigeria-tanzania-and-brazil. Accessed 30 May 2021

Leyte JD, Delaquis E, Van Dung P, Douxchamps S (2021). Linking Up: The Role of Institutions and Farmers in Forage Seed Exchange Networks of Southeast Asia. Human Ecology. https://doi.org/10.1007/s10745-021-00274-5

Maredia MK, Shupp R, Opoku E, Mishili F, Reyes B, Kusolwa P, Kusi F, Kudra A (2019) Farmer perception and valuation of seed quality: evidence from bean and cowpea seed auctions in Tanzania and Ghana. Agric Econ 50:495–507. https://doi.org/10.1111/agec.12505

McEwan MA, Almekinders CJM, Andrade-Piedra JL, Delaquis E, Garrett KA, Kumar L, Mayanja S, Omondi BA, Rajendran S, Thiele G (2021a) Breaking through the 40% adoption ceiling: Mind the seed system gaps. A perspective on seed systems research for development in One CGIAR. Outlook Agric 50:5–12. https://doi.org/10.1177/0030727021989346

McEwan MA, Spielman DJ, Gatto M, Abdoulaye T, Hareau G (2021b) User guide to seed regulatory framework analysis and implications for vegetatively propagated crops. User guide. RTB No. 2021-2. CGIAR Research Program on Roots, Tubers and Bananas (RTB), Lima, Peru https://cgspace.cgiar.org/handle/10568/111347

McEwan MA, Spielman DJ, Okello J, Hareau G, Bartle B, Mbiri D, Atieno E, Omondi BA, Wossen T, Cortada L, Abdoulaye T, Maredia M (2021c) Exploring the regulatory space for improving availability, access and quality of vegetatively propagated crop seed: potato in Kenya. RTB working paper. No. 2021-1. CGIAR Research Program on Roots, Tubers and Bananas (RTB), Lima, Peru: https://cgspace.cgiar.org/handle/10568/110925

McGuire S, Sperling L (2011) The links between food security and seed security: facts and fiction that guide response. Dev Pract 21(4–5):493–508. https://doi.org/10.1080/09614524.2011.562485

Mrisho LM, Mbilinyi NA, Ndalahwa M, Ramcharan AM, Kehs AK, McCloskey PC, Murithi H, Hughes DP, Legg JP (2020) Accuracy of a smartphone-based object detection model, PlantVillage Nuru, in identifying the foliar symptoms of the viral diseases of cassava–CMD and CBSD. Front Plant Sci 11:1964. https://doi.org/10.3389/fpls.2020.590889

Mulesa TH, Dalle SP, Makate C, Haug R, Westengen OT (2021) Pluralistic seed system development: a path to seed security? Agronomy 11(2):372. https://doi.org/10.3390/agronomy11020372

Mulugo L, Kyazze BF, Kibwika P, Kikulwe E, Omondi AB, Ajambo S (unpublished data) Farmer selection preferences for banana planting materials in central Uganda

Mulugo L, Ajambo S, Kikulwe E (2021) User guide to the four-square method for intervening in root, tuber and banana seed systems. RTB User Guide No. 2021-3. CGIAR Research Program on Roots, Tubers and Bananas (RTB), Lima, Peru. https://doi.org/10.4160/9789290605805

Navarrete I, Almekinders CJM, Lopez V, Borja RM, Oyarzún P, Andrade-Piedra JL, Struik P (2019) Diversos actores, variedades, fuentes y transacciones en los sistemas de semilla de papa en Cotopaxi. In: Rivadeneira J, Racines M, Cuesta X. Artículos del Octavo Congreso Ecuatoriano de la Papa. Ambato, Ecuador. Junio 27, 28, 2019. pp 135–136. https://hdl.handle.net/10568/107872

Nkengla-Asi L, Omondi AB, Che-Simo V, Assam E, Ngatat S, Boonabaana B (2020) Gender dynamics in banana seed systems and impact on banana bunchy top disease recovery in Cameroon. Outlook Agric 49(3):235–244. https://doi.org/10.1177/0030727020918333

Nyadanu D, Opoku-Agyeman M (2015) Survey of diversity and production of yams in four communities in Southern Ghana. Afr J Agric Res 10(24):2453–2459. https://doi.org/10.5897/AJAR2014.9468

Ogero K, Pamba N, Walsh S (2015) Review and learning workshop on Marando Bora (Quality Vines), sweetpotato seed systems case study in the Lake Zone, Tanzania. Proceedings of a workshop held in Mwanza, Tanzania, 27–28 January 2015

Ogero K, Kreuze J, McEwan MA, Luambano N, Bachwenkizi H, Garrett KA, Andersen K, Thomas-Sharma S, van der Vlugt R (2019) Efficiency of insect-proof net tunnels in reducing virus-related seed degeneration in sweet potato. Plant Pathol 68:1472–1480. https://doi. org/10.1111/ppa.13069

Okello JJ, Jogo W, Kwikiriza N, Muoki P (2018a) Motivations and cognitive models associated with decentralized seed multiplication: experiences from biofortified sweetpotato vine multipliers in Kenya and Ethiopia. J Agribus Develop Emerg Econ 8(4):626–641. https://doi. org/10.1108/JADEE-06-2017-0058

Okello JJ, Lagerkvist CJ, Kakuhenzire R, Parker M, Schulte-Geldermann E (2018b) Combining means-end chain analysis and goal-priming to analyze Tanzanian farmers' motivations to invest in quality seed of new potato varieties. Br Food J 120(7):1430–1445. https://doi.org/10.1108/ BFJ-11-2017-0612

Okello JJ, Zhou Y, Barker I, Schulte-Geldermann E (2019) Motivations and mental models associated with smallholder farmers' adoption of improved agricultural technology: evidence from use of quality seed potato in Kenya. Eur J Dev Res 31(2):271–292. https://doi.org/10.1057/ s41287-018-0152-5

Ouma T, Kavoo A, Wainaina C, Ogunya B, Kumar PL, Shah T (2019) Open data kit (ODK) in crop farming: an introduction of mobile data collection methods in seed yam tracking in Ibadan, Nigeria. J Crop Improv 33(5):605–619. https://doi.org/10.1080/15427528.2019.1643812

Pircher T, Obisesan D, Nitturkar H, Asumugha G, Ewuziem J, Anyaegbunam H, Azaino E, Akinmosin B, Ioryina A, Walsh S, Almekinders CJM (2019) Characterizing Nigeria's cassava seed system and the use of planting material in three farming communities. RTB working paper. No. 2019-1. International Potato Center, Lima, 28 p. https://doi.org/10.4160/2309658 6RTBWP20191

Rajendran S, Kimenye L, McEwan MA (2017) Strategies for the development of the sweetpotato early generation seed sector in Eastern and Southern Africa. Open Agric 2(1):236–243. https:// doi.org/10.1515/opag-2017-0025

Rajendran S, McEwan MA (2018) Using the sustainable early generation seed business analysis tool (SEGSBAT) for a financially sustainable seed business. CIP Science week 2018, Nairobi, Kenya

Rajendran S, McEwan MA (2021a) User guide to the sustainable early generation seed business analysis tool (SEGSBAT). RTB user guide no. 2021-8. CGIAR Research Program on Roots, Tubers and Bananas (RTB), Lima, Peru. https://doi.org/10.4160/9789290605836

Rajendran S, McEwan MA (2021b) The sustainable early generation seed business analysis tool (SEGSBAT). International Potato Center, Dataverse. https://doi.org/10.21223/FROU6W

Remington T, Maroko J, Walsh S, Omanga P, Charles E (2002) Getting off the seeds-and-tools treadmill with CRS seed vouchers and fairs. Disasters 26:316–328. https://doi. org/10.1111/1467-7717.00209

Reynolds T, Gutman J (1988) Laddering theory, method, analysis, and interpretation. J Advert Res 28(1):11–31

Rutkoski J (2019) A practical guide to genetic gain. Adv Agron 157:217–249. https://doi. org/10.1016/bs.agron.2019.05.001

Sartas M, Schut M, Proietti C, Thiele T, Leeuwis C (2020) Scaling readiness: science and practice of an approach to enhance impact of research for development. Agric Syst 83. https://doi. org/10.1016/j.agsy.2020.102874

Simbare A, Sane CAB, Nduwimana I, Niyongere C, Omondi BA (2020) Diminishing farm diversity of East African Highland bananas in banana bunchy top disease outbreak areas of Burundi—The Effect of Both Disease and Control Approaches 16. https://doi.org/10.3390/ su12187467

Sikkema A (2020) The N2Africa formula works (almost) everywhere. Resource: https://resource. wur.nl/en/show/The-N2Africa-formula-works-almost-everywhere-.htm

Sperling L (2008) When disaster strikes: a guide to assessing seed system security. International Center for Tropical Agriculture, Cali. http://ciat-library.ciat.cgiar.org/Articulos_Ciat/sssa_manual_ciat.pdf. Accessed 31 May 2021

Sperling L, Gallagher P, McGuire S, March J, Templer N (2020) Informal seed traders: the backbone of seed business and African smallholder seed supply. Sustainability 12:7074. https://doi.org/10.3390/su12177074

Spielman DJ, Gatto M, Wossen T, McEwan MA, Abdoulaye T, Maredia M, Hareau G (2021) Regulatory options to improve seed systems for vegetatively propagated crops in developing countries. IFPRI Discussion Paper. https://doi.org/10.2499/p15738coll2.134441

Tadesse Y, Almekinders CJM, Schulte RPO, Struik PC (2017a) Understanding farmers' potato production practices and use of improved varieties in Chencha, Ethiopia. J Crop Improv 31(5):673–688. https://doi.org/10.1080/15427528.2017.1345817

Tadesse Y, Almekinders CJM, Schulte RP, Struik PC (2017b) Tracing the seed: seed diffusion of improved potato varieties through farmers' networks in Chencha, Ethiopia. Exp Agric 53(4):481. https://doi.org/10.1017/S001447971600051X

Thiele G, Dufour D, Vernier P, Mwanga ROM, Parker ML, Schulte Geldermann E, Teeken B, Wossen T, Gotor E, Kikulwe E, Tufan H, Sinelle S, Kouakou AM, Friedmann M, Polar V, Hershey C (2020) A review of varietal change in roots, tubers and bananas: consumer preferences and other drivers of adoption and implications for breeding. Int J Food Sci Technol 56:1076–1092. https://doi.org/10.1111/ijfs.14684

Thomas-Sharma S, Abdurahman A, Ali S, Andrade-Piedra JL, Bao S, Charkowski AO, Crook D, Kadian M, Kromann P, Struik PC, Torrance L, Garrett KA, Forbes GA (2016) Seed degeneration in potato: the need for an integrated seed health strategy to mitigate the problem in developing countries. Plant Pathol 65(1):3–16. https://doi.org/10.1111/ppa.12439

Thomas-Sharma S, Andrade-Piedra JL, Carvajal Yepes M, Hernandez Nopsa J, Jeger M, Jones R, Kromann P, Legg J, Yuen J, Forbes GA, Garrett KA (2017) A risk assessment framework for seed degeneration: Informing an integrated seed health strategy for vegetatively propagated crops. Phytopathology 107:1123–1135. https://doi.org/10.1094/PHYTO-09-16-0340-R

Urrea-Hernandez C, Almekinders CJM, van Dam YK (2016) Understanding perceptions of potato seed quality among small-scale farmers in Peruvian highlands. NJAS – Wageningen J Life Sci 76:21–28. https://doi.org/10.1016/j.njas.2015.11.001

USAID (2013) Enabling Agricultural Trade (EAT) project. Seed CLIR Tanzania Pilot Report, Place, USAID

van Etten J, de Sousa K, Aguilar A, Barrios M, Coto A, Dell'Acqua M, Fadda C, Gebrehawaryat Y, van de Gevel J, Gupta A, Kiros AY et al (2019) Crop variety management for climate adaptation supported by citizen science. Proc Natl Acad Sci U S A 116(10):4194–4199. https://doi.org/10.1073/pnas.1813720116

van Wijk M, Hammond J, Gorman L, Adams S, Ayantunde A, Baines D et al (2020) The Rural Household Multiple Indicator Survey, data from 13,310 farm households in 21 countries. Sci Data 7:46. https://doi.org/10.1038/s41597-020-0388-8

Walker TS, Alwang J, Alene A, Ndjuenga J, Labarta R, Yigezu Y, Diagne A, Andrade R, Andriatsitohaina RM, De Groote H, Mausch K, Yirga C, Simtowe F, Katungi E, Jogo W, Jaleta M, Pandey S, Charyulu DK (2015) Varietal adoption, outcomes and impact. In: Walker TS, Alwang J (eds) Crop improvement, adoption, and impact of improved varieties in food crops in sub-Saharan Africa. CGIAR Consortium of International Agricultural Research Centers and CAB International. 388–405 pp. ISBN-13: 978 1 78064 401 1 (CABI)

Walker BA, Olson JC (1991) Means-end chains: connecting products with self. J Bus Res 22(2):111–118

Watkins L (2010) The cross-cultural appropriateness of survey-based value(s) research: a review of methodological issues and suggestion of alternative methodology. Int Mark Rev 27(6):694–716. https://doi.org/10.1108/02651331011088290

White H, Phillips D (2012) Addressing attribution of cause and effect in small n impact evaluations: towards an integrated framework. International Initiative for Impact Evaluation (3ie).
Working paper 15

Wossen T, Spielman DJ, Abdoulaye T, Kumar PL (2020) The cassava seed system in Nigeria:
opportunities and challenges for policy and regulatory reform. CGIAR Research Program on
Roots, Tubers, and Bananas (RTB) working paper no. 2020-2. International Potato Center,
Lima, Peru. https://doi.org/10.4160/23096586RTBWP20202

Xing Y, Hernandez Nopsa JF, Andrade-Piedra JL, Beed FD, Blomme G, Carvajal Yepes M, Coyne
DL, Cuéllar WJ, Forbes GA, Kreuze JF, Kroschel J, Kumar PL, Legg JP, Parker M, Schulte-
Geldermann E, Sharma K, Garrett KA (2020) Global cropland connectivity: a risk factor for
invasion and saturation by emerging pathogens and pests. BioScience 70:744–758. https://doi.
org/10.1093/biosci/biaa067

Chapter 12
Securing Sweetpotato Planting Material for Farmers in Dryland Africa: Gender-Responsive Communication Approaches to Scale Triple S

Margaret A. McEwan, Tom A. van Mourik, Mihiretu C. Hundayehu, Frezer Asfaw, Sam Namanda, Issahaq Suleman, Sarah Mayanja, Simon Imoro, and Prince M. Etwire

Abstract Triple S (Storage in Sand and Sprouting) is a root-based system for conserving and multiplying sweetpotato planting material at the household level. In sub-Saharan Africa, farmers predominantly source planting material by cutting vines from volunteer plants that sprout from roots left in the field from a previous crop. However, it takes 6 to 8 weeks after the rains start to produce enough vines for

M. A. McEwan (✉)
International Potato Center, Regional Office for Africa, Nairobi, Kenya
e-mail: M.McEwan@cgiar.org

T. A. van Mourik
Royal Tropical Institute (KIT), Amsterdam, The Netherlands
e-mail: t.v.mourik@kit.nl

M. C. Hundayehu
International Potato Center, Hawassa, Ethiopia
e-mail: M.Cherinet@cgiar.org

F. Asfaw
International Potato Center, Addis Ababa, Ethiopia
e-mail: F.Asfaw@cgiar.org

S. Namanda · S. Mayanja
International Potato Center (CIP), Kampala, Uganda
e-mail: S.Namanda@cgiar.org; S.Mayanja@cgiar.org

I. Suleman
International Potato Center, Kumasi, Ghana

S. Imoro
International Institute for Tropical Agriculture, Ibadan, Nigeria

P. M. Etwire
Council for Scientific and Industrial Research, Savanna Agricultural Research Institute, Tamale, Ghana

© The Author(s) 2022
G. Thiele et al. (eds.), *Root, Tuber and Banana Food System Innovations*,
https://doi.org/10.1007/978-3-030-92022-7_12

planting material, and normally these vines are infected by sweetpotato diseases and pests carried over from previous crops. Where rainfall is unpredictable, farmers can use Triple S to take advantage of the whole growing season, planting and harvesting early to obtain food, higher yields, and income. Triple S facilitates household retention and adoption of new sweetpotato varieties, notably the beta-carotene-rich, orange-fleshed varieties. Triple S PLUS is the combined innovation package of core Triple S components and complementary components used to scale the innovation. These included good agricultural practices, different storage containers, local multiplication and sales of planting material, and a multimedia communication strategy for training and extension to encourage the uptake of Triple S. Components were at different levels of scaling readiness. This chapter explores evidence from Ethiopia and Ghana (2018–2019) on the extent to which exposure to different communication channels and their combinations influenced the uptake of Triple S PLUS by male and female farmers, the partnering arrangements that supported this, and the resulting changes in food security. We discuss implications for future scaling initiatives.

12.1 Introduction and Background

Agricultural production in many countries in sub-Saharan Africa (SSA) is dominated by smallholder farmers, and sweetpotato is among the most widely grown and consumed food crops (Low et al. 2009). The crop has several advantages for resource-poor households. It produces reliable yields even with minimum inputs, and it is rich in carbohydrates and micronutrients. The orange-fleshed sweetpotato (OFSP) varieties are especially rich in provitamin A (Low et al. 2017). The sweetpotato has proven its value in post disaster recovery due to its early maturity. But the crop's advantages are only realized if the vegetatively propagated planting materials are available for timely and early planting. In areas with a long dry season, the most common way farmers source planting material is from roots that were left in the field from a previous crop, which then sprout at the start of the rains. However, it takes about 6 to 8 weeks to produce enough vines for planting material, which may have accumulated sweetpotato virus diseases, weevils, and other pests, affecting future yields and quality of the harvested roots. This results in a shortage of timely, quality planting material at the start of the rains. Some households may have access to water or to land near wetlands, which can be used to conserve planting material over the dry season. But many households prioritize this land for high-value horticultural crops rather than for conserving sweetpotato planting material. Furthermore, female and resource-poor farmers may have poor access to these wetlands.

12.1.1   The Potential of Triple S

Triple S (Storage in Sand and Sprouting) is system that allows households to store sweetpotato roots during the dry season and use them to multiply vines to use as planting material. Triple S can be used for any sweetpotato variety. The method requires less labor than maintaining vines throughout the dry season, requires little watering, minimizes exposure to pests and diseases, and produces clean and early planting material. This can increase the productivity of sweetpotato and extend the period in which it can be cultivated. As rainy seasons become shorter and less predictable, farmers can use Triple S to take advantage of the whole growing season, planting right at the start of the rains, to harvest early and obtain food, higher yields, and income. Triple S can be used in areas with a dry season of 4 months or longer. The warm, semiarid and tropical savanna climate zones in Africa are potential areas for using Triple S. A key contribution of this innovation is to facilitate household retention and adoption of new sweetpotato varieties, notably the beta-carotene-rich, orange-fleshed varieties.

12.1.2   Validation of Triple S and Bottlenecks for Scaling

Building on farmer seed management practices (i.e., using vines which resprout from roots after the rains), the Triple S technology was first validated in Uganda and Tanzania between 2009 and 2010 (Namanda 2012; Namanda et al. 2013). Farmers who piloted Triple S testified to its potential to solve the chronic shortage sweetpotato planting material[1] at the start of the rainy season. Between 2015 and 2017, training and testing was conducted with additional varieties and in places with dry seasons up to 9 months long, in Ethiopia, Kenya, Malawi, Mozambique, Nigeria, Ghana, and Burkina Faso. The project "Scaling Sweetpotato Triple S PLUS – gender responsive options for quality planting material, higher yields and extended shelf life for storage roots" supported the scaling of Triple S in Ethiopia and Ghana in 2018–2020. There had been wide distribution of OFSP planting materials in many countries in SSA, but in areas with a long dry season, farmers still struggled to retain the new varieties. Triple S could help address this challenge, but there were bottlenecks to scaling. These included overcoming gender-based constraints to access information and developing appropriate partnering strategies. In many contexts, sourcing and managing sweetpotato seed is the responsibility of women. Triple S is a knowledge-intensive technology which requires training and information at key steps throughout the season. However, extension services often overlook or do not appropriately target female farmers who may have low literacy levels, little time to attend trainings and limited access to extension agents; thus women are at a disadvantage in being able to acquire the knowledge necessary to try out a

---

[1] In this chapter planting material, vines, cuttings, and seed are used synonymously.

technology. Even if they can access information, women may have limited access to productive resources, such as land and equipment, to put their knowledge into practice. Thus, effective scaling would require gender-responsive communication materials, channels, and innovations to address these constraints. Moreover, as a research organization, for the Triple S PLUS innovation package to be institutionalized and scaled, we at the International Potato Center (CIP) needed to identify new types of partners to work with, who could effectively engage with both men and women. Therefore, as part of the scaling initiative, the project team developed the Triple S PLUS innovation package to address these bottlenecks, so that the technology could be scaled to enable the long-term and sustainable adoption of improved orange-fleshed sweetpotato varieties. Triple S PLUS is the combined package of the core Triple S system, plus the complementary components that were used to scale the innovation. These included good agricultural practices (GAPs), options for men and women to use the stored roots as seed or as food for a longer part of the year, and a multimedia information and communication strategy to train farmers.

The chapter starts by illustrating how an assessment of the level of scaling readiness (Sartas et al. 2020a) of the components of Triple S PLUS informed our theory of change, scaling approach, and partnering strategy in Ethiopia and Ghana. It shows how we changed our partnering strategy as we found that the scaling readiness of certain components was not as high as originally assessed. We then describe how exposure to different types and combinations of communication channels influenced the use of Triple S by male and female farmers and their perceptions about the efficacy of the different communication channels. We argue that while we were able to achieve relatively high levels of participation of women in demonstrations and in video screening events, there was still a gender gap in early use of the components of Triple S. This may have been due to several other constraints such as the type of communication tools used, differences between women's and men's access to starter seed and intra-household decision-making. The chapter concludes with a discussion on the implications of these findings for future scaling of Triple S PLUS and agricultural innovations in general.

12.2   The Triple S PLUS Innovation Package and Scaling Readiness

12.2.1   Triple S Cycle

Figure 12.1 shows the Triple S cycle, which starts with production and selection of medium-sized healthy roots. The seed roots are sourced from a crop produced by following recommended management protocols (Namanda et al. 2013) and Triple S training guidelines: good agricultural practices for the production of sweetpotato roots, negative selection (roguing), positive selection (growing plants are monitored and the best ones are pegged for selection at harvest), and management of weevils

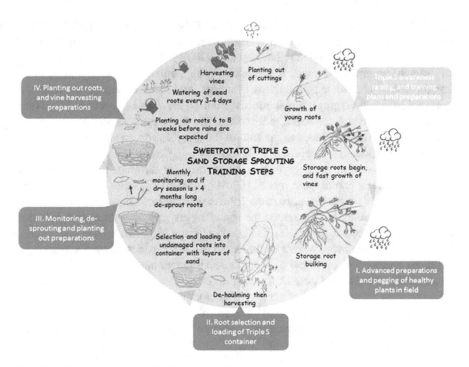

Fig. 12.1 The Triple S cycle (Stathers et al. 2017)

through remaking ridges (hilling up) to cover the cracks in the soil through which weevils can enter (Stathers et al. 2017). The roots are then carefully harvested, checked for pests, mechanical damage and disease symptoms, and then the selected roots are cured,[2] graded, and arranged in layers of sand in a container (plastic or locally available alternative). The size of the container varies depending on how many roots are stored, ranging from old basins and jerry cans to larger, purpose-built structures such as pits and sandboxes built from mud bricks. The roots are stored over the dry season, usually for 4 to 6 months. During storage, the roots' metabolic rate slows down, but eventually the roots start to sprout. Farmers check the roots on a regular basis to de-sprout if necessary and remove any spoiled roots. At 6 to 8 weeks before the anticipated start of the rains, farmers plant out the sprouted roots in a root seed bed which is protected from grazing animals. Farmers irrigate the root seed bed until the start of the rains when they can harvest the vines to use as planting material for root production. Over three harvests, each seed root can produce 30–40 cuttings, so 100 seed roots (stored in sand in two small basins) can produce about 12,000 cuttings, enough to plant a field of almost 0.4 hectares.

---

[2] Curing is done by either de-haulming (cutting the foliage off) the plant 2 weeks before harvesting, or after harvesting the roots are placed in shade under a tree, covered with banana leaves, or a mat (but not plastic sheets). Curing helps to thicken the skin and heals or prevents minor wounds. This reduces likelihood of rotting during root storage.

What constitutes a meaningful and viable innovation package depends on the context, which implies that the innovation package can change over time and is likely to differ across locations (Sartas et al. 2020b). During piloting and validation in different countries, changes were made to the Triple S technology to fit local environmental conditions, production practices, and socio-economic contexts. In northern Ethiopia (Tigray), larger roots were used for Triple S because they were more likely to survive the seven to 9-month-long dry season. Also, planting out in the root seed bed was delayed to the start of the rainy season, to align with the local sweetpotato production calendar for planting sweetpotato. Similarly, the composition of an innovation package may need to vary for different types of farmers according to their context and means (Glover et al. 2016; Sartas et al. 2020b). In Malawi, farmers wanted to store roots for use as food in addition to seed (Abidin et al. 2016), so adaptation included the use of larger structures such as sandboxes and stepped pits, which were later also used in Ghana (Fig. 12.2).

"Innovation readiness" refers to the demonstrated capacity of an innovation to fulfill its promise or contribute to specific development outcomes in specific locations. The innovation progresses from an untested idea to a fully mature practice being used by farmers or other value chain actors without input by the researcher. "Innovation use" indicates the level of use of the innovation or the package by a project, partners, and society. "Scaling readiness" of an innovation is a function of innovation readiness and innovation use. Table 12.1 defines for each level of technology readiness and use. At the start of the scaling intervention, the technology readiness of Triple S was between stages seven and eight on this scale and technology use between 0 and 2 (Tables 12.2 and 12.3).

Fig. 12.2 Preparation of stepped pit (l) and sand box (r) Navrongo, Ghana. (Photo credit: P. Abidin)

Table 12.1 Definition of each stage of technology readiness and use

Stage	Technology readiness (simplified)	Technology use (simplified)
0	Idea	Project partners
1	Hypothesis	Innovation network (rare)
2	Basic research	Innovation network (some)
3	Basic model	Innovation network (common)
4	Formulated working model	Innovation systems (rare)
5	Working model	Innovation systems (some)
6	Working application	Innovation systems (common)
7	Proof of application	Livelihood system (rare)
8	Incubation	Livelihood system (some)
9	Ready	Livelihood system (common)

Source: (Sartas 2018)

Table 12.2 Example of unpacking core components and understanding scaling readiness: Ghana, March 2018

Triple S PLUS core components	Technology readiness	Technology use	Scaling readiness	Decision: improve, reorient, outsource
Negative selection (rouging) and positive selection of healthy plants and pegging	7	0	7	*Improve* knowledge of farmers and emphasize these practices during training and demonstration
Root selection and loading of Triple S container	8	2	10	*Improve*: promote use through training and demonstration, suggest diversity of possible sizes and types of containers according to means and needs. Suggest additional protective structure for stepped pit container
Checking, de-sprouting and removing rotten and weevil infested roots	8	0	8	*Improve* knowledge of farmers and emphasize this practice during training and demonstration
Root seedbed (preparation, planting, and management)	8	2	10	*Improve* knowledge of farmers and emphasize fencing and regular irrigation of root seedbeds

NB. Scoring followed scale in Table 12.1

12.2.2 Triple S PLUS Innovation Package

In preparation for scaling, we (the co-authors as the project team) had included GAPs and gender-responsive communication materials as complementary components to form the Triple S PLUS innovation package. The Ethiopia and Ghana country teams used the scaling readiness tools to unpack the Triple S PLUS innovation

Table 12.3 Example of unpacking complementary components and understanding scaling readiness: Ghana, March 2018

Triple S PLUS complementary components	Technology readiness	Technology use	Scaling readiness	Decision: improve, reorient, outsource
Quality seed production system for pre-basic (PBS) and basic seed from NARI	8	4	12	*Improve* linkage of DVMs to quality PBS. (Crops Research Institute [CRI] and Savannah Agricultural Research Institute [SARI])
Good agricultural practices package available for root production	8	2	10	*Improve* farmer knowledge through training of field agents and in Triple S PLUS farmer training and demos
Adapted and translated gender-responsive training tools	3	0	3	*Improve:* add visuals and develop video and radio scripts in local languages to ensure training tools reflect diverse gender and cultural context and storage containers. Pilot the tools to get feedback on effectiveness
Multichannel communication approach	1	9	10	*Improve:* develop inventory of communication tools and short communication strategy
Training video that shows Triple S core components	1	0	1	*Outsource* to a local communication company and collaborate with farmers, extension, and research to develop videos
Provide different storage size options for multiple uses of stored roots	8	2	10	*Improve* farmer knowledge through training of field agents and Triple S PLUS farmer training and demos and by showing a diversity of storage options in the videos
Available improved varieties	9	5	14	*Improve* linkages and feedback to breeders (CRI and SARI)
Workplans adapted to engage with existing extension system (women's savings and loans groups)	1	0	1	*Reorient* workplans with partners so that their training tools and communication channels are more gender inclusive
Farmers and NGOs buying vines from DVMs and seed companies every two to four crop cycles	8	5	13	*Improve* linkage DVMs and seed company to quality PBS. (CRI and SARI)

(continued)

Table 12.3 (continued)

Triple S PLUS complementary components	Technology readiness	Technology use	Scaling readiness	Decision: improve, reorient, outsource
Strategy to involve early innovators and champion households based on gender responsibilities	6	0	6	*Reorient* with partners to ensure household-based approach (i.e., not just one spouse)
Cooking demonstrations for demand creation for OFSP	9	7	16	*Outsource* to produce a video explaining the nutritional benefit of OFSP and use in different local dishes
Video (and radio) for demand creation for OFSP	4	0	4	*Outsource* as part of partnering strategy: video was disseminated by extension, NGOs, agricultural colleges; local radio stations contracted

NB. Scoring followed scale in Table 12.1
Source: This table was developed by key team members (CIP and partners) during the first regional workshop on scaling readiness McEwan et al. 2018 (adapted) Theory of Change and Scaling Strategy Using Gender-Responsive Communication Approaches

package, i.e., the core technology and complementary components needed for successful scaling. Our assessment and discussions resulted in an increase in the complementary components, and so the final Triple S PLUS innovation package included (1) a quality seed production system through the National Agricultural Research Institutes (NARIs) and decentralized vine multipliers (DVMs), (2) good agricultural practices, and (3) a gender-responsive, multimedia information, and communication approach.

Videos and demonstrations on how to use the OFSP roots were also considered important. Initially there was no funding for this component, although later partners were able to include it. Experience had shown that farmers and researchers could adapt the core components of Triple S in different environmental and social contexts, and so the technical readiness was high. But not all complementary components had been tested in different contexts, and so the technical readiness and extent of use varied by component and country. Tables 12.2 and 12.3 show the core and complementary components of the Triple S innovation package identified for Ghana. The technology readiness of the core components scored high, between seven and eight out of ten (Table 12.2). This reflected the extensive research into the biophysical elements of Triple S, such as the use of different sizes of roots, types of storage media, and containers. However, technology use scored low (between zero and two), because use had remained within a project sphere under the control of the International Potato Center (CIP).

At that point independent use of Triple S by other organizations had only happened in Uganda (personal observation, Sam Namanda). Table 12.3 illustrates the actions that were agreed upon after discussing the scaling readiness of the complementary components. This highlighted areas where improvement was required, e.g., (1) emphasis on negative and positive selection during training and demonstrations, (2) partnering strategies and work plans that would need to be reoriented to ensure that both spouses could receive training, and (3) outsourcing, e.g., production of training videos. Conducting the scaling readiness exercise revealed that in Ghana we did not have easy to understand, illustrated, gender-responsive communication tools in the local language that reflected diverse gender and cultural contexts. So, these had to be developed and tested.

The process of unpacking scaling readiness also helped us realize that we had not yet addressed what would be required for sustained use of Triple S beyond the project context. These included enabling factors related to marketing, policy, value chain linkages, service provision, attitude and behavior change, education, and infrastructure. For example, Triple S is a technical innovation, intended to increase productivity and availability of surplus fresh roots for sale. But without access to a vibrant market for fresh roots, farmers would not be interested in adopting the Triple S technology. To demonstrate this, because the OFSP market was poorly developed in both countries, it was necessary to link with other projects that increased demand for sweetpotato by raising awareness about its health benefits and improving options for processing OFSP into popular convenience foods such as bread, crisps, and yoghurt. Therefore, cooking demonstrations, video, and radio messages were developed to create demand and promote local processing options. Understanding the level of scaling readiness of the different components informed our theory of change, scaling approach, and partnership strategy. The same process for assessing technology readiness and use was followed in Ethiopia. There were some differences in scoring which reflected previous in-country experiences with using the technology and context-specific approaches for partnerships and scaling.

Our initial theory of change proposed that face-to-face training (with practical demonstrations) was effective but also costly when done at scale, and diffusion of technology would not always occur spontaneously (Glover et al. 2016). Therefore, multimedia communication including demonstrations, video, and radio would be used for training on Triple S. Messages translated into local languages would be tailored to sensitize and train next users such as researchers, agricultural training institutions, extension, policymakers, community-based development partners, and farmers as end users of Triple S. A cascade training strategy would result in the improved technical capacity of extensionists and Triple S champion households to be the main source of information and to support community-level uptake. Considering differences in literacy rates between men and women, an audio and visual approach would mean that information would be better understood by women and would ensure wider exposure to the Triple S method with spillover through farmer-to-farmer sharing.

If farmers appreciated the benefits of the Triple S method, extensionists would include it in their activities, and agricultural training institutions would add it to curricula, which would contribute to longer term institutional capacity for the uptake of Triple S. Greater access to clean seed in time for early planting, when the soil is moister, would enable better crop growth and higher yields. Improving production and productivity would allow the sweetpotato to contribute more to household food security, nutrition, and incomes.

12.2.3 Developing Gender-Responsive Communication Materials

The communication tools included: visual training materials using cartoons with few words in large format, on A1 size flipcharts and farmer leaflets (Ethiopia only), demonstrations, and video-based extension and radio programs (Fig. 12.3). Content for the different communication tools was developed through a collaboration between research and extension agents, with feedback from farmers. Videos were developed using the zooming in-zooming out (ZIZO) method, a participatory approach combining a script that describes the innovation with farmers who show how they use the new idea, and explain why they use it (Van Mele 2006). Videos were produced in English, after which scripts were translated into local languages using a voice-over. In Ghana, four short videos were produced that described all the steps of Triple S, except the complementary component of GAPs. In Ethiopia, one video was produced that described the whole Triple S cycle, including GAPs. The video screenings were facilitated by extension staff who were trained in video-based extension, e.g., allowing time and encouraging questions and providing repeat

Fig. 12.3 Multimedia communication (clockwise from upper left): screening videos, filming videos, and flipcharts and other written material for farmers

sessions. The radio programs were developed by local and regional FM stations based on key messages in the video scripts. The programs were broadcast in local languages at key points in the Triple S seasonal cycle. Farmers could call in to these talk shows and ask questions of invited agricultural specialists.

12.2.4   Partnering for a Gender-Responsive Approach

The partnering approach to scaling Triple S was three-pronged (Tables 12.4 and 12.5). First, the project collaborated with large, nutrition-sensitive agricultural development projects and programs to achieve short-term but large-scale exposure of target populations to the Triple S PLUS innovation package. These projects and programs were implemented by non-governmental organizations (NGOs) and

Table 12.4 Partners and their roles in the Triple S scaling initiative in Ghana

Organization, project	Country, region	Partner type	Role in scaling initiative
Global Communities, Resilience in Northern Ghana (RING)	Ghana, Northern Region	Rural and agricultural development project, collaboration with district assemblies	Human and logistical resources to reach out to farmers at scale
Mennonite Economic Development Associates (MEDA), Generating Rural Opportunities for Women (GROW)	Ghana, Upper East Region	Rural and agricultural development project, collaboration with local NGOs	Human and logistical resources to reach out to farmers at scale
Ministry of Food and Agriculture (MoFA), regional departments	Ghana, Northern and Upper East Regions	Government extension, network of extension agents	Human and logistical resources to reach farmers at scale. Link to central government
Savanna Agricultural Research Institute (SARI-Ghana)	Ghana, Northern and Upper East Regions	Agricultural research, sweetpotato breeding program	Provide training, technical support, and quality pre-basic seed to vine multipliers
Crops Research Institute (CRI)	Ghana, Ashanti Region	Agricultural research, sweetpotato breeding program	Provide training, technical support, and quality pre-basic seed to vine multipliers
Damongo Agricultural College	Ghana, Northern Region	Vocational training institute for agriculture technicians (extension agents)	Test and adopt Triple S and sweetpotato modules in curriculum, support advocacy at national level
Integrated Water and Agricultural Development (IWAD)	Northern Region	Seed and agricultural produce company	Support in production of planting material, after videos created large (unforeseen) demand

Table 12.5 Partners and their roles in the Triple S scaling initiative in Ethiopia

Organization, project	Partner type	Role in scaling initiative
Southern Nations, Nationalities and Peoples' Region (SNNPR) Bureau of Agriculture and Natural Resource Development (BoANRD)	Government extension	Facilitation and logistics for training of trainers (ToT) at regional and cascading activities at district and village levels
People in Need (PIN)	NGO	Facilitate training to member of the Healthy Living Club
Quality Diet for Better Health Project (QDBH)	Project funded by European Union to promote orange-fleshed sweetpotato to 15,000 households in 3 districts	Financial and logistic support
Sodo and Shire Agriculture, Technical, Vocational Education and Training Institutes (ATVET)	Train agricultural development agents	Integrate Triple S technology in the ATVET curriculum
Southern Agricultural Research Institute (SARI-Ethiopia)	Agricultural research, sweetpotato breeding program	Provide technical support in training and provide clean planting material to vine multipliers

government extension[3] staff working at different administrative levels. Second, sustainability was built into the activities by partnering with agricultural colleges, government extension authorities, and the ministries of agriculture to encourage integration of the Triple S PLUS package into national and local work plans and training curricula. Third, partners identified and trained local Triple S champion households, i.e., both spouses (and decentralized vine multipliers and seed companies in Ghana) in the Triple S package to increase the long-term uptake of the package at the community level. Involving the NARIs as partners allowed for future research and the use of Triple S roots as part of the variety maintenance, multiplication, and dissemination strategy to introduce new sweetpotato varieties.

12.2.5 Implementation Modalities

Implementation started with a regional inception workshop for lead teams including researchers and selected partners from Ethiopia and Ghana (Wigboldus 2018). This workshop provided practical training on the scaling readiness approach to develop a first draft of the theory of change and the proposed scaling approach (Sartas et al. 2020a, b). This was followed by in-country start-up workshops led by the CIP

---

[3] In Ethiopia the Bureau of Agriculture and Natural Resource Development (BoANRD) works at regional, zonal, *woreda* (district), and *kebele* (village) levels. In Ghana the Ministry of Food and Agriculture (MoFA) works through District Assemblies.

scaling champions in Ghana and Ethiopia to adapt the theory of change and scaling approach to the country contexts (Cherinet and Asfaw 2018; van Mourik and Wigboldus 2018). The participants were project partners and relevant stakeholders, including NGOs managing nutrition-sensitive agricultural development programs, Ministry of Agriculture (MoA), community-based organizations (CBOs), national agricultural research institutes (NARIs), and pre-service and in-service agricultural training institutes and universities.

The main outputs from the country start-up workshops were the vision of success as a visual representation of the theory of change, including the core and complementary innovations, the partners, and the expected outcomes of the scaling process (Wigboldus et al. 2016) (Fig. 12.4a, b). This oriented the work plan with the roles and responsibilities of each of the partners, with a shortlist of low-hanging fruits and

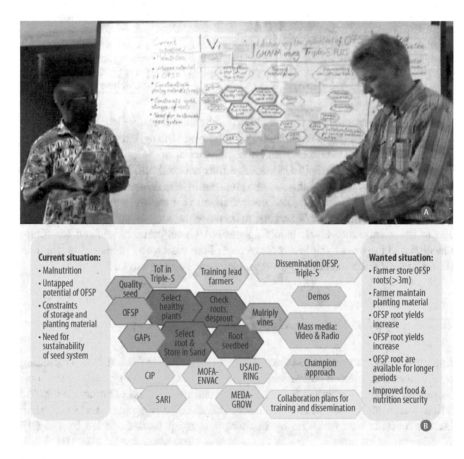

Fig. 12.4 (**a**) Developing a vision to unlock the potential of OFSP using Triple S in Ghana. This key stage encourages partners to add their input and to commit to a common vision and strategy for scaling. (**b**) Synthesis of the vision for impact, which combines elements of the core and complementary innovations, the theory of change and the approach to scaling, including the partners

immediate actions. Noting that women in the intervention areas may have lower literacy levels, which might limit participation in training events, a deliberate strategy to target women through partners' programs was incorporated to enhance exposure to and potential uptake of the technology. We recognized that both the processes of technical change and partnering would be dynamic, so we built in regular reflection steps (S. Wigboldus, personal communication), and "learning journeys" (www. africa.procasur.org) in the annual project cycle.

In Ghana, the cascade training worked at two levels, the CIP scaling champion implemented training of trainers (ToT) with three groups (two with NGO partners' field agents and one with MoFA extension agents), who then trained farmer groups and their leaders at community level (Fig. 12.5). In Ethiopia, the government extension structure required that staff should be trained at regional, through zonal to district and village levels. The CIP scaling champion trained BoANRD experts, research, and NGO partners (People in Need) at regional, zonal, and district level. Then zonal and district partners trained village development agents (DA). DAs trained champion households and village officials. At the community level, village officials and Triple S champion households shared the responsibility to train farmers.

The trainings covered the core and complementary practices (Tables 12.2 and 12.3) of Triple S. In Ghana, the two NGO partners with large agricultural development projects initially focused their agricultural training programs on women's Village Savings and Loan Association (VSLA) groups. However, to increase

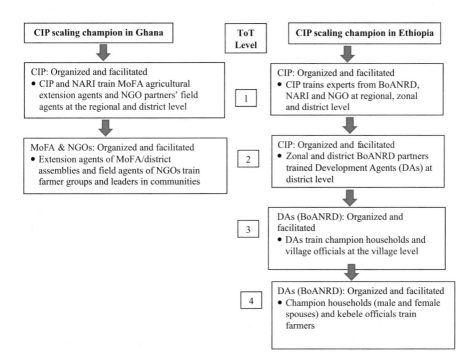

Fig. 12.5 Cascade training on Triple S PLUS innovation package in Ghana and Ethiopia

participation by men, the scaling project encouraged husbands to also join the trainings. In Ethiopia, the NGO partner, People in Need (PIN), disseminated Triple S as part of Healthy Living Club (HLC) initiatives that targeted caregivers of young children, who were mostly women. Local FM radio stations were selected based on the language used and coverage in the target areas. The broadcasts were accessible to those in households which owned a radio or to people who could listen in with neighbors and friends.

12.3 Methods for Assessment of Gender-Responsive Multimedia Communication Tools and Channels

We wanted to pilot and evaluate the effectiveness of the different types of communication tools in changing knowledge, attitudes, and practices on GAPs and Triple S in Ghana and Ethiopia. We also wanted to assess how the different communication channels were gender-responsive and their ability to encourage women to participate and use the innovation. We used a quasi-experimental design. Three levels of promotion intensity were tested in each country. "Core" farmers in communities in Ghana were exposed to the three communication channels, i.e., face-to-face training, video, and radio; in Ethiopia they were exposed to face-to-face training using flipcharts, farmer leaflets, and radio. "Spillover" farmers were exposed to video and radio, and "informed" farmers were exposed to radio only. The radio programs were broadcast in all intervention areas, so in theory farmers who were in either of the first two treatment categories could also hear about Triple S on the radio.

Quantitative surveys in both countries assessed how these different levels of promotion intensity (communication channels) influenced gender-differentiated changes in awareness, knowledge, and early uptake (i.e., use of Triple S in the previous 12 months). Data were also collected to assess any association between technology use and selected sweetpotato output variables (e.g., production or sales of roots). Both spouses were interviewed together for general household information and implementation of GAPs, and then they were interviewed separately on specific Triple S knowledge, attitudes, and practices. The survey also assessed household decision-making about planting, eating, and selling sweetpotatoes and using proceeds from the sale. In Ghana, the sample size was 294 households from 24 communities in eight districts across the two regions (Upper West and Northern). In Ethiopia, 745 households in three districts (Hawassa Zuria, Misrak Badewacho, and Mirab Abaya) participated.

Qualitative insights were gained through separate focus group discussions for men and women to gain a deeper understanding of gendered perceptions of the communication tools. Study communities were selected according to the communication channels they had been exposed to. In Ethiopia, 124 respondents (64 men, 60 women) and in Ghana, 235 respondents (110 men, 125 women) participated in the study. An index was developed to compare women and men's perception of the efficacy of different communication channels. We defined "efficacy" as the perceived level of ability of the channel to deliver the Triple S information to the project beneficiaries for their eventual use. The index was based on five indicators, i.e., (1) ease of access to the channel; (2) ease of understanding the information; (3) suitability of the time of day, and suitability of the season of the year when the information was relayed; (4) ease of sharing; and (5) ease of applying the information obtained. Respondents used a five-point Likert scale (5 = very easy or suitable; 1 = very difficult or unsuitable) to score each indicator. Each indicator was accorded a weight of 0.2 and the maximum score being 1.00.

In Ghana, during data analysis, it became apparent that the quasi experimental setup had not maintained discrete treatments for the core, spillover, and informed communities. Many farmers in communities that were not supposed to have been exposed to training and demonstration (spillover and informed) did attend a training and demonstration. Conversely, many farmers who were supposed to have received training or watched videos, in fact, had not. It was difficult to keep the treatments separate, in part because the communities were near each other, so curious farmers could visit other villages and attend their communication events as well. So, we analyzed our data by communication channels that farmers reported to have been exposed to, rather than the ones that the project had intended. This resulted in the following five categories: "not exposed" (i.e., not exposed to information on Triple S PLUS through radio, video, or demonstration), "radio" (i.e., exposed to the information through a neighbor, radio, or a neighbor and radio), "demonstration" (i.e., exposed to demonstration, radio and demonstration, a neighbor and demonstration, or a neighbor and demonstration and radio), "video" (i.e., exposed to video, radio and video, neighbor and video or neighbor, video and radio), and "video with demonstration" (i.e., exposed to demonstration, video, and any combination of the other channels), discussed in Sect. 12.5. In Ethiopia, farmers were classified in four categories: (1) face-to-face training using flipcharts, farmer leaflets, demonstrations, and radio, (2) video and radio, (3) radio only, and (4) no exposure.

In both countries the percentage of male and female farmers using individual GAPs and Triple S practices was analyzed. Two indices were constructed: the first for household level "use of GAPs" and the second for individual, sex-disaggregated "use of Triple S." We classified the use of at least three out of the four recommended practices (1) crop rotation, (2) timely weeding, (3) uplifting, and (4) remaking ridges as "use of GAPs." For Triple S, the index was based on the use of at least three of four practices: (1) identify and tag healthy plants, (2) sort roots before storage, (3) store in a container or structure with sand, and (4) check the roots in storage. While planting out sprouted roots in seed beds is also part of the technology, the timing of the survey in both countries did not allow us to assess the use of this practice.

12.4 Findings on Gender-Responsive Communication Channels and Early Uptake of Triple S

12.4.1 Participation in Trainings and Events

In Ghana, 148 trainers (26% women) from partners were initially trained in Triple S and GAPs and later cascaded the training to women and men farmers. Trainers reached 36,866 participants (60% women) through 2041 training and demonstration events. The video screenings reached 42,799 persons (65% women) through 846 video screening events. While it was impossible to measure the number of farmers who listened to the different radio programs, the stations reported much feedback from listeners, with 2544 documented calls into the programs (17% from women callers). Though distribution and sale of vines was not an initial goal of the project, early communication activities unleashed a large, unexpected demand for OFSP planting material, and through a coordinated effort with vine multipliers and a private seed company, over 20,000 farmers (~80% women) were able to buy vines.

In Ethiopia, 487 (32% female) BoANRD, NGO staff, and Triple S champion households participated in the cascade training at regional, district, and village levels. Each champion household was then responsible for training 50 households, reaching 13,693 farmers (39% female). An additional 18 BoANRD staff (all male) were trained in video-based extension, which was extended to 2463 farmers (10% female).

12.4.2 Early Uptake of Triple S and GAPs

Table 12.6 summarizes reported use of Triple S (three out of four practices) in the previous 12 months.

In Ghana, by the end of the 2-year intervention, 48% of the men and 35% of the women who had been exposed to any of the communication channels reported that they had used Triple S. These figures were much higher than for people who had not been exposed to any of the information (7% of the men and 5% of the women). Sixty-two percent of households which had been exposed to any communication channel reported using GAPs, compared to just 33% of those which had not been exposed to any information reported using GAPs. In other words, in Ghana use of the practices was fairly high among people who had received the information through any channel.

In Ethiopia, 24% of males ($n = 327$) and 7% of females ($n = 319$) had started using Triple S at the time of the survey. Three percent of both men and women who were in one of the treatment groups reported that they had used three out of four Triple S practices, compared to none of those who had not received information. 84% of households which had been exposed to any communication channel reported using GAPs, while 76% of households which had not been exposed were using GAPs.

Table 12.6 Summary data from Ghana and Ethiopia end-line surveys showing sex-disaggregated use of Triple S[a] and household use of GAPs[b] by treatment or no treatment

Ghana	Treatment $n = 181$	No Treatment $n = 95$
Using three out of four Triple S practices[a]	87 (48%)	7 (7%)
Using three out of four Triple S practices	64 (35%)	5 (5%)
Using three out of four GAPs[b]	113 (62%)	31 (33%)
Ethiopia	*Treatment*	*No treatment*
Ethiopia men	*n = 327*	*n = 395*
Using three of four Triple S practices	11 (3%)	0%
Ethiopia women	*n = 319*	*n = 396*
Using three of four Triple S practices	9 (3%)	0%
Ethiopia household	*n = 317*	*n = 395*
Using three of four GAPs	266 (84%)	299 (76%)

[a]The Triple S index was based on the use of at least three of four practices: (1) identify and tag healthy plants, (2) sort roots before storage, (3) store in a structure with sand, and (4) check the roots in storage
[b]GAPs index was based on the use of at least three of four practices: (1) crop rotation, (2) timely weeding, (3) uplifting, (4) remaking ridges

In both countries, those who received no information had the lowest rates of uptake of both Triple S and of GAPs. In Ethiopia the uptake of GAPs was much higher (76% to 94%) than the use of Triple S (0% to 7%). However, it is likely that GAPs were already widely used before the project intervention. The very low uptake of Triple S leaves little room for any difference between men and women. In Ghana, the uptake of GAPs was higher (33% to 62%) than that of Triple S (5% to 48%). Men had higher uptake than women (48% for men versus 35% for women) who were exposed to any communication channel.

We then compared uptake of Triple S by communication channel.

In Ghana, we grouped farmers into five categories of exposure to information: (1) not exposed, (2) radio only, (3) demonstration, (4) video, and (5) video and demonstration. More men used Triple S (at least three out of four practices) than women (Figs. 12.6 and 12.7). For instance, 63% of men exposed to both video and demonstration used Triple S, compared to 45% of the women. Use of Triple S was somewhat higher for men than for women, across the communication channels.

Treatments 4 (video) and 5 (video plus demo) were more effective at encouraging farmers to adopt Triple S practices, with one exception. For storage in sand, demonstration was especially important, suggesting that some practices that involve a lot of manipulation of different materials (getting the sand, putting it in the container, burying the roots) may be most effectively taught with a demonstration. For most other practices, the videos were as effective or a demonstration or more so. The most intensive treatment (5), combining demonstration and video, is particularly efficient in encouraging both male and female farmers to apply Triple S, although it is also the most expensive.

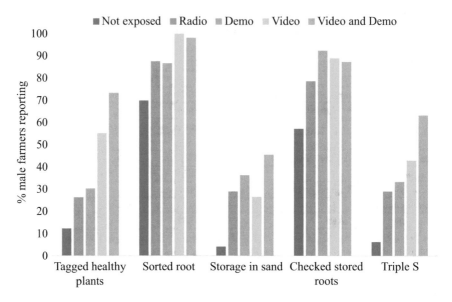

Fig. 12.6 Ghana: percentage of male farmers applying Triple S practices disaggregated by communication channel. (Source: Etwire et al. 2020)

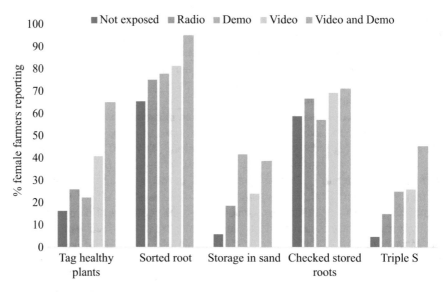

Fig. 12.7 Ghana: percentage of female farmers applying Triple S practices disaggregated by communication channel. (Source: Etwire et al. 2020)

In Ethiopia, uptake of Triple S was assessed across three communication channels, i.e., (1) radio with face-to-face training (using flip charts, handouts, and demonstrations), (2) video and radio, and (3) radio only, and a fourth group of those who had not been exposed to information on Triple S. The end-line survey showed that a higher proportion of respondents reported that they been trained in Triple S through the videos (87% of men and 36% women) than by face-to-face training (57% men, 20% women) or radio (22% men, 6% women). In the nontreatment group, few (1% men, 1% women) reported that they had been trained in Triple S.

Spouses were interviewed separately for Triple S specific practices. Use of clean planting material scored highest both by men (93%) and women (91%) reached through video.

For Ethiopia, while the number of farmers that reported that they had started using Triple S was low (121 men and 32 women), Figs. 12.8 and 12.9 show that men applied Triple S practices more than women. For instance, 26% of men exposed to face-to-face training used Triple S (i.e., at least three out of four practices) compared to 11% of the women. Moreover, no women exposed to the other communication channels reported using Triple S. Men may use Triple S after watching videos, but for women, face-to-face training is still critical for uptake.

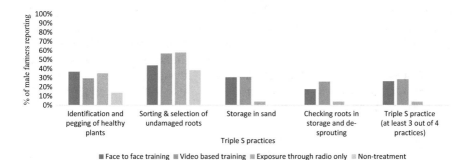

Fig. 12.8 Ethiopia: percentage of male farmers who started using Triple S (*n* = 131) by communication channel. (Source: Asfaw et al. 2020)

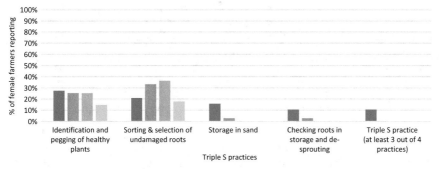

Fig. 12.9 Ethiopia: percentage of female farmers who started using Triple S (*n* = 31) by communication channel. (Source: Asfaw et al. 2020)

12.4.3   Changes in Food Security

The end-line surveys made a preliminary assessment of any association between technology use and selected variables such as sweetpotato production and sale. In Ghana, 67% of households surveyed in the intervention area reported an increase in area under sweetpotato cultivation. 81% of households reported an increase from the sales of roots and vines. The average household reported income over the preceding rainy season of GHC 70 (about USD 15) and GHC 180 (USD 34) from sales of sweetpotato vines and roots, respectively (Table 12.7). Farmers who applied GAPs tended to earn more from roots and vines. Nonusers of either Triple S or GAPs sold more roots but earned the lowest income from sweetpotato, while farmers who applied both GAPs and Triple S earned the highest income, perhaps because they were able to harvest earlier and capture higher prices. As a result of increased sales and conservation of roots, these farmers may also consume less roots (Table 12.7) (Etwire et al. 2020).

The end-line survey found no significant relationship between sweetpotato output (root production) and the use of GAPs and Triple S. This could be due to the general constraints that farmers faced in accessing planting materials irrespective of whether they were exposed to different communication channels and irrespective of whether they applied GAPs and Triple S or not. In addition, data was collected immediately after the end of project activities and therefore did not allow for complete manifestation of the effects of technology use in future cropping seasons (Etwire et al. 2020).

Table 12.7 Use of GAPs and Triple S vs sweetpotato production and use in Ghana

Quantitative variable	Nonusers	Triple S	GAPs	GAPs and Triple S	Sample mean
Income from vines (GHC)	4.00	15.00	81.20	130.00	79.24
	8.00	*1.00*	*127.00*	*171.54*	*133.72*
Income from roots (GHC)	148.68	148.57	200.61	215.94	187.72
	149.45	*102.38*	*228.44*	*196.74*	*186.51*
Roots sold (kg)	190.91	108.50	203.97	184.24	183.32
	265.49	*84.94*	*429.13*	*216.92*	*288.12*
Roots consumed (kg)	108.08	109.62	80.93	87.59	92.50
	159.10	*102.36*	*96.52*	*118.85*	*123.62*
Roots conserved as seed (kg)	37.09	46.67	57.82	41.38	45.04
	31.66	*33.45*	*155.30*	*39.06*	*79.28*
Roots given as gift (kg)	48.33	29.25	30.79	45.69	40.41
	81.60	*29.64*	*48.05*	*62.05*	*60.45*
Roots for other purpose (kg)	24.00	31.80	5.71	30.22	22.79
	34.47	*27.18*	*6.83*	*66.15*	*49.44*

Figures in italics are averages and those in italics are associated standard deviations
GHC Ghana Cedi (5.25 GHC = 1 USD)
Source: Etwire et al. (2020)

Table 12.8 Households (HH) reporting vine and root sales in previous 12 months by different communication channels in selected villages, SNNPR, Ethiopia

Indicators		Face-to-face training % HH	Video-based training % HH	Exposure through radio only % HH	Nontreatment % HH
% households selling vines		8	8	15	27
Changes in sweetpotato vine sales	Gone up	67	57	55	63
	Stayed the same	17	0	9	17
	Gone down	17	29	9	17
	Don't know	0	14	27	4
% households selling roots		61	49	40	48
Changes in sweetpotato root sales	Gone up	72	58	69	62
	Stayed the same	9	12	21	24
	Gone down	19	30	10	14

Source: Asfaw et al. (2020)

In Ethiopia, there was insufficient time since the Triple S training cycle had been completed to assess an increase in root production. Therefore, farmers were asked about trends in area under sweetpotato over the previous 12 months; 61% of farmers receiving face-to-face training reported an increment. This is statistically significant when compared with farmers receiving video-based training (33%). Few households in any group reported that their area under cultivation had gone down (Asfaw et al. 2020, data not shown). Table 12.8 presents the percentage of households selling vines and roots in the previous 12 months. The percentage selling vines is lower than those selling roots. About 61% of households exposed to face-to-face training reported root sales. For both vines and root sales, households reported that sales had gone up in the recall period. Average income from sweetpotato sales (vine and root) was the following: face-to-face training, USD 105; video training, USD 76; radio training, USD 105; and nontreatment, USD 63. The result is similar for both treatment and nontreatment groups, possibly because there was continuous rainfall in both short and long rainy seasons in all project districts.

12.4.4 Gender-Based Perspectives on Different Communication Tools and Channels

In Ethiopia women identified four communication channels through which they obtained information (face-to-face, video, radio, and HLCs), with men adding meetings and the farming radio program "8208." In Ghana, men and women identified face-to-face, video, radio, and personal communication (from friends, neighbors) channels and in Upper West Region "talking books" which were the extension method used by the NGO implementing partner. We used the 5-point index described in Sect. 12.4 (ease of access, ease of understanding, etc.) to compare women's and men's perception of the efficacy of different communication channels (Figs. 12.10, 12.11, and 12.12. Weighted scores of 0.80 and above rated as adequate for each indicator (Mayanja et al. 2020).

Ease of access: Respondents in both countries perceived face-to-face channels such as demonstrations to be the most accessible. Women in both countries and men in Ghana, Upper West Region, perceived radio and video as difficult channels to access due to the lack of radio sets, batteries, and power outages, and they were dependent on the extension officer to repeat the video events (Fig. 12.10). On the other hand, men in Ghana, Northern Region (Fig. 12.11), and Ethiopia (Fig. 12.12) reported full access to radio (twice as much as women) and slightly higher access to video as compared to women.

Ease in understanding information: Information relayed via face-to-face channels was scored as adequately understood by both men and women in both countries. In Ghana, personal communication from friends and neighbors was rated to be very adequate for men in both regions and only for women in the Upper West Region. In Ghana, ease of understanding information through video was rated as slightly below adequate for men in both regions, inadequate by women in Upper West Region (Fig. 12.10), but adequate for women in the Northern Region. In Ethiopia, both men and women reported that information from videos was not easy to understand. Surprisingly, men in Ghana, Northern Region, rated information transmitted via radio as much harder to understand than women did – with women scoring adequate (Fig. 12.11).

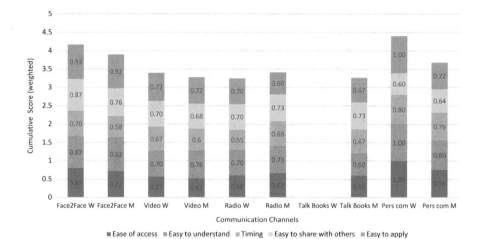

Fig. 12.10 Assessment of efficacy of communication channels by women and men in Upper West Region, Ghana. (Source: Mayanja et al. 2020)

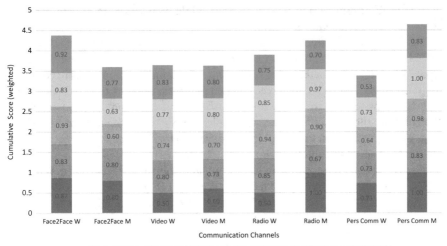

Fig. 12.11 Assessment of efficacy of communication channels by women and men in Northern Region, Ghana. (Source: Mayanja et al. 2020)

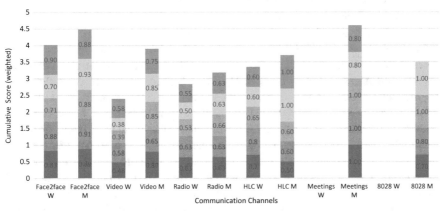

Fig. 12.12 Assessment of efficacy of communication channels by women and men in Ethiopia. (Source: Mayanja et al. 2020)

Timing (appropriate time of the day and suitable time of year): In Ethiopia, men gave a score of adequate for the timing of face-to-face, video, and meetings (Fig. 12.12). Women in Ethiopia and men from the Ghana, Upper West Region (Fig. 12.10), gave a score of inadequate for the timing of all information channels. Men from the Northern Region, Ghana, also perceived timing for all the information channels inadequate except for radio and personal communication (Fig. 12.11). In Ghana, Northern Region, women perceived face-to-face and radio channels to have relayed information at a suitable time, while men scored radio and personal communication as adequate.

Ease of sharing: Information received through face-to-face channels was perceived to be easy to share by women and nearly made the mark for men from the Upper West Region, Ghana (Fig. 12.10). This was similar for women from Northern Region, Ghana, who also perceived information from radio as easy to share. Men from the same region also selected radio and personal communication as adequate. Women from Ethiopia did not find information through any channel as easy to share (Fig. 12.12), while men identified face-to-face, video, HLC, and meetings as channels that delivered information which was easy to share.

Ease of application: In Ethiopia, information obtained face-to-face was perceived to be easy to apply by both men and women (Fig. 12.12). Men also scored information from HLC, meetings, and farmer radio as easy to apply. In Ghana there were similar ratings for face-to-face channels for both men and women. Women in Upper West Region scored personal communication adequate as did men in Northern Region. Information from video was perceived to be easy to apply by both men and women from the Northern Region (Fig. 12.12).

Mayanja et al. (2020) report on focus group discussions which provided additional insights for the different channels for this project. Radio was a popular communication channel in both countries. One advantage mentioned by men and women was that agricultural information can be transmitted quickly through radio and can be accessed by many households at the same time, wherever people happened to be. They also said that the information was likely to be up-to-date information and that issues could be clarified by calling in to the station. Women noted that radio motivates them because they can learn a lot from other farmers while also obtaining information on sweetpotato in general, Triple S, and prevention of pests. They appreciated information from other towns and countries related to different planting methods and use of improved seed. Women mentioned that they could use their phones to listen to radio, especially if the broadcast was in the local language and at a suitable time. This made radio versatile since it could be used anywhere at any time. Women also noted that people found it easier to share information from the radio which could then benefit women without radios. The disadvantages mentioned related to limited access to a radio set, power outages, and battery costs. At times, Triple S information via radio was not clear, and farmers found it hard to put such information into practice. This was compared to face-to-face training where they could ask questions when they were not clear. This challenge was noted in both countries.

Video was appreciated as a new and alternate means of relaying agricultural information. Men in Ghana likened video to being in a classroom where all the steps of the Triple S were clearly explained. This highlighted the power of video to relay practical information on Triple S. In Ethiopia, men noted that information from video was easy to share with others; while women in Northern Ghana appreciated the information obtained because they found it easy to apply. Indeed, for women and men in Suke Lambussie and women in Chogsia West, video was voted the most preferred communication channel. Respondents from these areas noted that messages communicated via video were easy to understand and inspirational, given that they saw what other farmers had experienced with the technology and everything screened could not be easily forgotten. Women in Ethiopia mentioned disadvantages, including difficulties in accessing the video shows as well as timing, which could conflict with household chores. Women from the Upper West Region, Ghana, noted that sometimes the information in the videos was not clear, which made it difficult for them to absorb the information, especially if there was no question and answer session. Another disadvantage noted by most women in both countries was the program was short and the shows were not available as and when they were needed. Nonetheless, the innovative nature of the channel was highly appreciated.

Face-to-face was the most highly rated channel in both countries, mainly because it enabled farmers to merge both practical and theoretical aspects of the technology, making it easier to understand and apply. Women appreciated this channel because information was relayed in the local language, and they could ask questions on the spot. Another advantage was that one could ask the same question repeatedly (unlike with other channels) to obtain a clear understanding on difficult technical issues. Face-to-face was also appreciated because it synched with the cropping calendar. Disadvantages included the tendency to lean towards theoretical aspects during trainings. Women in Ethiopia noted that this made it harder to implement the technology as sometimes they left the training with insufficient knowledge to apply Triple S on their own.

12.5 Drawing Lessons on Gender-Responsive Communication Channels

The multimedia communication approach and our partnering strategy contributed to reducing the gender gap (especially in Ghana) in knowledge transfer with a higher proportion of women farmers reached (60% for demonstrations and 65% for video screenings) than the 5% cited for conventional extension initiatives (UNDP 2012). This is also higher than the 30% participation rate of women reported for the demand-led extension approach of CABI's Plantwise program in Ghana (Williams and Taron 2020).

In Ethiopia, the World Bank (2010) reports that women's participation in extension is on average about 20% but with high regional variations ranging from 2% in Afar to 54% in Tigray. While recognizing that the use of a technology often requires a joint decision by husband and wife, there are signs of early uptake of Triple S. In both countries there is higher uptake in the group reached through video (in combination with demonstration in Ghana). Men often favor the novel video-based extension approach, but women prefer face-to-face interactions which allow more opportunities to ask questions and for practical demonstrations. This is reflected in the lower uptake rates of women compared to men exposed to video-based extension. However, as women's uptake of GAPs and Triple S practices are lower than for men in general, other constraints are also at play.

12.5.1 Country Context and Early Uptake of GAPs and Triple S

The total number of farmers exposed to Triple S through the different treatments was 16,156 farmers in Ethiopia compared to 79,655 in Ghana. Moreover, a higher proportion of male and female farmers reported that they had used Triple S (three out of four practices) in Ghana compared to Ethiopia. This reflects differences in context, climate, and partnering strategies between the two countries. For example, in Ethiopia, the agricultural season of 2018–2019 was unusually wet in some of the intervention areas, so farmers did not see the urgent need to start practicing Triple S. Moreover, in Ethiopia, the extension agents in the government system were responsible for multiple crops, whereas in Ghana the NGO staff in nutrition-sensitive agriculture programs were focused on fewer crops.

In both countries, the 2-year intervention period was short, especially for a technology with an annual cycle. In the first year, planting material was in short supply, so few roots were available to store in sand. We overestimated the readiness of this component and should have treated it as a core rather than complementary component (Table 12.2) to ensure stronger engagement with the NARIs in each country to guarantee sufficient production of early generation seed. In the second year in Ethiopia, roots were harvested and selected for Triple S in October–November 2019, and the survey was conducted the following month (December 2019 to January 2020), too soon to learn if farmers were using the final Triple S step of planting out the sprouted roots in seed beds.

In Ghana, video-based trainings and demonstrations (either separately or in combination) had positive effects on the use of GAPs and Triple S. Males living in wealthier regions with larger farms were more likely to use the technology. In Ethiopia, more male and female farmers turned up for the video screenings than was planned and higher than the face-to-face training. It is possible that the selected farmers listed as beneficiaries (and sampled for the end-line survey) were not actually invited to the face-to-face training. It is also possible that the face-to-face trainings were poorly attended or that many farmers who were not on the beneficiary list went to watch the videos. The novelty of the videos attracted a larger audience, especially because they were screened in the village (rather than at the Farmer Training Center). Fewer participants (only 25 farmers) were intended to watch videos, while 50 were assigned to each face-to-face training, and those events may have been under-attended. Abate et al. (2019) showed that video-based extension method was more effective than the traditional method of extension when disseminating teff, wheat, and maize technologies in Ethiopia.

Literacy levels and access to radio influenced the effectiveness of flipchart-mediated and radio-based communication. In Ethiopia, literacy rates were 72% for men and 53% for women reached by face-to-face training with flipcharts and printed materials. Farmers with access to radio sets were more likely to have been exposed to Triple S through radio programs (64%) and the least likely to have attended face-to-face training (22% of men and 6% of women).

A good proportion of male and female farmers who were exposed to Triple S and GAPs did not take up either of the technology packages. More time is needed, as individuals usually go through five stages prior to adopting a technology, i.e., awareness, gaining adequate knowledge, making a decision to adopt, trying the technology, and then deciding to continue using it or not (Rogers 1995). Technology adoption is complex, and individuals form perceptions of a technology which influence their decisions to adopt (Straub 2009). The attributes of a technology, how long it has been around, the communication channels used to share it, and the social system of adopters all influence adoption (Straub 2009; Hermans et al. 2021). Risk considerations and associated constraints can also limit widespread adoption of technologies (Mukasa 2018). For knowledge-intensive technologies such as tissue culture for banana, successful adoption required major changes in farming practices and a good understanding of the technology (Kabunga et al. 2012). So, adoption is not just about getting the technology right but about making it fit with social and institutional components and the perceptions of different types of farmers (Glover et al. 2019), which in the case of Triple S was through the multimedia communication strategy.

12.5.2 Gender-Based Perspectives on Communication Tools and Channels

We have gained insights on gender-based perceptions about the different communication channels used and how they influence the type and number of participants reached and the quality of knowledge transfer. While we show that using the different communication channels reduced the gender gap for sharing information on Triple S, women's access to information was still lower than men's in Ghana and lower still in Ethiopia. Women were more disadvantaged than men when we used communication channels that required reading or access to radio sets. Despite the novelty of video-based extension, most women and men in both countries preferred face-to-face communication and rated the efficacy of this channel as adequate. Women found face-to-face training easier to understand, and it was easier to ask questions and to learn by seeing or doing. Demystifying a technology from the theory into practice may still need face-to-face sessions for women (Mudege et al. 2017), with two-way communication between knowledge receivers and senders. Women appreciated the demonstrations not just for the technical information but also because they could catch up on the news and community issues at these social networking events. Such gatherings are safe places where women share, mentor, and encourage one another to address challenges related to farming and personal well-being (Jones et al. 2017). Given the lower literacy levels for women, such channels will still be preferred and relevant for them.

Men appreciated video and radio as alternate channels of information. In Ghana, they likened video to being in class, showing how authoritative this channel can be in relaying information on the technology. Witnessing the powerful testimonies of other farmers using the technology was valuable, especially for men who anticipate cash benefits. This further confirms the importance of digital platforms to strengthen delivery and uptake of technology (Aker 2011). Videos could help reduce the high costs associated with face-to-face communication. However, this benefit will depend on how the challenges associated with video can be addressed. There is a digital gender divide, as women have less access to ICTs, as evidenced in Ethiopia (see also Ogato 2013). In Ethiopia, women perceived the efficacy of radio and video as below adequate, while men ranked them much higher. Women in both countries noted lack of ownership of radio sets as a limiting factor to tuning in. Kyazze et al. (2012) also noted that lack of money to buy radios and batteries limited women's access to weather information. Societal gender norms affect women's ability to access agricultural services (McCormack 2018). It is important to understand household dynamics that influence access to radio sets before designing radio information packages. Strategies such as community radios or talking books could be explored as an add-on to household radio sets.

Besides access to radio sets and video screenings, respondents in both countries explained that the timing of the programs conflicted with domestic chores. This restricted their ability to understand Triple S since they only heard part of the program. This has also been noted in Tanzania, where women had to limit their time for listening to educational programs to attend to household work and childcare (Poulsen et al. 2015). Repeated broadcasts of radio shows and videos could help to alleviate this challenge. Extension programs should also schedule these programs at times that are the most convenient for women. For example, women may have more time to listen to radio early in the morning, and video screenings may be easier to attend in the early evening, but this should be determined with the women in each community.

In some communities, men and women did not access the same communication channels. For example, in Ethiopia, women did not mention meetings as a source of information. Also, in communities where the farming radio program "8208" was mentioned (and rated as a highly effective communications channel), only men listened to it. It is usually more difficult for women to attend meetings especially those held outside the community. Cultural norms can restrict women's access to events where information is relayed, but it is easier for women to access information services in the village, where their childcare and household responsibilities are (Gumucio et al. 2020). The scaling readiness of the video and radio components may have been higher for an audience of male farmers, but we needed more research on existing social norms to understand how to design communication channels for women.

12.6   Way Forward: Implications for Continued Scaling of Triple S PLUS

12.6.1   Embedding Triple S in Sweetpotato Vine Distribution Programs

Various training and dissemination activities have continued and will go on well after the Triple S PLUS project. Communication and training tools such as radio programs and videos (including video projector kits) are available and continue to be used. For example, the training videos from Ghana have been disseminated via USB sticks, are still available online on the AgTube (now EcoAgtube) channel, and have been viewed over 35,600 times in 2 years (e.g., van Mourik 2018). In both Ethiopia and Ghana, NGOs that promote OFSP as part of a nutrition-sensitive intervention have come to consider the Triple S PLUS innovation package, in combination with processing and value chain activities, as an option to increase the sustainability of their intervention and catalyze lasting change. In Ghana, several district assemblies have adopted OFSP and the Triple S package as a priority and budgeted for the activity in their annual agricultural development plans. At the national level, subsidized sales of quality OFSP planting material are part of the presidential Planting for Food and Jobs agricultural development program; although the Triple S innovation package is not specifically mentioned here, quality planting material is a prerequisite for successfully applying. Modules for training on OFSP, production, processing, storage, and Triple S PLUS have also been incorporated into the curricula of agricultural colleges in Ethiopia and Ghana, which will teach it to the extension agents and field technicians of tomorrow.

12.6.2   Partnering Capacities and Capabilities

Partners have different capacities to contribute to scaling. For instance, NGOs with large, well-funded but time-limited projects can expose more farmers to the innovations and permit for measurable early uptake at scale. However, other types of partners are needed to bring about more long-term and systemic changes, such as the Ministry of Agriculture and agricultural colleges. The short-term impact achieved by NGOs and community interventions, if well documented and communicated, can convince government actors to adopt appropriate policies and training. Some elements of the scaling strategy implemented by the scaling champions that contributed to this were (1) quarterly scaling reflection and planning meetings in-country; (2) cascade training of field and extension agents, sequenced to the Triple S yearly cycle; (3) yearly regional exchange visits with partners from Ethiopia, Ghana, and a regional team; (4) learning journeys to project activities, with clear learning objectives, information gathering, and presentation of conclusions and action points; (5) assessing perceptions and early uptake of Triple S with men and women farmers and communities; and (6) communicating findings to key actors, using briefs, blogs, and articles (International Potato Center 2019). Partners also have different capacities for training and disseminating innovations, monitoring, and evaluation and quality control of interventions. The scaling champions were able to provide continuous training and technical support which were essential to make sure that a many farmers were reached but also to make sure that the information provided was of good technical quality, understandable, and followed up with support where needed.

12.6.3 Reflecting on Enabling Factors for Scaling Triple S

We started this chapter recognizing that several enabling factors related to infrastructure, attitude and behavior change, marketing and value chains, and policy are critical for scaling innovation packages. As we look to the future, several changes may be needed to enhance the Triple S scaling process.

The communication strategies stimulated so much interest in sweetpotato in Ghana that 20,000 people bought vines. We used multimedia communication channels as part of the scaling strategy because demonstrations and face-to-face extension are expensive. Farmers like face-to-face communication, but the results also show that multimedia can reach more people, e.g., in Ghana 36,866 people were reached through 2041 training events, but videos reached slightly more, 42,799 at just 846 screenings. Videos can reach a lot more people, about three times as many per session, yet the early uptake rates are comparable. Future research should undertake detailed analysis of the cost effectiveness of the different channels.

Continued work on understanding gender-responsive communication channels in different sociocultural contexts is also required. Research on local social norms can inform the design and implementation of gender-responsive communication tools. Separate channels may be appropriate, or it may be necessary to ensure equitable opportunities to access the same channels. Face-to-face training allows women to participate in talks and in practical demonstrations while using social networks to enhance information sharing. Men are drawn to video-based extension, which is increasingly relevant as digital platforms become a new norm. But the events need electricity, devices to show videos, and sensitivity to the right time and place to screen videos so that women can attend.

There are several technical adaptations in progress. Sweetpotato breeders have started to include "sproutability" as a key trait in their breeding programs and to consider the use of sprouted roots for disseminating new varieties. As climate becomes increasingly unpredictable, there is greater interest in using root-based seed interventions. As farmers have started to shift their attitudes and behavior, commercial seed producers need to test the technical and economic feasibility of Triple S technology using larger quantities of stored roots.

As we worked with partners, we realized that truly achieving impact and scale would require more interventions than the Triple S technology. For instance, in our vision of impact, partners made it clear that awareness raising of the nutritional advantages of OFSP and supporting innovation in processing and marketing of OFSP products were essential to increase demand for the roots, continued interest in increasing OFSP production, and the relevance of Triple S. This highlights the importance of the policy context which enables growth in the sweetpotato sub-sector and will continue to be an enabling factor and driver for the need for and uptake of innovative seed technologies.

Acknowledgments The authors would like to thank the many women and men farmers and decentralized vine multipliers who have started to use Triple S, adapting it to their own needs and contexts, together with the colleagues and partners who are supporting them. We sincerely appreciate the support and encouragement from colleagues in the CGIAR Research Program on Roots, Tubers and Bananas Scaling Fund and Program Management Unit. The chapter has benefited from suggestions by the external reviewers and editors.

This research was undertaken as part of, and funded by, the CGIAR Research Program on Roots, Tubers and Bananas (RTB) and was supported by CGIAR Fund Donors. Funding support for this work was provided by Bill & Melinda Gates Foundation through grants to SASHA (ID: OPP1019987) and SweetGAINS (ID: OPP121332), European Union-funded "Sustained Diet Quality Improvement with Climate-smart, Nutrition-smart Orange-fleshed Sweetpotato in Southern Nations, Nationalities and Peoples' Region (SNNPR), Ethiopia," known as Quality Diets for Better Health (QDBH). (Contract number FOOD/2016/380-038), USAID/OFDA – Extending orange-fleshed sweetpotato availability for vulnerable households through good agricultural practices and postharvest storage (1337-USAID).

In-kind support was provided by Crops Research Institute (CSIR-CRI), Ghana; Damongo Agricultural College, Northern Region, Ghana; Greater Rural Opportunities for Women (GROW) Project, implemented by Mennonite Economic Development Associates (MEDA) and partners, and Integrated Water Management & Agricultural Development Ghana Limited (IWAD), seed production company, Northern Region, Ghana; People in Need (PIN), Regional Departments of Agriculture of the Northern and Upper West Regions, Ghana; Savanna Agricultural Research Institute (CSIR SARI), Ghana; Sodo and Shire ATVET, Ethiopia; SNNPR Bureau of Agriculture and Natural Resource Development, Ethiopia; Southern Agricultural Research Institute, SNNPR, Ethiopia; Technologies for African Agricultural Transformation (TAAT) "Orange-Fleshed Sweetpotato Compact"; and USAID Resiliency in Northern Ghana (RING) Project, implemented by Global Communities and partners.

References

Abate GT, Bernard T, Makhija S, Spielman DJ (2019) Accelerating technical change through video-mediated agricultural extension: evidence from Ethiopia, vol 1851. International Food Policy Research Institute

Abidin PE, Kazembe J, Atuna RA, Amagloh FK, Asare K, Dery EK, and Carey E (2016) Sand storage, extending the shelf-life of fresh sweetpotato roots for home consumption and market sales

Aker JC (2011) Dial "A" for agriculture: a review of information and communication technologies for agricultural extension in developing countries. Agric Econ 42(6):631–647. https://doi.org/10.1111/j.1574-0862.2011.00545.x

Asfaw F, Cherinet M, and McEwan MA (2020) Scaling Sweetpotato Triple S PLUS – gender responsive options for quality planting material, higher yields, and extended shelf life for storage roots. End line Survey Report. CGIAR Research Program on Roots, Tubers and Bananas (RTB), International Potato Center

Cherinet M, Asfaw F (2018) Report for RTB Scaling Fund. Scaling Triple S PLUS in Ethiopia: country start up and planning workshop, April 2018. Hawassa, SNNPR, Ethiopia

Etwire, P. M., Imoro, S., Suleman, I. & Van Mourik, T (2020) Gender responsive extension tools in Northern Ghana: effectiveness of a multi-channel communication approach on knowledge, attitudes and practices around orange-fleshed sweetpotato innovations. CGIAR Research Program on Roots, Tubers and Bananas (RTB), International Potato Center

Glover D, Sumberg J, Andersson JA (2016) The adoption problem; or why we still understand so little about technological change in African agriculture. Outlook Agric 45(1):3–6. https://doi.org/10.5367/oa.2016.0235

Glover D, Sumberg J, Ton G, Andersson J, Badstue L (2019) Rethinking technological change in smallholder agriculture. Outlook Agric 48(3):169–180. https://doi.org/10.1177/0030727019864978

Gumucio T, Hansen J, Huyer S, Van Huysen T (2020) Gender-responsive rural climate services: a review of the literature. Clim Dev 12(3):241–254

Hermans TDG, Whitfield S, Dougill AJ, Thierfelder C (2021) Why we should rethink 'adoption' in agricultural innovation: empirical insights from Malawi. Land Degrad Dev 32(4):1809–1820. https://doi.org/10.1002/ldr.3833

International Potato Center (CIP) (2019) Triple S technology: Availing clean sweetpotato planting material at the on-set of rains 1 p. https://hdl.handle.net/10568/105600

Jones N, Holmes R, Presler-Marshall E, Stavropoulou M (2017) Transforming gender constraints in the agricultural sector: the potential of social protection programmes. Glob Food Sec 12:89–95

Kabunga NS, Dubois T, Qaim M (2012) Heterogeneous information exposure and technology adoption: the case of tissue culture bananas in Kenya. Agric Econ 43(5):473–486. https://doi.org/10.1111/j.1574-0862.2012.00597.x

Kyazze FB, Owoyesigire B, Kristjanson PM, Chaudhury M (2012) Using a gender lens to explore farmers' adaptation options in the face of climate change: results of a pilot study in Uganda. CCAFS Working Paper 26. CGIAR Research Program on Climate Change, Agriculture and Food Security (CCAFS), Copenhagen, Denmark

Low J, Lynam J, Lemaga B, Crissman C, Barker I, Thiele G, Namanda S, Wheatley C, Andrade M (2009) Sweetpotato in Sub-Saharan Africa. In: Loebenstein G, Thottappilly G (eds) The Sweetpotato. Springer, Dordrecht, pp 359–390. https://doi.org/10.1007/978-1-4020-9475-0_16

Low JW, Mwanga ROM, Andrade M, Carey E, Ball A-M (2017) Tackling vitamin A deficiency with biofortified sweetpotato in sub-Saharan Africa. Glob Food Sec 14:23–30. https://doi.org/10.1016/j.gfs.2017.01.004

Mayanja S, Suleiman I, Imoro S, van Mourik T, Asfaw F, Cherinet M, and McEwan M (2020) Gender responsive communication tools and approaches for scaling the Triple S Technology in Ethiopia and Ghana. Retrieved from https://hdl.handle.net/10568/110029

McCormack C (2018) Key factors in the use of agricultural extension services by women farmers in Babati District, Tanzania

McEwan M, van Mourik T, Namanda S, Mihiretu C, Suleman I, Asfaw F (2018) Scaling Sweetpotato Triple S PLUS – gender responsive options for quality planting material, higher yields and extended shelf life for storage roots. Baseline Report on Scaling Readiness at April 2018. International Potato Center, Nairobi

Mudege NN, Mdege N, Abidin PE, Bhatasara S (2017) The role of gender norms in access to agricultural training in Chikwawa and Phalombe, Malawi. Gend Place Cult 24(12):1689–1710. https://doi.org/10.1080/0966369X.2017.1383363

Mukasa AN (2018) Technology adoption and risk exposure among smallholder farmers: panel data evidence from Tanzania and Uganda. World Dev 105:299–309. https://doi.org/10.1016/j.worlddev.2017.12.006

Namanda S (2012) Current and potential systems for maintaining sweetpotato planting material in areas with prolonged dry seasons: a biological, Social and economic framework. University of Greenwich

Namanda S, Amour R, Gibson R (2013) The triple S method of producing sweet potato planting material for areas in Africa with long dry seasons. J Crop Improv 27(1):67–84

Ogato GS (2013) The quest for gender responsive information communication technologies (ICTs) policy in least developed countries: policy and strategy implications for promoting gender equality and women's empowerment in Ethiopia. Int J Inf Technol Business Manage 15(1):23–44

Poulsen E, Sakho M, McKune S, Russo S, Ndiaye O (2015) Exploring synergies between health and climate services: assessing the feasibility of providing climate information to women farmers through health posts in Kaffrine, Senegal. CCAFS Working Paper no. 131. CGIAR Research Program on Climate Change, Agriculture and Food Security (CCAFS), Copenhagen

Rogers E (1995) Diffusion of innovations, 4th edn. Free Press, New York

Sartas M (2018) Scaling readiness in action. Presentation at RTB scaling fund kick off workshop, Nairobi, March 2018. Scaling readiness in action - Google Slides

Sartas M, Schut M, van Schagen B, Velasco C, Thiele G, Proietti C, and Leeuwis C (2020a) Scaling readiness: concepts, practices, and implementation. CGIAR research program on roots, tubers and bananas (RTB). January 2020, pp 217

Sartas M, Schut M, Proietti C, Thiele G, Leeuwis C (2020b) Scaling readiness: science and practice of an approach to enhance impact of research for development. Agr Syst 183:102874. https://doi.org/10.1016/j.agsy.2020.102874

Stathers T, Namanda S, Agili S, Cherinet M, Njoku J, McEwan M (2017) Guide for trainers - Sweetpotato planting material conservation triple S method: sand, storage, Sprouting. International Potato Center, Nairobi

Straub ET (2009) Understanding technology adoption: theory and future directions for informal learning. Rev Educ Res 79(2):625–649. https://doi.org/10.3102/0034654308325896

UNDP (2012) Gender, climate change and food security. In: Gender and climate change Africa policy. Brief 4. United Nations Development Programme, New York. Accessed 15 Feb 2021

Van Mele P (2006) Zooming-in zooming-out: a novel method to scale up local innovations and sustainable technologies. Int J Agric Sustain 4(2):131–142. https://doi.org/10.1080/1473590 3.2006.9684796

van Mourik T (2018) Storage in sand and sprouting of the orange-fleshed sweetpotato (Triple-S). Video. EcoAgtube. https://www.ecoagtube.org/content/storage-sand-and-sprouting-orange-fleshed-sweetpotato-triple-s

van Mourik T, Wigboldus S (2018) Report of the "Scaling sweetpotato Triple-S PLUS startup workshop", Hotel Radach, Tamale, Ghana, 12 and 13 March 2018. International Potato Center, Tamale, Ghana

Wigboldus S (2018) RTB Scaling Fund Kick-off Workshop. March 2018. Nairobi, Kenya

Wigboldus S, Klerkx L, Leeuwis C, Schut M, Muilerman S, Jochemsen H (2016) Systemic perspectives on scaling agricultural innovations. A review. Agron Sustain Dev 36:46. https://doi.org/10.1007/s13593-016-0380-z

Williams FE, Taron A (2020) Demand-led extension: a gender analysis of attendance and key crops. J Agric Educ Ext 26(4):383–400. https://doi.org/10.1080/1389224X.2020.1726778

World Bank, International Food Policy Research Institute (2010) Gender and governance in rural services: insights from India, Ghana, and Ethiopia. Retrieved from Washington DC, USA. https://doi.org/10.1596/978-0-8213-7658-4 Accessed 15 Feb 2021 World Document

Chapter 13
Revolutionizing Early Generation Seed Potato in East Africa

Elmar Schulte-Geldermann (ID), **Rogers Kakuhenzire** (ID), **Kalpana Sharma** (ID), and **Monica Parker** (ID)

Abstract Poor access to healthy, high-yielding planting materials hampers potato production in East and Central Africa (ECA). The need to improve the quality and increase the quantity of seed potato available to farmers has been the basis of previous efforts in the subregion. One bottleneck in the seed value chain is the low quantity of early generation seed (EGS) for further multiplication. To break this bottleneck, the International Potato Center (CIP) and local partners introduced two rapid multiplication technologies (aeroponics and rooted apical cuttings) and an improved conventional system (sand hydroponics). These three technologies differ in terms of multiplication rates, investment costs, profitability, required skills, infrastructure, risks, and linkages to the rest of the seed value chain, with its actors, policy environment, plus supply, and demand. The three introduced technologies have helped to increase the supply of certified or high-quality seed in the region over the last decade. However, for successful scaling, the technologies have to be carefully selected based on their situation and their natural and economic environments.

13.1 Introduction

Improved potato production in sub-Saharan Africa can be a pathway out of poverty. The potato grows quickly, is high-yielding, and makes more efficient use of water than do many other food crops. The potato also makes fairly efficient use of capital, making it a smallholder cash crop of the future for the densely populated East and

E. Schulte-Geldermann (✉)
Bingen University of Applied Science, Bingen am Rhein, Germany
e-mail: e.schulte-geldermann@th-bingen.de

R. Kakuhenzire
International Potato Center, Addis Ababa, Ethiopia
e-mail: r.kakuhenzire@cgiar.org

K. Sharma · M. Parker
International Potato Center, Nairobi, Kenya
e-mail: kalpana.sharma@cgiar.org; m.parker@cgiar.org

© The Author(s) 2022
G. Thiele et al. (eds.), *Root, Tuber and Banana Food System Innovations*,
https://doi.org/10.1007/978-3-030-92022-7_13

Central African highlands, with a high potential of improving livelihoods. The potato is also grown by women in East Africa, although female farmers often face barriers to obtaining land, capital, and seed (Mudege and Demo 2016).

13.1.1 Status of Seed Systems in East Africa

Potato yields are relatively low in the tropical highlands of East and Central Africa, stagnating at around 10 tons per ha for the past decades (FAOSTAT 2020). Increase in production has been achieved by area expansion rather than by productivity. The low yields have largely been attributed to increasing incidences of pests and diseases, particularly late blight, bacterial wilt, nematodes, and viruses, most of which are seed-borne. There is an inadequate supply of clean (healthy) seed, and farmers have poor access to quality seed and to improved varieties, contributing to a huge yield gap (Gildemacher et al. 2009; Harahagazwe et al. 2018b; Schulte-Geldermann et al. 2013; Thomas-Sharma et al. 2017).

Formal seed systems involve specialized activities of the seed value chain governed by an official regulatory environment, which is weak in many countries of East and Central Africa. The lack of high-quality, inexpensive, early generation seed (EGS) remains a major bottleneck in the seed supply chain (Muthoni and Kabira 2014; Negash 2014; Thiele et al. 2011). In the target region, only about 0.1–3% of the seed potatoes used are certified (Ferrari et al. 2017; Kaguongo et al. 2014). In the region, Kenya has the most advanced seed system with an established certification scheme. However, despite a great increase in the supply of certified seed over the last decade (430 tons in 2009 vs. 10,600 tons in 2018) production still only reached a third of the actual demand for certified seed which is estimated at around 10% of the total area planted (National Potato Council of Kenya 2018).

In East and Central Africa semiformal and informal seed systems provide more than 90% of the seed used; they follow various exchange and dissemination patterns. The semiformal seed system in Kenya, referred to as the "clean seed system," uses an onward multiplication of certified seed for one to three seasons by locally recognized informal seed producers, who usually have bigger farms than the average smallholder farmer has and produce an estimated 2–5% of the seed potato supply (Kaguongo et al. 2014). This seed has not been tested, but it is likely to be healthier than farm-saved planting material or other sources from the informal market. However, most smallholders use seed from their own farms, from neighbors, from local markets, or from ware potato traders. These traders drive varietal change as they distribute small ware potatoes of market-demanded varieties as seed to farm communities (Gildemacher et al. 2009; Kaguongo et al. 2014).

Hence the formal seed systems are still not able to meet the estimated demands, and smallholder farmers have poor access to quality seed. Low yields impede profitable farming and confine smallholders to semi-subsistence agriculture (Okello et al. 2017), limiting small farm sector development and entrenching food insecurity and poverty.

Projects intended to improve seed potato systems have been an important component of agricultural development in East Africa for several decades with limited success. Top-down approaches failed because they focused too narrowly on technology, without adequately considering local circumstances and demands (Almekinders et al. 2019a, 2019b; Ferrari et al. 2017). Seed is still a priority for agricultural interventions, with projects of different shapes and scales aiming to increase seed security, food and nutritional security, and agricultural productivity or to reduce poverty. Most interventions strive to make quality seeds and improved varieties more available to farmers. The high cost of quality seed is a bottleneck (Okello et al. 2018). Certified seed is about three times the price of ware potato and would account for up to 60% of the total variable production costs, limiting its use by smallholders.

13.1.2   Enhancing the Supply of Early Generation Seed (EGS)

The supply of affordable, high-quality seed can be increased by improving the efficiency in EGS production, by introducing rapid multiplication technologies (RMTs) and improved conventional technologies. New RMTs must consider the limited resources of the target countries, the local seed sector potential, and the seed customers (ware potato farmers). Multiplying seed in fewer generations reduces the risk of acquiring degenerative diseases, mainly viral and bacterial. Two RMTs, aeroponics and rooted apical cuttings, and an improved conventional system (sand hydroponics) have been introduced to public and private sector seed producers in order to increase early generation seed production. The technologies were selected and designed specifically for these enterprises and validated and improved during development.

Additional reasons for introducing the improved EGS technologies were (i) the faster introduction of new varieties, giving farmers quicker access to genetic gains, (ii) the quicker revitalization of popular varieties and lost seed stocks from a limited number of healthy propagules, and (iii) fast bulking of foreign germplasm and the introduction of varieties into other regions or countries (Struik and Wiersema 1999). Besides improving seed health, removing the bottleneck of poor access to genetic gains has been another major reason for promoting technologies that increase the capacity to produce EGS, to give smallholder farmers faster access to varieties that tolerate abiotic and biotic stresses (Parker et al. 2019; Thiele et al. 2020).

However, a remaining challenge for large-scale uptake is the poor linkages between the formal and the informal seed systems, which would allow for a dynamic increase in supply and demand. Thomas-Sharma et al. (2017) and Forbes et al. (2020) suggest linking the intervention with a more integrative on-farm seed quality improvement strategy. Such a strategy should help farmers manage their own seed stocks, while accessing a regular supply of high-quality seed, which would rely on EGS.

Increased EGS production would reduce the land required for seed multiplication. The region has relatively high pressure of soil- and seed-borne diseases, hence limited land suitable for seed multiplication. More disease-free seed can also be produced by reducing the number of field generations from five to seven to two to four through rapid EGS multiplication. However, producing quality seed requires a set of different skills at various levels, from planning and management of seed production to farm operations, including postharvest handling. Early generation seed production also requires highly skilled technical staff and significant capital investment, which was a constant bottleneck with previous public sector initiatives. Attracting local private investment in RMTs while strengthening the capacity of the public sector and generating efficient public-private partnership (PPP) models has been a way to break the seed bottleneck.

This chapter describes the success and limitations from experiences with three EGS multiplication technologies (aeroponics, sand hydroponics, and rooted apical cuttings), which have been introduced and promoted in East and Central Africa.

13.2 Suitable Technologies for Early Generation Seed Potato Production

The potato seed value chain begins with clean (disease-free) micro-plants that are first reproduced in test tubes by micro-propagation. These micro-plants are then transplanted in a greenhouse or screenhouse to produce the first generation of tubers, called "breeder" or "early generation seed" (EGS). The second and third generations ("foundation" seed) are propagated in carefully selected, disease-free fields. Until recently, certified seed was produced by another two to four field multiplication rounds. This procedure is long, has relatively low multiplication rates of seed potato, and is based on producing EGS with conventional pot-substrate technologies.

Several interventions in sub-Saharan Africa aimed to produce EGS faster with adapted RMTs. Many RMTs exist, but the most common in developing countries are micro-propagation (plantlets and micro-tubers), cuttings (single-node, tuber sprout, axillary, leaf-bud, apical), aeroponics, and hydroponics.

Micro-propagation has been used in developing countries since 1980s (Dodds 1988; Naik and Karihaloo 2007). In vitro techniques allow many micro-tubers to be produced per plantlet, but they have a longer dormancy period than normal tubers. They must also be planted in greenhouses or in net houses. Micro-tubers also suffer from high storage losses, and they may not be suitable for certain climates (Naik and Karihaloo 2007). The vitality of seed tubers from micro-propagated potato plants can be enhanced by producing mini-tubers, which are less delicate and can be directly planted in open fields or in net houses (Wiersema et al. 1987). However, plantlets usually produce fewer than ten mini-tubers (Ritter et al. 2001; Singh et al. 2010). More mini-tubers per plantlet can be produced by repeated tuber harvesting where potato plants are carefully lifted from the soil and replanted after removing tubers above a critical size (Ritter et al. 2001). This technique however is labor-intensive, disrupts plant growth, and may cause early plant senescence, often resulting in few tubers per plantlet.

Potato mini-tuber production from in vitro propagated plantlets permits faster multiplication of seed potato than conventional methods and reduces the number of generations of open field bulking (Farran and Mingo-Castel 2006). However, there is a need to improve the multiplication rates further, especially in the tropical highlands where previous seed potato production technologies have done little to improve the quantity of affordable, quality seed for resource-poor farmers.

In recent decades, soil-less technologies have been developed to produce mini-tubers in controlled environments to increase the multiplication rate through repeated mini-tuber harvesting. Such technologies include hydroponics (Muro et al. 1997; Tibbitts and Cao 1994) and aeroponics (Chang et al. 2008; Farran and Mingo-Castel 2006). Other technologies such as rooted cuttings from apical shoots (Van Minh et al. 1990) and improved conventional systems such as sand hydroponics have been developed and validated in particular for low-middle-income countries.

In cooperation with local partners, the International Potato Center (CIP) introduced and validated the EGS potato technologies in the target region: aeroponics in 2008, sand hydroponics in 2012 and rooted apical cuttings in 2016.

13.2.1 Aeroponic Technology

Aeroponics is a method of growing crops, particularly vegetables, in an air-nutrient mixture without soil or any other porous medium. Aeroponics has been adapted to produce large quantities of potato mini-tubers (Otazú 2010) in South Korea, China, Africa, Spain, and Latin America. The technology started with sophisticated equipment and relatively low yields Farran and Mingo-Castel 2006; Ritter et al. 2001; Soffer and Burger 1988), but after 2006, CIP had improved the yields and adapted aeroponics for developing countries (Otazú and Chuquillanqui 2008).

In aeroponics, plant roots are grown suspended in a dark box that is closed or semi-closed. Roots are sprayed with a nutrient-rich solution (Otazú 2010; Singh et al. 2010). The nutrients and improved aeration improve the growth of the root system (Farran and Mingo-Castel 2006; Otazú 2010) (Fig. 13.1).

The investment in the aeroponic units in African countries became possible by 2008, thanks to the support of the US Agency for International Development (USAID), German Agency for International Cooperation (GiZ), Irish Aid, and others. The technology was crucial for the successful Three Generation (3G) project in three sub-Saharan countries and for the release and dissemination of new late blight-resistant varieties in Kenya (Landeo et al. 2009). However, the technology requires training, electricity, an adequate greenhouse, a safe source of water, and robust in vitro plants.

Fig. 13.1 Aeroponic boxes with potato plants (left) and tuber production in boxes (right) at Kevian Farm ltd. in Kenya. (Photo credits M.L. Parker)

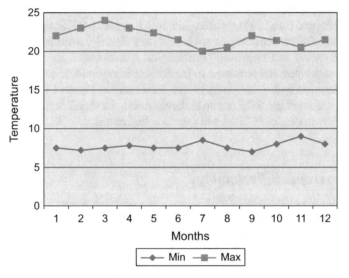

Fig. 13.2 Stable temperatures in Nyahururu, Kenya (2300 masl), permit year-round potato seed production by aeroponics

When the weather is too warm, plants grow foliage but few tubers. In cold weather, plants are low yielding. Temperature can be artificially controlled, but this increases production costs. In a few places, aeroponics can perform well all year, as in Nyahururu, Kenya (Fig. 13.2), where stable temperatures allow year-round production.

In some places, even though adequate greenhouses were built, frequent interior temperatures of over 30 °C for more than 4 hours caused setbacks, including pathogen proliferation. Greenhouses at higher elevations, above 2200 masl, produced more tubers (Schulte-Geldermann 2013; Demo et al. 2015; Atieno and Schulte-Geldermann 2016; Kakuhenzire et al. 2017, Sharma and Atieno 2020). Temperatures can also be regulated with low-cost methods such as shade, ventilation, and production during the cooler months.

In aeroponics, harvesting can last up to 5 months, creating uneven sprouting. At the end of the season, the mini-tubers are not of uniform dormancy or physiological age. Planting mini-tubers with large differences in physiological age results in crops with large variations in important yield parameters such as date of emergence, stem numbers, canopy growth pattern, maturity date, total yield, and tuber size distribution. This can cut yields in half (Lung'aho et al. 2013) and make mini-tuber production unprofitable. This can be managed with plant growth regulators (PGR) (Tsegaw 2006; Van Ittersum and Scholte 1993). Aeroponics requires judicious management to ensure that mini-tubers are in the correct physiological stage at the time of planting.

Aeroponics requires proper training over at least one season. The best learning experience involves the trainees in building the modules, followed by lectures. This worked well, except in places with a rapid staff turnover. In many cases, new personnel were not trained appropriately, which put several units at risk. The in vitro plantlet to mini-tuber multiplication rates of the aeroponic units differed largely by environmental, management, and genotypes (varieties) between 1:20 and 1:70 (Kakuhenzire et al. 2015; Atieno and Schulte-Geldermann 2016; Kakuhenzire et al. 2017, Muthoni et al. 2017; Harahagazwe et al. 2018a, b

Profitability of aeroponics Labarta (2013) compared the profitability of three mini-tuber production systems: (1) conventional pot system, (2) aeroponics with a regular power supply and a backup generator, and (3) aeroponics using a solar power supply and a backup generator. All of them produced mini-tubers of the same quality, but they had different levels of investment, variable costs, and multiplication rates of mini-tubers. Aeroponics reduced the unit cost of each mini-tuber from 0.60 USD to 0.10 or 0.12 USD. All three techniques require a high initial investment (Table 13.1).

The calculation in Table 13.1 presents the level of investment required and assumed that each technology last for ten seasons (5 years). The variable cost include the cost of plantlets, substrates, nutrients, water, electricity, labor, and soil sterilization where needed.

Table 13.1 Profitability of mini-tuber production under three different production schemes (in USD)

	Conventional pot	Aeroponics with backup generator	Aeroponics with solar power and backup generator
Initial investment	31,053	27,310	32,841
Variable cost (season)	3146	1220	1315
Discounted total cost	57,288	37,482	43,809
Mini-tuber per plantlet	7	26.4	26.4
Unit cost per mini-tuber	0.60	0.10	0.12

Source: Labarta (2013)

All of them produced mini-tubers of the same quality but using different levels of investment and variable cost and showing different multiplication rates of mini-tubers. For the profitability analysis, we use the average multiplication rate found for the base conventional pot system (seven mini-tubers per plantlet) and for both aeroponic systems (26.4 mini-tubers per plantlet).

In the public sector, where conventional pot mini-tuber production is dominant, the initial investment is often not accounted in the production cost, and they can barely cover the variable cost that is highly driven by the cost of sterilizing the soil (Labarta 2013).

13.2.2   Rooted Apical Cuttings Technology

Rapidly growing young vegetative tissue of potato can be cut and rooted in a number of ways (Bryan et al. 1981). Rooted apical cuttings have long been used in SE Asia (Vander Zaag and Escobar 1990), particularly in Vietnam (Van Minh et al. 1990). This technique is being introduced into sub-Saharan Africa to provide a simple, effective technique for multiplying early generation seed as alternative to mini-tubers (Parker 2019).With rooted apical cutting (RAC) technology; seedlings are produced vegetatively, with cuttings taken from plantlets in the screenhouse every two to three weeks (Fig. 13.3). The cuttings are planted into plugs for rooting with a substrate of coconut sawdust, clean subsoil, and sterilized decomposed manure. After 6 weeks, the cuttings are fully rooted, and they are transplanted to the field. Rooted apical cuttings are transplanted right away in the field, thereby saving one generation, as mini-tubers are no longer needed (Parker 2019). The cuttings can also be placed in flower boxes, but they must be planted within 24 hours or kept in a protected location until planting. However, mini-tubers are more versatile; they can be stored until ready to plant and are easy to transport.

Fig. 13.3 Rooted apical cutting production in a screenhouse at Stokman Rozen Ltd., Naivasha, Kenya (left), and harvested tubers derived from RAC field multiplication (right). (Photo credits: Benson Kisinga and Monica L. Parker)

Rooted apical cuttings have high multiplication rates after only 5 months in the screenhouse. Investment costs are lower than for aeroponics. Electricity is needed only for the tissue culture lab. There is no need to break tuber dormancy, and no storage facility is required. This makes seed potato available to farmers one season sooner.

A manual by Nyawade and Parker (2020) describes this technique in detail for any size of farm. Integrating cuttings in seed systems greatly expands opportunities in seed businesses of any scale (e.g., small and medium enterprises, youth groups, farmer groups, entrepreneurial farmers) at different points along the seed production chain. Cuttings are penetrating the seed system, and the opportunities they present are in validation in Kenya and Uganda to scale out the technology through diversified partnership models and uses.

13.2.3   Sand Hydroponic Technology

Unlike aeroponics and rooted apical cuttings, which are new technologies, sand hydroponics is an improved conventional technique that lowers the cost of quality EGS. Sand hydroponics is relatively easy to handle, and it costs less than conventional multiplication technologies. Sand hydroponics uses a greenhouse with wooden boxes, plastic pots, polythene bags filled with sterilized river sand, and growing plants (Naik and Karihaloo 2007). This new hydroponic system is independent from electricity. The system uses gravity to distribute the nutrients dissolved in water, and sand replaces the conventional substrate. It combines the principles of hydroponics and fertigation (chemical fertilizer dissolved in irrigation water) (Fig. 13.4).

Fig. 13.4 Potato seed multiplication using sand hydroponics in pots, Holeta, Ethiopia

Sand hydroponics is an open (run-to-waste) system in which the nutrient solution is not recycled, greatly reducing the likelihood of diseases (unlike aeroponics where a disease can spread quickly in recycled nutrient media). Besides wooden beds, sand hydroponics can work with any available container. Sand hydroponics is a versatile technology, and existing resources can be adapted for seed potato multiplication.

Substrate must be sterilized to kill the pathogens that can cause plant disease. This requires an expensive boiler and fuel. If the substrate does not get hot enough, diseases may occur. If the substrate is overcooked, toxic manganese may be released and damage the plants. In the conventional system, the substrate is made of mixtures of peat moss, compost, dark soil, and sand. Some of these are hard to obtain and expensive. However, with sand hydroponics, the sand is sterilized with ordinary sodium hypochlorite bleach, so sterilization is easier, and cheaper without the risk of toxic releases. Before planting, sand is treated with boiling water to eliminate pathogens. Bleach is affordable and efficient as disinfectant, and it can also be used to treat sand (Otazú 2008). Sand can be reused several times, but after each harvest, it must be cleaned of any plant debris, root pieces, and mini-tubers to avoid pathogen buildup.

Unlike some soils, sand allows the free expansion of tubers for good tuber formation. Other than the aeroponic units, the system does not require electricity and can be managed by relatively unskilled workers, because water and nutrients flow by gravity.

Aeroponics can produce some mini-tubers that are too small (under 5 g) to plant in the field, and seed multipliers are reluctant to use them. Sand hydroponics produces bigger mini-tubers and slightly more than in conventional pots (11 vs. 8.9), which also helps to reduce seed costs (Mbiri et al. 2015). Sand hydroponics is a good technology for places where electricity, trained staff, and investment capital are limited.

13.3   Strengths and Weaknesses of Introduced EGS Production Technologies

The introduction of EGS multiplication technologies into sub-Saharan Africa has been successful, but no technology is suitable for all conditions. Nevertheless, all three technologies when operated properly, in the right environment, have helped to increase early generation seed production (Sect. 13.5). The rooted apical cutting technology produces certified seed in less time. Up to generation four, its multiplication rate up is about 20 times faster than with conventional seed production and about three times faster than with aeroponics (Table 13.2).

Each technology has its strengths and weaknesses, so some may be more suitable for certain natural or social environments (Table 13.3). It is important to choose the right technologies for each business type and to develop a business plan which

Table 13.2 Certified seed production cycle by various potato EGS technologies

	Screenhouse production (G1)		Breeder seed (G2)	Foundation seed (G3)		Certified 1 seed (G4)	
	per in vitro plant	Months	# seed tubers	# seed tubers[a]	Months	# seed tubers[a]	Months
Rooted apical cuttings	120 rooted cuttings	5	1080[b]	10,800	17	108,000 [3]	23
Aeroponics	35 mini-tubers	8	280[c]	2800	20	28,000	26
Sand hydroponics	10 mini-tubers	8	80[c]	800	20	8000	26
Conventional	8 mini-tubers	8	64[c]	640	20	6400	26

[a]Assumed multiplication rate of 10
[b]Assumed multiplication rate of 12 at 75% of survival
[c]Assumed multiplication rate of 8

Table 13.3 Strengths and weaknesses of EGS potato production technologies

Technology	Strengths	Weaknesses
Aeroponics	Effective RMT with high multiplication rates of 1:20–80 Needs fewer generations of seed potato multiplication in the field, lowering costs Seed can be harvested at seed size (from 5 to 30 grams) since the fertilizing sprays that are applied to the roots allow the plant to grow without interrupting its vegetative cycle of up to 180 days Costs are about one-quarter those of a conventionally grown tuber Requires much less water and fertilizer than conventional systems, minimizing fertilizer residues seeping into ground water Excellent circulation of air, which strengthens the roots Uses less greenhouse space Attracts private sector engagement	High investment Requires uninterrupted power supply Consumables and equipment are often unavailable on the local market Lack of synchronization of aeroponic production with downstream activities due to sequential harvests (normally every 2 weeks) Small and fragile mini-tubers require cold storage facility Requires well-trained staff Risks of contamination within the boxes, causing plant health problems Sensitive to high temperatures Difficult to clean properly inside the boxes Leakages of nutrients in system raise production costs Metal equipment causes rust within the boxes Long growth cycles, which normally end with less productive periods

(continued)

Table 13.3 (continued)

Technology	Strengths	Weaknesses
Sand hydroponics	Mini-tubers of high quality and larger than with conventional and aeroponic technologies No need for expensive substrate steaming or boiling Easy to establish and run Substrate can be recycled Suitable for resource-poor settings	Difficulty to determine the optimal nutrient use efficiency Sand disinfestation requires a lot of clean water to wash out the sterilization agent Labor-intensive when pots are used Yield lower than with RMTs Nutrients and equipment are often unavailable on the local market
Rooted apical cuttings	High multiplication rates (cuttings) High field multiplication rates (1:12–18) Not dependent on electricity First field-grown seed produced in one season less than with other technologies Stable production across varieties Doesn't require cold storage Fast dissemination of new varieties Attracts private sector investment Excellent opportunity for diversifying business operations of existing tissue culture labs	Cuttings cannot be stored Sensitive to cold temperatures Output market is difficult to create and manage due to cutting harvests carried out over time Labor-intensive Bulky for transport from greenhouses to farmers' fields Needs irrigation when transplanted in the field
Conventional pot system	Easy to manage with stable yield in mini-tubers (backup technique when conditions for RMTs are not optimal) Acceptable tuber size and high survival rates when planted	Low mini-tuber per plantlet ratio High requirement of greenhouse space Risk of contaminations with soil-borne diseases Substrate expensive and not recycled High cost of substrate sterilization Labor-intensive

Source: adapted from Harahagazwe et al. (2018a)

includes further multiplication steps and the point of sale. Decision support tools such as the risk assessment framework for investments in aeroponics (Andrade-Piedra et al. 2019) are useful for choosing the right technology. The potato varieties, staff, funding, and the possibility of using existing buildings and resources should be considered on an individual basis before investing. The seed producer should anticipate seed multiplication efficiency in order to assess the cost of goods sold (COGS) as well as the demand from seed customers (farmers) for specific varieties.

13.4 Scaling Up Improved Early Generation Multiplication Technologies

Producing quality seed requires different skills at various levels, from planning and management of seed production, through to farm operations, including postharvest handling. This is particularly true when introducing these early generation seed innovations. Aeroponics and rooted apical cuttings require skilled technical staff and high capital investment, which are usually limited in national programs. Improving infrastructure and capacity for high-grade seed production is resource- and knowledge-intensive. It requires investments in facilities, equipment, and inputs, which is risky because of the uncertainty about market size. Technical support is essential at the start of seed enterprises. But the private sector is increasingly realizing the market potential for seed potato and has begun to invest in seed production.

13.4.1 Partnerships for the Innovation: Engaging Local, Private Enterprises in the Seed Potato Business

A major thrust of the interventions was to attract private sector engagement in early generation seed, as in most countries EGS potato was solely produced by the public sector. These partnerships were established for the first time during the project "Tackling the Food Price Crisis in Eastern and Central Africa with the Humble Potato: Enhanced Productivity and Uptake through the '3G' Revolution." The USAID-funded project (generally referred to as "3G") was implemented in Kenya, Rwanda, and Uganda from 1 October 2008 to 30 June 2011. The strategy involved delivering low-cost, quality seed to growers in three generations of field multiplication, rather than the conventional five to seven generations (hence "3G" for three generations). The 3G seed strategy envisaged producing many mini-tubers through one generation of a rapid multiplication technology, thus allowing bulking of sufficient seed in fewer field generations. This reduced both the cost of production and prevented the buildup of diseases in the field (Atieno and Schulte-Geldermann 2016).

CIP worked with several private companies and public organizations to deliver seed of improved varieties to smallholder farmers. Large-scale private farms were attracted to produce quality seed through a 50:50 cost sharing for constructing aeroponic facilities and by linking them to markets. CIP advertised through the local media for private sector companies interested in producing breeder seed. CIP supplied in vitro plantlets to the companies and offered technical support to produce mini-tubers. The private companies produced the mini-tubers using both the traditional soil-based method and aeroponics. Kenya Plant Health Inspectorate Service (KEPHIS), the authority responsible for seed certification, ensured the inspection of the seed for pests and diseases.

While supporting the private sector's production of early generation seed, the interventions continued to increase the capacity of national agricultural research institutes, which still played a vital role in producing EGS. In the early stages of establishing a seed potato system, the national programs provided a base to private enterprises and assured them of business opportunities in seed production.

Crucially, the project helped private and public enterprises develop business plans to select the most appropriate and cost-effective RMTs (aeroponics, rooted stem cuttings, sand hydroponics, or other improved conventional systems) for producing mini-tubers (G1). The project funded a cost-share scheme to support capital investments in screenhouses and equipment for producing mini-tubers. Backstopping of seed producers focused on creating awareness of emerging pathogens, adherence to the plant health standards, and improving working relations with the regulatory body. The project trained entrepreneurs and farmers in seed quality management, seed use and storage, and accessing new varieties in order to contribute to the uptake of new varieties and improved practices.

13.4.2   Quality Control: Enable Seed Quality Control and Disease-Monitoring Schemes

One bottleneck was quality assurance and the certification system, which are vital to develop a seed system based on guaranteed quality standards, and of equal importance for seed producers and users. A formal seed quality control system generally involves seed certification by national regulatory agencies; however, many African countries either do not have such systems, or they are not implemented. Of the target countries, only Kenya has an established and functional certification scheme, mandated to KEPHIS. But recent massive increases in seed production are taxing KEPHIS's ability to implement timely, customer-friendly services, because its human capacity is limited, and its inspection technologies and schemes are outdated. Despite its problems, Kenya's quality control is still far ahead of the other countries in the region, where the National Agricultural Research Systems (NARS) are basically testing the seed they produce and only to a certain level. This discourages specialized private investment in quality seed production, as buyers doubt the product's quality.

To strengthen formal seed certification schemes, it is vital to develop and introduce easy-to-use and affordable technologies for rapid, low-cost quality control or better seed health assessments. Examples include pocket diagnostics such as lateral flow devices, heel-end coring devices for rapid disease detection, and molecular methods based on polymerase chain reaction for confirmation tests (i.e., LAMP assay). In Rwanda, Malawi, Tanzania, and Uganda, where formal seed certification schemes for seed are in the planning stages, technical support and training in new methods will be critical. Standards for seed quality such as defined thresholds for degenerative seed-borne diseases need to be aligned to local conditions. Though time-consuming, a strategy that involves constant stakeholder reviews has been the most useful way to improve the quality and quantity of certified seed. An alternative semiformal quality control system named Quality Declared Planting Material (FAO 2010), based on visual inspections, has been introduced to countries like Ethiopia, Malawi, Rwanda, Tanzania, and Uganda where formal certification systems are not operational.

In both Kenya and Uganda, rooted apical cuttings have been included into the national seed certification protocol for potato. In both countries RACs are classified as breeders' seed (stock seed) in the same category as mini-tubers.

13.4.3 Stimulating Demand for Potato Seed and Making It More Accessible to Smallholders

The demand for high-quality seed is key for attracting investment in early generation seed and for developing a vital seed sector. Most potato farmers are smallholders with limited cash. As seed is a large investment, most farmers save their own seed or access it from other local sources. The accessibility of certified seed is also limited because it is often produced in regions with poor roads and because a 1-hectare potato field requires about 2 tons of seed tubers, which are perishable and may be lost. For these reasons, certified seed has little penetration into the local seed system (Gildemacher et al. 2009). Recycling farm-saved seed potato leads to a decline in productivity due to seed degeneration (Gildemacher et al. 2009; Schulte-Geldermann et al. 2013). In order to provide smallholder farmers with better seed sources, CIP and partners investigated the efficacy of establishing a network of decentralized seed multipliers to increase the supply and reduce the price of seed. However, there was not enough clean land suitable for producing healthy seed tubers. These difficulties have been reported in Ethiopia and Kenya, where bacterial wilt and potato cyst nematodes (the latter only in Kenya) have been widely distributed by local seed multipliers (Abdurahman et al. 2017; Mburu et al. 2020). Even though seed from local multipliers outperformed the farm-saved seed significantly, the risk of spreading pests and diseases throughout the seed system has to be considered (Andersen et al. 2019).

Another option is to advise farmers to invest in certified seed for a smaller area of their farm, in combination with improved management (Thomas-Sharma et al. 2016). To curb seed degeneration and lower the cost of certified seed, farmers can buy small amounts of certified seed and bulk it themselves using on-farm seed selection and management. One study showed that with little investment in high-quality certified seed, farmers were able to more than double their yields in only 1 year. After the first year, participating farmers continued to source small amounts of certified seed. However, in some cases local availability remained an issue (Ochieng 2021).

Local extension services increased awareness with on-farm demonstration trials. CIP's USAID-funded 3G and Better Potato for a Better Life projects distributed massive amounts of certified seed to smallholders, including new, farmer- and market-preferred varieties in affordable, low-risk units (5 to 25 kg bags). Other strategies to improve accessibility included seed banks (based on commodity loan systems), seed fairs, input loans (e.g., in Kenya, Equity Bank and Kenyan Women Finance Trust), input insurance systems (e.g., Kilimo Salama by the Syngenta Foundation for Sustainable Agriculture), and strengthening seed value chain linkages with contract farming. In Kenya an information and communication technology (ICT)-based information portal, Viazi Soko ("potato market"), was set up and is maintained by the National Potato Council of Kenya, which includes detailed information on the traits and availability of seed varieties. Other activities generated knowledge about and linkages to seed distribution channels.

13.4.4   Scaling Strategy and Key Partnerships for Scaling

Successful scaling requires assessing institutional arrangements, the business environment and capacities of the public and private sectors, the natural environment, supply and demand, and the general enabling environment for the comparative advantage of the innovation. Scaling-improved EGS technologies depend on tissue culture labs and tissue culture material of the varieties to be produced. Nurseries or seed multipliers must be identified who are willing to invest in producing cuttings or mini-tubers. Markets must be developed by raising awareness among seed customers. Regulatory authorities must be engaged to ensure that seed produced from cuttings and mini-tubers from aeroponics is eligible for certification.

Critical Bottleneck for Further Scaling of Rooted Apical Cuttings
Nurseries must be close to potato producing areas. Technical backstopping is a further bottleneck because few of nursery operators know how to produce apical cuttings.

Critical Bottleneck for Further Scaling of Aeroponics
Aeroponics also faces a high risk for failure if conditions are not favorable and the capacities are inadequate (Atieno and Schulte-Geldermann 2016; Kakuhenzire et al. 2017). CIP developed a risk analysis tool based on the lessons learned from previous interventions. This tool provides a total score by weighing different factors according to their impact on performance of aeroponics and applying scores from 1 (very good) to 4 (poor). With total scores ranging between 21 and 30, aeroponics should operate with no major problems. From 31 to 40, aeroponics can be implemented with caution, and with scores of over 40, aeroponics is likely to fail. Table 13.4 summarizes an assessment conducted in 2013 by Otazu and Schulte-Geldermann (unpublished; method described in Andrade-Piedra et al. 2019).

Table 13.4 Risk analysis performance in selected sites where aeroponics is operating (green = no major problems, yellow = implement with caution, red = likely to fail)

Location*	Electricity	Water	Climate	Facility	Plants	Training	Total score
Tigoni, KE	5 × 3 = 15	3 × 1 = 3	4 × 2 = 8	3 × 1 = 3	3 × 1 = 3	3 × 2 = 6	38
Kisima, KE	5 × 1 = 5	3 × 1 = 3	4 × 1 = 4	3 × 1 = 3	3 × 2 = 6	3 × 2 = 6	27
Molo, KE	5 × 1 = 5	3 × 1 = 3	4 × 1 = 4	3 × 1 = 3	3 × 2 = 6	3 × 3 = 9	30
Nairobi, KE	5 × 1 = 5	3 × 1 = 3	4 × 3 = 12	3 × 2 = 6	3 × 1 = 3	3 × 1 = 3	32
Holeta, ET	5 × 5 = 25	3 × 1 = 3	4 × 2 = 8	3 × 1 = 3	3 × 1 = 3	3 × 1 = 3	43
Mbeya, TZ	5 × 3 = 15	3 × 3 = 9	4 × 2 = 8	3 × 3 = 9	3 × 2 = 6	3 × 2 = 6	53
Gisozi, BU	5 × 3 = 10	3 × 1 = 3	4 × 1 = 4	3 × 1 = 3	3 × 2 = 6	3 × 4 = 12	38
Lichinga, MZ	5 × 3 = 15	3 × 2 = 6	4 × 2 = 8	3 × 1 = 3	3 × 1 = 3	3 × 2 = 6	41

[a]KE Kenya, ET Ethiopia, TZ Tanzania, BU Burundi, MZ Mozambique

This assessment indicated that locations with poor electricity service are most likely to fail or get low numbers of mini-tubers. Frequent power cuts of more than 2 hours are a major limiting factor for scaling the technology. The cost for running a backup generator would greatly increase the cost of mini-tuber production.

Climate is difficult to change. Other factors are more within immediate human control. For instance, in Tanzania, a new dependable backup generator with an automatic starting up system was installed after the assessment. Improved management and facilities might change a risky operation to one that could operate with caution.

13.4.5 Example for a Scaling Approach of Rooted Apical Cuttings and Robust Varieties of Potato in Uganda

As mentioned above, rooted apical cuttings have been successfully implemented in Kenya staring from a small pilot intervention with a private sector partner. CIP aimed to scale this technology to other countries in the region such as Uganda. Within a project funded through the RTB scaling funds, a systematic scaling approach was developed for the intervention in Uganda, described below.

The scaling project team used the concept of "scaling readiness to assess the innovation packages". "Innovation readiness" refers to the demonstrated capacity of an innovation to fulfill its contribution to development outcomes in specific locations. This is presented in nine stages showing progress from an untested idea to a fully mature proven innovation. "Innovation use" indicates the level of use of the innovation by the project, partners, and society. This shows progressively broader levels of use beginning with the intervention team who develop the innovation to its widespread use by users who are completely unconnected with the team or their partners. "Scaling readiness" of an innovation is a function of innovation readiness and innovation use. Table 13.5 provides summary definitions for each level of readiness and use adapted from Sartas et al. (2020).

Description of Innovation Package   Starter material is crucial for seed multiplication. Mini-tubers normally serve as starter material, producing five to ten tubers per unit. Apical cuttings are a faster alternative, producing 10 to 20 or more tubers per unit. There are two stages in apical cutting systems: producing apical cuttings in a screenhouse and then changing hands to a seed producer or farmer to plant the apical cuttings in the field to produce seed tubers. Seed produced from apical cuttings can be saved on-farm for a few seasons with little risk of quality loss. Apical cuttings offer opportunities for seed production in areas without enough land for traditional seed bulking and crop rotation. Apical cuttings accelerate new varieties quickly. Farmers can also plant apical cuttings directly to produce seed on-farm. In this context a scaling pathway at different levels of the seed potato value chain has been developed and is described in Fig. 13.5.

Present Level of Innovation Readiness and Innovation Use   For a systematic monitoring and evaluation of the scaling process and success of the technology in Uganda, the project team used a scaling tool (Sartas et al. 2020) that defines each stage of innovation readiness and use scale. Figure 13.6 represents the starting point

Table 13.5   Summary definition of levels of innovation readiness and use (Sartas et al. 2020)

Stage	Innovation readiness	Innovation use
1	Idea	Intervention team
2	Basic model (testing)	Direct partners (rare)
3	Basic model (proven)	Direct partners (common)
4	Application model (testing)	Secondary partners (rare)
5	Application model (proven)	Secondary partners (common)
6	Application (testing)	Unconnected developers (rare)
7	Application (proven)	Unconnected developers (common)
8	Innovation (testing)	Unconnected users (rare)
9	Innovation (proven)	Unconnected users (common)

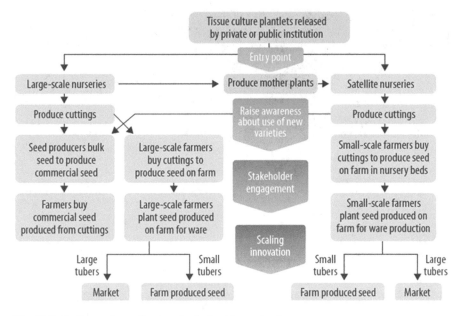

Fig. 13.5 Scaling pathway for rooted apical cutting technology

of the innovation package components (technologies, products, services, and institutional arrangements) when interventions began in Uganda in December 2018 based on feedback from private, public, and NGO stakeholders.

Most components started at a low level of readiness and use and progressed rapidly in both readiness of innovation and use scale within 2 years (Fig. 13.6) due to demand from users and stakeholders.

13.5 Outcomes from Scaling the Innovation in East and Central Africa

13.5.1 Impact of New Technologies and Partnership Models on the Production of EGS

The dissemination of technologies and the impact on early generation seed production have been studied by Harahagazwe et al. (2018a). Of the 18 institutions studied, 9 belong to governments, 1 is a parastatal, and 8 are private. The production of EGS potato in sub-Saharan Africa is still dominated by the conventional technique. Over 500,000 in vitro plantlets can be accommodated at one time under protective structures for mini-tuber production in the 7 study countries, and more than 7.5 million

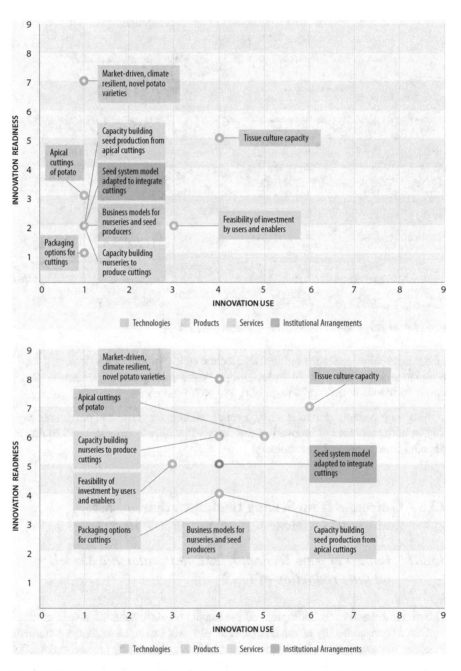

Fig. 13.6 *Readiness* versus use at introduction (top) and 2 years after (below) introduction of apical cuttings and robust potato varieties in Uganda

mini-tubers can be produced in a single season. This is enough seed to plant over 150 ha per season for the first field generation, compared with only 10 ha per season in 2008—using mini-tubers locally produced, i.e., 93,500 mini-tubers in Uganda, Rwanda, Malawi, Kenya, and Ethiopia (Demo et al. 2015) and 148,034 mini-tubers in Burundi. There was, however, no mini-tuber production in Tanzania in 2008 (Kakuhenzire et al. 2015) as shown in Fig. 13.7. In the region, 49% of mini-tubers derive from conventional technique, 30% from sand hydroponics, and 21% from aeroponics (Harahagazwe et al. 2018a). Additionally, more than 650,000 rooted apical cuttings are produced in Kenya and about 300,000 in Uganda enough to plant 13 and 6 ha for further multiplication, respectively.

This means that, as in Latin America (Mateus-Rodríguez et al. 2013), the conventional technique remains the most commonly used technology to produce mini-tubers. As one seed expert told Harahagazwe et al. (2018a), "the conventional technology remains the easiest and best known way of producing clean seed potato when working conditions and funding are suboptimal."

Nonetheless, the introduction of new technologies seems to have triggered interest in EGS production and contributed to the significant increase in it since 2008 (Fig. 13.7).

Despite this huge capacity of EGS infrastructure in the seven countries, the actual production remains below expectations. The total number of mini-tubers produced in 2017 is only 65% of the mini-tubers that could be produced in a single

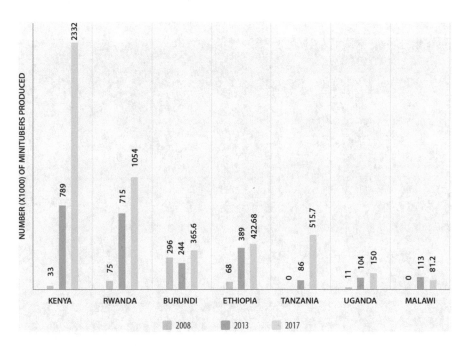

Fig. 13.7 Mini-tuber production over time in seven SSA countries. (Adapted from Demo et al. (2015). 2017 data were provided from Harahagazwe et al. (2018a))

season of 3 to 6 months, indicating inefficiencies in the management of the units, in particular by some public sector institutes. Public sector aeroponic units in Kenya, Ethiopia, and Uganda only produce at about 30% of their capacity, while some are not operational at all. However, results from Tessema and Dagne (2018) report that one-third of the 600,000 mini-tubers produced in Ethiopia were coming from aeroponic units in the period from 2011 to 2014 suggesting a better performance prior to the study conducted by Harahagazwe et al., in 2018. In contrast, private sector aeroponic units in Kenya produce at about 90% of their capacity. Aeroponics is challenging to operate but offers a high potential when conditions and capacities are adequate. This should be considered when deciding to scale this technology, which is best suited to enterprises with the capacity to invest in infrastructure and which already have experience with similar operations (e.g., greenhouses) in the flower and horticultural sectors.

Nonetheless, data collected in seven African countries show that the capacity for mini-tuber production is significantly increasing (Fig. 13.8), in part as a result of the investments in RMTs, including the implementation of the CIP project funded by USAID, Irish Aid, GiZ, and the Government of Finland in Ethiopia, Kenya, Malawi, Rwanda, Tanzania, and Uganda (Demo et al. 2015).

Rooted apical cuttings are starting to be used in the potato sector; the number of rooted apical cuttings sold in Kenya is rapidly increasing from about 70,000 in 2017 to 265,615 in 2018 and 337,418 in 2019 (unpublished data provided by Stokman Rozen Ltd.). In 2019 about 46% of the cuttings were ordered by 44 private

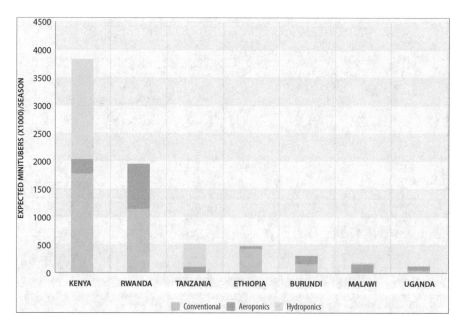

Fig. 13.8 Capacity of facilities used for potato mini-tuber production in seven African countries using three techniques

enterprises (farmers and commercial seed multipliers) and 54% by two projects (CIP and an NGO), which increased to 71% of sales to private enterprises in 2020 out of 586,000 cuttings. Cuttings are penetrating the seed system, and their opportunities for being scaled out are being validated in Kenya.

13.5.2 Impact on Availability of Certified Seed: Results from Kenya

The impact of the new techniques on the availability of certified seed is difficult to support with data as only Kenya has a reliable certification system and documentation of amounts produced. In many countries such as Ethiopia, Rwanda, and Uganda, foundation seed produced by programs is distributed to trained farmers, farmer groups, or seed cooperatives, which are not engaging in certification systems, so production is not officially documented. In Ethiopia, for instance, the semi-formal Quality Declared Seed standards have been introduced for the crop in 2016 as an intermediate quality control system based on local visual inspection; however, the amounts produced are not systematically documented. Therefore, the focus in this section will be on Kenya.

Since 2010 Kenya showed a clear trend of increasing amounts of certified seed. From only 1000 tons in 2005 and a drop to 300 tons in 2008 (due to political unrest) provision of certified seed rose to 2500 tons in 2010. This increase can be solely attributed to the introduction of RMTs in 2000–2009, as well as the further increase to 5600 tons annually in 2015 and 2016 (compare Fig. 13.9). Increases in the following years have been achieved in part due to the local multiplication of imported foundation seed from Europe. However, about 70% of the certified seed has been

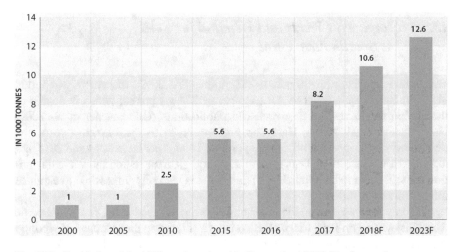

Fig. 13.9 Certified seed (in 1000 tons) produced in Kenya since 2000 (F = forecast)

produced using locally produced EGS (data from KEPHIS, amounts of certified seed in 2019 by producer). The most important local producers of certified seed, Kisima Farms Ltd. and the parastatal Agricultural Development Cooperation (ADC) produce 3400 tons and 2500 tons annually, respectively. Both produce EGS locally. Kisima Farms is producing its EGS through aeroponics and sand hydroponics and purchases rooted apical cuttings (RAC) from Stokman Rozen Ltd. Hence, all of the certified seed at Kisima Farms has been produced from EGS produced with the recently introduced technologies. The EGS production at ADC, however, is still dominated by conventional pot systems, accounting for about 80% of their mini-tuber production. In sum, currently about 4000 tons, or 40% of certified seed, is produced using EGS derived from the introduced technologies.

The impact of the recently introduced RAC technology on certified seed production is difficult to estimate, as most cuttings are directly purchased by farmers and further multiplied on-farm without certification. Nonetheless, the impact of the technology could be even bigger than with previous techniques. Based on the production data from 2019 (see previous paragraph) and assuming a field survival rate of 70% with two further field multiplications before using the seed to produce ware potatoes, the production would be enough to plant about 640 ha. This would be equivalent to 1280 tons of seed which is about 11% of the current provision of certified seed in the country, hence adding a significant amount of high-quality seed to the sector.

Initial results with pilot farms and farmer groups indicate huge yield gains after two rounds of on-farm multiplication. These yield increases are similar to those when using certified seed but with reduced costs. This technology could contribute significantly to improve seed quality at the farm level and nationally. The technology would also be a perfect fit for multiplying true potato seed from hybrid potatoes, a technology that promises to revolutionize potato breeding and seed systems.

13.5.3   Impact on Yields and Demand for Seed at Smallholder Level

Total yield impacts are difficult to determine. FAOSTAT inaccurately reported yields in Kenya from 2008 to 2012 at around 20 tons per ha, which is double of those reported in several peer-reviewed publications (Gildemacher et al. 2009; Harahagazwe et al. 2018b; Haverkort and Struik 2015). In following years, the reported yields dropped significantly to around 14 tons per ha in 2014 and 2015 and to 9 or 10 tons since 2016. Data from other countries in the region show less fluctuation but still seem to be unreliable. Therefore, it is basically impossible to estimate impacts at larger scales.

The only reliable data on the impact of high-quality seed at the farm level come from surveys and on-farm trials. Okello et al. (2017) found that adopters of certified seed (N = 167) had a yield about 2.5 tons per ha higher (11.5 vs. 8.7) than non-adopters (N = 241). The survey however was conducted during a season with low rainfall which might have dampened yields. The adopters did not plant 100% of their fields with certified seed every season but reused the seed for two to three seasons before purchasing new certified seed. This might be a good tactic for balancing investment risk and yield improvements and probably portrays a realistic scenario for potential demand.

Participatory research on 18 farms in Kenya between 2010 and 2011 evaluated the effects of different seed qualities on yield and crop heath. When using recommended management practices, certified seed can double the yield from farm-saved seed (Schulte-Geldermann et al. 2013). However, this yield gain can only be achieved by properly managing plant nutrition and pests and diseases. With suboptimal management, e.g., lower rates of fertilizer and reduced chemical control of late blight, the yield gain was usually 30–75% higher than farm-saved seed. That means that the investment in high-quality seed, which amounts to 1200 to 1500 USD per ha, needs to be protected by additional investments in farm inputs, often difficult for cash-constrained farmers.

Ochieng (2021) indicate that with current farm-gate ware potato prices, if farmers planted only certified seed, yields would have to be about 18 tons per ha to achieve the same profit as with farm-saved seed, which yields on average 8 tons per ha. That suggests that investing in certified seed is risky, which explains the reluctance of many farmers to invest in it.

However, alternative strategies such as purchasing 5% of certified seed, bulking it on land where potatoes haven't been planted for a long time, at close spacing and conducting positive selection allow for replacing the old seed stock with better seed within three seasons. With this strategy, yields were just 10–15% below those obtained with certified seed but at much lower costs. Compared to the use of farm-saved seed, this strategy required a yield increase of only 2.5 tons per ha to cover the additional investment. This was achieved easily at all eight farms involved in the study. The integrated strategies have a profit margin about 10–25% higher than when using certified seed alone (Ochieng 2021).

Farmers need to be capacitated on the best management options, in order to break the vicious cycle of declining yields brought about by using farm-saved seed, while also reducing the high cost and risk of planting only certified seed. A bottleneck is dissemination, as seed producers are usually reluctant or unable to deal with the logistics of distributing small quantities to many individual customers. Organizing farmers into groups or cooperatives to buy larger quantities would be a viable option. This has been documented in a study (110 farmers) in Rwanda comparing access to high-quality seed by individual farmers and by members of cooperatives. The cooperatives were able to purchase certified seed from the Rwanda Agricultural Board and multiplied it one further season before it was distributed to all members. Only about 10% of the individual farmers accessed high-quality seed. This resulted in huge differences in yields between the two groups. About 40% of the cooperative members achieved yields higher than 20 tons per ha, and 90% harvested more than 15 tons per ha. About 80% of individual farmers achieved yields of below 15 tons per ha, and 60% had yields below 10 tons (Uwimana 2020).

Demand for high-quality seed could also be bolstered by distributing it in smaller packages, e.g., in bags of 5 to 25 kg, even though this is below the standard size of 50 kg defined in local seed laws. Smaller seed bags would improve smallholders' access to quality seed (and to validate its quality on their own farms), allowing for the growth of the seed sector and creating business opportunities based on RMTs.

13.5.4   *Contribution to Gender and Social Inclusiveness*

The technologies indirectly contributed to gender and social inclusiveness by reducing the price for certified seed. Mudege and Demo (2016) and Mudege et al. (2020) concluded that improved technologies that are affordable and available in local markets are generally friendly for vulnerable groups, including women. In Kenya, for instance, the price for a 50 kg bag of generation four seed prior to the introduction of the EGS technologies in 2008 has been 3300 or 3500 KSH (33 or 35 USD) and dropped to between 2500 and 3000 KSH (25 and 30 USD) in 2020, hence making it more available to resource-poor farmers.

In particular, rooted apical cuttings are becoming more accessible as they can easily be traded in small quantities of 100–200 cuttings at an affordable price of 10–15 KSH (0.09 to 0.14 USD) per cutting, which can be multiplied on the small parcels typically farmed by women and youths (Fig. 13.10). Investing in rooted cuttings for seed production could be interesting for women and youth as little land is required, and profit margins are high.

Fig. 13.10 Women farmers in Kenya multiplying seed from rooted apical cuttings. (Photo credits Benson Kisinga CIP)

13.6 Conclusion

Rapid multiplication technologies such as aeroponics, rooted apical cuttings, and sand hydroponics save costs and have a place in the seed business in Eastern Africa. RMTs have contributed significantly to the increased supply of certified or high-quality seed in the region over the last decade. However, technologies have to be carefully selected for their situation and for underlying conditions including climate. Highly productive technologies such as aeroponics and rooted apical cuttings place high demands on management and infrastructure while requiring suitable business models that go beyond producing EGS. Both technologies are well suited for businesses that already manage tissue culture or screenhouses, e.g., in the horticultural and floricultural sectors. For these enterprises, both technologies offer viable opportunities to diversify their business and earn an additional stream of income. The less productive sand hydroponic technology reduces costs while slightly improving production over the conventional pot system, contributing to lowering seed production costs. Its relatively ease of management is an opportunity for less skilled and resource-constrained enterprises and public sector institutes to produce EGS. A careful assessment (e.g., a business plan) is required before choosing the best technologies for each situation.

Acknowledgments This research was undertaken as part of, and funded by the CGIAR Research Program on Roots, Tubers, and Bananas (RTB) and supported by the CGIAR Trust Fund contributors. Funding support for this work was provided by BMZ/GIZ, Government of Finland, Irish Aid, USAID.

References

Abdurahman A, Griffin D, Elphinstone J, Struik PC, Schulz S, Schulte-Geldermann E, Sharma K (2017) Molecular characterization of Ralstonia solanacearum strains from Ethiopia and tracing potential source of bacterial wilt disease outbreak in seed potatoes. Plant Pathol 66(5):826–834. https://doi.org/10.1111/ppa.12661

Almekinders CJ, Walsh S, Jacobsen KS, Andrade-Piedra JL, McEwan M, De Haan S, Kumar L, Staver C (eds) (2019a) Why interventions in the seed systems of roots, tubers and bananas crops do not reach their full potential. Food Secur. First Online 23 Jan 2019, 20 p. ISSN 1876-4525

Almekinders CJ, Beumer K, Hauser M, Misiko M, Gatto M, Nkurumwa AO, Erenstein O (2019b) Understanding the relations between farmers' seed demand and research methods: the challenge to do better. Outlook Agric 48(1):16–21

Andersen KF, Buddenhagen CE, Rachkara P, Gibson R, Kalule S, Phillips D, Garrett KA (2019) Modeling epidemics in seed systems and landscapes to guide management strategies: the case of sweet potato in Northern Uganda. Phytopathology 109(9):1519–1532. https://doi.org/10.1094/PHYTO-03-18-0072-R

Andrade-Piedra JL, Kromann P, Otazu V (2019) Manual for seed potato production using aeroponics. In: Ten years of experience in Colombia, Ecuador and Peru. International Potato Center (CIP), National Institute of Agricultural and Livestock Research (INIAP), Colombian Agricultural Research Corporation (CORPOICA), Quito, Ecuador. ISBN 978–92–9060-504-1, 267 p

Atieno E, Schulte-Geldermann E (2016) Public-private partnerships to multiply seed potato in Kenya. Case studies of roots, tubers and bananas seed systems, 2016-3

Bryan JE, Jackson MT, Malendez NG (1981) Rapid multiplication techniques for potatoes. International Potato Center, Lima, Peru, p 20

Chang DC, Park CS, Kim SY, Kim SJ, Lee YB (2008) Physiological growth responses by nutrient interruption in aeroponically grown potatoes. Am J Potato Res 85(5):315

Demo P, Lemaga B, Kakuhenzire R, Schulz S, Borus D, Barker I, Woldegiorgis G, Parter ML, Schulte-Geldermann E (2015) Strategies to improve seed potato quality and supply in sub-Saharan Africa: experience from interventions in five countries. In: Potato and sweetpotato in Africa: transforming the value chains for food and nutrition security. CABI, Wallingford, pp 155–167

Dodds JH (1988) Tissue culture technology: practical application of sophisticated methods. Am Potato J 65:167–180. https://doi.org/10.1007/BF02854450

FAO (2010) Quality declared planning material. Protocols and standards for vegetatively propagated crops. In: FAO plant production and protection paper 195, Rome, p 126

FAOSTAT (2020) Food and Agriculture Organization of the United Nations. FAOSTAT Statistical Database. FAO, Rome: http://www.fao.org/faostat/en/#data/QC. Accessed December 2020

Farran I, Mingo-Castel AM (2006) Potato mini-tuber production using aeroponics: effect of plant density and harvesting intervals. Am J Potato Res 83(1):47–53

Ferrari L, Fromm I, Jenny K, Muhire A, Scheidegger U (2017) Formal and informal seed potato supply systems analysis in Rwanda. In: Tropentag 2017: future agriculture: social-ecological transitions and bio-cultural shifts, Bonn. September 20–22, 2017

Forbes GA, Charkowski A, Andrade-Piedra J, Parker ML, Schulte-Geldermann E (2020) Potato seed systems. In: Campos H, Ortiz O (eds) The potato crop. Springer, Cham, pp 431–447

Gildemacher PR, Kaguongo W, Ortiz O, Tesfaye A, Woldegiorgis G, Wagoire WW, KakuhenzireR KPM, Nyongesa M, Struik PC, Leeuwis C (2009) Improving potato production in Kenya, Uganda and Ethiopia: a system diagnosis. Potato Res 52(2):173–205

Harahagazwe D, Andrade-Piedra J, Parker M, Schulte-Geldermann E (2018a). Current situation of rapid multiplication techniques for early generation seed potato production in Sub-Saharan Africa International Potato Center, CGIAR Research Program on Roots, Tubers and Bananas (RTB) RTB Working paper no. 2018-1

Harahagazwe D, Condori B, Barreda C, Bararyenya A, Byarugaba AA, Kude DA, Lung'aho C, Martinho C, Mbiri D, Nasona B, Ochieng B, Onditi J, Randrianaivoarivony JM, Tankou CM, Worku A, Schulte-Geldermann E, Mares V, de Mendiburu F, Quiroz RQ (2018b) How big is the potato (Solanum tuberosum L.) yield gap in Sub-Saharan Africa and why? A participatory approach. Open. Agriculture 3(1):180–189

Haverkort AJ, Struik PC (2015) Yield levels of potato crops: recent achievements and future prospects. Field Crop Res 182:76–85

Kaguongo W, Maingi G, Barker I, Nganga N, Guenthner J (2014) The value of seed potatoes from four systems in Kenya. Am J Potato Res 91:109–118. https://doi.org/10.1007/s12230-013-9342-z

Kakuhenzire R, Temu N, Kwigizile O, Valkonen J, Shayo T, Msemwa J (2015) Seed potato development project in Tanzania (2012–2015). Final Project Report. International Potato Center, Uyole, Tanzania, p 44

Kakuhenzire R, Tibanyendera D, Kashaija IN, Lemaga B, Kimoone G, Kesiime VE, Otazu V, Barker I (2017) Improving minituber production from tissue-cultured potato plantlets with aeroponic technology in Uganda. Int J Agric Environ Res 3:3948–3964

Labarta RA (2013) Possibilities and opportunities for enhancing the availability of high-quality seed potato in Ethiopia: lessons from the successful 3G project in Kenya. Proceedings of the National Workshop on Seed Potato Tuber Production and Dissemination, 12-14 March 2012, Bahir Dar, Ethiopia. ISBN: 978-99944-53-87-x. pp. 21–34

Landeo J, Barker I, Otazú V (2009) Novel approaches to promote and diffuse new Potato varieties in Kenya. 15th Triennial of the Symposium of the International Society for Tropical Root Crops. Proceedings

Lung'aho C, Mbiyu M, Nyongesa M, Otieno S, Onditi J, Schulte-Geldermann E (2013) Are dormancy management and physiological age the Achilles' heel of aeroponic minituber production in seed potato value chains in SSA? In: Transforming potato and sweetpotato value chains for food and nutrition security. 9. Triennial Congress of the African Potato Association, Naivasha. 30 Jun-9 Jul 2013

Mateus-Rodríguez JR, de Haan S, Andrade-Piedra JL, Maldonado L, Hareau G, Barker I, Chuquillanqui C, Otazú V, Frisancho R, Bastos C, Pereira AS, Medeiros CA, Montesdeoca F, Benítez J (2013) Technical and economic analysis of aeroponics and other systems for potato mini-tuber production in Latin America. Am J Potato Res 90(4):357–368

Mbiri D, Schulte-Geldermann E, Otazu V, Kakuhenzire R, Demo P, Schulz S (2015) An alternative technology for pre-basic seed potato production–sand hydroponics. Potato and Sweetpotato in Africa: Transforming the Value Chains for Food and Nutrition Security. p. 249

Mburu H, Cortada L, Haukeland S, Ronno W, Nyongesa M, Kinyua Z, Bargul J, Coyne D (2020) Potato cyst nematodes: a new threat to potato production in East Africa. Front Plant Sci 11:670. https://doi.org/10.3389/fpls.2020.00670

Mudege N, Demo P (2016) Seed potato in Malawi: Not enough to go around. In: Andrade-Piedra J, Bentley JW, Almekinders C, Jacobsen K, Walsh S, Thiele G (eds) Case studies of roots, tubers and bananas seed systems. CGIAR Research Program on Roots, Tubers and Bananas (RTB), Lima, RTB Working Paper No. 2016-3. ISSN 2309-6586., pp 151–167

Mudege NN, Escobar SS, Polar V (2020) Gender topics on potato research and development. Potato Crop 475

Muro JV, Goñi JL, Lamsfus C (1997) Comparison of hydroponic culture and culture in a peat/sand mixture and the influence of nutrient solution and plant density on seed potato yields. Potato Res 40:431–438

Muthoni J, Kabira J (2014) Multiplication of seed potatoes in a conventional potato breeding programme: a case of Kenya's national potato programme. Aust J Crop Sci 8(8):1195

Muthoni J, Mbiyu M, Lung'aho C, Otieno S, Pwaipwai P (2017) Performance of two potato cultivars derived from in vitro plantlets, mini-tubers and stem cuttings using aeroponics technique. Int J Horticult 7(27):246–249

Naik PS, Karihaloo JL (2007) Micropropagation for production of quality potato seed in Asia-Pacific. Asia-Pacific Consortium on Agricultural Biotechnology, New Delhi, p 54

National Potato Council of Kenya (2018) Background and seed potato challenges. In: Kaguongo W (ed) Seed potato stakeholders' meeting. PanAfric Hotel, Nairobi

Negash K (2014) Rapid multiplication techniques (RMTs): a tool for the production of quality seed potato (Solanum tuberosum L.) in Ethiopia. Asian. J Crop Sci 6(3):176–185

Nyawade S, Parker M (2020) Planting and management of potato rooted apical cuttings: a field guide. Extension brief

Ochieng BO (2021) Bacterial wilt and virus disease management in potato production systems in Kenya Dissertation for the acquisition of the degree Doktor der Agrarwissenschaften (Dr. agr. sc.). Submitted to Faculty of Organic Agricultural Sciences at University of Kassel, January 2021

Okello JJ, Zhou Y, Kwikiriza N, Ogutu S, Barker I, Schulte-Geldermann E, Atieno E, Ahmed JT (2017) Productivity and food security effects of using of certified seed potato: the case of Kenya's potato farmers. Agricult Food Secur 6(1):1–9

Okello JJ, Lagerkvist CJ, Kakuhenzire R, Parker M, Schulte-Geldermann E (2018) Combining means-end chain analysis and goal-priming to analyze Tanzanian farmers' motivations to invest in quality seed of new potato varieties. Br Food J

Otazú V (2008) Steam sterilization of greenhouse substrates (in Spanish). In: Alternativas al uso del bromuro de metilo en la producción de semilla de papa de calidad. International Potato Center (CIP). Documento de trabajo 2007–2, Lima, pp 15–25

Otazú V (2010) Manual on quality seed potato production using aeroponics. International Potato Center, p 44

Otazú V, Chuquillanqui C (2008) Quality seed potato production by aeroponics. (in Spanish). In: Alternativas al uso del bromuro de metilo en la producción de semilla de papa de calidad. International Potato Center (CIP). Documento de trabajo 2007–2, Lima, pp 35–45

Parker ML (2019) Production of apical cuttings of potato. International Potato Center, Lima. https://doi.org/10.4160/9789290605195

Parker ML, Low JW, Andrade M, Schulte-Geldermann E, Andrade-Piedra J (2019) Climate change and seed systems of roots, tubers and bananas: the cases of potato in Kenya and Sweetpotato in Mozambique. In: Rosenstock T, Nowak A, Girvetz E (eds) The climate-smart agriculture papers. Springer, Cham, pp 99–111. https://doi.org/10.1007/978-3-319-92798-5_9

Ritter E, Angulo B, Riga P, Herran C, Relloso J, San Jose M (2001) Comparison of hydroponic and aeroponic cultivation systems for the production of potato minitubers. Potato Res 44(2):127–135

Sartas M, Schut M, Proietti C, Thiele G, Leeuwis C (2020) Scaling readiness: science and practice of an approach to enhance impact of research for development. Agric Syst 183:102874

Schulte-Geldermann E (2013) Tackling low potato yields in Eastern Africa: an overview of constraints and potential strategies. Proceedings. National Workshop on Seed Potato Tuber Production and Dissemination. Bahir Dar (Ethiopia), 12-14 Mar 2012, (Ethiopia), Ethiopian Institute of Agricultural Research (EIAR), Amhara Regional Agricultural Research Institute (ARARI), International Potato Center ISBN 978-99944-53-87-x. pp 72–80.

Schulte-Geldermann E, Wachira G, Ochieng B, Barker I (2013) Effect of field multiplication generation on seed potato quality in Kenya. Proceedings of the National Workshop on seed potato tuber production and dissemination, 12-14 March 2012, Bahir Dar, Ethiopia. ISBN: 978-99944-53-87-x, pp 81–90

Sharma K, Atieno EO (2020) Dissemination of climate-smart consumer-demanded potato varieties in Kenya. Second Annual Progress Report to Syngenta Foundation for Sustainable Agriculture. January 1st–December 31st 2019, Lima

Singh S, Singh V, Singh B, Panday SK (2010) Aeroponics for potato seed production. In Council Agric Res Newsletter 16:1–2

Soffer H, Burger DW (1988) Studies on plant propagation using the aero-hydroponic method. In: Symposium on High Technology in Protected Cultivation 230. pp. 261–270

Struik PC, Wiersema SG (1999) Seed potato technology. Wageningen Academic Publishers, Wageningen, p 383. https://doi.org/10.3920/978-90-8686-759-2

Tessema L, Dagne Z (2018) Aeroponics and sand hydroponics: alternative technologies for pre-basic seed potato production in Ethiopia. Open Agric 3(1):444–450

Thiele G, Labarta R, Schulte-Geldermann E, Harrison G (2011) Roadmap for investment in the seed potato value chain in Eastern Africa. Lima (Peru), International Potato Center, p 27

Thiele G, Dufour D, Vernier P, Mwanga RO, Parker ML, Schulte Geldermann E, Teeken B, Wossen T, Gotor E, Kikulwe E, Tufan HA, Sinelle S, Kouakou AM, Friedmann M, Polar V, Hershey C (2020) A review of varietal change in roots, tubers and bananas: consumer preferences and other drivers of adoption and implications for breeding. Int J Food Sci Technol 56:1076–1092. https://doi.org/10.1111/ijfs.14684

Thomas-Sharma S, Abdurahman A, Ali S, Andrade-Piedra JL, Bao S, Charkowski AO, Crook D, Kadian M, Kromann P, Struik PC, Torrance L (2016) Seed degeneration in potato: the need for an integrated seed health strategy to mitigate the problem in developing countries. Plant Pathol 65(1):3–16

Thomas-Sharma S, Andrade-Piedra J, Carvajal Yepes M, Hernandez Nopsa JF, Jeger MJ, Jones RAC, Kromann P, Legg JP, Yuen J, Forbes GA, Garret KA (2017) A risk assessment framework for seed degeneration: informing an integrated seed health strategy for vegetatively propagated crops. Phytopathology 107(10):1123–1135

Tibbitts TW, Cao W (1994) Solid matrix and liquid culture procedures for growth of potatoes. Adv Space Res 14:427–433. https://doi.org/10.1016/0273-1177(94)90332-8

Tsegaw T (2006) Response of paclobutrazol and manipulation of reproductive growth under tropical conditions. PhD Thesis University of Pretoria

Uwimana J (2020) Evaluation of the quality and availability of potato seed to respond to the dynamics of farmer's needs and production increase in communities of Musanze district, Rwanda. Master Thesis at Bingen University of Applied Sciences, Bingen am Rhein, Germany, March 2020

Van Ittersum MK, Scholte K (1993) Shortening dormancy of potatoes by haulm application of gibberellic acid and storage temperatures regimes. Am Potato J 70:7–9

Van Minh T, Van Uyen N, Vander Zaag P (1990) Potato (*Solanum spp.*) production using apical cuttings and tuberlets under three contrasting environments of Vietnam. Am Potato J 67(11):779. https://doi.org/10.1007/BF03044529

Vander Zaag P, Escobar V (1990) Rapid multiplication of potatoes in the warm tropics: rooting and establishment of cuttings. Potato Res 33:13–21

Wiersema SG, Cabello R, Tovar P, Dodds JH (1987) Rapid seed multiplication by planting into beds micro tubers and in vitro plants. Potato Res 30(1):117–120

Chapter 14
Transforming Yam Seed Systems in West Africa

Norbert Maroya, Morufat Balogun ⓘ**, Beatrice Aighewi** ⓘ**,**
Djana B. Mignouna ⓘ**, P. Lava Kumar** ⓘ**, and Robert Asiedu** ⓘ

Abstract The availability of clean planting materials and functional seed regulatory systems is indispensable for fostering a sustainable seed yam system. The Yam Improvement for Income and Food Security in West Africa (YIIFSWA) project of the International Institute of Tropical Agriculture (IITA) developed the capacity of National Agricultural Research Institutes (NARIs) in their use of standardized Temporary Immersion Bioreactor (TIB) and Vivipak (VP) systems for high-ratio propagation and post-flask handling of yam breeder seed plantlets. Foundation seed was enhanced by supporting five private seed companies in Nigeria and three in Ghana. They were equipped with aeroponic and hydroponic technologies for foundation seed tuber production using single-node vine seedlings. For certified seed, seed yam out-growers were trained in good agronomic practices and entrepreneurship for certified seed tuber production using the adaptive yam minisett technique (AYMT). New certification standards were established for various classes of seed produced using different propagation methods and quality assurance procedures in Ghana and Nigeria. The capacity of the national regulatory organizations in both countries was enhanced to implement seed quality control and certification. Increased public sensitization and advocacy were done to raise awareness among

N. Maroya (✉) · P. L. Kumar · R. Asiedu
International Institute of Tropical Agriculture (IITA), Ibadan, Nigeria
e-mail: N.Maroya@cgiar.org; L.Kumar@cgiar.org; r.asiedu@cgiar.org

M. Balogun
International Institute of Tropical Agriculture (IITA), Ibadan, Nigeria

University of Ibadan, Department of Crop Protection and Environmental Biology,
Ibadan, Nigeria
e-mail: M.Balogun@cgiar.org

B. Aighewi
International Institute of Tropical Agriculture (IITA), Abuja, Nigeria
e-mail: B.Aighewi@cgiar.org

D. B. Mignouna
International Institute of Tropical Agriculture (IITA), Cotonou, Benin
e-mail: D.Mignouna@cgiar.org

© The Author(s) 2022 421
G. Thiele et al. (eds.), *Root, Tuber and Banana Food System Innovations*,
https://doi.org/10.1007/978-3-030-92022-7_14

relevant stakeholders to enhance the uptake of the seed propagation technologies and ensure a smooth interaction between the public and private sectors. This chapter summarizes the accomplishments of YIIFSWA in Ghana and Nigeria and the spillover impact on the yam belt of West Africa and beyond. The key lessons could inform the design and implementation of more effective seed projects, especially for vegetatively propagated crops.

14.1 Introduction

Yam (*Dioscorea* spp.) is a clonally propagated, high-value crop, cultivated for its underground edible tubers and planting materials. It is an essential source of income and food for millions of smallholder farmers, processors, and consumers in West Africa. Yams have been cultivated since 11,000 BC in West Africa (Coursey 1976). In the region, yam cultivation covers over 8.1 million hectares (ha), with a total annual production of over 67 million tonnes[1] (FAO 2020). More than 92% of the world's yams are produced in five countries (Nigeria, Ghana, Côte d'Ivoire, Benin, and Togo) in the yam belt of West Africa. Ghana and Nigeria alone account for 77%. The average per capita consumption of yam in major producing countries ranges from 193 kcal a day in Togo to 502 kcal per day in Côte d'Ivoire. The crop also contributes to much more protein to the region's diet than the more widely grown cassava and even more than meat protein (FAO 2021). Yam is also a culturally important crop in the region, used in many key ceremonies such as yam festivals, weddings, chieftaincy ceremonies, and sacrifices to the gods (Nweke 2016).

Although yam production is only 38% of cassava, its total crop value is slightly higher, making it the most valuable food crop in Africa (FAO 2021). In terms of the value of production (income to farmers), yam is far ahead of the other five main food commodities (maize, rice, cassava, sorghum, and millet) in Nigeria (Table 14.1).

Table 14.1 Comparison of average production value of five food crops in Nigeria from 1995 to 2016

Crops	Area × 1000 (ha)	Yield (T/ ha)	Prod × 1000 tons	Price US$/ton	Prod value Million US$	Percentage to yam
Yam	3328	9949	31,898	398	12,694	100%
Cassava	4100	10,339	41,086	157	6462	51%
Maize	4425	1572	6918	334	2311	18%
Millet	4050	1163	4976	297	1480	12%
Sorghum	6303	1174	7401	301	2226	18%
Rice	2387	1651	3976	364	1448	11%

Source: FAOSTAT 2018

---

[1] 1 tonne (t) = 1000 kgs

The annual production value of yam is over US$12.7 billion in Nigeria and US$ 1.4 billion in Ghana. If quality seed of improved varieties were used with a 30% yield increase, the production value could increase to US$17.9 billion in Nigeria and US$1.9 billion in Ghana. Those increases would benefit primarily millions of small-holder farmers for whom yam is a cash crop. Of the yam produced in both countries, about 30% is allocated for seeds, 42% is sold at market, and the remaining 28% consumed at home (Mignouna et al. 2015).

Yam is the only vegetatively propagated crop in sub-Saharan Africa (SSA) with a regular cash-based seed system, and farmers expect that they will need to replace at least some of their seed each year (Nweke 2016). In Igalaland in Kogi, Nigeria, farmers buy seed yam each growing season rather than producing their own, seeking quality judged by physical inspection for symptoms of damage by pests and diseases (Nweke 2016).

The unavailability of quality seed is largely due to the traditional methods of seed yam production in West Africa. The first traditional method – milking – harvests physiologically immature tubers at between 60% and 70% of the growing season without destroying the feeding roots system. This early harvest of ware yam is consumed and sold at market (Aighewi et al. 2015). Before total senescence, the parental plant redevelops small new yam tubers used as seeds for the following planting season. The second traditional method of seed production uses small whole tubers from varieties that produce multiple tubers per stand or by sorting small tubers from a ware crop. This method is risky if the tubers are small and infected with pests and diseases in the field (Nweke 2016). The third traditional method – setts – involves cutting mature ware tubers into small portions (100–250 g) (Aighewi et al. 2002). These three traditional methods are slow, with a multiplication ratio of about 1:6, compared to some cereals which multiply at 1:200 (Mbanaso et al. 2011). As such, these methods cannot supply seed in sufficient quantity and quality needed for real growth of the yam sector throughout the yam belt of West Africa.

The prioritized constraints of yam production formed the basis for the interventions of the first phase of the Yam Improvement for Income and Food Security in West Africa (YIIFSWA) project. This work was funded by the Bill & Melinda Gates Foundation and managed by the International Institute of Tropical Agriculture (IITA) from 2011 to 2016 in Ghana and Nigeria and is continuing as a second phase through the end of 2021. The Project seeks to develop and establish a functional, commercial seed yam seed system in the two countries to benefit smallholder farmers through timely and affordable access to high-quality seed yam tubers of improved varieties.

14.2 Technologies for Breeder and Foundation Seed Yam Production

14.2.1 Tissue Culture and Temporary Immersion Bioreactor System

The YIIFSWA project developed and standardized tissue culture-based heat therapy combined with meristem tissue culture for cleaning yam of viruses at a 73% success rate for yam mosaic virus (YMV) (Balogun et al. 2017a). YIIFSWA also developed the Temporary Immersion Bioreactor System (TIBS) to scale up the propagation of breeder plantlets from which foundation and certified seed are generated (Balogun et al. 2017b). The cleaned plantlets constitute stocks for the rapid multiplication of superior varieties, ensuring that the virus is not passed on to subsequent generations. 25 genotypes of yam, including improved and local varieties, were cleaned of diseases using these approaches.

Plantlets from TIBS are of higher quality than those from conventional tissue culture (CTC), as they are more vigorous and resilient to post-flask acclimatization. Due to more efficient process control, large batches are handled more easily for scale-up propagation with lower risks of mix-ups. The propagation ratio in TIBS was five to six per plantlet every 8–10 weeks (Balogun et al. 2017b) compared to three to four every 12–16 weeks with CTC. In addition, the rate of subculturing in TIBS was 100 cuttings per person hour, double that of CTC, reducing the cost of labor as well, and the use of liquid without agar/geltrite reduces medium cost by 50%. All these advantages curried favor for the establishment of a laboratory for TIBS at NRCRI in Umudike, Nigeria, in 2019 (Fig. 14.1).

Fig. 14.1 The NRCRI TIBS laboratory at Umudike, Nigeria

For post-tissue culture operations, for acclimation to the outside environment, the plantlets from the TIBS are hardened in 50:50 carbonized rice husks in topsoil using perforated Vivipak that allows for adequate ventilation (Balogun et al. 2017b).

Technologies for breeder seed production were transferred to the National Center for Genetic Resources and Biotechnology (NACGRAB) and the National Root Crops Research Institute (NRCRI) in Nigeria and the Crops Research Institute (CRI) and Savanna Agricultural Research Institute (SARI) in Ghana through the provision of equipment and supplies for backup electrical power, TIBS, post-flask handling and documentation, and initial clean stock of planting materials. Solar/inverter systems were installed at NARIs to mitigate interruptions in power supply in operating the TIBS. The National Agricultural Research Institutes (NARIs) are now providing clean breeder planting materials to private seed companies for foundation seed production.

Each TIBS can produce at least 50 plantlets every 2 months, so up to 76,500 stock plantlets per year can be produced in the laboratory. When plantlets are introduced into aeroponic or hydroponic systems, combined with vine cutting, production is increased at least 100 times, up to 7.65 million (i.e., 6.12 million breeder planting materials with a buffer of 20% reserved against possible losses). Following installation of the TIBS at NARIs in 2019, 11,132 plantlets were produced from January through April 2020 (Table 14.2). This figure was projected to reach 100,188 by the end of 2020 at a multiplication ratio of three per cycle of 4 months. NARIs have also distributed clean seed yam among stakeholders and are better positioned to prime the seed value chain (Table 14.3).

14.2.2 The Aeroponic System for Seed Yam Production

Laboratory production of seed is continuous (Balogun et al. 2017b), while farmers' demand for seed yam is seasonal. Thus, aeroponic and hydroponic systems were developed for time- and cost-effective agronomic and quality-assured production (Table 14.4). The single-node-derived vine seedlings are generated from aeroponics and hydroponics after 3–4 months of plantlet growth. After 30–45 days in the nursery, the seedlings are transplanted to the field, maintained for 6 months, and harvested during December to February when most farmers harvest their ware yams and need seed tubers to replant (or expand) their harvested areas. Plantlets from TIBS are planted with aeroponics or hydroponics in the screenhouse, thus circumventing the use of soil and restricted seasonal production cycles (due to tuber dormancy and access to rainwater) because vine seedlings and tuber production in the screenhouse is continuous. Up to 300 vines were cut per plantlet of TIBS grown aeroponically (Maroya et al. 2014b, 2017), while the drip system hydroponics saves on electricity (Balogun et al. 2018, 2020). Up to 100 single-node vine cuttings were made per TIBS plantlet using hydroponics after 8 weeks of growth, with 95% rooting success followed by field planting. The seed tubers produced ranged from 5 g to 220 g per plant after 5 months of hydroponic growth.

Table 14.2 Early generation seed production by NARIs from January to April 2020

NARIs	Products	Quantity
NACGRAB	Plantlets in TIBS	657
	Plantlets in conventional tissue culture vessels	2307
	Directly hardened plants from TIBS or conventional tissue culture	401
	Vine cutting seedlings from hardened in vitro plants in trays	1687
	Subtotal	*5052*
NRCRI	Plantlets in TIBS	1850
	Plantlets in conventional tissue culture vessels	412
	Directly hardened plants from TIBS or conventional tissue culture	2025
	Subtotal	*4287*
CRI	Plantlets in TIBS	8535
	Plantlets in conventional tissue culture vessels	3025
	Directly hardened plants from TIBS or conventional tissue culture	9196
	Subtotal	*20,756*
SARI	Plantlets in TIBS	90
	Vine cutting seedlings from hardened in vitro plants in trays	3846
	Micro-tubers from hardened in vitro plants or vine cuttings	4199
	Plantlets hardened in Vivipak	156
	TIBS plantlets through Vivipak to pots	120
	Single nodes in pots	7318
	Subtotal	*15,729*
Grand total		*45,824*

Aeroponic yam propagation was started under the YIIFSWA project in January 2013 at IITA in Ibadan, Nigeria (Maroya et al. 2014a, b). Genotypes of both *D. rotundata* and *D. alata* were successfully propagated with aeroponics using both pre-rooted and fresh vine cuttings. Yam mini-tubers harvested from aeroponics varied from 0.2 g to 110 g, depending on the genotype, harvest age, and the composition of the nutrient solution (Maroya et al. 2014c).

Three types of planting materials are generated with aeroponics: mini-tubers harvested underneath the aeroponic boxes, the aerial bulbils on both water yam and white yam varieties, and one-node cuttings from vines. All these planting materials had a propagation rate of over 90%.

The single-node vine cuttings from aeroponic plants performed the best with an average of 200–300 cuttings per plant in 4–6 months. A *Manual for Clean Foundation Seed Yam Production Using Aeroponics System* is available to private seed companies (Maroya et al. 2017).

Table 14.3 Distribution of early generation planting materials of yam among public and private institutions in Ghana and Nigeria in 2019

Institution receivers		Distributors			
		IITA	CRI	NACGRAB	TOTAL
Public	NRCRI	2170		1000	3170
	SARI	1192	980		2172
	CRI	900			900
	NACGRAB	127			127
	UI	250			250
	Subtotal	4639	980	1000	6619
Private	PS NUTRAC	400		1200	1600
	BIOCROPS	150			150
	Hikma Farms		592		592
	Fosuah Food		588		588
	Strategic seeds	450			450
	GoSeed	856			856
	Iribov		420		420
	Others		2492	1000	3492
	Subtotal	1856	4092	2200	8148
Grand Total		*6495*	*5072*	*3200*	*14,767*

NB. *UI* University of Ibadan, *PS NUTRAC* PS Nutraceuticals International Ltd., *BIOCROPS* Biocrops Biotechnology Company Ltd.

Table 14.4 Model cropping calendar for seed and ware yam production showing non-seasonality in production based on novel high-ratio propagation technologies

Jan		Feb	March	April	May		June	July	Aug	Sep	Oct	Nov	Dec
Generation of clean stock plantlets in laboratory													
Scale-up propagation of breeder stock plants in TIBS in the laboratory													
Post-flask acclimatization of breeder plants in the screenhouse													
Production of clean breeder mother plants in aeroponics and hydroponics in screenhouse													
Single-node vine seedling production from clean aeroponic/ hydroponic breeder mother plants in nurseries					Field planting of breeder 1 vine seedlings under rain-fed conditions for foundation 1 seed tuber production								
Single-node vine breeder 2 tuber production from clean aeroponic/hydroponic breeder 1 mother plants in nurseries													
Dormancy of breeder and foundation seed tubers					Field planting of breeder and foundation seed tubers under rain-fed conditions for foundation and certified seed production, respectively								
Dormancy of foundation and certified seed tubers					Field planting of foundation and certified seed tubers under rain-fed conditions								

In 2017, YIIFSWA-II project funded five selected seed companies with USD 30,000 each, half of the cost to build a screenhouse and an aeroponic system. Unfortunately, contrary to the letter of agreement that provided construction specifications, the seed companies used inferior materials to construct their screenhouse and other cost-cutting changes contrary to the guidelines. Consequently, plant mortality was very high, and the plants could not grow big enough to allow vines to be cut.

These undesired alterations also led to overheating in the screenhouse and in the nutrient solution tank. To remedy this issue, the project staff developed an air-conditioning cooling system that kept the nutrient temperature between 22 °C and 26 °C. All the materials needed for the nutrient tank cooling system can be sourced locally in Ghana and Nigeria. Eventually, the seed companies asked IITA to help them build a standard aeroponic unit with a 10 KVA solar energy panel (Fig. 14.2). All the materials for constructing this unit, including the solar panels, were purchased locally.

By 2020, all the private seed companies both in Ghana and Nigeria were producing first generation of foundation (FS1) seed and second generation of foundation (FS2) seed yam tubers, simultaneously using single-node vine seedlings under irrigation for FS1 and adaptive yam minisett technique (AYMT-see Sect. 14.3.2) under rain-fed conditions for FS2. Generally, the seed companies used out-growers to produce the FS2 seed yam tuber. These companies would supply certified seed entrepreneurs, out-growers, some NGOs, and local governments with tubers. NGOs and local governments typically requested more seeds than could be supplied.

The major persisting challenge for the aeroponic screenhouse is the heat. Temperatures can reach 42 °C and above during the warm season of January to March. The project used two approaches to keep screenhouses cool enough to avoid damaging the plants in AS:

1. An automated screenhouse cooling system with showers and misters (Fig. 14.3). The system kicks in when the temperature reaches 32 °C and stops when it decreases to 28 °C.
2. Adjusting the dimensions of the screenhouse to 5 m high, 8 m wide, and 30 m long and providing a double roof to facilitate air circulation.

Fig. 14.2 The solar panels (left) and the aeroponic unit with cooling powerhouse for Da-Allgreen Seeds (right)

Fig. 14.3 An aeroponic screenhouse cooled by showering and misting systems

14.2.3 Potential of Semi-autotrophic Hydroponics

Semi-autotrophic hydroponics (SAH) is a robust, low-cost technique for efficient and rapid multiplication of clonally propagated crops (see Chap. 15 in this book on EGS in cassava). SAH is a high-ratio multiplication of true-to-type, virus-free plants of tissue culture-derived material. SAH was originally developed for potato multiplication by a company in Argentina called SAHTecno LLC. SAHTecno licensed the technology to IITA and IITA modified it to suit propagation of cassava through the Building an Economically Sustainable Integrated Cassava Seed System in Nigeria (BASICS) project and later with yam through the Yam Breeding Program and AfricaYam Project. The BASICS project helped design and establish commercial early generation seed (EGS) enterprises known as GoSeed at IITA and Umudike Seeds at NRCRI.

For yam, the SAH facility has the potential to complement the other propagation technologies. The current SAH lab at IITA-Ibadan can produce 600,000–720,000 seedlings per year (50,000–60,000 per month) within a laboratory space of 39.5 m^2 (Fig. 14.4). The technology can generate one million plants per year within 50 m^2, making it suitable for commercial seed production and enhanced multiplication in breeding programs. Apart from the fast multiplication rate and low production cost, SAH offers less than 2% plant loss with no contamination. Tissue culture (TC) plantlets introduced as single nodal cuttings (SNC) into SAH boxes develop new shoots that are ready for cutting after 30 days. The mother plants then regenerate new shoots in 14 days or fewer. By conservative estimate, a rack of 40 test tube plantlets can plant 10 SAH boxes, which translates into 150 boxes within 120 days. Requests for SAH seedlings have been on a steady increase from February to August 2019 due to the attractive look of the SAH yam plantlets. So far, the SAH lab had generated USD 9965 from seedlings sold in 2019 (Table 14.5).

Fig. 14.4 Staff at the SAH laboratory at IITA-Ibadan manage the production of SAH plantlets under controlled conditions

Table 14.5 SAH yam plantlets distribution from IITA-Ibadan lab to end users between February and August 2019

SAH plantlet buyers	Number of plantlets purchased or granted	Period of purchase	Purpose of purchase	Amount paid in Naira (N)*
Da All-Green Seeds Ltd.	4000	Feb–July 2019	Seed production	100,000
PSN International Ltd.	70,250	Feb–-Aug 2019	Seed production	3,000,000
Lecturer UI	2250	Jul–Aug 2019	Seed production	67,500
Ecoprime	5000	Jul 2019	Seed production	125,000
ART	30,000	Aug–Sept 2019	Seed production	270,000
Internal IITA various uses	39,180	Feb–Aug 2019	Trials and demos	Free
GoSeed/Ecoprime	18,000	Jul–Aug 2019	Out-grower tests	Free
Total	168,680	Feb–Sept 2019	Various uses	3,562,500

Source: SAH lab record; * 1$ = 360 N

14.3   Seed Yam Actors Strengthened on Seed Production and Commercial Seed Marketing

14.3.1   Training in Production of Early Generation Seed Yam

The training on breeder and foundation seed yam production organized by IITA included the breeder and foundation seed modules and the day-to-day management of aeroponics (Maroya et al. 2017) and TIBS (Balogun et al. 2017b). The modules included step-by-step procedures for establishing pathogen-free cultures, scale-up propagation in TIBS, and post-flask management of plantlets. 49 trainees from NARIs were trained on breeder seed production between 2017 and 2020, and another 49 trainees from private sector on foundation seed.

The Human Resource Capacity Development on breeder and foundation seed yam production include the breeder seed module and the day-to-day aeroponic system management. The breeder seed training module included:

- Step-by-step procedures of establishment of pathogen-free cultures: virus elimination using heat therapy, practical aspects of virus elimination using thermotherapy, medium preparation, explant collection and disinfection, culture initiation, meristem excision, thermotherapy, and plantlet regeneration
- Step-by-step procedure of scale-up propagation in SETIS Temporary Immersion Bioreactors: indexing for endophytes, setting up of the SETIS bioreactor unit, medium preparation, autoclaving vessels and instruments, introduction of explants to TIBs in the laminar flow hood, assembly on the shelf, changing medium in bioreactor cultures, and cleaning bioreactor vessels
- Post-flask management of plantlets: practical aspects on hardening, prehardening medium (PHM) preparation in TIBS, subculturing plantlets into PHM, preparing acclimatization chamber and transfer of plantlets, and identifying hardened plantlets
- Documentation, including creation of spreadsheets, use of bar codes, and tracking the cost of production

In partnership with Context Global Development, business advisory support was provided to NARIs in 2019 to develop business plans for breeder seed yam production using TIBS. These were based on an institution-specific, cost-tracking templates for financial and operating models.

14.3.2   Private Seed Companies Strengthened for Seed Business and Commercial Seed Marketing

The private seed companies were selected through a due diligence process and were supported to develop business plans based on the project promoted seed production technologies. After review, eight businesses with the soundest plans were engaged in the YIIFSWA-II project. The business plan format used for the private seed companies was an FAO document titled *Small-Scale Seed Enterprise: Start-up and Management* (2007).

The NRCRI and IITA developed an effective and affordable propagation technique, the yam minisett technique (YMT), for yam growers to produce their seed yam (IITA 1985). Using this technique, the multiplication ratio increases from 1:5 in traditional methods to 1:30 (Orkwor et al. 2000). A refined model, the adaptive yam minisett technique (AYMT) (Odu et al. 2016), was introduced under YIIFSWA project to strengthen quantity and quality assurance in the yam seed system, especially for farmer seed growers in Ghana and Nigeria, and research has led to further improvements (Aighewi et al. 2020). Since seed was mainly obtained from dual crops of ware and seed yam, the AYMT was promoted at the start of the project to introduce the notion of producing sole crops of seed yam for commercial purpose, while research on high-ratio propagation techniques for yam was going on.

The project established partnerships with several non-governmental organizations (NGOs) in major yam-growing regions of Ghana and Nigeria to scale out the AYMT as well as to encourage the commercial production of seed yam. The NGOs received training of trainers on the technique before they conducted several training activities for extension officers and individual farmers or cooperative seed yam growers with backstopping from IITA. At the start of the project in 2012, one NGO worked in Nigeria, Missionary Sisters of the Holy Rosary (MSHR), and, in Ghana, Catholic Relief Services (CRS). Over 3 years, 35,135 farmers were trained on the AYMT in Ghana and Nigeria. Based on recommendations of an external evaluation of the project, three more NGOs in Nigeria and two in Ghana were added.

These NGOs trained an additional 43,625 farmers during the last 2 years of the project, bringing the total to 78,760 (40% women) (Table 14.6). Considering that yam is generally considered a man's crop in West Africa, the high percentage of women in the training was encouraging as few women are involved in the value chain.

Table 14.6 Farmer seed growers trained by YIIFSWA through the AYMT demonstrations in Ghana and Nigeria

Year	Number of farmer seed growers trained		
	No. male	No. female	Total
2012	1555	605	2160
2013	9700	6542	16,242
2014	9562	7171	16,733
2015	17,793	11,488	29,281
2016	8961	5383	14,344
Total (5 years)	47,571	31,189	78,760
Percentage	60.4%	39.6%	100%

14.3.3   Challenges to Farmers Capacity Development in Seed Production and Resolution

After 4 years of training farmers and extension officers, project staff realized that most farmers were not convinced that seed yam production could be more profitable than ware yam, even though all the farmers knew how expensive it was to buy seed yam (as much as 60% of production costs). We found the farmers had difficulty grasping three concepts:

1. Specialized seed yam production improves quality and yield.
2. Seed yam is more profitable to produce than ware yam.
3. Seed yam should be produced on ridges.

Having isolated these barriers, we invited farmers and extension staff from Ghana and Nigeria to Illushi, Edo, in Nigeria; seed yam production is the primary economic activity. They learned seven key ideas to encourage more seed yam production:

1. A new method of staking using pieces of bamboo (pyramidal staking) to minimize the cutting of trees for staking.
2. Small tubers (up to 50 g) that were normally discarded by the visiting farmers were displayed and sold at the market.
3. Seed yams were planted on ridges and growing well.
4. Illushi seed yam farmers select and tag apparently healthy plants from which seed yam will be harvested and good tubers selected to plant the next crop.
5. The quality of seed yam sold in the market was evidence of a good selection of disease- and pest-free seed tubers after harvest.
6. Seed tubers in the whole market were well-protected from the sun, a major cause of rot, using palm fronds to cover the market area.
7. Minisett technology is a profitable business venture, and farmers are now ready to adopt it at a commercial scale.

Farmers and extension staff indicated that site visit had been more valuable than earlier trainings. Farmers trusted the information that came from practicing farmers more than they trusted "theories" from researchers. Illushi had many seed yam farms 3–5 ha in size and a well-structured 1-hectare market that prohibited the sale of anything but seed yam.

14.3.4   Ware Yam Demonstrations Using Good Agronomic Practices and Improved Varieties to Drive the Demand for Released Varieties

Awareness of improved varieties may drive demand for improved seed. Baseline studies conducted at the start of the YIIFSWA project showed average yam yields of 9.4 tonnes (t)/ha in Nigeria and 7.0 t/ha in Ghana, for local varieties were less than 25% of the yield of improved released varieties, which ranged from 30 to 40 t/ha (Mignouna et al. 2014a, b). Yet, farmers were not aware of the 19 improved yam varieties that had been selected by plant breeders in Nigeria between 2001 and 2010. Adoption failed because there was no formal yam seed system to rapidly multiply large quantities of quality seed for farmers at an affordable price.

The yields for improved varieties can be further increased if farmers adopt better seed quality and good agronomic practices. To sensitize farmers on the superiority of improved varieties, seed tubers of selected released varieties [two *Dioscorea rotundata* (TDr 89/02665 and TDr 95/19177) and one *D. alata* (TDa 98/01176)] bred by IITA were planted in 80 on-farm demonstration plots to compare with the best local varieties by state in Nigeria (Oju-Iyawo in Oyo, Nwaagba in Enugu, Hembakwase in Benue, Meccakusa in the Federal Capital Territory (FCT), Nasarawa, and Niger) (Fig. 14.5). Farmers were identified to establish the demonstration plots in consultation with the National Root Crops Research Institute (NRCRI)-Umudike, NGOs, and local farmers who had participated in previous project activities. A deliberate effort to involve women in the demonstrations resulted in a proportion of 89% men and 11% women (Table 14.7).

A questionnaire to participating farmers revealed, on average, 23 years' experience with yam cultivation. About 46% of farmers claimed said their yam production had increased over the past 5 years (mostly due to an increase in the area cultivated), 36% noted said their productivity was decreasing, and 18% reported fluctuations due to climate change, increased pest and disease infestations, declining soil fertility, low yield potential of local varieties, and labor scarcity. Farmers said they selected yam varieties that resulted in high yields, big tubers, and good performance in dry years, with high market demand, early maturity, and white flesh that does not oxidize.

The project wrote a protocol describing good agricultural practices to follow in establishing seed plots. After planting, plots were monitored regularly by the IITA technical staff and the seed companies (including farmers' groups linked to the plots) to assess the growth and development of the crop.

At each locality, one of the best demo plots was selected for field days at the peak of the crop's vegetative growth and again at harvest. About 5 months after planting, the crop foliage was assessed, and 58% of the farmers preferred the improved variety TDr 95/19177, while 32% preferred TDa 98/01176. Ten percent were undecided between the local variety and TDr 89/02665. Although most ranked the improved varieties better during the vegetative stage, some farmers ranked the local varieties best, claiming that they were more certain of its tuber

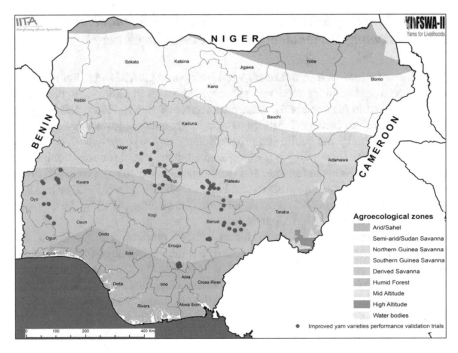

Fig. 14.5 Improved yam varieties were sown in 80 demonstration plots across six states in Nigeria

Table 14.7 Number of male and female farmers involved in the demonstrations for ware yam production

State	Benue	Enugu	FCT	Nasarawa	Niger	Oyo	Total
Male	14	6	9	14	15	13	71
Female	1	0	1	2	0	5	9
Total	15	6	10	16	15	18	80

quality. It was only after harvest that this group of farmers selected the improved varieties when they could confirm the tuber yield and quality. Although the *D. alata* variety, TDa 98/01176, had the highest mean yield of 19.1 t/ha (maximum of 34.6 t/ ha), the most preferred variety was *D. rotundata* TDr 95/19177, with an average yield of 18.0 t/ha (maximum of 31.5 t/ha). The farmers' best local variety yielded 12.1 t/ha. All the farmers who participated in the study were convinced of the improved varieties' potential and requested more seed to plant. The three improved and released varieties TDr 95/19177, TDr 89/02665, and TDa 98/01176 were later named by farmers as Kpamyo ("provides wealth"), Asiedu (in honor of the scientist who bred many of the yam varieties), and Swaswa ("fast," i.e., provides food and wealth fast).

14.3.5   Creating Awareness for the New Seed Technologies Among Stakeholders of the Yam Value Chain

Apart from the few ware yam farmers, seed growers and NARIs were not aware of recent developments in the seed yam subsector, so an awareness-raising campaign was carried out to (1) explain the economic impact of yam in Ghana and Nigeria; (2) advocate for an increased priority to yam in government investment plans for sustained funding by the public and private donors; and (3) popularize YIIFSWA's improved seed yam technologies for increased adoption by private seed companies and commercial ware yam farmers. The campaign engaged policymakers and other stakeholders with more emphasis given to Nigeria, where yam is prioritized as a key commodity in the 2016–2020 Agricultural Promotion Policy (APP) by the Ministry of Agriculture and Rural Development.

For sustainability of the improvements of seed production technologies, the YIIFSWA project assigned the role of connecting the yam value chain actors to the seed regulatory services: the National Agricultural Seed Council (NASC) in Nigeria and the Plant Protection and Regulatory Services Directorate (PPRSD) in Ghana. The public sector has the responsibility of regulating formal seed production, which needs the patronage or services of other stakeholders to thrive.

Stakeholders of the yam value chain were brought together to brainstorm on how to form partnerships and linkages to establish coordinated networks. The new business opportunities in seed yam production (for yam producers, processors, marketers, and transporters) using available new technologies to produce high-quality seed yam tubers of released varieties were demonstrated to intensify yam production for economic gain and food security as well as to facilitate a demand-driven seed system. The participants included researchers, public and private seed yam producers, ware yam dealers (producers, marketers, exporters, processors, input dealers, service providers, yam commodity associations and transporters), the Ministry of Agriculture's Yam Value Chain Department, and consumers and other relevant institutions. Recommendations were made to overcome the challenges for each value chain actor such as formalizing sales by weight, develop improved storage systems, improve knowledge sharing among stakeholders, improve extension education to farmers, build capacity and equip certification officers, and encourage the private sector to provide extension services, among others.

14.4   Advances in Seed Quality Assurance and Seed Certification

The formal yam seed system being developed in Ghana and Nigeria is promoting two generations of breeder seeds (BS1; BS2) followed by two generations of foundation seeds (FS1; FS2) and finally three generations of commercial seeds (CS1,

CS2, and CS3). The chronology accepted by all project partners is BS1 → BS2 → FS1 → FS2 → CS1 → CS2 → CS3.

14.4.1 Managing Pests and Diseases Through Quality Seed

Pests and diseases are major bottlenecks that reduce yam tuber yield and quality (Mignouna et al. 2019). Severe leaf diseases can significantly reduce yields. If the quality of the tubers is affected, then they lose commercial value. The most important diseases in West Africa are yam mosaic virus (YMV) and yam mild mosaic virus (YMMV). A third disease, yam anthracnose disease (YAD), is caused by the fungus *Colletotrichum gloeosporioides.* Nematode pests to yam include *Scutellonema bradys, Meloidogyne incognita,* and *M. javanica* and rot fungi, *Botryodiplodia theobromae* and *Aspergillus niger.* All these endemic threats are carried from one generation to another by recycling infected planting materials. Pests such as nematodes, mealybugs, and scale insects are transferred from the field on tubers into storage where they continue to multiply and cause damage, leading to higher tuber losses and reduced market value (Ampofo et al. 2010; MEDA 2011). Lack of pest- and disease-free seed tubers is a significant constraint to yam production in West Africa. In the absence of durable resistance, the use of pest- and disease-free seed yams, often referred to as "clean" seed yams, is the best alternative for enhancing the yields and quality of yam.

Various efforts have been made to improve the quality of seed yams, including the selection of healthy-looking yams and the treatment of seed yam tubers (setts and minisetts) with insecticides (e.g., chlorpyrifos) and fungicides (e.g., mancozeb) as part of the AYMT. These methods are useful to eliminate fungi, nematodes, and insects but ineffective against viruses. The YIIFSWA project made the first concerted effort to generate virus-free seed yams. The procedure, detailed in the previous section, involves in vitro production of virus-free plants used as sources for high-ratio propagation of virus-free seed yams (Aighewi et al. 2015; Balogun et al. 2017a). Plants are propagated hydroponically; then vines are cut from them to generate rooted vines for direct planting in the field or to produce mini-tubers in screenhouses.

Before the YIIFSWA project, the seed yam system in West Africa was mainly informal and entirely market-driven (Nweke et al. 2011). There was no quality control, and pest- and disease-infected tubers moved back and forth from field to storage. With YIIFSWA, a formal seed system was established in Ghana and Nigeria to control quality and provide certifications for all categories of seed yam (breeder, foundation, and certified seed) by the regulatory agencies.

14.4.2 Seed Certification for Sustainable Production of Quality Seed Yam

Seed certification schemes have been introduced to regulate the production and marketing of quality seeds and to reduce the use of diseased planting materials. For vegetatively propagated crops in sub-Saharan Africa, seed certification is emerging as an important tactic to prevent the further perpetuation of pathogens and pests in planting materials, to reduce disease inoculum in the fields, to prevent losses due to virus diseases, and ultimately to contribute to sustainable and profitable production (Almekinders et al. 2019; Bentley et al. 2018). There was no seed certification for most vegetatively propagated crops, including yams, until the 2000s in Ghana and Nigeria, when the first schemes were established to certify seed yams produced as part of an IFAD-funded root and tuber improvement program. These early schemes were tailored for the certification of the foundation seed (FS) and the certified seed (CS) tubers produced in open fields. The yam certification schemes were later adopted as part of the ECOWAS harmonized seed regulation procedures, which included standards for breeder seed (BS), FS, and CS.

The official seed yam certification schemes stipulate parameters to produce quality seed, which include:

1. Requirements for registration and accreditation of seed producer as BS, FS, or CS
2. Registration of seed production fields
3. Parameters for source seed used for seed production
4. Parameters for agronomic management of seed fields
5. Maximum threshold levels for notified pest and diseases in the seed production field

The seed yams can be reused for a maximum of seven generations: two generations each for BS (BS1 and BS2) and FS (FS1 and FS2) and three generations for CS (CS1, CS2, and C3). However, this procedure was challenged by the lack of virus-free source seed, so seed production remained mainly informal. Furthermore, existing seed certification procedures did not meet the needs of yam propagation material generated using newly established high-ratio propagation methods such as TIBS, aeroponics, hydroponics, and SAH. To overcome these challenges, revised certification procedures have been established by YIIFSWA-II and the CGIAR Research Program on Roots, Tubers, and Bananas' (RTB) initiatives, coordinated by IITA with national program partners in Nigeria (NASC) and Ghana (PPRSD).

14.4.3 Adapting Certification Procedures to Products of New Propagation Technologies

Emerging commercial seed yam enterprises fostered by the new wave of high-ratio propagation technologies require fit-for-purpose certification schemes. Revised seed yam certification procedures were developed after in-depth consultations between government regulators in Ghana and Nigeria and seed producers from public and private sector organizations, NGOs, and IITA in 2018 and 2019. The revised seed yam certification scheme aims to certify seed yams of registered varieties, which are true-to-type and meet the phytosanitary status concerning notified pathogens and pests, and ensure the supply of virus-free seed yams for ware yam production (Table 14.8).

The revised certification scheme introduces a new class of seed termed "nucleus seed (NS)" as the progenitor of BS. NS is allowed to reuse stock in perpetuity as long as they remain true-to-type and disease-free. BS and FS seed can be reused for an additional generation each, while the CS can be used for three generations.

Table 14.8 Basic parameters for seed yam quality assurance

	Nucleus seed (NS)	Breeder seed (BS)	Foundation seed (FS)	Certified seed (CS)
Propagation	Laboratory (TC, TIBS)	Laboratory (TC, TIBS, SAH) and screenhouse (aeroponics, hydroponics, SAH, and vine propagation)	Screenhouse (aeroponics, hydroponics, SAH, and vine propagation) and field (mini-tubers and minisetts)	Field (mini-tubers and minisetts)
Registration of seed producer	Required	Required	Required	Required
Source seed verification	Required	Required	Required	Required
Isolation distance (for field propagation)	NA	NA	5 meters	5 meters
Number of inspections	Lab accreditation	Screenhouse accreditation	Three inspections (1. preplanting; 3 to 4 months stage; at harvest)	Three inspections (1. preplanting; 3 to 4 months stage; at harvest)
Maximum number of generations	Perpetual[a]	Two generations (BS1 and BS2)	Two generations (FS1 and FS2)	Three generations (CS1, CS2, and CS3)
Virus (mosaic virus)	0	0	<10% of plants with symptom severity score (SSS) 2[b]	<20% of plants with SSS 2
Fungi (anthracnose etc.)	0	<10% of plants with symptom severity score 2	<10% of plants with SSS 3	<20% of plants with SSS 2
Nematode and fungal damage (tuber rot)	0	0	<10% of plants with SSS 2	<20% of plants with SSS 2

NA not applicable
[a]If material remains true-to-type with zero incidence of pest and diseases
[b]On a scale of 1 to 5 where 5 = most severe

The NS are produced in vitro in the laboratory or controlled environments with TIBS, which allows perpetual propagation of disease-free seed yams. Most BS is produced in laboratory using TIBS or SAH and in screenhouses with aeroponics and hydroponics. FS and CS are grown in screenhouses or open fields depending on the business model. Seed entrepreneurs (SE) and local seed producers are required to register seed fields, and the fields are inspected three times for suitability of field, varietal purity, and plant health.

14.4.4   Tools for Enabling Seed Certification

Tools and procedures have been established to assure seed qualities which are useful for seed producers and seed regulators.

Determination of genetic purity Morphological keys and DNA markers have been established to determine the identity of yam cultivars.

Seed health testing Tools for the reliable detection of yam viruses have been established based on polymerase chain reaction (PCR), loop-mediated isothermal amplification (LAMP), recombinase polymerase amplification (RPA), and rolling circle amplification (RCA) methods for the reliable and sensitive detection of yam viruses of importance to seed production (Bömer et al. 2019; Nkere et al. 2018; Silva et al. 2018).

Seed tracker The Yam Seed Tracker (YST) was developed as an innovative ICT solution to enable the digital integration of various value chain actors of the formal seed yam systems (Ouma et al. 2019), including institutions and entrepreneurs formally recognized to produce and market seed of released varieties. The YST is usable on any Internet-enabled device and offers real-time tracking of the seed production database, generates geographic maps, and offers analytics. The tool has been tailored to the regulatory procedures used for seed quality assurance and certification in Ghana and Nigeria. YST is suitable for registering seed producers and seed fields, organizing field inspection, and certification of seed lots. The tool has been adopted by the NASC of Nigeria as the national e-certification platform, making it easy for seed yam producers to ensure compliance with regulatory procedures and has been piloted to register seed yam value chain actors and for seed certification in Nigeria. Seed Tracker is one of the 11 tools of the RTB Seed System Toolbox.[2]

---

[2] https://www.rtb.cgiar.org/seed-system-toolbox/: The toolbox (11 tools) for seed systems of roots, tubers, and bananas that brings together proven strategies and tools that work to successfully diagnose, plan, and develop new vegetative seed systems.

14.4.5 Promotion of Seed Quality Assurance for Mainstreaming Seed Systems

The procedures for seed yam quality assurance by state seed certification agencies were new in West Africa. Therefore, YIIFSWA-II, AfricaYam, RTB, and other initiatives have made efforts to generate awareness about seed certification among growers, buyers, and seed yam users through the Yam Forums established in Ghana and Nigeria. The forum is comprised of all the seed value chain actor ware producers, traders, processors, input suppliers, policymakers, regulators, government agencies, private sector, NGOs, and representatives from the Ministry of Agriculture and serves as an excellent platform for sharing new advancements and benefits of adoption. Training courses have been organized in seed yam certification, virus diagnostics, and the use of YST for the seed certification officers from Ghana and Nigeria.

14.5 Outcomes and Value Added to Farmers Including Women and Youth

14.5.1 The Framework Established for Sustainable Value Creation in the Yam Sector

The YIIFSWA project has created products, services, and solutions in the yam sector by providing new technologies, organizing training events, and finding solutions that address challenges of the sector, to improve food security. However, there is still need (1) to produce yam differently using new high-ratio propagation techniques (HRPTs) and (2) to use these new technologies to bring yam to consumers and increasing efficiencies in the food chain.

Despite the COVID-19 pandemic during 2020, the early generation seed system (EGS) performed better than in 2019, generating 68% of the expected EGS of 5.8 million seeds. The performance is very high (309%) in terms of breeder seed production but below expected target levels for foundation seed production (65%).

In Nigeria, this is the first time that National Root Crops Research Institute (NRCRI-Umudike) has a facility established for breeder seed yam production using TIBS. It generated up to 22,663 seeds production level above its set target of 21,840 seeds. NACGRAB reached 98% of the 2020 seed yam production target of 8736. As for Ghana, in 2020, CRI was able with assistance from YIIFSWA-II to achieve a seed yam production level of 350% of its target of 21,840 seeds. SARI with its first TC Laboratory since inception attained 399% of its target of 8736 breeder seed plantlets.

In total, 221,735 breeder seeds were produced by IITA and NARIs in 2020. This total exceeded the target production by 309%. In terms of nucleus stock plantlets, production was higher in TIBS than conventional TCs at all NARIs. Nucleus stock was provided by IITA to NRCRI and SARI to enhance production before the COVID-19 travel constraints. Highest production of breeder seeds was through Vivipak and vine cuttings.

For foundation seed production, a total of 3,631,284 seed tubers were produced in 2020. The project partnered with eight commercial seed companies in Ghana (3) and Nigeria (5). The NARIs supplied the private seed companies with breeder seed to allow them generate foundation seed. However, it was observed that the seed companies of Nigeria are not performing well in term of single-node vine cuttings. Some of them (Da-Allgreen and PS Nutraceuticals) generated enough vines with their hydroponics and aeroponics but lost the single-node vines during the process of rooting: only 30% survived. With this low level of foundation seed production, the SE were not able to receive foundation seed from private seed companies to produce certified seed.

14.5.2 Cleaned Seed Delivery, Yield Gains, and Reduced Poverty

The most important challenge of the seed yam value chain is the scarcity of high-quality seed yam. This obstacle undermines farmers' ability to gain higher yields and generate sustainable incomes. Most farmers are not familiar with producing sole crops of seed yam due to the widespread traditional practices of obtaining seed yam from ware crops.

Based on the proven benefits of high-quality seed yam, IITA and its partners are establishing a robust seed yam system in Ghana and Nigeria that uses a market-based, integrated approach to deliver clean seed yam of improved varieties to at least 320,000 smallholder farmers by 2021. After the first phase of the project, the mean productivity advantage was about 18% in Nigeria and 13% in Ghana. The AYMT adoption rate was about 18% in Nigeria and 43% in Ghana. The private seed companies are being encouraged and backstopped for production of high-quality of quantities of target varieties for specific market segments.

Quality assurance standards of seed yam of improved varieties are resulting in increased productivity at the farm level. High-quality breeder seed is gradually improving productivity through lower losses to viruses and nematodes. Three improved yam varieties (*Asiedu*, *Kpamyo*, and *Swaswa*) are being promoted by the project and proving to be up to 38% more productive than local varieties.

Research by YIIFSWA-II has left a legacy of better technologies to address far-reaching yam concerns. At the end of the first phase of the project, the rural poverty had been reduced by 10% in Ghana and Nigeria (25,040 and 119,117 households, respectively). The project will move an estimated one million people out of poverty by 2037 in West Africa, and more than 96 million people are expected to benefit from yam technologies in all the yam-producing countries (Mignouna et al. 2020).

14.5.3 Trust of End Users and Informed Decision-Making for Sustainable Stakeholder Enterprises

Activities such as field days and demonstration plots were organized to educate certified SEs, local seed producers, and seed company out-growers on good agronomic practices required for quality seed yam production. The integration of women in the operation of demonstration sites was strongly encouraged to make informed decisions on how to decrease the constraints and increase the benefits to female and male farmers.

14.5.4 Sustainable Seed Yam System

Prior to interventions, the yam sector had a fragmented seed value chain, but the YIIFSWA project has upgraded portions that cover activities from the use of plant genetic resources to the marketing and/or distribution of improved yam varieties to farmers.

Through scaling out the high-quality seed production technologies developed in YIIFSWA, seed companies, in partnership with national regulatory agencies, delivered foundation and certified seed yam tubers to seed entrepreneurs. Private seed companies were trained to understand pricing, contracting, and demand forecasting. Demand forecasting is critical given the high downstream demand variability for each variety and the length of time between when foundation seed is produced and when demand for certified seed is satisfied. Further training is ongoing to:

- Strengthen capacities of the seed yam value chain actors to support the development of commercially viable seed businesses. Business models developed with seed yam value chain actors are providing good practice models for further replication.
- Disseminate the results of YIIFSWA (improved varieties and clean seed) through scaling to flow commercially beyond the project areas. YIIFSWA-II outputs (such as high-quality breeder seed production using TIBS and clean foundation seed production by private companies using aeroponics, virus diagnostic tools, and seed quality certification systems) are being appropriated, adapted, and extended to other places in the world where yam is an important crop.[3]
- Ensure efficient partnerships and coordination between seed actors along the value chain. Quality control and quarantine regulators, seed producers, seed sellers, seed yam buyers, and yam processing businesses are collaborating effectively. Record-keeping, sharing of information, and accountability are facilitating

---

[3] www.wecrop.com.au. Multinational company, source of 100% organic tropical food staples, for all lovers of yam, taro, cassava, and plantain (Accessed on 10 January 2021).

seed traceability and creating an enabling environment for seed yam businesses (Aighewi et al. 2017) to thrive even beyond the project's life.

14.5.5 Advocacy and Awareness Creation

The project's advocacy and awareness creation activities were aimed at communicating the results of YIIFSWA and are being scaled in YIIFSWA-II to sustain investments in yam and to raise investments to reach new areas in Ghana and Nigeria. It is expected that by the end of the project, the activities under this component should have raised at least $2 million.

14.5.6 Gender Mainstreaming in the Production and Commercialization of Seed Yam Systems

Gender is an important consideration for looking at how social norms and power structures differentially impact the lives and opportunities available to groups of men and women. Gender and diversity were built into the project through several strategies, tools, and resources to ensure that project activities specifically benefit women. These strategies included organizational gender mainstreaming and gender assessments aiming at increasing awareness and gender equality.

Women were provided various opportunities to earn income in commercial yam seed production and distribution. Of the eight private commercial seed companies are partnering with YIIFSWA-II, one in Nigeria and one in Ghana are female-owned. Moreover, 13% of staff contracted by these private companies in the seed yam value chain were women as heading toward the gender mainstreaming target of 40%.

An assessment was conducted in Ghana and Nigeria in 2019 to understand and address gender issues in seed yam production, delivery, and access. The study used a gender-sensitive questionnaire for individuals and focus group discussions to assess the role and status of gender in the seed yam sector and to understand the gender dynamics within yam production in the study area. States/regions and communities/districts were selected purposively, while male and female respondents (farmers) were randomly selected. In total, 226 male and 220 female individual respondents were interviewed, while separate focus groups comprised of a total of 55 men and 53 women were interviewed from 6 states in Nigeria. In Ghana, the groups of a total of 37 women and 165 men were interviewed from 3 regions.

Results revealed that decision-making for all the farm operations are controlled primarily by men and that men carried out most of the activities related to yam production (fertilizer application, harvest, storage, marketing). In some rare cases, the household head was a woman, and, in such situations, they were either divorced or widowed.

Table 14.9 Seed yam production and objectives for growing seed yam in Nigeria

	Gender	Response	Benue	Enugu	FCT	Nasarawa	Niger	Oyo	Average
Produce seed	Female	No	32.5	16.2	6.3	50	36.5	3.8	24
		Yes	17.5	32.4	43.8	5.3	12.2	43.9	26
	Male	No	30	6.8	0	31.6	27	4.6	17
		Yes	20	44.6	50	13.2	24.3	47.7	33
Objective of growing seed	Female	Planting	47.5	21.6	39.6	55.3	47.2	41.7	42
		Sales	2.5	27.1	10.4	0	1.4	6.1	8
	Male	Planting	50	18.9	35.4	39.5	43.3	28.8	36
		Sales	0	32.4	14.6	5.3	8.1	23.4	14

Forty percent of the female respondents interviewed in Ghana and 26% in Nigeria reported that they produced seed yam, while 56% of the male respondents in Ghana and 33% in Nigeria said the same. In this gender study, seed yam in Nigeria is perceived as a nascent business because many farmers get their seeds from the previous season to plant and buy only a few for planting. Some leftover seeds are sold seed after planting by other farmers. Table 14.9 presents the situation of seed yam production and utilization in the project states in Nigeria.

In terms of gender participation in yam production, about 46% and 24% of men and women, respectively, reported full participation in Nigeria. Similarly, in Ghana, about 92% and 21% of men and women, respectively, reported full-time participation. Those women who participated full-time were either divorced or widowed. Few women owned their yam farms.

Concerning access to financial resources (such as lending), fewer people in Nigeria reported having access compared to those in Ghana. In Nigeria, only 15% and 13% of men and women, respectively, said they could access bank assistance while those numbers in Ghana were 32% and 30% of men and women, respectively.

14.6 Lessons Learned From Scaling Improved Seed Yam Production Technologies

The lessons learned included the following:

- Overall, provision of key infrastructure for the private seed companies, training as a component of the mechanism for delivery of the new technologies to next users, strong partnerships (especially with national regulatory agencies), and learning through monitoring and evaluation were important to progress.
- We underestimated the time needed to establish formal yam seed systems in countries with underdeveloped seed certification for clonally propagated crops, which required developing, validating, and scaling out high-ratio technologies; building capacity of key actors; and producing clean stock materials of registered varieties.

- Control of high temperatures in screenhouses of the aeroponic systems (using methods developed in the project) is critical, as is the timing of field planting of rooted single-node vine cuttings in relation to ambient weather conditions.
- Closer interaction of project staff with the private seed companies at the early stages of their engagement could have forestalled the setbacks that resulted from their ignoring some of the specifications for construction and operation of their aeroponic systems.
- Business plans for efficient and sustained breeder seed production using the novel technologies must pay attention to the costs and access to laboratory reagents, skilled technical capacity, and power (electric and/or solar).

14.7 Conclusions, Recommendations, and Emerging Issues

14.7.1 Conclusions

The aim of YIIFSWA project has been to transform seed yam systems by facilitating availability and access to high-quality seed of improved yam varieties in yam-growing areas of Ghana, Nigeria, and beyond. High-ratio propagation technologies and seed health management techniques developed during the first 5 years (2011–2016) were scaled out during the second phase (2017–2021). Thus, the elements for sustainable formal seed systems for yam are in place in Ghana and Nigeria:

- Temporary Immersion Bioreactor System (TIBS) have been established in laboratories at two NARIs in each country to produce breeder seed. Eight private seed companies have been provided with technical training and infrastructure for aeroponic (AS) and hydroponic (HS) systems to produce foundation seed. Seed entrepreneurs have been trained in commercial seed yam production using the adapted yam minisett technology (AYMT).
- Improved seed health management methods that incorporate virus indexing and virus elimination techniques (including heat therapy combined with meristem tissue culture) have been established. The regulatory bodies in Ghana and Nigeria have been trained and provided with equipment for yam virus diagnostics, and protocols are available for seed yam quality assessment and certification.
- The COVID-19 pandemic limited many planned interactions among implementing agencies and potential beneficiaries, but actors in the seed sector have been trained to facilitate scaling the seed production technologies. Technology uptakers were first trained in specifications and procedures for producing seed yam. These trainees can be service providers in scaling the production technology to new or existing actors. Seed producers were then trained in daily management of a seed production system. SOPs and a manual of procedures have been produced and disseminated.

- Males are more involved in yam production and benefit more from all factors of production than females in Ghana and Nigeria. Active efforts have been made through the project to enhance gender equity by involving women at more stages of yam systems.

14.7.2 Recommendations

Additional research is needed on the following topics:

- Studies of the purchasing power of farmers for adequate pricing and recurrent purchases to encourage replacement of their degenerated planting materials.
- Fine-tune processes in the new propagation technologies, as well as associated agronomy, postharvest handling, and storage of seed tubers (especially micro- and mini-tubers), and produce SOPs to facilitate adoption.
- Assess the cost-effectiveness of the use by seed companies of aeroponic, hydro- ponic, or semi-autotrophic hydroponic systems in foundation seed production under their conditions.
- Identify the conditions or methods to break seed tuber dormancy to allow plant- ing shortly after harvest as may be necessary to speed up seed production cycles.
- Estimate the rate of seed degeneration as the basis for deciding optimum timing for replacement of seed stocks at various stages along the seed value chain.
- Conduct ex ante analysis of the prices for quality seeds of newly released variet- ies in the formal seed system relative to those of traditional varieties in the infor- mal and later the formal, system. For seed potato, Almekinders et al. (2019) noted that certification requirements increased the cost of planting materials prohibitively.

Scaling should focus on the following areas:

- On-farm demonstrations of the superiority of high- over poor-quality seed yams in terms of productivity, postharvest losses, and overall profitability when used in producing ware yams.
- Engage with major yam processors, especially companies who are linked or could be linked with networks of ware tuber producers, to expand opportunities for commercial seed producers to supply healthy seeds of their desired varieties to the producers.
- Continue advocacy for policy and institutional support of the nascent formal seed sector, and target more women and the youth in efforts to promote commer- cial production of quality seed.
- Advocate for increased transparency in, and enforcement of, seed quality control and certification to limit competition in the seed markets between high-quality seeds and cheaper but poor-quality or contaminated seeds.

- Monitoring of uptake of certified seed production and foster stronger links between seed entrepreneurs and local seed producers to other key actors in the seed sector.

14.7.3   Emerging Issues

The establishment and growth of a formal seed system for yam in Nigeria and Ghana creates some issues that need to be addressed going forward. According to Nigerian seed regulations, a seed company cannot sell seeds in more than one class. Ahead of enforcement of this regulation in the formal yam seed sector, the seed companies need to identify and use the combination of inputs and growth conditions that maximize the production of seeds with the desired attributes (e.g., weight) for the class they target to specialize in. In addition, breeder seed tubers are currently produced by public institutions (NARIs) in Ghana and Nigeria. Fair prices for these seeds are yet to be determined and applied consistently for their sale to the private sector. Sustainable breeder seed production and sale need to be established as a business, for example, through private companies affiliated with the public breeding institutions. Alternatively, the national governments could provide temporary subsidy to the public institutions to continue to perform this function while the formal sector gets established.

There is need for much increased awareness of the emerging formal seed sector by yam growers' associations and actors in the informal sector, especially marketers, who have operated for decades. Likewise, the few improved varieties being promoted in the formal seed system are yet to earn recognition in the markets in West Africa. The high-ratio propagation technologies can be applied to registered traditional varieties that are already popular with farmers as requested by some seed companies. This will build confidence of the seed companies in the profitable use of the technologies at large scale, strengthen their links with their clients, and bring the health benefits from the new technologies to large numbers of farmers through clean seeds of already popular varieties.

The exciting opportunities offered by the formal sector calls for strong linkage with the yam breeding programs in West Africa and regular feedback toward the maintenance of a pipeline for new and superior market-preferred varieties that would sustain demand for certified seeds.

Acknowledgments   This research was undertaken as part of, and funded by the CGIAR Research Program on Roots, Tubers, and Bananas (RTB) and supported by the CGIAR Trust Fund contributors. Funding support for this work was provided by Bill & Melinda Gates Foundation.

References

Aighewi B, Akoroda M, Asiedu R (2002) Seed yam (Dioscorea rotundata Poir.) production, storage, and quality in selected yam zones of Nigeria. AJRTC 5(1):20–23

Aighewi BA, Asiedu R, Maroya N, Balogun M (2015) Improved propagation methods to raise the productivity of yam (Dioscorea rotundata Poir.). Food Sec 7:823–834. https://link.springer.com/article/10.1007/s12571-015-0481-6

Aighewi B, Maroya N, Mignouna DB (2017) Key for a profitable and sustainable seed yam business enterprise: business plan and market development with record keeping for seed yam farmers. YIIFSWA working paper series, No. 8. International Institute of Tropical Agriculture, Ibadan. https://hdl.handle.net/10568/82693

Aighewi B, Maroya N, Mignouna D, Aihebhoria D, Balogun M, Asiedu R (2020) The influence of minisett size and time of planting on the yield of seed yam (Dioscorea Rotundata). Exp Agric 56(3):469–481. https://doi.org/10.1017/S0014479720000095

Almekinders C, Walsh S, Jacobsen K, Andrade-Piedra J, McEwan M, de Haan S, Lava K, Staver C (2019) Why interventions in the seed systems of roots, tubers and bananas crops do not reach their full potential: a reflection based on literature, and thirteen case studies. Food Security 11(1):23–42. https://doi.org/10.1007/s12571-018-0874-4

Ampofo JKO, Kumar L, Seal SE (2010) Integrated crop management for sustainable yam production. In: yam research for development in West Africa – working papers. IITA-BMGF Consultation Documents, IITA, pp 46–80

Balogun M, Maroya N, Augusto J, Ajayi A, Kumar L, Aighewi B, Asiedu R (2017a) Relative efficiency of positive selection and tissue culture for generating pathogen-free planting materials of yam (*Dioscorea spp.*). Czech J Genet Plant Breed 53(1):9–16. https://doi.org/10.17221/117/2016-CJGPB

Balogun M, Maroya N, Taiwo J, Ossai C, Ajayi A, Kumar L, Pelemo O, Aighewi B, Asiedu R (2017b) Clean breeder seed yam tuber production using temporary immersion bioreactors. IITA, Ibadan, p 66. ISBN 978–978-8444-89-3

Balogun M, Maroya N, Ossai C, Ajayi A, Aighewi B, Asiedu R (2018) Breeder seed yam production from soil to soilless systems: yam hydroponics. In: Proceedings: 18th Triennial Symposium of the International Society for Tropical Root Crops. 22–25 October 2018, International Center for Tropical Agriculture (CIAT) Cali, Colombia. www.istrc.org

Balogun M, Maroya N, Aighewi B, Ossai C, Ajayi A, Taiwo J, Lava K, Mignouna D, Asiedu R (2020) Evolvement and advances in the hydroponics system for clean seed yam production. YIIFSWA Working Paper Series No 8, International Institute of Tropical Agriculture, Ibadan, p 15

Bentley JW, Andrade-Piedra J, Demo P, Dzomeku B, Jacobsen K, Kikulwe E, Kromann P, Lava K, McEwan M, Mudege N, Ogero K, Okechukwu R, Orrego R, Ospina B, Sperling L, Walsh S, Thiele G (2018) Understanding root, tuber and banana seed systems, and coordination breakdown: a multi-stakeholder framework. J Crop Improv 32(5):599–621

Bömer M, Rathnayake AI, Visendi P, Sewe SO, Sicat JPA, Silva G, Kumar PL, Seal SE (2019) Tissue culture and next-generation sequencing: a combined approach for detecting yam (Dioscorea spp.) viruses. Physiol Mol Plant Pathol 105:54–66. https://doi.org/10.1016/j.pmpp.2018.06.003

Coursey DG (1976) The origins and domestication of yams in Africa. In: Harlan JR (ed) Origins of African plant domestication. Mouton, La Hague, pp 383–408

FAO (2007) Small-scale seed enterprise. Guidelines & business skills for seed producers in Afghanistan. http://www.fao.org/3/a1516e/a1516e00.htm

FAO (2018) Food and Agriculture Organization of the United Nations. On-line and multilingual database. http://faostat.fao.org/

FAO (2020) Food and agriculture Organization of the United Nations. On-line and multilingual database Available at: http://faostat.fao.org/. Accessed 10 Jan 2020

FAO (2021) Food and Agriculture Organization of the United Nations. On-line and Multilingual Database, New food balances. Updated 08 February 2021. http://www.fao.org/faostat/en/#data/FBS

IITA (1985) Root and tuber improvement program. Agriculture research highlights 1981–1984. IITA, Ibadan

Maroya N, Balogun M, Asiedu R (2014a) Seed yam production in an aeroponics system: a novel technology. YIIFSWA Working Paper Series No 2 (Revised). Yam Improvement for Income and Food Security in West Africa. International Institute of Tropical Agriculture, Ibadan, p 20. http://yiifswa.iita.org/wp-content/uploads/2017/09/Seed-Yam-Production-in-an-Aeroponics-System-A-Novel-Technology.pdf

Maroya N, Asiedu R, Lava K, Mignouna D, Lopez-Montes A, Kleih U, Phillips D, Ndiame F, Ikeorgu J, Otoo E (2014b) Yam improvement for income and food security in West Africa: effectiveness of a multi-disciplinary and multi-institutional team-work. J Root Crops 40(1):85–92. https://hdl.handle.net/10568/76119

Maroya N, Balogun M, Asiedu R, Aighewi B, Lava K, Augusto J (2014c) Yam propagation using 'Aeroponics' technology. Ann Res Rev Biol 4(24):3894–3903. SCIENCEDOMAIN international. https://doi.org/10.9734/ARRB/2014/11632

Maroya N, Balogun M, Aighewi B, Lasisi J, Asiedu R (2017) Manual for clean foundation seed yam production using aeroponics system. IITA, Ibadan, p 68. ISBN 978-978-8444-88-6. https://hdl.handle.net/10568/91020

Mbanaso ENA, Egesi CN, Okogbenin E, Ubalua AO, Nkere CK (2011) Plant biotechnology for genetic improvement of root and tuber crops. In: Amadi CO, Ekwe KC, Chukwu GO, Olojede AO, Egesi CN (eds) Root and Tuber Crops Research for food security and empowerment

MEDA (2011) Ghana yam market, subsector and value chain assessment. Mennonite Economic Development Associated (MEDA), Canada, p 71

Mignouna BD, Abdoulaye T, Alene AD, Asiedu R, Manyong VM (2014a) Characterization of yam-growing households in the project areas of Nigeria. International Institute of Tropical Agriculture (IITA), Ibadan, p 92

Mignouna BD, Abdoulaye T, Alene AD, Asiedu R, Manyong VM (2014b) Characterization of yam-growing households in the project areas of Ghana. International Institute of Tropical Agriculture (IITA), Ibadan, p 95

Mignouna DB, Abdoulaye T, Akinola A, Alene A, Nweke F (2015) Factors influencing the use of selected inputs in yam production in Nigeria and Ghana. J Agric Rural Dev Trop Subtrop 116(2):131–142

Mignouna DB, Kumar L, Coyne D, Bandyopadhyay R, Ortega-Beltran A, Bhattacharjee R, De Koeyer D (2019) Identifying and managing plant health risks for key African crops: yam, taro, and cocoyam. In: Neuenschwander P, Tamò M (eds) Critical issues in plant health: 50 years of research in African agriculture. Burleigh Dodds Science Publishing, Cambridge, UK

Mignouna DB, Adebayo AA, Abdoulaye T, Alene AD, Manyong V, Maroya NG, Aighewi B, Kumar L, Balogun M, Lopez-Montes A, Rees D, Asiedu R (2020) Potential returns to yam research investment in sub-Saharan Africa and beyond. Outlook Agric 49:1–10. https://doi.org/10.1177/0030727020918388

Nkere CK, Oyekanmi J, Silva G, Bömer M, Atiri GI, Onyeka J, Maroya NG, Seal SE, Kumar L (2018) Chromogenic detection of yam mosaic virus by closed tube reverse transcription loop-mediated isothermal amplification (CT-RT-LAMP). Arch Virol 163(4):1057–1061. https://doi.org/10.1007/s00705-018-3706-0

Nweke F (2016) Yam in West Africa: food, money, and more. Copyright 2016 Michigan State University. ISBN: 978-1-61186-187-79 (pbk.); ISBN: 978-1-60917-474-3 (ebook:PDF)

Nweke F, Akoroda M, Lynam J (2011) Seed Systems of Vegetatively Propagated Crops in sub-Saharan Africa. Workshop convened by the Bill & Melinda Gates Foundation in Nairobi, November 12–17, 2010. p 98

Odu BO, Coyne D, Lava K (2016) Adapting a yam seed technique to meet farmers' criteria. In: Andrade-Piedra J, Bentley JW, Almekinders C, Jacobsen K, Walsh S, Thiele G (eds.) Case studies of roots, tubers and bananas seed systems. CGIAR research program on roots, tubers and bananas (RTB), Lima: RTB Working Paper No. 2016-3. ISSN 2309-6586, p 244

Orkwor GC, Asiedu R, Ekanayake IJ (2000) Food yams: advances in research. IITA, Ibadan and NRCRI, Umudike, p 249

Ouma T, Kavoo A, Wainaina C, Ogunya B, Kumar L, Shah T (2019) Open data kit (ODK) in crop farming: mobile data for seed yam tracking in Ibadan, Nigeria. J Crop Improvement 33(5):605–619. https://doi.org/10.1080/15427528.2019.1643812

Silva M, Oyekanmi J, Nkere CK, Bömer M, Kumar L, Seal SE (2018) Rapid detection of poty-viruses from crude plant extracts. Anal Biochem 546:17–22. https://doi.org/10.1016/j.ab.2018.01.019

Chapter 15
Commercially Sustainable Cassava Seed Systems in Africa

James P. Legg ⓘ, Elohor Diebiru-Ojo ⓘ, David Eagle,
Michael Friedmann ⓘ, Edward Kanju, Regina Kapinga ⓘ, P. Lava Kumar ⓘ,
Sanni Lateef ⓘ, Stephen Magige, Kiddo Mtunda ⓘ, Graham Thiele ⓘ,
Juma Yabeja, and Hemant Nitturkar ⓘ

Abstract Cassava is an important crop in sub-Saharan Africa for food security,
income generation, and industrial development. Business-oriented production systems require reliable supplies of high-quality seed. Major initiatives in Nigeria and
Tanzania have sought to establish sustainable cassava seed systems. These include
the deployment of new technologies for early generation seed (EGS) production;
the promotion of new high-yielding and disease-resistant varieties; the updating of
government seed policy to facilitate enabling certification guidelines; the application

J. P. Legg (✉) · R. Kapinga · J. Yabeja
International Institute of Tropical Agriculture (IITA), Dar es Salaam, Tanzania
e-mail: j.legg@cgiar.org; r.kapinga@cgiar.org; j.yabeja@cgiar.org

E. Diebiru-Ojo · P. L. Kumar · S. Lateef
International Institute of Tropical Agriculture (IITA), Ibadan, Nigeria
e-mail: e.diebiru-ojo@cgiar.org; L.Kumar@cgiar.org; l.sanni@cgiar.org

D. Eagle
Mennonite Economic Development Associates (MEDA), Waterloo, Ontario, Canada
e-mail: deagle@meda.org

M. Friedmann · G. Thiele
CGIAR Research Program on Roots, Tubers and Bananas (RTB), led by the International
Potato Center, Lima, Peru
e-mail: g.thiele@cgiar.org

E. Kanju
International Institute of Tropical Agriculture (IITA), Uganda, Kampala, Uganda
e-mail: e.kanju@cgiar.org

S. Magige
Mennonite Economic Development Associates (MEDA), Dar es Salaam, Tanzania
e-mail: smagige@meda.org

K. Mtunda
Tanzania Agricultural Research Institute (TARI), Tumbi, Tabora, Tanzania

H. Nitturkar
FAO, Riyadh, Saudi Arabia

G. Thiele et al. (eds.), *Root, Tuber and Banana Food System Innovations*,
https://doi.org/10.1007/978-3-030-92022-7_15

453

of ICT tools, Seed Tracker and Nuru AI, to simplify seed system management; and the establishment of networks of cassava seed entrepreneurs (CSEs). CSEs have been able to make profits in both Nigeria (US\$ 551–988/ha) and Tanzania (US\$ 1,000 1,500/ha). In Nigeria, the critical demand driver for cassava seed businesses is the provision of new varieties. Contrastingly, in Tanzania, high incidences of cassava brown streak disease mean that there is a strong demand for the provision of healthy seed that has been certified by regulators. These models for sustainable cassava seed system development offer great promise for scaling to other cassava-producing countries in Africa where there is strong government support for the commercialization of the cassava sector.

15.1 Introduction to Cassava and Cassava Seed in Africa

15.1.1 The Importance of Cassava in Africa

Cassava was introduced to the Gulf of Guinea in West Africa from Brazil in the sixteenth century. Later introductions were made to coastal areas of East Africa in the eighteenth century, and the crop diffused gradually into the interior. Cassava's value as a food security staple was recognized during the early part of the twentieth century, and its cultivation was widely encouraged. However, it was during this period – the 1920s and 1930s – that the crop faced its first major disease challenge as epidemics of cassava mosaic virus disease (CMD) spread to all of the major production zones, leading to substantial yield losses.

Cassava is a semi-perennial shrub (*Manihot esculenta* Crantz), which grows best in well-drained soils in regions with tropical temperature regimes where there is well-distributed annual rainfall of >800 mm. Once the crop is established, however, it is tolerant of periods of drought and will yield in soils which are unsuitable for more nutrient-demanding crops such as maize. Since the 1990s, increased levels of governmental support have led to expanded cassava production in several African countries. Total production in Africa increased from 70.3 million tonnes in 1990 to 192.1 million tonnes in 2019, a 173% increase, although most of this growth resulted from an expansion in the cultivated area (+151%) rather than from higher yields (+8.6%) (FAOSTAT 2021). These changes have led to a shift in the global dynamic of cassava cultivation, as Africa's share of total production has increased from 46% in 1990 to more than 63% in 2019. Although this change is partly a result of the need for food production to keep pace with population increase, it has also arisen from an increasing recognition of the industrial value of the cassava crop. Rapid urbanization has opened up new markets for processed cassava products, and there have been increasing efforts to strengthen cassava value chains. The increased commercialization associated with these changes has provided new opportunities to modernize the seed systems that are a key component of efficient cassava value chains.

15.1.2 The Fundamental Importance of Cassava Seed

Although cassava plants produce true seeds under favorable environmental conditions, these are genetically heterogeneous and may not germinate readily. By contrast, stem cuttings are clones of the parent plant, retaining desired characteristics. Cuttings sprout readily in moist soils, are relatively easy to handle, and have become the primary propagule for establishing a new cassava crop. However, there are several drawbacks to this form of vegetative propagation. Firstly, the cuttings are bulkier and more prone to deterioration than true seeds, and, secondly, there is a high likelihood that pathogens that infected the parent stock will be carried directly via the cuttings into the newly established plants. This is particularly true for viruses. In Africa, the two most important biotic production constraints are CMD and CBSD (cassava brown streak disease), both of which are propagated by stem cuttings as well as through transmission by a whitefly insect vector. CMD is distributed wherever cassava is grown in Africa, while CBSD occurs in East, Southern, and Central Africa, although it continues to spread further westward (Legg et al. 2015). Both diseases can be effectively managed with clean planting material of resistant varieties, but few farmers currently have access to this material. This situation provides a clear justification for investments in improving seed systems with the aim of the sustainable delivery of high-quality seed to farmers. This will necessarily involve efforts to shift from entirely informal seed systems, where farmers share or trade planting material with no regulation of the provenance, characteristics, and health of the material, toward more formal seed systems, where the source, identity, and quality of planting material are regulated. An extensive monitoring survey of cassava varieties and landraces being grown in Nigeria, using DNA fingerprinting to ascertain the identity of the varieties in the field, showed that about 20% of respondents erroneously thought they were growing a landrace when it was an improved variety, and about 13% thought the opposite (Wossen et al. 2017). This exemplifies the importance of being able to source seed that is true-to-type of the desired variety. Experience from Nigeria has also shown that high-yield potential and other quality traits are attributes of cassava seed that can encourage investment in commercial seed systems (Bentley et al. 2020b).

15.1.3 Social Background to Cassava Seed Systems

Informal seed systems continue to predominate for most crops in most countries of sub-Saharan Africa. McGuire and Sperling (2016) estimated that 90.2% of seed is accessed through informal systems in smallholder farming communities, particularly those in sub-Saharan Africa. For cassava, the balance is even more heavily skewed toward informal systems. Even in countries where there have been large-scale interventions to promote more formal seed systems, the quantity of formal seed produced still remains less than 1% of the total volume of seed planted by farmers. Both women and men are active in cassava production, processing, and marketing, although in Nigeria, men play a greater role in planting, crop management, and harvesting, while women are primarily responsible for farm and household-level processing (Nweke et al. 2002). Up to 30% of men and women indicate that they sometimes purchase cassava planting material, although main sources are farmers' own fields or those of their neighbors or friends (Teeken et al. 2018). Gender studies on cassava production systems in both West and Southern Africa have demonstrated clearly, however, that men play an increasingly dominant role as production, processing, and marketing systems increase in scale and become more commercialized (Forsythe et al. 2015). These trends have highlighted the importance of recognizing gender effects as more systematic efforts are made to increase the commercial potential of cassava seed systems.

15.1.4 Why a Commercially Sustainable Seed System for Cassava Is Critical

There have been efforts made over several decades to develop cassava varieties in Africa that are high-yielding and disease-resistant and meet the quality requirements of farmers. These have frequently been coupled with large-scale seed dissemination schemes (Dixon et al. 2003). However, prior to 2010 there was little attention given to the sustainability of seed dissemination programs, meaning that when support for the program came to an end, the provision of improved seed ceased. Furthermore, it was also widely considered that disseminating seed of disease-resistant varieties was enough, whether or not the seed was of good quality. More recently, however, it has become widely recognized that improved seed programs can only be sustained if commercial mechanisms are incorporated. A significant proportion of cassava farmers in Africa already pay for cassava seed, predominantly through the informal sector, and as the sub-sector becomes increasingly business-oriented, this proportion will increase. Commercial seed systems grounded on the formal sector are developing rapidly for major grain, legume, and vegetable crops in Africa, as well as for potato, which provides a useful comparator example for cassava in view of its similar vegetative propagation and bulkiness. Commercially sustainable seed systems for cassava, which deliver certified seed of improved varieties, offer great potential benefits to farmers, including increased productivity resulting from low disease and pest levels, as well as an assurance that what they pay for is what they get. Commercial seed systems that supply high-quality products also provide substantial new income-generating opportunities for farmers. Furthermore, structured commercial cassava seed systems that provide quality assurance may enhance dissemination of improved varieties to smallholder farmers and inform breeding programs on varietal demand and trait preferences, which in turn will result in a more dynamic and productive cassava economy.

15.2 The Essential Elements of a Commercially Sustainable Seed System for Cassava

15.2.1 Markets for Cassava Products

Vibrant markets for cassava products are an essential driver for the development of sustainable seed systems. Cassava is consumed as food or feed or is used to make industrial products like starch and biofuels. Although the starchy roots are the main food product, the leaves are also consumed in certain regions. The Cassava Monitoring Survey in Nigeria (Wossen et al. 2017) showed that cassava is used mainly for food within the household and as a cash crop, with about 50% of the cassava produced sold for cash income, primarily for processing into food products. The demand for cassava roots is increasing due to more food manufacturing to meet demand in growing urban populations. This includes industrial products such as starch and flour but predominantly processed food products such as gari and fufu (Bentley et al. 2020a). As trade in seed grows, certification will be increasingly required for commercial seed production as there will be a clear benefit to planting material that is of high quality and of known provenance of specific improved varieties (Thiele et al. 2021). This sets the stage for establishing a sustainable seed system. This review presents two contrasting market demand situations for high-quality cassava seed. In Nigeria, where disease problems are relatively minor, there is strong demand for high-yielding farmer-preferred varieties such as TME 419. In Tanzania, however, the significant losses incurred by farmers as a result of infection of local varieties by the virus diseases CBSD and CMD means that there is strong willingness to pay for healthy seed of disease-resistant varieties. Studies in Tanzania and Uganda suggest that farmers are willing to pay a price premium of between 8 and 25 percent for certified stems relative to uncertified material (MEDA 2016).

15.2.2 Availability of Early Generation Seed (EGS) for Seed Entrepreneurs

Formal seed systems, where seed is certified, have been effective for cereal crops such as hybrid maize, but these have rarely been applied to vegetatively propagated crops such as cassava until recently. Large-scale cassava seed production programs have worked intermittently through donor driven initiatives since the 1990s. In seed multiplication projects during this period, an approach of primary, secondary, and tertiary multiplication of cassava stems was applied by the public sector and NGOs. The outbreak of CBSD in East Africa in the 2000s meant that it was very difficult to multiply healthy planting material using existing approaches, and there was a recognition that more formal systems for seed quality control were required. This led to the inclusion of cassava among the seed crops that were officially certified in countries such as Nigeria and Tanzania, although it was recognized that seed certification for cassava would be much more challenging than that for cereals, since cassava seed is bulkier and perishable.

Two large Bill & Melinda Gates Foundation (BMGF) funded projects – BASICS in Nigeria and BEST in Tanzania – have provided impetus both for the completion of policies and regulations relating to the certification of cassava, as well as for activating EGS production efforts (Bentley et al. 2020b). This has involved recruiting a range of public and private organizations for EGS production and delivery. By 2021, several other countries had also implemented similar types of activities, including Burundi, the Democratic Republic of Congo (DRC), Rwanda, and Uganda. In addition, common cassava standards were established as part of the ECOWAS harmonized seed regulations adopted by the West African member states.

EGS production for cassava is still in its infancy, with formal seed accounting for less than 1% of the total, and it is vital that government and development agencies continue to nurture this evolving system. Furthermore, awareness raising and advocacy campaigns on the value of quality seed and seed replacement are important requirements for creating demand for quality seed, which is pivotal for a successful cassava EGS system. Although many African countries have seed laws that provide guidelines for the certification of seed of vegetatively propagated crops, in most cases these are unrealistically stringent and are not applied. A key new requirement for establishing effective commercial seed systems has therefore been the revision of certification guidelines so that they serve both to enhance seed quality while at the same time enabling seed producers to succeed. Costs and benefits associated with these revised certification requirements should be quantified and integrated into business models for cassava seed entrepreneurs.

15.2.3   Productive, Disease-/Pest-Resistant and Farmer-Preferred Varieties

Acquisition of new varieties – whether for their pest/disease resistance or higher yield potential – is the key incentive for farmers to purchase cassava seed. This means that establishing and maintaining a strong breeding pipeline is a key requirement for sustainable cassava seed systems. Productive, disease-/pest-resistant and farmer-preferred varieties are developed through breeding. Breeding is complex and time-consuming as new germplasm needs to be tested for qualitative and quantitative traits in multiple environments. This process may take up to 10 years before varieties with the desired traits can be released. Adoption is enhanced by involving farmers in on-farm evaluations. Since CMD and CBSD are the two most important disease constraints affecting cassava production, the development of varieties resistant to both is a key breeding priority. Requirements for virus-resistant varieties are less stringent in parts of Africa not yet affected by CBSD, which includes all of West Africa as well as the western part of Central Africa. In these regions, breeding efforts therefore place a greater emphasis on yield and the quality characteristics of both fresh and processed roots.

15.2.4 Technologies for Rapid Propagation

Tissue culture The multiplication throughput of clonal crops such as cassava has been significantly increased by using in vitro propagation methods (Ng 1992). These have also helped to introduce virus-tested, true-to-type plantlets into seed systems. Tissue culture activities for cassava are well established in several African countries, such as Nigeria, Ghana, Kenya, Rwanda, and Tanzania. The use of cassava tissue culture plants for commercial production of cassava seed has not been successful, however, particularly in countries such as Nigeria, in spite of the high demand for seed of improved cassava varieties. This is mainly due to the relatively sophisticated laboratory requirements and associated high cost of individual plantlets, as well as the difficulties that users often face in the post-flask management procedures. There has therefore been a need to explore other rapid propagation technologies for early generation seed.

Mini-cutting propagation The traditional method of cassava propagation entails the use of a 20–25 cm cutting that is cut from a mature stem that is usually 1 to 3 meters in length. Each planting stake can have five to ten nodes depending on the variety. Some varieties have nodes that are close together, while in other varieties the nodes are far apart. In the 1990s, breeders sought to explore the use of shorter planting materials (mini-cuttings) to obtain more planting materials for breeding trials. This led to the emergence of nodal cutting technology, where stems are cut into smaller pieces, about 5 cm long and which usually have two nodes. The technology has not been widely used for commercial seed production, so little is known about its performance when tailored toward quality seed production. Nevertheless, when used to multiply planting material for trials, the small size of the cuttings frequently results in many drying out and subsequently failing to sprout when planted. Also, the productivity of both the vegetative part and roots is hampered due to reduced transportation and distribution of assimilates in the plants. A further drawback is the labor required for preparing large volumes of nodal cuttings. For these reasons, mini-cutting propagation has not been widely used in cassava seed systems.

Semi-autotrophic hydroponics (SAH) SAH was developed by the SAHTECHNO Company (Argentina), to enhance the production of potato plantlets for greenhouse or field transplantation. In 2016, the International Institute of Tropical Agriculture (IITA) adapted SAH technology for rapid propagation of cassava and subsequently for yam. SAH offers the benefit of tissue culture grade quality planting material at lower cost and under "low-tech" conditions. SAH makes use of the autotrophic capacity of the plant to grow in a friendlier environment, where there is more effective gas exchange and where growing conditions are more natural (Rigato et al. 2000). SAH is a technology focused on the mass propagation of virus-free plants derived from tissue culture. It is low-cost and easy to set up for commercial seed production and enhanced multiplication in breeding programs. SAH has recently been adapted for cassava propagation in Nigeria. Currently, there are five SAH labs operating in Nigeria: one each at IITA in Ibadan and the National Root Crops Research Institute (NRCRI) in Umudike and three at private companies. SAH laboratories have also been established in other African countries, including the Democratic Republic of Congo (DRC), Malawi, Tanzania, and Zambia.

15.2.5   Seed Quality Assurance and Certification

Guidelines for quality assurance   Cassava has not traditionally been included as part of formal systems for seed quality assurance. This is partly because there is less international trade of cassava, and it has less commercial value than many other crops. This applies in all parts of the world where the crop is grown, including countries where there is a high degree of commercialization, such as South and Southeast Asia. In Africa, regional efforts to harmonize seed quality regulations, such as those included in COMESA's seed trade harmonization regulations of 2014, have not included cassava. The CMD and CBSD virus disease pandemics of the 1990s and 2000s, however, highlighted the need to have mechanisms for the production and trade in cassava seed that was free of these two diseases. Models for seed quality certification were piloted by national seed certification agencies in Tanzania and Nigeria and have subsequently been improved and scaled out to several other countries, including Burundi, DRC, and Rwanda. Specific details of the protocols used in each country vary, but all cases address the key elements of varietal purity, pest/disease incidence levels, virus testing, isolation distance, and the number of ratoon crops allowed.

Technologies and infrastructure for seed quality assessment including lab and field diagnostics   Seed certification systems normally include field inspections, where visual assessments are made of pest/disease incidence, as well as some laboratory testing for higher seed categories (pre-basic/breeder seed). Most national seed certification authorities have existing human and physical capacity for field and lab assessments, which are primarily used for the major seed crops, such as the cereals, oilseeds, and grain legumes. This means that there is at least some potential to adapt these capacities to implement cassava seed quality assurance. However, seed inspectors will need specific training on cassava. More inspectors may need to be hired, and decentralized, in order to provide enough inspections for new cassava seed entrepreneurs. Such efforts need to ensure that the seed certification systems are affordable and reliable. For example, in Tanzania, decentralization efforts focused on training existing extension staff as authorized seed inspectors (ASIs) who are able to conduct basic certification assessments on behalf of the regulatory authority (Tanzania Official Seed Certification Institute, TOSCI). This has greatly reduced costs since ASIs are based throughout the seed production zones (Douthwaite 2020). New digital tools, such as PlantVillage Nuru, are available for field-based disease/pest identification for cassava, although visual inspections continue to be the main method used by the ASIs. Laboratory-based virus diagnostics may be required at higher levels of the seed system, particularly in countries where the inconspicuous CBSD occurs. Novel diagnostic methods are likely to become more widely used as they reduce virus testing costs and are easier to conduct in lightly equipped laboratories. Currently, new methods such as loop-mediated isothermal amplification PCR (LAMP) are being validated with regulatory authorities in several countries.

Electronic information systems in support of certification, business registration, and marketing Cassava seed production involves a chain of interconnected activities between various institutions and entrepreneurs formally recognized to produce and market different classes of certified seed of released varieties. The lack of awareness about seed regulations, cumbersome procedures, and the high cost of seed certification are some of the challenges to implementing seed certification. To simplify this process and integrate the cassava seed production value chain, IITA developed the "Seed TrackerTM" ICT app (www.seedtracker.org/cassava) that digitally links multistage seed value chain actors. The app is usable on any Internet-enabled device, including mobile phones, tablets, and personal computers. It can be used to collect and organize seed production data and for real-time tracking of seed production. It generates maps, offers secure accounts for data input and retrieval, and provides a diverse set of analytics (Fig. 15.1). The app features customized digital data collection forms with offline data collection allowing it to be used in areas without an Internet connection, and it relays data remotely when the device gets connected to Wi-Fi. The application is tailored to national seed certification systems and provides support for registration of seed producers, registration of seed fields, field inspection, and certification of seed lots and generates barcoded labels for traceability.

The National Agricultural Seed Council of Nigeria (NASC) in 2017 and TOSCI in 2018 adopted Seed Tracker for e-certification of cassava. Seed certification officers were trained to perform e-seed certification and manage seed quality assurance activities using Seed Tracker. In 2019, NASC adopted Seed Tracker as a national e-certification platform and expanded its use to all crops. Seed Tracker offers a "one-stop shop" for information on seed quantity, location, and variety, enabling buyers to know when and where certified seed is available. It therefore supports the adoption and impact of improved varieties developed through breeding programs. Real-time reporting and the simplification of the administrative processes associated with certification mean that Seed Tracker can help seed system stakeholders overcome many of the problems traditionally associated with formal seed certification programs. Regulatory agencies are also using experience gained from other

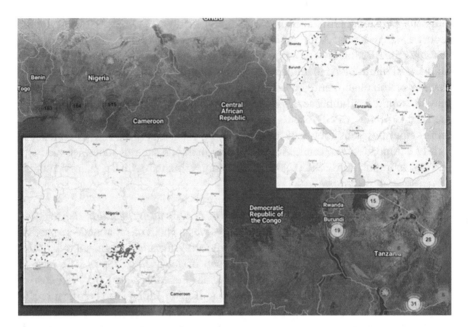

Fig. 15.1 Distribution of commercial cassava seed producers registered on Seed Tracker in Nigeria and Tanzania

more commercialized crops to develop ICT systems for the digital tagging of seed packages in order to prevent the sale of counterfeit seed.

15.2.6   *Enabling Environment for Business Development*

There are several important requirements for creating an enabling environment for commercial cassava seed system development. These include the free flow of information and data among all system participants, availability and access to sources of high-quality seed that targets demand, strong markets for cassava products, a business-friendly agricultural policy with ready access to business development services, and the system-wide application of a seed regulatory system that promotes quality while remaining affordable.

Information flow   Seed producers should be aware of market demand, of available varieties to meet that demand and of the crop management requirements for producing high-quality cassava seed. Root producers need information about the relative benefits to be gained from improved seed. Processors should have access to information about sources of roots, their quality traits, and markets for processed products. Finally, for effective quality control, seed regulators require a thorough understanding of the network of seed producers with easy, real-time access to information about the producers and their seed crops.

Availability and access to seed In order for a cassava seed system to flourish, seed producers at all levels, as well as root producers, need to have easy access to sources of quality seed. Unlike seed of cereal crops or grain legumes, cassava seed is not typically sold through markets or input dealers. Informal systems prevalent in most cassava-producing countries in Africa are dominated by local exchange with neighbors and relatives, with occasional roadside selling. Newly emerging commercial seed systems for cassava are addressing the issue of access and availability by enhancing information systems to openly share seed information such as variety, quantity available, price, and location.

Strong markets Cassava producers are paying greater attention to the commercial potential of the crop as new processing businesses and market outlets are set up, most notably in Nigeria. Where cassava is sold, the importance of productivity is increasingly recognized, leading to increased farmer demand for improved high-yielding varieties and a greater willingness to invest in inputs such as certified seed. Therefore, commercial seed systems are unlikely to succeed for cassava, unless there is a commitment to establish and sustain output markets for fresh or processed cassava products.

Business-friendly agricultural policy Business policy can have a critical impact on the likelihood of success for agricultural business, including seed production. Common impediments to success can include bureaucracy that can make it difficult to set up and run a business, weak government support for the agricultural sector, lack of access to finance, and poor infrastructure in agricultural areas. In countries where commercial cassava seed systems have not yet been established, business training may be required to support cassava seed entrepreneurs setting up businesses for the first time. Cassava seed entrepreneurs should also have ready access to the range of business development services required to establish and maintain their new seed businesses, including access to inputs, marketing advice, telephone and Internet infrastructure, as well as financial services.

Practical seed regulatory system Seed policy needs to be market-responsive, widely known, and implemented well. The first step for many countries will be defining guidelines for affordable seed quality regulation. The threats posed by vegetatively propagated viruses, such as those causing CMD and CBSD, as well as an increasing demand for greater productivity, have encouraged the implementation of formal systems for seed quality assurance in several countries. However, these certification systems must enhance the quality of cassava seed, be applied in a sustainable way, and be practically scaled beyond a project context, with affordable tariffs that are appropriate for the business plans of seed producers at each level of the system.

15.3 African Experiences of Establishing Sustainable Cassava Seed Systems: Case Studies from Nigeria and Tanzania

15.3.1 Building a Sustainable Integrated Seed System for Cassava in Nigeria (BASICS)

The context: cassava in Nigeria in the 2000s Nigeria is the world's largest producer of cassava (59 million t/yr). About 90% of cassava in the country is processed as food, 9% for industrial flour or starch, and less than 1% of the cassava output is exported. For food, most is produced and processed locally into gari (grated, fermented, and toasted to a dry granular flour), fufu/akpu (fermented wet paste then cooked with additional water to make a thick porridge), or lafun (dried chips ground into flour and cooked into a thick porridge). Almost all processing is done by women at home or in specialized processing centers in rural areas.

Up until the early 1990s, cassava cultivation was mostly based on traditional low-yielding cultivars (with average yields of 7–10 t/ha) and manual processing. Since the early 2000s, many states in Nigeria have witnessed greater attention by different actors to the promotion of cassava as an industrial crop, with the objectives of diversifying farmers' incomes, enhancing foreign exchange earnings and increasing employment opportunities. This process was boosted further between 2002 and 2010 with the implementation of a Presidential Initiative that aimed at promoting cassava for the purposes of poverty alleviation and improved food security. The development of cassava markets in several states during this time enhanced the adoption of high-yielding cultivars, with on-farm yield potential up to 25 t/ha, resulting in increasing mechanization of labor-intensive processing stages (grating, pressing, roasting, drying, and milling). The impact of these developments was positive, although productivity in Nigeria (8.2 t/ha – FAOSTAT 2021) remains significantly below the potential. This highlights the need for further efforts to establish a sustainable system for delivering quality seed of high-yielding cassava varieties to farmers.

Brief history of cassava seed production initiatives in Nigeria In spite of the importance of the cassava crop for livelihoods in Nigeria, the seed system remains largely informal. Some of the key constraints include an insufficient and unsustainable supply of EGS, a poorly functioning national variety release system, policies that limit access to publicly develop improved varieties by private seed companies, and the widespread occurrence of seed in markets which is not true-to-type.

Formal plant breeding for cassava started in Nigeria in the 1970s, although there were few large-scale efforts to disseminate improved varieties until the early 2000s after which projects began to play the major role in cassava seed delivery (Okechukwu and Kumar 2016). Examples included the cassava mosaic disease (CMD) project from 2004 to 2007, the Cassava Enterprise Development Program

(CEDP) (2008–2010), the UPoCA Project (2008–2010), the CFC West Africa Project (2008–2012), and Cassava: Adding Value for Africa (CAVA) (2008–2019). These projects distributed disease-resistant varieties to farmers, who then distributed to community seed gardens, which multiplied seed for smallholders. Seed gardens and multiplication work were project-funded activities. None of these projects had a commercial focus for seed delivery, so as the projects came to an end, seed multiplication and dissemination also stopped.

The first project to attempt more formal commercialization of the cassava seed business was the sustainable cassava seed system (SCSS) project (2012–2015), which was the first to work with Village Seed Entrepreneurs (VSEs). From 2016, this approach was carried forward and expanded through the project entitled "Building an Economically Sustainable Integrated Cassava Seed System in Nigeria" (BASICS), managed by the CGIAR Research Program on Roots, Tubers, and Bananas (RTB). This was the first initiative in Nigeria to attempt to develop a comprehensive sustainable system for the supply of improved cassava seed. Some of the key partners in this work were NRCRI, NASC, IITA, Catholic Relief Services (CRS), Context Global Network (USA), and Fera Science Limited (UK). All partners worked with VSEs and private sector processing companies to commercialize cassava seed. By the time the grant to BASICS had come to an end in early 2020, significant progress had been made to build the breeder, foundation, and commercial seed levels necessary to create an economically sustainable seed system in Nigeria. But there were still kinks in the system and fragility in the new institutions. Most notably, efficient and practical EGS production approaches had not been established, the cassava seed regulatory system had just been implemented at pilot level, and there was very little geographical coverage of areas targeted by the sustainable seed production effort. Additionally, IITA and NRCRI had established two new private seed companies. These were GoSeed and Umudike Seeds, respectively, and were still in the early stages of becoming fully self-sustaining by the end of the BASICS project. In order to cement progress already achieved in building sustainable cassava seed systems, BMGF funded another phase – BASICS-2 – which aimed to strengthen the work in Nigeria and link it with similar activities in Tanzania (the BEST project) to consolidate the progress made to date, address the weaknesses, and begin the process of marketing this novel model to other governments and donors for replication.

The business case for commercial seed production, markets, and demand Prior to the implementation of the SCSS and the BASICS projects in Nigeria, the view was that very few farmers saw a business opportunity in cassava seed production. BASICS, building on the SCSS experience, showed that seed entrepreneurs can produce and sell certified seed of improved varieties, generating profits for both the seller and the buyer. The project documented the profitability of commercial seed production by the top 50% of the VSEs in different states in Nigeria. These VSEs in Benue State and South-South regions made average net profits of US$ 988 and US$ 551 per ha respectively (Table 15.1). Some VSEs in Benue State followed the ratooning strategy, and they obtained a net profit of US$ 1588 at the end of the two

Table 15.1 Average profitability of cassava commercial seed production by VSEs per ha in different regions of Nigeria. (Typical sale prices of cassavas stems/bundle: Foundation Seed US$ 2, Commercial Seed US$ 1–1.5, informal (roadside) unverified seed US$0.5–1, although prices vary by region and market demand)

Production costs (US$/ha)	VSEs Benue State	VSEs Abia, Imo, Akwa Ibom, and Cross Rivers States
Land (prevailing lease cost – actually paid or opportunity cost)	19	70
Seed (foundation seed – 100 bundles/ha)	153	114
Labor (planting, weeding, harvesting, etc.)	238	506
Inputs (including herbicides and other pesticides, fertilizer)	112	205
Miscellaneous (fuel, repairs, transport, utilities, certification, fees, etc.)	78	50
Total production costs (US$/ha)	600	945
Revenues		
Stem yield (bundles/ha)	300	412
Total revenue from stem sales (US$/ha)	566	916
Root yield (t/ha)	15	15
Total revenue from root sales (US$/ha)	1022	580
Total revenue (US$/ha)	1588	1496
Summary		
Total production cost (US$/ha)	600	945
Total revenue (US$/ha)	1588	1496
Net profit (US$/ha)	988	551

seasons. Most of the entrepreneurs sold their cassava stems at the farm gate, followed by sales at the local market; little seed reached distant markets.

Both the VSEs and the cassava varieties they were producing needed to be promoted via various channels including demo plots alongside main roads in various locations, market day promotions, advertisements in banners and flyers in local languages, radio spots, and a public service call-in system that provided free, on-demand information. As these were supported through the BASICS project, going forward, such commercialization activities and costs need to be incorporated into the business plan of the seed companies and farmer associations selling certified seed. In addition, a website (www.cassavastems.com) was launched providing information about the project, the VSEs with their location and contact details, as well as the varieties available.

A review of the more successful VSEs showed that the viability of commercial seed production is enhanced when the producer can dedicate larger areas for production (>1 ha) and that smaller VSEs need to be organized into an association that can support training, access to inputs, and markets by bringing in economies of scale. Moreover, a survey of 218 cassava farmers who bought certified seed from VSEs in 2017–2019 showed that the main reason for buying stems was to access new varieties and to be sure of their identity. These criteria were considered much

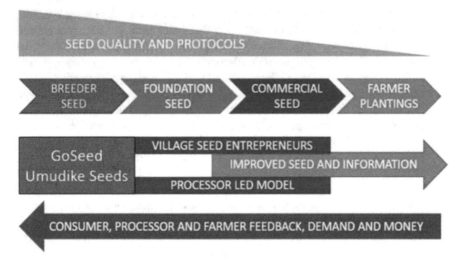

Fig. 15.2 Integrated approach for the seed value chain

more important than the state of health of the certified material. Nevertheless, NASC certification did mean that the VSEs' trustworthiness and prestige was enhanced. It was observed early on that women VSEs face more challenges in accessing credit and inputs. Consequently, the BASICS project placed a special focus on building capacity and promoting social capital and networking among female seed entrepreneurs.

Approaches to achieving a sustainable supply of quality seed Through the SCSS and BASICS projects in Nigeria, a sustainable supply of quality EGS and certified seed was established following different approaches. The integrated value chain approach that this encompasses is illustrated in Fig. 15.2.

(i) *EGS system*: An adequate and reliable supply of healthy and virus-free breeder and foundation seed of improved varieties is the starting point for a sustainable seed system. IITA and NRCRI in Nigeria are the mandated institutions for these seed sources. In a pioneering intervention, the BASICS project helped IITA and NRCRI establish the private limited companies, GoSeed and Umudike Seeds, respectively. These companies produce and sell EGS of improved varieties on a commercial basis. With the assistance of Sahel Consulting Agriculture and Nutrition Ltd (Sahel), financially viable business plans were developed for both these entities, and they started implementing the plans in 2019. GoSeed has established a network of contract farmers to produce breeder and foundation seed, to increase capacity and expand the geographic reach of EGS. Both companies certify their seed through NASC following updated certification protocols, ensuring seed health as well as varietal identity. Foundation seed is sold to commercial seed producers, and the companies are now expanding their commercialization efforts by promoting new improved varieties. GoSeed and Umudike Seeds are linked to the breeding programs of IITA and NRCRI, and as new varieties are tested with farmers and processors, end user experiences and requirements for improvements are fed back to the breeding programs. These two companies are also expanding marketing activities to promote the improved varieties and certified seed.

(ii) *Certified seed system*: Certified seed is produced from foundation seed and can be multiplied through ratooning a maximum of three times by certified seed producers, subject to approvals by NASC inspectors. Certification occurs at breeder seed, foundation seed, and certified seed levels. Certification standards at the highest levels of EGS production are the most stringent, as these impact the entire seed system at national level. Standards at VSE level are more lenient to ensure that seed producers at this level succeed but also so that the inspections are affordable. Cassava farmers purchase certified seed for root production. Every planting season, there is a large demand for planting material. Traditionally, this has been met by farmers' own saved seed; from friends, family and neighbors; through dissemination campaigns from governments and projects; via informal markets; or from formal markets (Pircher et al. 2019). The formal seed market is almost negligible. According to a NASC Annual Report in 2017, only 117,722 bundles (0.0003% of the total cassava seed market) of certified cassava seed were produced and sold. Given this situation, the BASICS project developed two models of certified seed production: the village seed entrepreneur (VSE) model and the processor-led model (PLM), described below. The initial scaling target of the certified seed system is to meet 2% of the cassava seed requirement for Nigeria, which would be equivalent to 5–10% of the nation's formal seed production.

(iii) *VSE model*: The project selected farmers, both young and old, men and women, trained them on cassava agronomy, quality assurance for meeting certification requirements, and business aspects of the seed business. VSEs were selected according to nine criteria, mostly relating to experience of agriculture as a business and willingness to adopt innovations. Criteria relating to assets were relaxed for women farmers to avoid structural bias. Approximately 30% of the VSEs were women. Responding to the demand for improved varieties in their vicinity, the VSEs grew these varieties and marketed them to farmers nearby. In order to enhance sustainability, the project organized the VSEs into a network that continued the training and promotion of the member VSEs as well as marketing of new improved varieties. The profit motive for such seed entrepreneurs ensures sustainability.

(iv) *PLM*: Traditionally cassava processors bought the roots needed for their factories in open markets, and this meant that they got a mixture of varieties, mostly local varieties, thus potentially reducing the factory's profitability as well as not fully meeting the processing demands for particular quality traits in the roots. The BASICS project selected cassava processors who demonstrated a strong interest in piloting the PLM model and developing business-oriented systems of seed dissemination. The best varieties for each processor were identified by setting up demonstration trials at the processor sites and processing the roots in their factory. The seed of these varieties was then supplied to a network of cassava outgrowers linked to the processor, with a root buyback agreement. In this manner, the PLM model looks to have the farmers benefit from the higher yields obtained from certified seed of improved varieties along with a guaranteed market. The processors benefit from getting an assured quantity and quality of cassava roots for processing. The increased net profitability ensures the sustainability of the PLM.

New varieties delivered and technologies established to increase the volume and efficiency in the supply of high-quality seed Through collaboration between the NextGen and BASICS projects, five new varieties were released in Nigeria in 2020. NextGen is a BMGF-funded project which aims to modernize cassava breeding in Africa using cutting-edge tools. All five varieties were targeted to meet the product profiles for the country which range from industrial applications, primarily high-quality cassava flour (HQCF) and starch, to fresh roots, gari, and fufu. Varieties with different optimal end uses were promoted using a market segmentation approach (e.g., each variety was marketed to the customers most likely to use it). In addition, a naming event was organized to rename the varieties from conventional breeding code names (mostly made up of numbers and easy to forget), to more farmer-friendly names. The renaming of the varieties was a strategy aimed at enhancing the promotion and delivery of the new varieties to farmers. Historically, when new varieties are released, it can take several years of field multiplication before the variety reaches grassroots farming communities. Difficulties experienced in accessing planting materials of new varieties can discourage farmers from cultivating them. Novel rapid multiplication techniques, such as SAH, combined with improved systems for the rapid dissemination of cassava seed, have given farmers much faster access to the new varieties. This has been further enhanced by establishing the EGS companies – IITA GoSeed and Umudike Seeds – and farmers can now access seed of new varieties within 2–3 years of their initial release. The EGS companies in addition to the production and commercialization of the varieties are also helping independent seed producers across the country increase the production of foundation seed. This will bolster sustainability of seed demand and supply of cassava seed in a structured manner.

Assuring the quality of cassava seed from breeder seed to community level NASC is the government agency responsible for regulating the production, marketing, and trade of seed in Nigeria, as well as for implementing national seed policy and regulations. NASC has adopted a three-tier system for cassava seed production, comprising breeder seed (BS), foundation seed (FS), and certified seed (CS). Cassava seed quality assurance involves control of the generations of seed and seed quality, as defined by its genetic purity and health status. BS is the first seed generation produced under the supervision of plant breeders, and FS is the progeny of BS, and both may be recycled twice. CS may be multiplied three times. The quality parameters considered during each cycle include genetic purity, isolation distance, field management (weediness), and incidences of notified pests and diseases which must

Table 15.2 Maximum pest and disease thresholds for Nuclear Seed (NS), BS, FS, and CS for cassava in Nigeria

	Maximum threshold			
Parameter	NS	BS	FS	CS
Off type	0	0%	≤3%	≤5%
[2]Cassava mosaic disease	0	0%	≤2%	≤2%
[1]Cassava anthracnose disease	0	≤3	≤3	≤3
[1]Cassava bacterial blight	0	≤3	≤3	≤3
[2]Scale insects and termites	0	≤5%	≤5%	≤10%
[1]Cassava mealybug	0	≤3	≤3	≤3
[1]Cassava green mite	0	≤3	≤3	≤3
[1]Cassava white or brown spot	0	≤3	≤3	≤3

Minimum isolation distance 3 m for BS, FS and CS; not applicable for NS
At least 7 months old for stem harvesting (stem girth of 1.5 to 2.5 cm)
About 1 m stems; 50 stems per bundle
No recycling limit for NS, two generations for BS and FS, and three generations for CS

[1]Mean severity score per field (using a 1–5 scale ranging from 1 = no symptoms to 5 = very severe)
[2]Percent incidence

be lower than tolerance levels set for each class (Table 15.2). If seed does not meet the required standards, it is rejected and cannot be sold as certified seed.

This system for cassava seed certification functions effectively in Nigeria, although the limited supply of tissue culture material to start the process has been a bottleneck. The adoption of SAH since 2016 helped to address this situation. In addition, a new NASC-approved seed class (nuclear seed, NS) was recently added. NS is produced from TC or SAH material and is used as the source for BS. NS seed is not limited by seed cycle, as long as the planting material remains genetically pure and true-to-type, and the material remains entirely free of notified pests and diseases.

As some VSEs do not meet all of the eligibility requirements for CS production, NASC allows them to be registered and produce CS seed as Community Seed Producers.

NASC organizes a minimum of three inspections. The first is prior to field establishment to verify seed source and field site. The second is 3–4 months after planting, and the third inspection is at the time of harvest. NASC may propose additional inspection visits, as well as the collection of samples – especially from EGS fields for testing of cassava mosaic viruses, which are endemic in Nigeria. Viruses that cause CBSD are not known to occur in Nigeria; however, NASC, IITA, and cassava stakeholders have prioritized CBSD prevention since this disease poses a major threat to cassava production in Nigeria. NASC has therefore established facilities for testing for cassava brown streak viruses and has trained inspectors in recognizing symptoms and how to respond in case the disease is introduced to the country.

Due to the limited number of seed certification officers, timely certification has been a bottleneck. To overcome this challenge, NASC introduced the use of "licensed seed inspectors (LSIs)" who are third party agents trained and approved to perform CS field inspection and to collect samples for testing under NASC supervision. Forty LSIs from four Nigerian states were commissioned in 2020.

To enhance seed quality compliance and enable e-seed certification, NASC adopted Seed Tracker in 2019 as the official e-certification tool for cassava. This ICT application has been tailored to NASC seed regulations to enable seed producer registration, seed field registration, inspections, digital data collection, and issuance of certificates and labels for eligible producers and seed fields.

Alternative "light touch" quality assurance models have been proposed for cassava seed systems in Nigeria that maintain certification-oriented measures for EGS but which recommend less regulation for community-level seed production and a primary focus on the delivery of seed of new varieties with known provenance (Wossen et al. 2020).

15.3.2 Building an Economically Sustainable Seed System for Cassava in Tanzania (BEST)

The context. Cassava in Tanzania in the 2000s Tanzania is the fifth largest cassava producer in Africa, with an annual production for 2019 of nearly 8.2 million tons (FAOSTAT 2021). Although yields declined from 1997 (8.6 t/ha) to 2015 (5.4 t/ha), largely due to CMD and CBSD epidemics, they have subsequently recovered – reaching 8.3 t/ha in the latest year for which data are available (2019). The 2002–2003 National Sample Census of Agriculture indicated that 24% of Tanzanian farmers grow cassava and that the production is concentrated in the southeast and northwestern parts of the country. Most farmers market only a small proportion of their output, although there are a relatively small number of medium- and large-scale farmers who sell most of their harvest. In Mtwara, the region in Tanzania that is most dependent on cassava, only 17% of the output is marketed. More fresh cassava might be sold if the roots were not so perishable. Cassava in Tanzania was heavily impacted by the severe CMD pandemic which spread through northwestern production zones from 1998 to 2010 (Legg 1999, 2010), compounded by outbreaks of CBSD in the same areas from 2006 to the present day (2022). In view of the current importance of cassava in Tanzania, any reduction in productivity arising from a disease or pest outbreak can lead to decreased food security, reduced incomes, higher food prices, and subsequently to social tension. To tackle emerging disease and pest attacks, farmers need access to high-quality seed of newly developed varieties with multiple pest and disease resistance. This highlights the particular importance of "clean seed" systems in Tanzania.

Brief history of cassava seed production initiatives in Tanzania Cassava seed systems have been functioning in an informal way ever since the crop was introduced to Tanzania. Seed exchange among farmers across villages, towns, and even countries in the region is a normal practice in the informal seed system. Most cuttings are obtained from friends, relatives, or neighbors without payment. Some in-kind sales have also occurred in addition to free exchange. In the informal seed system farmers may select seed from vigorous plants in their own fields or obtain seed of favored varieties from trusted sources, but there is no systematic approach to managing quality. Although informal approaches may work under normal circumstances, they can often result in rapid and widespread declines in productivity where cassava is affected by diseases such as CMD and CBSD, which are readily propagated through stem cuttings.

It has been estimated that less than 2% of cassava planting material in Tanzania is quality seed (AGRA 2016), although the current proportion of certified planting material by 2022 may be less than 1%. Previously, between 1995 and 2000, the Southern Africa Root Crops Research Network (SARRNET) introduced a semiformal, three-tier seed multiplication scheme. Cassava was multiplied at primary, secondary, and tertiary levels. At the primary level, the sites were at or near research stations for easy supervision and monitoring by scientists. At the secondary level, the sites were established and managed by extension staff, NGOs, faith-based groups, and some individual farmers. Tertiary multiplication sites were mainly smaller, more numerous, and managed by farmers but usually backstopped by scientists, extension staff, or NGOs.

From 2007 to 2012, the BMGF-funded Great Lakes Cassava Initiative (GLCI) project operated across six countries in East and Central Africa including Tanzania (Walsh 2016). The project brought together national and international research, government, and nongovernmental organizations in a single network tasked with producing and delivering disease-tolerant cassava seed to 1.15 million smallholder farming families. GLCI partners collectively learned and applied four innovative seed system approaches: (1) decentralized production and dissemination, (2) quality management protocols, (3) targeted dissemination and traceability in the seed system, and (4) disease surveillance, including source site field testing with virus diagnostics to control CBSD spread. This semiformal seed dissemination system promoted quality of seed but was not formally administered by government regulatory institutions.

In 2014, TARI piloted public private partnerships in cassava seed systems. Two private companies – Aminata seed company and CBS Arusha – were involved in producing basic seed and TC plantlets respectively. TARI piloted pre-basic, field-based seed production at the station for closer supervision. From 2012 to 2017, scientists from IITA, TARI, and the Mennonite Economic Development Associates (MEDA) increased their efforts to advocate for a commercialized, clean (healthy) cassava seed system and implemented several projects to address this challenge. Currently, there is a strong pipeline of formally released superior varieties that flow into an organized seed system with registered cassava seed entrepreneurs, where quality is controlled by seed inspectors through a legally enforced but producer-friendly seed certification process. Moreover, the Government of Tanzania is providing strong policy-level support for the modernized seed system as part of its overall National Cassava Development Strategy which was officially launched in 2021 (IITA 2021).

The business case for commercial seed production, markets, and demand Demand for cassava seed production comes from farmers who are growing cassava roots that they consume or sell to markets. The market in Tanzania comes primarily from home consumption and fresh root sales. Increasingly, commercial markets are developing with small-scale HQCF production and, more recently, with larger-scale investments being made into flour and starch production and even nascent opportunities with industrial applications such as ethanol production and export of dried cassava. The Tanzanian demand is amplified by the high disease pressure of CMD and CBSD which dramatically reduces farm productivity and increases the farmers' need for improved, disease-resistant, and high-yielding varieties. The commercial cassava seed system is the mechanism for these farmers to access the improved varieties that have been released by the government.

The commercial seed production model based on cassava seed entrepreneurs (CSEs) distributed in a hub-and-spoke model with quality control oversight is a profitable and sustainable model to reach cassava seed farmers. The development of the models in the BEST Cassava project has demonstrated that farmers are willing to pay for improved cassava seed that has been certified. The average profitability of CSEs from 2018 to 2020 ranged from US$1000 to 1500 per ha which demonstrates a positive business case, and it is a key indicator for a commercially sustainable seed system. Farmer business associations have been set up to support CSE networks with the aim of gradually taking over training and support functions which are currently being delivered through the BEST Project. In addition, the Tanzania Cassava Producers and Processors Association (TACAPPA) is a newly created independent organization which aims to provide long-term advocacy and support for cassava farmers, processors, and traders, including CSEs. This and allied government initiatives to foster the commercial development of cassava in Tanzania are providing a strong business-oriented environment which will enhance the sustainability of the emerging cassava seed system.

Approaches to achieving a sustainable supply of quality seed The cassava seed system in Tanzania was developed using a formal, commercialized approach that factors in quality assurance and profitability for the CSEs to ensure a cost-effective and sustainable system. Since cassava seed is bulky and perishable to transport, the BEST Cassava project has developed the seed system based on a hub-and-spoke model to foster effective seed delivery. It includes a smaller number of pre-basic (equivalent to BS) and basic (equivalent to FS) CSEs, who then sell cassava seed in subsequent steps to certified and quality declared seed (QDS) CSEs who ultimately sell to root producing farmers (Fig. 15.3). Cultural norms and traditional patterns of land ownership mean that there have been difficulties in encouraging gender balance among CSEs. Currently women comprise about 30% of CSEs, although the proportion is higher at the community level and lower among basic seed producers. Under the follow-on BASICS-2 project, however, there are twin goals to nearly double the

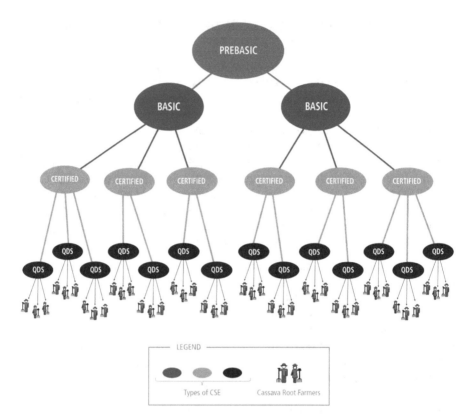

Fig. 15.3 Hub-and-spoke model of the commercialized cassava seed system in Tanzania. CSE - cassava seed entrepreneur, QDS - quality declared seed

number of CSEs while at the same time increasing representation of women to 50% of the total.

(i) *Pre-basic/basic seed.* The production of pre-basic seed in Tanzania is overseen by TARI. The role of this level is to produce initial seed stock of the released varieties that will then feed into the commercial seed system for distribution. The original model for pre-basic seed production was to take small amounts of lab-tested, virus-free stock from tissue culture, harden it off in a screenhouse, plant it in the field for one cycle of field multiplication, and sell mature cassava cuttings to basic CSEs who would then multiply one more cycle of seed and sell it to the next level – certified CSEs. This model is evolving with more advanced techniques being used at both the pre-basic and basic levels to increase the capacity and production. Pre-basic producers now use a "two-node" rapid multiplication step in the screenhouse and aim to incorporate SAH capability. Basic CSEs are also developing their own screenhouse capability for "two-node" rapid multiplication.

(ii) *Certified seed.* Certified CSEs purchase basic seed as their input and multiply a cycle of seed in the field before they sell to QDS CSEs. To maximize profitability of the model, certified CSEs vary ratooning strategies and planting densities (within the parameters laid out in the Tanzania certification protocols) to optimize their stem production. They also sell roots to supplement stem sales either using a piecemeal root harvest strategy or selling all the roots at the end of their ratooning cycle when it is time to refresh their fields with new basic seed stock.

(iii) *QDS seed.* QDS producers sell both their stem harvest to farmers and their root harvest to the market. To gain certification, QDS CSEs are inspected by decentralized ASIs who have been gazetted by TOSCI. The QDS CSEs also have the least stringent certification requirements of any level in the seed system. QDS producers plant at an optimum spacing for root production (1 m × 1 m) rather than for stem yield. Their farming strategy is driven more by root yield than by stem production.

(iv) *CSE business associations.* Business associations made up of CSE members have been formally organized and registered as legal entities in Tanzania. The role of these business associations is to support the CSEs in multiple aspects of making the CSE businesses sustainable such as registering their business, training on good agricultural practices, providing access to finance through a revolving loan fund, and marketing their products.

New varieties delivered and technologies established to increase the volume and efficiency in the supply of high-quality seed Tanzanian researchers have officially released 26 improved cassava varieties that are high-yielding, have high dry matter content, and that are tolerant to CMD and CBSD. Most of the new released varieties are not known by farmers, and it may take up to 10 years for a new variety to become popular. For example, variety Kiroba was officially released in 2004 but only became popular after 10 years. TARI has therefore collaborated with partners such as IITA, MEDA, and farmers' associations in efforts to promote new varieties through demonstration plots, farmer field days, seed fairs, annual agricultural shows, and TARI agribusiness expo events organized at research centers. As part of this effort, 109 plots to demonstrate released varieties were established in 2020 in the Lake Zone (30), Eastern Zone (61), Southern Zone (17), and Central Zone (1). Most of the varieties were selected primarily for yield and virus disease resistance, although they have diverse processing qualities. Targeting specific varieties for particular output markets as part of a market segmentation strategy is an important future objective of cassava value chain stakeholders in Tanzania.

Technologies that have been implemented in Tanzania to increase the volume and efficiency in the early stages of multiplication include tissue culture, mini-cuttings, the pencil stem technique, and SAH which has been piloted by a private company, KilimOrgano. TARI has implemented the use of mini-cuttings and the pencil stem technique for rapid propagation of new varieties. Stems are allowed to grow from hardened TC plantlets for about 8 weeks under screenhouse conditions.

Green mini-cuttings are then harvested to establish new plants under controlled conditions. With this method 1 plantlet can give up to 300 plants in 6 months.

Assuring the quality of cassava seed from pre-basic to community level Seed quality in Tanzania is controlled by TOSCI. As in Nigeria and elsewhere, the two primary considerations during seed quality inspections are the genetic purity of the material and the restriction of pest and disease incidences to within prescribed tolerance levels. Four seed classes are recognized under the Seed Act of the Government of Tanzania. These are pre-basic (PB), basic (B), certified (C), and quality declared seed (QDS). TOSCI inspects all PB, B, and C fields on at least two occasions and is mandated to inspect at least 10% of QDS fields twice. In view of the small number of trained TOSCI seed inspectors relative to the size of the country, a selected group of extension officers has been trained to conduct QDS inspections on behalf of TOSCI as ASIs. By the end of 2021, there were 130 officially gazetted ASIs in Tanzania.

Field inspections have traditionally been documented by filling out paper forms. Since 2020, however, inspectors have been able to conduct certification assessments digitally using tablets and the Seed Tracker platform. During assessment, 200 plants are examined in each of 5 quadrants of 40 plants for each hectare. The presence of major pests and diseases is scored (Table 15.3), either on the basis of the percentage of symptomatic plants or the average severity, which is assessed using a one to five scale in which one is unaffected and five is very severely affected. The two most significant differences in certification systems in Tanzania and Nigeria both arise from the greater importance of the virus diseases (CMD and CBSD) in Tanzania than in Nigeria. Required isolation distances are much greater in Tanzania (50 m to 300 m) than in Nigeria (3 meters), while tolerance levels for CMD are less

Table 15.3 Maximum pest and disease thresholds for PB, B, CS, and QDS for cassava in Tanzania

Parameter	Maximum threshold			
	PB	B	CS	QDS
Off type	0	0%	1%	1%
Minimum isolation distance	300 m	200 m	100 m	50 m
[1]Cassava mosaic disease	1%	2%	3%	10%
[1]Cassava brown streak disease	2%	4%	7%	10%
[2]Cassava bacterial blight	2.5	2.5	2.5	3.5
[1]Scale insects and termites	1%	2%	4%	8%
[1]Cassava mealybug	1	2%	4%	8%
[2]Cassava green mite	2.5	3.0	3.5	3.5

Harvest age (new crop: 8–18 months; ratoon crop 6–12 months)
Minimum cutting sizes (length: 20 cm; girth: 2 cm; nodes: 5)
Maximum ratoons (PB, B, CS: 2; QDS: 1); Maximum shoots per ratoon: 3
Minimum number of inspections: 2

[1]Percent incidence
[2]Mean severity score per field (using a 1–5 scale ranging from 1 = no symptoms to 5 = very severe)

restrictive in Tanzania (3% at CS level) than they are in Nigeria ($\leq$2% at the equivalent CS level).

Virus testing is only required for PB seed production where the sampling protocol provides for the testing of 2% of plantlets. Testing is only done for cassava brown streak viruses, and this is conducted by TOSCI using a real-time PCR protocol (Adams et al. 2013). PB producers pay TOSCI for this testing, and the charges are incorporated into the PB business plans and are factored into the price of the PB seed. Recent work has focused on the development of LAMP protocols for virus testing that will be cheaper and quicker and which will have the potential to be run in a decentralized network of less well-equipped laboratories.

Tanzania continues to strengthen its cassava seed system, and there are plans for the period 2022–2025 to increase the number of CSEs to more than 1000. The decentralized seed certification system that is being established should offer sustainable capacity to maintain the quality of seed produced through this system while at the same time improving the awareness of the importance of seed quality among all actors in the cassava sub-sector in the country.

15.4 Key Lessons Learned and Future Opportunities

Coordinated cassava seed multiplication initiatives have been running in many of the major cassava-producing countries of Africa for several decades. It is only in recent years, however, that more attention has been directed at establishing sustainable and commercially oriented seed systems for cassava. The largest initiatives of this type, described in this chapter, have been the BASICS project in Nigeria and the BEST project in Tanzania. Both have provided valuable lessons about the potential viability of business-focused cassava seed systems, and some of the key issues that have emerged are as follows:

15.4.1 Lessons Learned

(i) *Farmers are willing to buy seed.* Results from Tanzania in particular have demonstrated that there is a strong demand for high-quality cassava seed. In the early years of the BEST project, there was a greater reliance on selling to institutional buyers such as government district authorities or NGOs, but by 2020 more than 80% of certified cassava seed was being sold to individual farmers.

(ii) *Demand drivers differ.* Smallholder producers are willing to pay for certified stems of desired varieties, but the drivers of this demand differ between regions and countries. In Nigeria, farmers buy cassava stems mainly to obtain new varieties, with strong implications for seed enterprises throughout the value chain. In Tanzania, by contrast, the main driver for seed demand is the health status of the material, as virus diseases affect a large proportion of stems obtained from uncertified material of local varieties. Studies of repeat purchases of cassava seed are required, however, to confirm that accessing a new variety (in Nigeria) or managing disease (in Tanzania) will not curtail business opportunities for seed producers.

(iii) *Cassava seed is a viable business under the right conditions.* Data from both Nigeria and Tanzania have confirmed that it can be profitable to produce and sell certified cassava seed of improved varieties. In Nigeria, the larger VSEs in particular are profitable, particularly those who also sell roots. However, without new varieties, the lack of a clear yield premium for certified seed undercuts the business case for VSEs, making it much more challenging to create a sustainable business than in Tanzania. In Tanzania, combining roots sales with stem production has also been shown to be a key to success for CSEs, and profitability has increased gradually from 2018 to 2020.

(iv) *There are major limitations to the scaling of the VSE model in Nigeria because of the high setup and support costs.* Establishing networks of cassava seed producers requires significant start-up investments, which can limit the potential for scaling. Several options are being explored to overcome this challenge, including creating an apex organization for VSEs, and moving to much larger VSEs.

(v) *Improved private sector engagement will improve long-term viability.* Involving cassava processing companies as facilitators of the commercial seed system warrants further investigation as these enterprises are the ultimate beneficiaries of the system and investments from them may offset expenditures that might otherwise be borne by projects or governments.

(vi) *New varieties.* An efficient and productive breeding pipeline is required to ensure that new varieties become available to cassava seed producers on a regular basis. Stronger linkages with new varietal development in breeding are essential. Demand creation trials can help to identify the most suitable varieties for processors to scale out via their outgrowers. An essential part of the seed value chain is naming varieties and promoting them to highlight their specific uses and production conditions.

(vii) *Modern tools for rapid multiplication.* EGS production needs to apply modern multiplication technologies to increase the speed of dissemination of new varieties. This has been achieved through the use of tissue culture, screenhouse-based mini-cutting multiplication, and SAH. SAH labs can produce hundreds of thousands of plantlets and allow for real-time inventory management. The technique enables the rapid multiplication of promising new varieties in response to market demand. Processors in Nigeria showed an interest in investing in cassava seed units for backward integration to provide stems of varieties of interest to their own enterprises and for outgrowers. Valuable progress was made in the adaptation of SAH by processors, but the commercial case for this still needs to be comprehensively made.

(viii) *Spin-off companies for EGS production.* In Nigeria, GoSeed and Umudike Seeds were established as spin-off companies from IITA and NRCRI, respectively, and have got off to a promising start. Both have played an effective role in producing EGS seed on a commercial basis. A similar approach has also been taken by TARI in Tanzania. These spin-off companies have the potential to play an important role in linking breeders with commercial seed producers; however, it will be essential that they are run with an entirely commercial mindset separated from the research function.

(ix) *Organization of VSEs/CSEs.* Community-level VSEs/CSEs can benefit from being organized into associations which can provide support with finance, training, and quality assurance. The seed entrepreneurs require extensive training both in seed production and business practices. This can be done sustainably by organizing them into large networks, where participation in the network for a fee can allow the network to provide the needed services. This needs to be accompanied by marketing campaigns using different strategies to promote the new varieties, create awareness of the benefits of certified seed, and publicize the seed entrepreneurs.

(x) *Quality management programs are required to facilitate commercial seed systems.* Two of the most important requirements for farmers buying cassava seed are that the variety is known and that the stem cuttings are of high quality. Cassava seed quality assurance systems established in Tanzania and Nigeria have shown that certification systems can facilitate effective trade in cassava seed. However, these systems need to be tailored to the local environment. Where disease is less important, such as in Nigeria, inspections at community level need to be cheap and pragmatic enough to enable seed businesses to flourish. Where disease is more important, as in Tanzania, effective quality control is required down to QDS level, although decentralized inspection systems should be affordable and have quality standards that are realistically achievable. CSE/VSE associations can play an important role in facilitating cassava seed quality management at community level.

(xi) *Digital tools such as Seed Tracker can greatly enhance the efficiency of the seed system and support M&E.* Seed Tracker has proven to be highly effective in increasing the efficiency of the cassava seed certification work of TOSCI and NASC. In addition to permitting e-certification and remote registration of seed producers, the platform promotes the seed businesses of its users and allows the seed regulatory agencies to track performance throughout the system in real time.

(xii) *There are important scaling constraints that need to be defined* (Sartas et al. 2020). Currently, the formal cassava seed systems established in both Nigeria and Tanzania each produce less than 1% of the national requirements for cassava seed. Although seed producers have been able to successfully sell to local root producers, establishing a network of commercial cassava seed producers is relatively slow and expensive, requiring significant training. This is likely to be a key constraint in scaling commercial cassava seed systems, both in countries where there are established systems as well as in places where systems would need to be set up from scratch. Experience from other work on RTB crops conducted by the CGIAR demonstrates that the application of scaling readiness analyses will help to identify scaling bottlenecks and enhance the impacts of commercial cassava seed systems. A key constraint for scaling to other countries is the investment needed to establish practical and affordable certification systems and a network of seed entrepreneurs at the various levels of the system. Teams focused on scaling the commercial cassava seed system models described here will need to specify and quantify these costs.

Significant investments have been made in modernizing and building sustainable models for cassava seed systems in Africa over the last decade. Contrary to the widespread perception, cassava seed is extensively traded. Furthermore, newly established smallholder seed entrepreneurs have been successful in generating profits from cassava seed businesses. Although the factors driving demand for seed differ in different contexts (disease-resistant varieties in East Africa versus high-yielding varieties in West Africa), successes being achieved provide considerable hope for the successful scaling of approaches described in this chapter to other parts of Africa where cassava is an increasingly important climate resilient crop.

Acknowledgments This research was undertaken as part of the CGIAR Research Program on Roots, Tubers, and Bananas (RTB) and supported by *CGIAR Trust Fund contributors*. Funding support for this work was also provided by the Bill & Melinda Gates Foundation (BMGF).

References

Adams IP, Abidrabo P, Miano DW, Alicai T, Kinyua ZM, Clarke J, Macarthur R, Weekes R, Laurenson L, Hany U, Peters D, Potts M, Glover R, Boonham N, Smith J (2013) High throughput real-time RT-PCR assays for specific detection of cassava brown streak disease causal viruses, and their application to testing of planting material. Plant Pathol 62:233–242

AGRA (2016) Tanzania early generation seed study. https://pdf.usaid.gov/pdf_docs/PA00MR49.pdf. Accessed 1st July 2021

Bentley JW, Nitturkar H, Obisesan D, Friedmann M, Thiele G (2020a) Is there a space for medium-sized cassava seed growers in Nigeria? J Crop Improv 34:842–857

Bentley J, Nitturkar H, Friedmann M, Thiele G (2020b) BASICS phase I final report. December 23, 2020, p 78. https://doi.org/10.4160/9789290605690. Accessed 1st July 2021

Dixon AGO, Bandyopadhyay R, Coyne D, Ferguson M, Ferris RSB, Hanna R, Hughes J, Ingelbrecht I, Legg JP, Mahungu N, Manyong V, Mowbray D, Neuenschwander P, Whyte J, Hartmann P, Ortiz R (2003) Cassava: from a poor farmer's crop to a pacesetter of African rural development. Chronica Horticulturae 43:8–14

Douthwaite B (2020) Development of a cassava seed certification system in Tanzania: evaluation of CGIAR contributions to a policy outcome trajectory. International Potato Center: Lima. https://doi.org/10.4160/9789290605560. Accessed 1st July 2021

FAOSTAT (2021) FAO database. Food and Agriculture Organization of the United Nations, Rome, Italy. Available at http://www.fao.org/faostat/en/#data/QC. Accessed 16 Feb 2021

Forsythe L, Martin AM, Posthumus H (2015) Cassava market development: a path to women's empowerment or business as usual? Food Chain 5:11–27

IITA (2021) IITA partners in launch of Tanzania's national cassava strategy. https://www.iita. org/news-item/iita-partners-in-launch-of-tanzanias-national-cassava-strategy/. Accessed 1st July 2021

Legg JP (1999) Emergence, spread and strategies for controlling the pandemic of cassava mosaic virus disease in east and central Africa. Crop Prot 18:627–637

Legg JP (2010) Epidemiology of a whitefly-transmitted cassava mosaic geminivirus pandemic in Africa. In: Stansly PA, Naranjo SE (eds) *Bemisia*: bionomics and management of a global pest. Springer, Dordrecht-Heidelberg-London-New York, pp 233–257

Legg JP, Lava Kumar P, Makeshkumar T, Ferguson M, Kanju E, Ntawuruhunga P, Tripathi L, Cuellar W (2015) Cassava virus diseases: biology, epidemiology and management. Adv Virus Res 91:85–142

McGuire S, Sperling L (2016) Seed systems smallholder farmers use. Food Security 8:179–195

MEDA (Mennonite Economic Development Associates) (2016) Commercially sustainable, quality-assured, cassava seed distribution system in Tanzania: Pilot Innovation Project

Ng SYC (1992) Tissue culture of root and tuber crops at IITA. In: Thottappilly G, Monti LM, Mohan Raj DR, Moore AW (eds) Biotechnology: enhancing research on tropical crops of Africa. IITA, Ibadan, pp 135–141

Nweke FD, Spencer SO, Lynam JK (2002) The cassava transformation: Africa's best-kept secret. International Institute of Tropical Agriculture (IITA), Ibadan

Okechukwu R, Kumar PL (2016) Releasing disease-resistant varieties of cassava in Africa. In: Andrade-Piedra J, Bentley JW, Almekinders C, Jacobsen K, Walsh W, Thiele G (eds) Case studies of roots, tubers and bananas seed systems. CGIAR research program on roots, tubers and bananas (RTB), Lima: RTB Working Paper No. 2016-3. ISSN 2309-6586, p 244

Pircher T, Obisesan D, Nitturkar H, Asumugha G, Ewuziem J, Anyaegbunam H, Azaino E, Akinmosin B, Ioryina A, Walsh S, Almekinders C (2019) Characterizing Nigeria's cassava seed system and the use of planting material in three farming communities. Lima (Peru). CGIAR research program on roots, tubers and bananas (RTB). RTB Working Paper. No. 2019–1, p 28. www.rtb.cgiar.org. Accessed 1st July 2021

Rigato S, Gonzalez A, Huarte M (2000) Producción de plántulas de papa a partir de técnicas combinadas de micropropagación e hidroponía para la obtención de semilla prebásica (Potato plantlet production by means of combined micropropagation and hydroponic techniques to obtain prebasic seed). In: XIX Congreso de la Asociación Latinoamericana de la Papa, February 28th-March 3rd 2000, La Habana, Cuba. Proceedings, p 155

Sartas M, Schut M, Proietti C, Thiele G, Leeuwis C (2020) Scaling Readiness: science and practice of an approach to enhance impact of research for development. Agr Syst 183:102874. https:// doi.org/10.1016/j.agsy.2020.102874

Teeken B, Olaosebikan O, Haleegoah J, Oladejo E, Madu T, Bello A, Parkes E, Egesi C, Kulakow P, Kirscht H, Tufan HA (2018) Cassava trait preferences of men and women farmers in Nigeria: implications for breeding. Econ Bot 72:263–277

Thiele G, Dufour D, Vernier P, Mwanga ROM, Parker ML, Schulte Geldermann E, Teeken B, Wossen T, Gotor E, Kikulwe E, Tufan HA, Sinelle S, Kouakou AM, Friedmann M, Polar V, Hershey C (2021) A review of varietal change in roots, tubers and bananas: consumer preferences and other drivers of adoption and implications for breeding. Int J Food Sci Technol 56:1076–1092

Walsh S (2016) Responding to two cassava disease pandemics in East and Central Africa. In: Andrade-Piedra J, Bentley JW, Almekinders C, Jacobsen K, Walsh W, Thiele G (eds) Case studies of roots, tubers and bananas seed systems. CGIAR Research Program on Roots, Tubers and Bananas (RTB), Lima: RTB Working Paper No. 2016-3. ISSN 2309-6586, p 244

Wossen T, Girma GT, Abdoulaye T, Rabbi IY, Olanrewaju A, Alene A, Feleke S, Kulakow P, Asumugha G, Adebayo M A, Manyong V (2017) The Cassava Monitoring Survey (CMS) in Nigeria. https://cgspace.cgiar.org/handle/10568/80706

Wossen T, Spielman DJ, Abdoulaye T, Kumar PL (2020) The cassava seed system in Nigeria: opportunities and challenges for policy and regulatory reform. CGIAR Research Program on Roots, Tubers and Bananas (RTB), Lima, Peru. RTB Working Paper. No. 2020-2, ISSN: 2309-6586, p 37

Chapter 16
Building Demand-Led and Gender-Responsive Breeding Programs

Vivian Polar ⓘ, Béla Teeken ⓘ, Janet Mwende ⓘ, Pricilla Marimo ⓘ,
Hale Ann Tufan ⓘ, Jacqueline A. Ashby ⓘ, Steven Cole ⓘ,
Sarah Mayanja ⓘ, Julius J. Okello ⓘ, Peter Kulakow ⓘ, and Graham Thiele ⓘ

Abstract Gender-responsive breeding is a new approach to making sure modern breeding takes advantage of opportunities to improve gender equality in agriculture. Conventional research on the acceptability of modern varieties has scarcely addressed gender differences during adoption studies. Gender-responsive breeding starts from a different premise that adoption and social impact will be enhanced if gender is addressed at early stages of variety design and priority setting in breeding. However, until recently, there was no concrete way to integrate gender considerations into the practice of breeding. This chapter draws lessons for the future from three RTB breeding programs innovating with gender-responsive breeding with a focus on piloting novel tools. The new G+ tools are designed to help gender researchers and breeders make joint, evidence-based decisions about the significance of gender differences for customer targeting and trait prioritization in variety development. Their piloting in the context of each program's practice of gender-responsive breeding throws light on some valuable good practices that contributed to successful innovation.

V. Polar (✉)
CGIAR Research Program on Roots, Tubers and Bananas (RTB), led by the International Potato Center, Lima, Peru
e-mail: v.polar@cgiar.org

B. Teeken
International Institute of Tropical Agriculture (IITA), Ibadan, Nigeria
e-mail: b.teeken@cgiar.org

J. Mwende
University of Nairobi, Nairobi, Kenya
e-mail: janetmwende70@gmail.com

P. Marimo
Alliance Bioversity – CIAT, Kampala, Uganda
e-mail: p.marimo@cgiar.org

G. Thiele et al. (eds.), *Root, Tuber and Banana Food System Innovations*,
https://doi.org/10.1007/978-3-030-92022-7_16

16.1 Why Innovate with Gender-Responsive Breeding?

Although modern breeding has introduced varieties beneficial to farmers in high-potential environments and to those who can profitably use inputs to modify their environments (Ceccarelli and Grando 2007; Ribaut and Ragot 2019), it has been challenging for breeding programs to equitably reach low-income users, in particular poor men and women farmers, who may have different needs and priorities. In low-income farming, varietal change is usually slow because modern varieties often do not meet users' needs and preferences. About 35% of many new food crop varieties were adopted across sub-Saharan Africa (SSA) over the past 15 years, in contrast with about 60% in Asia and 80% in South America (Kimani 2017; Walker et al. 2015; Walker and Alwang 2015). Lower use of modern varieties among women farmers is a significant trend, reflecting, in part, unequal access to technology as well as differences in preferences (Ashby and Polar 2019; Wale and Yalew 2007). Gender inequality is a stumbling block to varietal adoption when women and men users have different trait preferences, because they face unequal costs and benefits from adoption and use. This challenge is particularly tough for root and tuber and banana (RT&B) crops with complex breeding requirements, some of which experience slow adoption of new varieties, while women make up a high proportion of poor growers and processors (Thiele et al. 2021). However, innovations that contribute to the modernization of breeding such as genotyping and phenotyping technologies, genomic resources, and analytics (Assefa et al. 2019; Ribaut and Ragot 2019; Watson et al. 2018; Yao et al. 2017) have created shortcuts and opportunities to address composites of traits that at first sight may have less economic value but often hold great significance for local populations, including poor men and women producers and other low-income value chain stakeholders.

H. A. Tufan
Cornell – GREAT, Ithaca, NY, USA
e-mail: hat36@cornell.edu

J. A. Ashby
Independent Consultant, Portland, OR, USA
e-mail: jacqueline.ashby@cantab.net

S. Cole
International Institute of Tropical Agriculture (IITA), Dar es Salaam, Tanzania
e-mail: s.cole@cgiar.org

S. Mayanja · J. J. Okello
International Potato Center (CIP), Kampala, Uganda
e-mail: s.mayanja@cgiar.org; j.okello@cgiar.org

P. Kulakow
International Institute of Tropical Agriculture (IITA), Ibadan, Nigeria
e-mail: p.kulakow@cgiar.org

G. Thiele
CGIAR Research Program on Roots, Tubers and Bananas (RTB), led by the International Potato Center, Lima, Peru
e-mail: g.thiele@cgiar.org

Product profiles (descriptions of the traits that users want in new varieties) contribute to effective breeding that meets customer demand (Kimani 2017) but they must go hand in hand with carefully segmented customer preferences (i.e., targeting specific groups of users (Thiele et al. 2021). Consumer preferences are important to consider in RT&B breeding because root, tuber, and banana crops are the most important staple foods in the humid tropics of SSA (Lebot 2019), produced, processed, and consumed by people with relatively low income in rural areas but also by urban consumers, creating an increasing demand (Bricas et al. 2016). Consumer preferences drive much of varietal change in RT&B crops, either through characteristics attractive to consumers that increase adoption (yam, cassava, potato) or unattractive traits that limit uptake (banana) (Kimani 2017; Polar et al. 2021; Thiele et al. 2021).

When breeding involves prioritizing plant traits that are valued quite differently by different types of consumers, the decision to select for one trait over another is also a decision to privilege one set of consumers and their preferences over another. Recognition of the social and gender dimensions of decisions about variety design is built into gender-responsive plant breeding. Such breeding ensures that gender differences in trait preferences are neither overlooked nor neglected when developing and disseminating new crop varieties (Orr et al. 2018). Additionally, understanding gender differences in trait preferences can help breeders identify opportunities for breeding new varieties that address gender-specific objectives for food, nutrition, or economic security.

Building gender-responsive breeding programs requires plant breeders to define which customers the breeding program intends to target and to design product profiles for varieties with an appreciation of what their choices mean for their program's impact on gender equality. This chapter analyses the experience of three breeding programs of the CGIAR Research Program on Roots, Tubers, and Bananas (RTB), innovating with gender-responsive breeding in conjunction with piloting the new G+ tools designed to assist gender-responsive customer targeting and product profiling. Section 16.1 describes the innovations undertaken by all three breeding programs to introduce gender-responsive breeding. Section 16.2 describes the practice of gender-responsive breeding and use of the new tools. The final section analyses what was learnt from innovating that can inform future efforts to build gender-responsive breeding teams.

16.1.1   The Innovations in Introducing Gender-Responsive Breeding

The innovations in introducing gender-responsive breeding are a series of tools, institutional changes and training strategies that lead plant breeders to formally and systematically query gender implications whenever they prioritize (1) the customer segments for targeting and (2) traits to include in their product profiles. The desired outcome of innovation, in piloting use of the G+ tools, was a formal commitment from the three breeding teams involved to continue this practice.

Gender-responsiveness does not mean a program breeds specifically "for women." Analysis using the G+ tools may clarify that gender differences are not important for the program. If, however, a program does identify gender differences in demand, then successful adoption of a gender-responsive approach should result in appropriate changes in customer profiles and/or product profiles. Therefore, while not a requirement of successful adoption of a gender-responsive approach, the discovery of new demands requiring selection for traits with a gender dimension is an important anticipated result.

The innovations described here, in the introduction of gender-responsive breeding by RTB program, developed and converged from different starting points over almost 5 years. Concern that agricultural research funded by international development donors should demonstrate how their work was contributing to gender equality in agriculture gathered new impetus with the publication of the World Bank Report *Gender Equality and Development* in 2012 (World Bank 2012). Major donors made gender-responsiveness a condition for disbursement of funding, spurring CGIAR in 2013 to introduce formal requirements for its programs to integrate gender into annual work and budget plans, monitored and reported to donors by the central CGIAR System Office and with budget approval conditioned on satisfactory progress.

In the past, CGIAR programs had not ignored the significance of gender inequality in their long-standing mission to reduce poverty, but this was the first time a formal and financial obligation to do so was institutionalized system-wide. However, in the 2 years after the requirement was introduced in 2013, plant breeders and the gender research specialists hired to implement integration had difficulty finding ways to work together that demonstrably enhanced breeding's gender-responsiveness. In response, and in view of the strategic and central importance of plant breeding to CGIAR's overall purpose and impact at that time, the System Office launched the CGIAR Gender and Breeding Initiative (GBI) in 2016 with the objective of developing common ground between breeders and gender researchers. The GBI aimed to foster the co-design of practical approaches and tools that would make it easy for breeders to identify desirable or undesirable features of varieties at different stages in the breeding pipeline, i.e., from the early stage of variety design through the subsequent stages leading to advanced testing, evaluation, and release.

Until the GBI, there was no systematic approach to applying gender analysis to variety design or product profile development in plant breeding on a routine basis (CGIAR and GBI 2017, 2018). Gender analysis of technology adoption in low-income countries has frequently observed gender bias in technology design and in how adoption decisions are made, often with inequitable outcomes (Doss and Morris 2001; Fisher and Kandiwa 2014; Peterman et al. 2014; Ragasa 2012; Teklewold et al. 2020; Udry 1996). However, this analysis was seldom keyed to specific plant traits and so had little impact on the practice of breeding. This left an important gap in research.

The GBI confronted this gap in 2016 with the first innovation of interest to this analysis, promoting dialog among plant breeders and gender researchers about needed changes in practice through a series of *cross-disciplinary "gender and breeding" workshops*. GBI's second innovation followed in 2016–2017: *placement of a gender researcher as a member of a breeding team*, co-financed by GBI and the team. In 2017, the System Office withdrew from research, and these functions, including the GBI, were divested to suitably qualified CGIAR Programs. From 2017 onward, RTB was recognized as a leader in social and gender analysis within the CGIAR and coordinated the GBI workshops. Since 2013 RTB had proactively included gender research into new projects, such as the NextGen Cassava Breeding project that involved collaboration with Cornell University and was associated with a project launched at this time: the Gender-Responsive Researchers Equipped for Agricultural Transformation (GREAT)[1] led by Cornell University. GREAT began implementing a third innovation: a *training model to confront deeply held norms about gender and research in crop breeding teams* by focusing on interdisciplinarity, attitude shifts, and changing practice for gender research in breeding programs.

These higher-level institutional innovations that broadened gender awareness and installed new capacity for gender research in breeding teams stimulated development of a new set of practical innovations: the G+ tools. With the broad participation of breeders and social scientists from across the CGIAR, GBI workshop participants identified "must-have" features of gender-responsive breeding, and critical input required from gender researchers at key decision points along the breeding pipeline. GBI drew on workshop findings to develop the G+ tools as a practical resource to help breeders and gender researchers realize this input.

Starting in 2018, GREAT training involved key individuals from the GBI organizing committees and introduced principles of gender-responsive crop improvement developed by the GBI (Ashby et al. 2018). In particular, the last two training cohorts (2020 and 2021) included exposure to G+ tools. The RTB programs discussed in this chapter participated in GREAT courses as fellows or mentors. The developers of GREAT courses consider them to have become more impactful since the introduction of GBI frameworks with their "must-have" features for gender-responsive breeding (Mascarenhas 2016) and the critical decisions for ensuring plant or animal breeding is gender-responsive (Ashby et al. 2018). The GREAT training became more applicable since the introduction of G+ tools.

As the G+ tools developed, the RTB gender team and RTB breeders began to trial them with breeding programs for cassava in Nigeria and sweetpotato and banana in Uganda. In parallel, and with the aim of gaining broader recognition for the G+ tools, RTB brought the tools to the attention of CGIAR's Excellence in Breeding Platform (EiB), responsible for coordinating breeding for the whole of the CGIAR and itself confronting the demand from donors for demonstrably gender-responsive programming. EiB agreed to co-sponsor the piloting of the G+ tools with bean breeding in Zimbabwe and cassava breeding in Nigeria.

These five breeding programs were selected because they already included a gender researcher and had substantial gender-related data. Piloting of the G+ tools in different crops aimed to generate feedback for adjusting the tools so they could be incorporated as innovations into a breeding program's regular operation.

---

[1] https://www.greatagriculture.org/

The G+ tools consist of three innovations, detailed below: the G+ Customer Profile tool, the G+ Product Profile Query tool, and a Standard Operating Procedure (SOP). Together they provide a procedure for incorporating the results of gender analysis into two key decisions that public sector plant breeders routinely make for variety development: (1) who is the intended customer for the breeding product? and (2) what are the important features of the breeding product intended for this customer?

- *The G+ Customer Profile* tool guides prioritization of customer segments that takes into account gender differences among the target customer population. An example of a segment defined with the tool could be "smallholder women cassava farmer-processors in southern Nigeria." If men and women express demand for the same varietal traits, a customer segment will include both. The tool helps to organize the evidence to decide which customer segments to prioritize from a socially inclusive and gender perspective. The tool also sheds light on the reasons to target breeding for important plant traits that men and women value differently (Orr et al. 2021).
- *The G+ Product Profile Query* tool assigns a "gender impact" score to each individual plant trait in a breeder's product profile. Scoring is similar to the nominal index that breeders often use to assign a value for disease tolerance to a variety. The tool helps organize the evidence for valuing individual plant traits from a gender perspective (Ashby and Polar 2021a). For example, the tool identifies trade-offs among traits women and men value differently, for example, when women prefer a low-yielding millet that adapts to their poor soils while men prefer a high-yielding variety that performs well on their more fertile plots (Weltzien et al. 2019).
- *The G+ Standard Operating Procedure* is a decision-support guide for using the other two tools in multidisciplinary teamwork (Ashby and Polar 2021b). It lays out a stepwise process for a team to use the results of gender analysis produced with the other tools.

Piloting the G+ tools involved knowledge sharing, capacity development, and a planning workshop, followed by tool application championed by at least one breeder, a gender specialist, and an economist or market-research specialist in each breeding team. A pilot version of the G+ tools was provided so each team could adapt the tools. All the teams shared experiences of piloting, documented improvements they tested during piloting, and wrote a formal review of the tools as feedback to RTB for tool adjustment. Finally, each team had the option to commit formally to routinely considering gender during future product design and advancement decisions.

Different actors and partners were engaged with different roles throughout the process. The core facilitating team from RTB conducted the capacity development and planning exercise with representatives from EiB and GREA and participants from the different research centers. EiB provided the customer and product profile template to CGIAR breeders and guided breeders on the stage-gate process that manages a product from design to delivery through a series of stages and decision-making gates. Given this pivotal role, the involvement of EiB was key in initiating and potentially mainstreaming the tools. The data processing and tool application were guided by the gender specialist from each case, with the participation of other social scientists and breeders. After the first cycle of tool implementation, the teams generated feedback to adjust the tools and the implementation process. In the final stage of piloting, the teams used the results from applying the G+ tools to decide about customer profiling and product design. This represented the formal inclusion of gender analysis in decision-making about breeding product and program design.

The following sections describe different experiences with RTB crops, reporting on two key dimensions of the experience: the practice of gender-responsive breeding that each program evolved and the piloting of the G+ tools. Gender-responsive breeding practice, levels of use for the G+ tools, and levels of good practice that enabled success for each program are compared in Tables 16.4 and 16.5 and discussed at the end of this section.

16.2 Experience 1: Cassava Breeding in Nigeria

16.2.1 Practice of Gender-Responsive Breeding

The NextGen Cassava Breeding project (NextGen Cassava) started operating in Nigeria in 2013, with the main objective of shortening the breeding cycle of cassava for the benefit of smallholder farmers.[2] This project included the two main partners in Nigeria that work together on cassava breeding: The International Institute of Tropical Agriculture (IITA) and National Root Crops Research Institute (NRCRI). The project staff included biological and social scientists who participated in the CGIAR cross-disciplinary workshops on gender and breeding and in GREAT training. The implementation of gender-responsive breeding in the IITA-NRCRI cassava program in Nigeria helped the breeding program organize gender-relevant data, identify important traits for sex-disaggregated value chain actors, and understand evidence gaps that need to be addressed through additional research. Work of NextGen Cassava and RTBfoods[3] project on food quality preferences received grants from GREAT. In 2016, the project added an interdisciplinary postdoctoral fellow specializing in gender who integrated gender analysis into IITA's participatory varietal evaluation methods (Teeken et al. 2020). This cooperation is recognized by the senior breeder as the starting point of the development of a cassava *cross-functional design team*[4] to advance products in the context of EiB's stage-gate process.

The contribution of the postdoctoral specialist in gender research was to analyze a large amount of sex-disaggregated data already collected by several projects to interpret its meaning for breeding objectives and trait prioritization. NextGen Cassava was set up with a survey division for understanding user preferences in a

---

[2] See https://www.nextgencassava.org/

[3] See https://rtbfoods.cirad.fr/

[4] The "cross-functional design team" as conceived by EiB is composed of downstream, market-oriented subject matter experts that provide guidance and data driven insights for product design.

gender-responsive way. Their work was complemented by a large-scale Cassava Monitoring Study (CMS), done by IITA in 2015 that collected sex-disaggregated data among 2500 households (Wossen et al. 2017), and generated important information on preferred and non-preferred characteristics at each step of processing (Bentley et al. 2017; Chijioke et al. 2020; Ndjouenkeu et al. 2020; Olaosebikan et al. 2019; Teeken et al. 2018). In parallel, sex-disaggregated data was collected to assess trait preferences related to cassava food quality characteristics by the RTBFoods project.[5]

The gender analysis using the CMS and initial research carried out under the NextGen Cassava project found gender differences in variety and trait preferences for cassava in Nigeria. Traits such as "easy to peel" and those related to "food product quality" were mentioned more frequently by women than by men, reflecting women's strong involvement in processing and trading (Teeken et al. 2018). The main recommendation was for the breeding program in Nigeria to prioritize cassava traits that ensure good-quality food products like fufu and gari-eba,[6] usually processed and traded by women. Subsequent studies with additional support from the RTBfoods project[7] confirmed that the color and texture of the dough-like products (eba, fufu) and the shininess/color of gari are important traits that should be considered when breeding cassava (Olaosebikan et al. 2019; Teeken et al. 2020). The convergence of findings and support from all these initiatives urged the breeding program to further explore user preferences related to processing and food quality.

16.2.2 Piloting the G+ Tools

Use of the G+ tools started through a knowledge sharing and planning workshop in March 2020. The cassava breeding program team realized that they had most of the required information and had already followed steps prescribed in the tools but in a less systematic manner. The program developed a product map from a review of literature and available data using the G+ Customer Profile Tool. The product map brought together information from different studies to highlight that 95% of the cassava in Nigeria is produced by smallholder farmers and 90% is processed at home or by small-scale processors who are mostly women (Forsythe et al. 2016). This supported the need to target the processing segment, due to the importance of processing and trading in the cassava value chain in Nigeria and the major role women play in carrying out these activities. Application of the G+ Customer Profile tool identified significant gaps in existing data, notably the trait preferences of value chain actors other than farmers and in parts of the value chain handled mostly by poor women. Data gaps included further understanding of preferences for food quality traits by different types of processors and retailers and how they translated into breeding traits.

---

[5] Led by the *Centre de coopération internationale en recherche agronomique pour le développement* (CIRAD).

[6] Eba is the dough-like product prepared by adding hot water to gari.

[7] See https://rtbfoods.cirad.fr/

The G+ Customer Profile tool helped the IITA-NRCRI cassava breeding program in Nigeria formulate a strong evidence base that supported the program's prioritization of the customer segment for the fufu and gari product profile. Prior assumptions could now be backed with more evidence. This stimulated the team to look systematically at how to translate preferred food product related characteristics into concrete traits that, through discussions with food scientists, could be made operational. Table 16.1 highlights gender-relevant traits and shows how traits

Table 16.1 Shift in traits of interest to the breeders' product profile based on gender analysis of preferences for cassava in Nigeria

Traits considered before gender-responsive cassava user preferences studies (until ca 2016)	Traits added to breeding selection after identification of gender differences (ca 2016–2018)	Additional traits currently under consideration –gender-relevant traits highlighted (2019–2020)

Cassava mosaic disease resistance
Plant type (erect)
Plant height (high for stems)
Branching height (high)
Fresh yield
Harvest index
Dry yield
Dry matter
	Fufu yield
	Gari yield
	L (brightness)
	B (yellowness)
		Cassava brown streak disease resistance
		Roots can be stored in the ground
		Early maturity
		Stability of dry matter
		Stem longevity
		Canopy closure to suppress weeds
		Big roots
		Multipurpose (fufu, gari, lafun, abacha)
		Food product color/browning
		Food product texture (after preparation and storage)
		Swelling when preparing food product
		How easy a variety releases its water during pressing (gari)
		Complete softening of roots for fufu during retting
		Few or no woody filaments in the root apart from the central fiber
		Ease of peeling
		    Easy to cut off the cortex by sliding under it with a knife
		    Undressing the cassava in which cortex loosens from the root
		Taste

prioritized previously and recently have been added to the product profile for gari and fufu of the cassava breeding program in Nigeria, facilitated by use of the G+ tools. Men and women mostly agree about the desirability of these traits but may weight them differently, depending on their role in the value chain. Piloting the G+ tools revealed the importance of analyzing preferences of gender-differentiated value chain actors. In 2020, for the first time, the breeding program formally made use of the findings and recommendations of gender research in the team decision to advance five candidate varieties for release.

As a result of their innovation with gender-responsive breeding, the cassava breeding program is currently modifying its customer and product profile templates to incorporate the gender-related information organized with the G+ tools and to support future product advancement decisions.

16.3 Experience 2: Sweetpotato Breeding in Uganda

16.3.1 Practice of Gender-Responsive Breeding

In Uganda, including gender in breeding design was a slow process that gained importance only due to donor demands. Sweetpotato breeders in Uganda recognized the important role women played in sweetpotato production since a study done in the early 1990s (Bashaasha et al. 1995). On-farm participatory plant breeding (PPB) trials that consulted the opinions of women and men farmers led to increased adoption and greater diffusion of an improved variety – NASPOT 11 (Gibson et al. 2008; Kiiza et al. 2012; Mwanga et al. 2011). Later breeding concentrated on improved nutrition from orange-fleshed sweetpotato (OFSP) varieties (Low et al. 2017), targeting women and children as beneficiaries, but was less concerned with gender relations in production. In 2016 the CIP sweetpotato breeding team participated in the multidisciplinary GBI workshops. Subsequently the team worked with a gender specialist on aspects of sweetpotato production and consumption, contributed to GREAT training and integrated gender into methods used by sweetpotato breeders for participatory varietal evaluation. National partners were vital contributors to further new studies that systematically analyzed preferences for traits by gender at farm and market levels. Both preferences at farm and market levels aimed at developing a sweetpotato product profile that considered gender differences in demand for raw and boiled or steamed sweetpotato, the most common form in which the crop is eaten in Uganda.

Traits desired by women and men along the sweetpotato value chain gained new attention by sweetpotato breeders. Studies that produced a gendered food map for raw and boiled/steamed sweetpotato showed that men and women had different quality preferences for raw and boiled/steamed sweetpotato roots, driven by gender norms and roles (Mwanga et al. 2020).

16.3.2 Piloting the G+ Tools

Piloting the tools enhanced the sweetpotato program's organization of gender-related information and motivated the development of a multidisciplinary team. Sweetpotato was a late entrant in the use of the G+ tools, starting in March 2020, but partners' studies that included gender differences had already done much of the foundational work. Some of these studies included the NARO-CIP Trait prioritization project (Sseruwu et al. 2015; Turyagyenda et al. 2015; Yanggen and Nagujja 2006), the RTBFoods gendered food mapping study (Asindu et al. 2020; Banda et al. 2021; Moyo et al. 2021; Mwanga et al. 2020), the trait prioritization and valuation analysis for sweetpotato conducted by AbacusBio[8] (Byrne et al. 2020), PPB and PVS trials (Gibson et al. 2008; Kiiza et al. 2012), the program on Sweetpotato Action for Security and Health in Africa (SASHA), and the Sweetpotato for Profit and Health Initiative (SPHI) (Mwanga et al. 2021). Long-established cooperation between the RTB program and breeders and social scientists of Uganda's National Agricultural Research Organization (NARO) was particularly important for getting up to speed with piloting the G+ tools.

Application of the G+ Customer Profile tool revealed that male and female actors along the sweetpotato value chain had different roles and responsibilities in production, processing, and marketing which drove interest in different traits. Women were interested in the cooking qualities of sweetpotato roots (such as taste, aroma), whereas men gave more attention to market-related traits (e.g., root size). Customer profiling revealed the importance of intersectional differences within a sex category, e.g., older and younger women had different root size and maturity preferences. The sweetpotato product map developed from a non-systematic literature review affirmed that 70% of the crop was eaten at home and it was mostly grown by women (66%) though men dominated large market transactions of up to 265 tons (Byrne et al. 2020; Echodu et al. 2019). Customer profiling identified reasons for the breeding program to consider gender in its product profiles and specifically, to target a customer segment whose priorities are consumption and processing of sweetpotato. At the same time, use of the G+ tools highlighted serious gaps in the available data on gender differences and the need for more up-to-date data collection.

A gender analysis of all the traits initially depicted in the product profile and traits that emerged from the literature review was conducted to assess their gender-responsiveness. This pointed to the need to adjust trait prioritization and product profiles to accommodate gender differences. For example, use of the G+ Product Profile tool to make this analysis drew breeders' attention to evidence of differences in variety preference where women mostly preferred local sweetpotato cultivars such as Okonynedo and Arakaraka and men preferred improved varieties. The analysis revealed that root yield and early maturity pose a potential conflict of interest between women and men. Even though it implies lower yields, early maturity is

---

[8] AbacusBio is a private company that provides services for agricultural innovation. https://abacus-bio.com/wp-content/uploads/2020/03/Case-study-Sweetpotato_compressed.pdf

often valued more highly by women producers. This is particularly true when gender norms deem they are responsible for providing food for the household, because it can help to relieve seasonal food shortages. On the same line, earliness can be highly valued because women commonly cultivate relatively smaller plots, and earliness may increase the returns to scarce land by opening up possibilities for intensification, e.g., relay cropping.

Men who trade sweetpotato value high yields and in commercialized production systems are less disposed to accept the trade-off between yield and early maturity, especially if they are not primarily responsible for putting food on the table. Men dominate the high-volume trade probably because they are able to exploit distant markets which are not accessible by women farmers who mostly sell their roots within their community. Differences in men's and women's market access influence their preferences for a variety: men looking for traits that address the needs of urban consumers, while women basing their preferences on the needs of the rural consumers. Sweetpotato commercialization has been shown to attract mostly men, and women may be displaced in the trade. Thus, prioritizing traits considered desirable for commercialization of sweetpotato by men has to consider the potential for creating disadvantages for women.

By using the G+ tools, the sweetpotato breeding team integrated gender into its ongoing analysis of demand and customer segmentation and consequently recognized the need to adjust trait prioritization. The G+ Product Profile tool added traits identified as gender-relevant to the existing product profile. These traits will inform future team discussions to assess if the product profiles need to be adjusted, once progress has been made in addressing the serious gaps detailed above in data on how gender affects use of the crop, trait preferences, and varietal choice.

16.4   Experience 3: Banana Breeding in Uganda

16.4.1   Practice of Gender-Responsive Breeding

The IITA-led, "Improvement of Banana for Smallholder Farmers in the Great Lakes Region of Africa" (the Breeding Better Bananas or BBB)[9] project was implemented from 2014 to 2019 to upscale existing breeding activities, build a breeding and selection pipeline, improve data management, increase the pace and efficiency of banana breeding. As part of the project, baseline research covering 1319 respondents was conducted in 2015–2016 in districts targeted for introducing hybrid banana varieties in Uganda (Luwero and Mbarara) and Tanzania (Bukoba, Meru, Moshi, and Rungwe). Representative, sex-disaggregated data were collected to describe characteristics of the target population and the demand for breeding products and banana varietal traits.

---

[9] http://breedingbetterbananas.org/

Explicitly incorporating gendered aspects in the breeding program in Uganda was not recognized as a priority at this time. However, staff from the BBB attended a CGIAR cross-disciplinary GBI workshop, and in November 2016, a CGIAR gender postdoctoral fellow was recruited and assigned to the BBB project specifically to focus on gendered trait preferences in the banana value chain. The postdoctoral fellow assessed the literature on gendered trait preferences in banana production, processing, and use (Marimo et al. 2020a) and conducted data analysis of the baseline survey under BBB (Marimo et al. 2019). Concurrently, teams from the Banana Program at the National Agricultural Research Laboratories (NARL) were participating in the GREAT program and gathering empirical evidence on gender dynamics in banana breeding (Nasirumbi-Sanya et al. 2018; Ssali et al. 2017). Additionally, the RTBFoods project contributed to building an evidence base as it assessed gendered preferences for food products in Uganda.

One of the key features of the work conducted by the BBB project was the close collaboration between Bioversity International, NARL, Tanzania Agricultural Research Institute (TARI), and IITA. This partnership emphasized the use of findings from a survey that collected sex-disaggregated data, evaluations of varieties by men and women farmers, and consumer acceptability tests to evaluate new hybrids before official release. The research involved a preference analysis exercise that allowed women and men farmers to visit the on-station trials and rate their most and least preferred varieties and a qualitative assessment of traits that farmers look for (or avoid) when selecting new banana varieties as well as farmer involvement in food preparation and taste tests of the hybrids that were under evaluation (Marimo et al. 2020b).

The dynamic interaction, collaboration, and production of joint outputs between the social scientists, gender researcher, and breeders provided an opportunity to discuss, brainstorm, and get a better understanding of the contribution and perspective of the different disciplines, an understanding of the breeding pipeline, and the history of breeding product profile development. The gender postdoctoral fellow and breeders from IITA and NARL attended a course together on inclusive breeding, produced joint publications, and joint presentations.

Unfortunately, between 2017 and 2020, when debate in the BBB was growing around the importance of gender analysis for the breeding program and when an evidence base was being developed, two of the lead breeders at NARL left the organization. This discontinuity hampered the development of an operational multidisciplinary team (Sanya et al. 2018). Additionally, the team was unable to attend the initial capacity development and planning meeting for piloting the G+ tools due to budget issues, further delaying implementation of the tools.

16.4.2 Piloting the G+ Tools

The experience of using the G+ tools for banana is still at an early stage. The departure of breeders who had participated in building the evidence base was a serious handicap. Staff turnover and difficulties replacing staff due to COVID-19 made it impossible to build multidisciplinary collaboration to use the G+ tools.

Nonetheless, application of the G+ tools by the gender researcher increased the banana breeding team's awareness of the importance of using gender-relevant and sex-disaggregated information on preferences to better identify and characterize priority traits to include in selections. The body of literature around banana trait preferences in Uganda has increased in the last years. Considerable information can be found from routine PVS trials (Akankwasa et al. 2013a, b, 2016; Ssali et al. 2010), in studies of agricultural technology and agribusiness advisory services (ATAAS) (Sanya et al. 2017, 2018, 2020; Ssali et al. 2017), more recent studies on end user preferences for the RTBFoods project (Akankwasa et al. 2020), and specific studies from the BBB project (Marimo et al. 2019, 2020a, b). Nevertheless, the use of the G+ tools suggested that more research is needed to make meaningful assessments and quantify implications of specific traits. For example, while using the G+ tools, the gender researcher found that both women and men mention the importance of the trait "ease of peeling" although women rank it higher (Akankwasa et al. 2020).

However, there are no studies that measure changes in labor input for peeling different varieties, only a subjective rating of whether a variety is easy to peel. This makes it difficult to make the G+ tool's assessment of whether "ease of peeling" affects drudgery. Although women rank "ease of peeling" more highly than men, as a preferred banana trait, there are no quantified standards for descriptors of banana fingers such as "big," "long," or "straight" that are associated with ease of peeling and are preferred. Without quantified standards, it is difficult to know whether a given variety is, for example, "straight" enough to satisfy a given users' preference or if it is likely to be rejected as "not straight enough" and thus, potentially, "difficult to peel." Also, until quantifiable standards are established, breeders cannot be sure if a preferred descriptor is correlated with a heritable trait.

In the Uganda banana experience, the use of G+ tools for gender-responsive breeding was hindered by the lack of data. There was a shortage of gender-relevant and sex-disaggregated data for customer profiling that made it difficult to do segmentation, targeting, and particularly trait valuation with a gender dimension. While there was some data on gender-differentiated trait preferences, the available information on gender relations in banana value chains in Uganda did not supply enough evidence or representative data to score traits for their positive or negative implications for men or women.

Another decisive handicap was the difficulty in reorganizing and operating a multidisciplinary team and the insufficient financial support for gender research from the projects reaching their end of phase. This experience underscores the value of teamwork in research for innovating in gender-responsive breeding. To understand the socioeconomic implications of divergent men's and women's trait preferences on breeding decisions, a scientific quantification of producers' banana descriptors and their correlation with heritable traits is needed. Nonetheless, the use of the G+ tools created an opportunity to build on existing teamwork and research on gender and trait preferences to conduct a deeper assessment of what gender analysis means for ongoing work within the breeding program. Despite the obstacles to applying the G+ tools with a multidisciplinary team, the specific banana trait preferences of women and men farmers were identified, and the gender roles for preparing banana-based products were determined. Preferences for some traits differed between women and men, while others did not. Quality and consumption-related attributes were regarded as the most important by both women and men. Traits valued by women not then included in the banana breeding profiles included agronomic attributes (e.g., adaptability to poor soils), processing traits, and social and cultural traits – plant parts which could be used for multiple purposes, (e.g., banana leaves to wrap food or roots for medicines; and size and shape attributes of fruit, e.g., uniform finger size, straight fingers for ease of peeling, and compact bunches for easy transport). Ease of peeling and short cooking time were desired traits mostly mentioned by women.

Innovation with gender-responsive breeding, despite serious lack of continuity in staffing that held up piloting of the G+ tools, has increased breeders' openness to revise product profiles considering newly identified traits valued by women and men.

16.5 Discussion

This section compares the experiences of the three RTB breeding programs implementing institutional innovations for gender-responsive breeding and applying new tools to help breeders integrate gender into customer profiling and product profile development. As a result of adopting the institutional innovations and G+ tools for gender-responsive breeding, all three programs systematically queried the gender implications of customer segments and the traits currently prioritized for selection. The cassava breeding program made a formal commitment to using gender analysis and the G+ tools, while the sweetpotato and banana programs are still discussing it.

All three programs identified traits with gender relevance that attained new recognition or gained more importance in their breeding objectives. At the time of writing, all three programs were in the process of incorporating one or more newly identified, gender-related traits into their breeding objectives or product profile. However, the programs experienced more difficulty with gendered customer profiling, as discussed further below.

All three programs successfully achieved the key objective of introducing gender-responsive breeding, i.e., that breeders systematically query gender implications whenever they decide on customer segments for targeting and prioritizing traits to include in their product profiles. This matters even if the conclusion of the enquiry is that there is no important gender difference in terms of trait priorities, because the enquiry reduces risk of overlooking gender inequality. In fact, in all three programs gender analysis found substantial agreement on some important trait preferences among men and women in different parts of the value chain, but the gender analysis gave new significance to breeding for aspects of the value chain that were economically vital to poor women.

A comparison of the three programs' work on gender-responsive breeding is shown in Table 16.2.

All the breeding teams implemented specific actions as part of the innovation package intended to enhance gender awareness and capability for gender-responsive breeding as summarized in Table 16.3: one or more members of the breeding team attended workshops designed to promote dialog between breeders and gender researchers and participated in gender training; each program installed new capacity in the team for gender analysis by appointing a gender researcher with a brief to promote cross-disciplinary interaction and who championed the piloting of the G+ tools.

Innovation was associated with four important good practices that evolved along similar lines in each program. Levels of four key good practices that contributed to success with innovating gender-responsive breeding are compared on a three-point scale (Table 16.4).

First, each gender researcher got to grip early on with the issue of how to make practical use of gender analysis to help breeders do their work more effectively. Each one encouraged the team to be more *gender-aware and receptive* to innovation by engaging directly in integrating gender into the program's methods for evaluating varieties with farmers. From the breeders' point of view, this had the benefit of becoming aware of gender-relevant data that might otherwise have escaped notice. Evaluating varieties with farmers generated dialog between them

Table 16.2 Results from working on gender-responsive breeding 2016–2020

Team's results from working on gender-responsive breeding	Cassava	Sweetpotato	Banana
Customer profiles and targets were queried from a gender perspective by the team	Yes	Yes	Yes
Trait preferences were queried by the team from a gender perspective	Yes	Yes	Yes
Team identified gender dimensions of some customer trait preferences or values	Yes	Yes	Yes
Team evaluated its trait priorities in the light of findings from gender analysis	Yes	Yes	Yes
Team added a new trait or category of traits as a result of gender analysis	Yes	Yes	Yes
Team formally committed to incorporating gender-responsiveness using G+ tools into its work	Yes	In progress	In progress

Table 16.3 Actions implemented to advance toward gender-responsive breeding 2016–2020

Team's practice of gender-responsive breeding	Cassava	Sweetpotato	Banana
Team took an active part in GBI cross-disciplinary workshops	Yes	Yes	Yes
Team incorporated a specialized gender researcher	Yes	Yes	Yes
One or more team members received or contributed to GREAT gender training	Yes	Yes	Yes
Gender specialist integrated gender into the program's varietal selection (PPB and/or PVS)	Yes	Yes	Yes
Team obtained and analyzed sex- disaggregated data	Yes	Yes	Yes

and program breeders, and when this dialog included a gender specialist, the breeder gained insights about the implications of gender differences in farmers' varietal choice.

Second, each gender researcher brought into focus the gender aspects of farmers' trait preferences whether from existing or new data and, in all cases, *the program invested in collecting or analyzing data* that threw new light on gender aspects of the value chain and variety choice. Third, in all three programs, the leaders took steps to promote *multidisciplinary teamwork* among breeders, the gender researcher, and other social scientists accessible to the program, especially through partnership. Finally, in all three cases, *partners were sought* who made vital contribution to obtaining and analyzing gender-relevant data and also to reinforcing the value of gender analysis for helping breeders to understand the demand for certain traits.

All three programs found that using the G+ tools helped the gender researcher and the breeding team to conduct a systematic process that reinforced the good practices described above. This provided a process for organizing and analyzing data to make sure that available evidence from gender analysis was actually used for customer targeting and product profiling. In each case, use of the tools highlighted some important data gaps, notably in relation to the trait preferences that carried weight at different points in the value chain where women either predominated or, in some instances, were at a disadvantage compared to men. This was important because a common outcome of the gender analysis conducted with the G+ tools was to identify or confirm the importance of prioritizing RTB-related quality traits that were highly valued by women, whether they processed for home consumption or for the market.

The values assigned to progress with the tools are presented in Table 16.5. To compare use of the G+ tools, a scale was devised for evaluating different levels of progress in the use of the core innovations (Table 16.5). Levels of good practice and progress with tool use are presented in Fig. 16.1.

Figure 16.1 displays the differences among programs in progress with use of the G+ tools and levels of good practice found in all three cases. The banana program is visibly restricted in terms of use of the G+ tools and levels of good

Table 16.4 Levels of good practices achieved by each case

Good practice	Cassava – Nigeria		Sweetpotato – Uganda		Banana - Uganda	
	Level[a]	Indicator description	Level[a]	Indicator description	Level[a]	Indicator description
Gender awareness and capability built in the team	3	Team took an active part in GBI cross-disciplinary workshops Team incorporated a specialized gender researcher One or more team members received or contributed to GREAT gender training	3	Team took an active part in GBI cross-disciplinary workshops Team incorporated a specialized gender researcher One or more team members received or contributed to GREAT gender training	2	Team took an active part in GBI cross-disciplinary workshops Team incorporated a specialized gender researcher One or more team members received or contributed to GREAT gender training Team commitment not yet developed due to staffing issues
Investment in obtaining and analyzing gender-relevant data	3	Gender specialist integrated gender into the program's varietal selection (PPB and/or PVS) Team obtained and analyzed sex-disaggregated data Team committed to further investment in gender data	3	Gender specialist integrated gender into the program's varietal selection (PPB and/or PVS) Team obtained and analyzed sex-disaggregated data Team committed to further investment in gender data	2	Gender specialist integrated gender into the program's varietal selection (PPB and/or PVS) Team obtained and analyzed sex-disaggregated data Team commitment not yet developed due to staffing issues

(continued)

Table 16.4 (continued)

Good practice	Cassava – Nigeria		Sweetpotato – Uganda		Banana - Uganda	
	Level[a]	Indicator description	Level[a]	Indicator description	Level[a]	Indicator description
Partnerships built that expanded capacity	3	Partners made critical input to gender capacity, data collection, and team decision-making	2	Partners made critical input to gender capacity and data collection, but team decision-making is yet to be implemented	2	Partners made critical input to gender capacity and data collection but team decision-making stalled due to staffing issues
Leaders fostered multidisciplinary teamwork and cooperation between biological and social scientists	3	Team formally committed to future use of G+ tools Team committed to inclusion of gender researcher in team decisions about breeding product design, evaluation, and release, as a model for future operation	2	Team not committed to future use of G+ tools at time of writing Team committed to inclusion of gender researcher in team decisions	1	Team not committed to future use of G+ tools at time of writing Team composition and role of gender researcher uncertain while new staff are recruited

[a]Levels of good practice: 0 = nonexistent; 1 = present but not actively enabling innovation; 2 = actively enabling innovation with gender-responsive breeding; 3 = actively enabling innovation that produces commitment to integrating gender into future variety design and product advancement after the project ends

practice. Above all, the diagram illustrates the critical importance of further investment in customer profiling and highlights how all three programs have run into difficulties in putting together a full, gendered customer profile due to the inadequacies of gender-relevant data. This is critical: if a breeding program is not clear about who are its priority customers, it is impossible to know whose trait preferences to compare in a gender analysis. The next section of the paper draws on the comparison to draw some lessons for future innovation with gender-responsive breeding.

Table 16.5 Levels of progress in use of the G+ tool innovations

Use of G+ tool	Cassava – Nigeria		Sweetpotato – Uganda		Banana - Uganda	
	Level[a]	Indicator description	Level[a]	Indicator description	Level[a]	Indicator description
G+ Customer Profile tool	2	Applied with existing data Report produced. Not used in discussions toward decision-making	2	Applied with existing data Report produced. Not used in discussions toward decision-making	1	Applied with existing data
G+ Product Profile Query tool	3	Applied with existing data Report produced Used for product profile development	2	Applied with existing data Report produced. Not used for product profile development	1	Applied with existing data
Standard Operating Procedure (SOP)	3	Used with the tools. SOP Gender Report produced Used for team discussions	2	Used with the tools SOP Gender Report produced Not used for team discussions	1	Used with the tools

[a]Levels of G+ Tool use: 0 = no use; 1 = desk review of tools; 2 = tool applied and report produced; 3 = results used to formally include gender in team decision-making

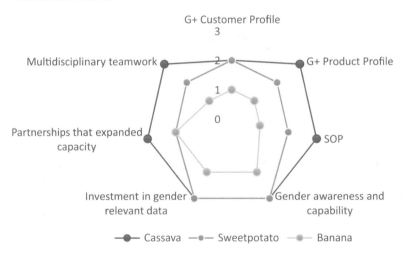

Fig. 16.1 Levels of progress in three breeding programs' use of the G+ tools and good practice of gender-responsive breeding. (*Source*: authors' interpretation)

16.6   Key Lessons Learned and Future Opportunities

Five lessons for innovating with gender-responsive breeding can be drawn from the experiences presented:

First, the *importance of a good foundation for gender-responsive breeding*. This was supplied by cultivating gender awareness though cross-disciplinary dialog, gender training targeted to breeding programs, and incorporation of a gender specialist into breeding teams. These innovations encouraged receptivity to teamwork among breeders and social scientists and enabled pioneering gender researchers to engage with practical aspects of variety development in a way that was useful to breeders.

Second, *hands-on, cross-disciplinary cooperation proved vital*. For example, the cassava program engaged earliest with cross-disciplinary dialog and rapidly enhanced its gender research capacity, developing an operational multidisciplinary team with active engagement of breeders. This program had effectively done the job involved in applying the G+ tools by the time these were made available and so achieved much progress in using the G+ tools to systematize work that was already well-advanced (Fig. 16.1). In comparison, the sweetpotato program started building a team later, and they made less progress with the G+ tools, despite having relatively abundant data for gender analysis. In the banana case, when the team lost the national breeders who had been closely involved in the cross-disciplinary dialog exercises and who had spurred the collection of gender-relevant data, teamwork could not develop, the gender researcher had to work in relative isolation, and this held up full application of the tools. It is important for leadership to pay close attention to how well the multidisciplinary breeding team operates as a mechanism for knowledge sharing, communication, and decision-making on important issues that enable gender-responsive decision-making and prioritization.

Third, *actively promoting multidisciplinary teamwork strengthens gender training*. Capacity building is a critical entry point for gender integration in agricultural development (Njuki 2016), and gender-responsive breeding as an innovation is no exception. In all the experiences described, gender capacity was enhanced with a dedicated gender specialist, but these researchers could only be as effective as the team in which they operated. GREAT training introduced the G+ Product Profile Query tool along with the concept of product profiles and the need for a clear understanding of demand from users of future varieties. The practical questions provided by the G+ tool for assessing "Do No Harm" lead the participants to immediate questions about their programs and to direct interactions between the social scientist and the breeders in each team. After training, teams gain a heightened sense of the need for interdisciplinary expertise to collect the necessary data for G+ tool application. Recent case studies on GREAT impact found that attending GREAT training was associated with changes in breeding programs, including their approach to on-farm testing, data collection, consumer testing, varietal evaluation, and fostering interdisciplinary teams. This was well appreciated by one of the RTB breeders who told the team's social scientist, "We are all breeders" (even the social scientist).

Fourth, *it is essential to invest in collecting data for gender analysis relevant to breeding*. To make sure the data is relevant, the gender analysis has to be planned with breeding objectives in mind. Latching onto any available sex-disaggregated data on a target population or even an adoption study is seldom fit for this purpose. The breeding teams presenting cases in this chapter already much data from different sources and collected through different methods and mechanisms, but all experienced the need to obtain additional data or to do further analysis. This involved not only demographic data but also biophysical data for breeders to understand better what plant traits are correlated with women's and men's preferences and descriptors. GREAT training has also highlighted the difficulty faced by most teams to find suitable evidence to complete tables in the G+ tools. As more than 80% of GREAT trainees are from NARES in SSA, this raises questions about potential scaling of the G+ tools in programs that may have less data.

Fifth, *using practical G+ tools helps to streamline teamwork and the assembly of data*, especially if tool application is hampered by the lack of data. Using the G+ tools stimulated further data collection and a demand for deeper gender analysis. All the cases experienced some challenges in assembling the data required for customer segmentation and profiling. In part this reflects the nature of customer profiling which involves much more than gender analysis. The G+ Customer Profile Tool proved useful in alerting the teams to gender-relevant data that are valuable for understanding demand. Most of the cases had access to large-scale, social surveys, conducted for other purposes such as baselines, impact, or adoption studies of varieties, but there was a significant shortage of foresight or forward-looking analysis and projections with gender content. Other data sources such as national statistics were not detailed enough to disaggregate individuals by gender or customer segments. The unevenness in the quality and quantity of available data requires full-time expertise in gender analysis. The paucity of gender-relevant and sex-disaggregated data at regional (agroecological) or national level underscored the need to include a gender dimension in future studies of customer demand and grower practices as well as variety adoption and impact.

These five lessons tell us a lot about the dos and don'ts of innovation with gender-responsive breeding. Above all, for future efforts to introduce gender-responsive breeding, the lessons highlight the importance of using the G+ tools in combination with institutional innovations to enable good practices to succeed.

Acknowledgments This research was undertaken as part of the CGIAR Research Program on Roots, Tubers, and Bananas (RTB), the CGIAR Excellence and Breeding Platform (EiB), and the CGIAR GENDER Platform and supported by CGIAR Trust Fund contributors. Funding support for this work was also provided by the Bill & Melinda Gates Foundation (BMGF).

The authors are thankful to Jillian Lenne and Gordon Prain for their comments and feedback on earlier drafts of this chapter. We appreciate Jeff Bentley's editorial support and Zandra Vasquez's assistance with formatting.

References

Akankwasa K, Ortmann GF, Wale E, Tushemereirwe WK (2013a) Determinants of consumers' willingness to purchase east African Highland cooking banana hybrids in Uganda. AJAR 8:780–791. https://doi.org/10.5897/AJAR12.1745

Akankwasa K, Ortmann GF, Wale E, Tushemereirwe WK (2013b) Farmers' choice among recently developed hybrid banana varieties in Uganda: a multinomial logit analysis. Agrekon 52:25–51. https://doi.org/10.1080/03031853.2013.798063

Akankwasa K, Ortmann GF, Wale E, Tushemereirwe WK (2016) Early-stage adoption of improved Banana "Matooke" hybrids in Uganda: a count data analysis based on farmers' perceptions. Int J Innov Technol Manag 13:1650001. https://doi.org/10.1142/S0219877016500012

Akankwasa K, Marimo P, Tumuhimbise R, Asasira M, Khakasa E, Mpirirwe I, Kleih U, Forsythe L, Fliedel G, Dufour D, Nowakunda K (2020) The east African highland cooking bananas 'Matooke' preferences of farmers and traders: implications for variety development. Int J Food Sci Technol. https://doi.org/10.1111/ijfs.14813

Ashby JA, Polar V (2019) The implications of gender for modern approaches to crop improvement and plant breeding. In: Sachs C (ed) Gender, agriculture and agrarian transformation. Routledge, London, pp 11–34

Ashby JA, Polar V (2021a) G+ product profile query tool. GBI, Lima

Ashby JA, Polar V (2021b) Standard operating procedure (SOP). GBI, Lima

Ashby JA, Polar V, Thiele G (2018) Critical decisions for ensuring plant or animal breeding is gender-responsive

Asindu M, Ouma E, Elepu G, Naziri D (2020) Farmer demand and willingness-to-pay for sweetpotato silage-based diet as pig feed in Uganda. Sustainability 12:6452. https://doi.org/10.3390/su12166452

Assefa T, Assibi Mahama A, Brown AV, Cannon EKS, Rubyogo JC, Rao IM, Blair MW, Cannon SB (2019) A review of breeding objectives, genomic resources, and marker-assisted methods in common bean (Phaseolus vulgaris L.). Mol Breed 39. https://doi.org/10.1007/s11032-018-0920-0

Banda L, Kyallo M, Domelevo Entfellner JB, Moyo M, Swanckaert J, Mwanga ROM, Onyango A, Magiri E, Gemenet DC, Yao N, Pelle R, Muzhingi T (2021) Analysis of β-amylase gene (Amyβ) variation reveals allele association with low enzyme activity and increased firmness in cooked sweetpotato (Ipomoea batatas) from East Africa. Jo Agric Food Res 4:100121. https://doi.org/10.1016/j.jafr.2021.100121

Bashaasha B, Mwanga ROM, Ocitti p'Obwoya C, Ewell PT (1995) Sweetpotato in the farming and food systems of Uganda: a farm survey report. CIP-NARO, Kampala

Bentley J, Olanrewaju A, Madu T, Olaosebikan O, Abdoulaye T, Wossen T, Manyong V, Kulakow P, Ayedun B, Ojide M, Girma G, Rabbi I, Asumugha G, Tokula M (2017) Cassava farmers' preferences for varieties and seed dissemination system in Nigeria: gender and regional perspectives. International Institute of Tropical Agriculture

Bricas N, Tchamda C, Martin P (2016) Les villes d'Afrique de l'Ouest et du Centre sont-elles si dépendantes des importations alimentaires? Cah Agric 25:55001. https://doi.org/10.1051/cagri/2016036

Byrne T, Santos B, Schurink A, Amer P, Martin-Collado D (2020) Trait prioritisation and valuation analysis for Sweetpotato in Central Uganda. AbacusBio

Ceccarelli S, Grando S (2007) Decentralized-participatory plant breeding: an example of demand driven research. Euphytica 155:349–360. https://doi.org/10.1007/s10681-006-9336-8

CGIAR, GBI (2017) Innovation in gender-responsive breeding (workshop report). Nairobi, Kenya

CGIAR, GBI (2018) Gender-responsive product profile development tool. Workshop report. November 12–13. Ithaca, USA (Report)

Chijioke U, Madu T, Okoye B, Ogunka AP, Ejechi M, Ofoeze M, Ogbete C, Njoku D, Ewuziem J, Kalu C, Onyemauwa N, Ukeje B, Achonwa O, Forsythe L, Fliedel G, Egesi C (2020) Quality attributes of fufu in South-East Nigeria: guide for cassava breeders. Int J Food Sci Technol 56:1247–1257. https://doi.org/10.1111/ijfs.14875

Doss CR, Morris ML (2001) How does gender affect the adoption of agricultural innovations? The case of improved maize technology in Ghana. Agric Econ 25:27–39. https://doi.org/10.1016/S0169-5150(00)00096-7

Echodu R, Edema H, Wokorach G, Zawedde C, Otim G, Luambano N, Ateka EM, Asiimwe T (2019) Farmers' practices and their knowledge of biotic constraints to sweetpotato production in East Africa. Physiol Mol Plant Pathol Crop Pathol Afr 105:3–16. https://doi.org/10.1016/j.pmpp.2018.07.004

Fisher M, Kandiwa V (2014) Can agricultural input subsidies reduce the gender gap in modern maize adoption? Evidence from Malawi. Food Policy 45:101–111. https://doi.org/10.1016/j.foodpol.2014.01.007

Forsythe L, Posthumus H, Martin A (2016) A crop of one's own? Women's experiences of cassava commercialization in Nigeria and Malawi. J Gender Agric Food Security 1:110–128

Gibson RW, Byamukama E, Mpembe I, Kayongo J, Mwanga ROM (2008) Working with farmer groups in Uganda to develop new sweet potato cultivars: decentralisation and building on traditional approaches. Euphytica 159:217–228. https://doi.org/10.1007/s10681-007-9477-4

Kiiza B, Kisembo LG, Mwanga ROM (2012) Participatory plant breeding and selection impact on adoption of improved sweetpotato varieties in Uganda. J Agric Sci Technol, ISSN 1939-1250. 2(5A):673–681. https://hdl.handle.net/10568/66545

Kimani PM (2017) Principles of demand-led plant variety design. In: Persley GJ, Anthony VM (eds) The business of plant breeding: market-led approaches to new variety design in Africa. CABI, pp 1–25

Lebot V (2019) Tropical root and tuber crops, 2nd edn. CABI

Low JW, Mwanga ROM, Andrade M, Carey E, Ball AM (2017) Tackling vitamin a deficiency with biofortified sweetpotato in sub-Saharan Africa. Global Food Security Food Security Governance Latin Am 14:23–30. https://doi.org/10.1016/j.gfs.2017.01.004

Marimo P, Karamura D, Tumuhimbise R, Shimwela MM, Van den Bergh I, Batte M, Massawe CRS, Okurut AW, Mbongo DB, Crishton R (2019) Post-harvest use of banana in Uganda and Tanzania: product characteristics and cultivar preferences of male and female farmers. (RTB Working Paper No. 3). RTB, Lima, Peru

Marimo P, Caron C, Van den Bergh I, Crichton R, Weltzien E, Ortiz R, Tumuhimbise R (2020a) Gender and trait preferences for banana cultivation and use in sub-Saharan Africa: a literature Review1. Econ Bot 74:226–241. https://doi.org/10.1007/s12231-020-09496-y

Marimo P, Nowakunda K, Aryamanya W, Azath H, Babley HF, Kazigye F, Khakasa E, Kibooga C, Kibura JK, Kindimba G, Kubiriba J, Magohe RB, Magona I, Masanja SR, Massawe C, Mbongo DB, Mgenzi AR, Mubiru DM, Mugisha JA, Mugume D, Mwabulambo BA, Namuddu MG, Ndagire L, Ngabirano W, Ngulinziva LD, Njau MC, Nsibirwa L, Nyemenohi S, Okurut AW, Shimwela MM, Swennen R, Urio PA, Van den Bergh I (2020b) Report on consumer acceptability tests of NARITA hybrids in Tanzania and Uganda (report)

Mascarenhas M (2016) The "must-have" features of gender-responsive plant or animal breeding (report). CGIAR Gender and Agriculture Research Network

Moyo M, Ssali R, Namanda S, Nakitto M, Dery EK, Akansake D, Adjebeng-Danquah J, van Etten J, de Sousa K, Lindqvist-Kreuze H, Carey E, Muzhingi T (2021) Consumer preference testing of boiled sweetpotato using crowdsourced citizen science in Ghana and Uganda. Front Sustain Food Syst 5. https://doi.org/10.3389/fsufs.2021.620363

Mwanga ROM, Niringiye C, Alajo A, Kigozi B, Namukula J, Mpembe I, Tumwegamire S, Gibson RW, Yencho GC (2011) 'NASPOT 11', a sweetpotato cultivar bred by a participatory plant-breeding approach in Uganda. HortScience 46:317–321. https://doi.org/10.21273/HORTSCI.46.2.317

Mwanga ROM, Mayanja S, Swanckaert J, Nakitto M, Zum Felde T, Grüneberg W, Mudege N, Moyo M, Banda L, Tinyiro SE, Kisakye S, Bamwirire D, Anena B, Bouniol A, Magala DB, Yada B, Carey E, Andrade M, Johannsmeier SD, Forsythe L, Fliedel G, Muzhingi T (2020)

Development of a food product profile for boiled and steamed sweetpotato in Uganda for effective breeding. Int J Food Sci Technol https://doi.org/10.1111/ijfs.14792

Mwanga ROM, Swanckaert J, da Silva Pereira G, Andrade MI, Makunde G, Grüneberg WJ, Kreuze J, David M, De Boeck B, Carey E, Ssali RT, Utoblo O, Gemenet D, Anyanga MO, Yada B, Chelangat DM, Oloka B, Mtunda K, Chiona M, Koussao S, Laurie S, Campos H, Yencho GC, Low JW (2021) Breeding progress for vitamin a, iron and zinc biofortification, drought tolerance, and sweetpotato virus disease resistance in sweetpotato. Front Sustain Food Syst 5. https://doi.org/10.3389/fsufs.2021.616674

Nasirumbi-Sanya L, Ssali RT, Akankwasa K, Nowankunda K, Barekye A, Namuddu MG, Kubiriba J (2018) Gender-differentiated preferences in breeding for new matooke hybrids in Uganda. In: Tufan HA, Grando S, Meola C (eds.) State of the knowledge for gender in breeding: case studies for practitioners. pp. 87–94

Ndjouenkeu R, Kegah FN, Teeken B, Okoye B, Madu T, Olaosebikan OD, Chijioke U, Bello A, Osunbade AO, Owoade D, Takam-Tchuente NH, Njeufa EB, Chomdom ILN, Forsythe L, Maziya-Dixon B, Fliedel G (2020) From cassava to gari: mapping of quality characteristics and end-user preferences in Cameroon and Nigeria. Int J Food Sci Technol 56:1223–1238. https://doi.org/10.1111/ijfs.14790

Njuki J (2016) Practical notes: critical elements for integrating gender in agricultural research and development projects and programs. Practical notes: critical elements for integrating gender in agricultural research and development projects and programs. https://doi.org/10.19268/JGAFS.132016.6

Olaosebikan O, Abdulrazaq B, Owoade D, Ogunade A, Aina O, Ilona P, Muheebwa A, Teeken B, Iluebbey P, Kulakow P, Bakare M, Parkes E (2019) Gender-based constraints affecting biofortified cassava production, processing and marketing among men and women adopters in Oyo and Benue States, Nigeria. Physiological and Molecular Plant Pathology, Crop Pathology in Africa 105:17–27. https://doi.org/10.1016/j.pmpp.2018.11.007

Orr A, Cox CM, Ru Y, Ashby JA (2018) Gender and social targeting in plant breeding (Working Paper). CGIAR

Orr A, Polar V, Ashby JA (2021) G+ customer profile tool. GBI, Lima

Peterman A, Behrman JA, Quisumbing AR (2014) A review of empirical evidence on gender differences in nonland agricultural inputs, technology, and services in developing countries. In: Quisumbing AR, Meinzen-Dick R, Raney TL, Croppenstedt A, Behrman JA, Peterman A (eds) Gender in agriculture: closing the knowledge gap. Springer Netherlands, Dordrecht, pp 145–186. https://doi.org/10.1007/978-94-017-8616-4_7

Polar V, Ram Mohan R, McDougall C, Teeken B, Mulema A, Marimo P, Yija J (2021) Examining choice to advance gender equality in breeding research. In: Pyburn R, van Eerdewijk A (eds) Advancing gender equality through agricultural and environmental research: past, present and future. IFPRI, Washington, DC

Ragasa C (2012) Gender and institutional dimensions of agricultural technology adoption: a review of literature and synthesis of 35 case studies

Ribaut JM, Ragot M (2019) Modernising breeding for orphan crops: tools, methodologies, and beyond. Planta 250:971–977. https://doi.org/10.1007/s00425-019-03200-8

Sanya LN, Kyazze FB, Sseguya H, Kibwika P, Baguma Y (2017) Complexity of agricultural technology development processes: implications for uptake of new hybrid banana varieties in Central Uganda. Cogent Food Agric 3:1419789. https://doi.org/10.1080/23311932.2017.1419789

Sanya LN, Ssali RT, Akankwasa K, Nowankunda K, Barekye A, Namuddu MG, Kubiriba J (2018) Gender-differentiated preferences in breeding for new matooke hybrids in Uganda. In: Tufan HA, Grando S, Meola C (eds.) State of the knowledge for gender in breeding: case studies for practitioners, pp. 87–94

Sanya LN, Sseguya H, Kyazze FB, Diiro GM, Nakazi F (2020) The role of variety attributes in the uptake of new hybrid bananas among smallholder rural farmers in Central Uganda. Agric Food Security 9:1. https://doi.org/10.1186/s40066-020-00257-7

Ssali RT, Nowankunda K, Barekye Erima R, Batte M, Tushemereirwe WK (2010) On-farm participatory evaluation of East African highland banana Matooke hybrids (Musa spp.). Acta Horticulturae, IV International Symposium on Banana: International Conference on Banana and Plantain in Africa: Harnessing International Partnerships to Increase Research Impact 585–591

Ssali RT, Narirumbi MG, Namuddu MG, Mayanja S (2017) Breeding cooking bananas: do men and women's trait preferences matter? (Technical Report). GREAT

Sseruwu G, Shanahan P, Melis R, Ssemakula G (2015) Farmers awareness and perceptions of alternaria leaf petiole and stem blight and their preferred sweetpotato traits in Uganda. J Plant Breed Genet 3:25–37

Teeken B, Olaosebikan O, Haleegoah J, Oladejo E, Madu T, Bello A, Parkes E, Egesi C, Kulakow P, Kirscht H, Tufan HA (2018) Cassava trait preferences of men and women farmers in Nigeria: implications for breeding. Econ Bot 72:263–277. https://doi.org/10.1007/s12231-018-9421-7

Teeken B, Agbona A, Bello A, Olaosebikan O, Alamu E, Adesokan M, Awoyale W, Madu T, Okoye B, Chijioke U, Owoade D, Okoro M, Bouniol A, Dufour D, Hershey C, Rabbi I, Maziya-Dixon B, Egesi C, Tufan H, Kulakow P (2020) Understanding cassava varietal preferences through pairwise ranking of gari-eba and fufu prepared by local farmer–processors. Int J Food Sci Technol. https://doi.org/10.1111/ijfs.14862

Teklewold H, Adam RI, Marenya P (2020) What explains the gender differences in the adoption of multiple maize varieties? Empirical evidence from Uganda and Tanzania. World Dev Perspect 18:100206. https://doi.org/10.1016/j.wdp.2020.100206

Thiele G, Dufour D, Vernier P, Mwanga ROM, Parker ML, Geldermann ES, Teeken B, Wossen T, Gotor E, Kikulwe E, Tufan H, Sinelle S, Kouakou AM, Friedmann M, Polar V, Hershey C (2021) A review of varietal change in roots, tubers and bananas: consumer preferences and other drivers of adoption and implications for breeding. Int J Food Sci Technol 56:1076–1092. https://doi.org/10.1111/ijfs.14684

Turyagyenda FL, Kankwatsa P, Muzira R, Kyomugisha M, Mutenyo H, Muhumuza JB (2015) Participatory agronomic performance and sensory evaluation of selected orange-fleshed sweet potato varieties in south western Uganda. Glob J Sci Front Res D Agric Veter 15:25–30

Udry C (1996) Gender, agricultural production, and the theory of the household. J Polit Econ 104:1010–1046. https://doi.org/10.1086/262050

Wale E, Yalew A (2007) Farmers' variety attribute preferences: implications for breeding priority setting and agricultural extension policy in Ethiopia. Afr Dev Rev 19:379–396. https://doi.org/10.1111/j.1467-8268.2007.00167.x

Walker TS, Alwang J (2015) Crop improvement, adoption and impact of improved varieties in food crops in Sub-Saharan Africa. CABI

Walker TS, Alwang J, Alene A (2015) Varietal adoption outcomes and impact, in: crop improvement, adoption and impact of improved varieties in food crops in Sub-Saharan Africa. CABI, pp 388–405

Watson A, Ghosh S, Williams MJ, Cuddy WS, Simmonds J, Rey MD, Asyraf Md Hatta M, Hinchliffe A, Steed A, Reynolds D, Adamski NM, Breakspear A, Korolev A, Rayner T, Dixon LE, Riaz A, Martin W, Ryan M, Edwards D, Batley J, Raman H, Carter J, Rogers C, Domoney C, Moore G, Harwood W, Nicholson P, Dieters MJ, DeLacy IH, Zhou J, Uauy C, Boden SA, Park RF, Wulff BBH, Hickey LT (2018) Speed breeding is a powerful tool to accelerate crop research and breeding. Nature Plants 4:23–29. https://doi.org/10.1038/s41477-017-0083-8

Weltzien E, Rattunde F, Christinck A, Isaacs K, Ashby J (2019) Gender and farmer preferences for varietal traits. In: Goldman I (ed) Plant breeding reviews. John Wiley & Sons, Ltd, pp 243–278. https://doi.org/10.1002/9781119616801.ch7

World Bank (2012) World development report 2012: gender equality and development. The World Bank, Washington, DC

Wossen T, Girma Tessema G, Abdoulaye T, Rabbi IY, Olanrewaju AS, Bentley J, Alene A, Feleke S, Kulakow PA, Asumugha GN, Abass A, Tokula M, Manyong VM (2017) The cassava monitoring survey in Nigeria: final report (Report). International Institute of Tropical Agriculture

Yanggen D, Nagujja S (2006) The use of orange-fleshed sweetpotato to combat vitamin A deficiency in Uganda: a study of varietal preferences, extension strategies and post-harvest utilization. Social Sciences Working Paper Series (CIP)

Yao N, Djikeng A, Shoham JL (2017) Visioning and foresight for setting breeding goals. In: Persley GJ, Anthony VM (eds) The business of plant breeding: market-led approaches to new variety design in Africa. CABI, pp 26–62

Part IV
Improving Livelihoods

Chapter 17
Scaling Readiness of Biofortified Root, Tuber, and Banana Crops for Africa

Jan Low (ID), Anna-Marie Ball, Paul Ilona (ID), Beatrice Ekesa (ID),
Simon Heck (ID), and Wolfgang Pfeiffer (ID)

Abstract This chapter describes the degree of readiness and use of biofortified root, tuber, and banana (RT&B) crops: sweetpotato, cassava, banana (cooking and dessert types), and potato. Efforts to develop and utilize orange-fleshed sweetpotato (OFSP), yellow cassava (VAC), and vitamin A banana/plantain (VAB) have been focused heavily in sub-Saharan Africa (SSA), where 48% of the children under 5 years of age are vitamin A-deficient. Iron-biofortified potato is still under development, and a recent study found high levels of bioavailability (28.4%) in a yellow-fleshed cultivar (Fig. 17.1). To date, adapted VAB varieties have been piloted in East Africa, and OFSP and VAC have scaled to 8.5 million households. The scaling readiness framework is applied to innovation packages underlying those scaling efforts to shed light on how scaling is progressing and identify remaining bottlenecks. Women dominate RT&B production in SSA, and women and young children are most at risk of micronutrient deficiencies; hence women's access to technologies was prioritized. Lessons learned from these scaling efforts are discussed, with

J. Low (✉)
International Potato Center (CIP), Nairobi, Kenya
e-mail: j.low@cgiar.org

A.-M. Ball
Independent Consultant (formerly HarvestPlus), Hartley Wintney, Hook, UK
e-mail: annamarieball2016@gmail.com

P. Ilona
HarvestPlus, Nigeria country office, Ibadan, Nigeria
e-mail: p.ilona@cgiar.org

B. Ekesa
Alliance Bioversity International-CIAT, Kampala, Uganda
e-mail: b.ekesa@cgiar.org

S. Heck
International Potato Center (CIP), Nairobi, Kenya
e-mail: s.heck@cgiar.org

W. Pfeiffer
Alliance Bioversity-CIAT, HarvestPlus, Washington, DC, USA
e-mail: w.pfeiffer@cgiar.org

© The Author(s) 2022
G. Thiele et al. (eds.), *Root, Tuber and Banana Food System Innovations*,
https://doi.org/10.1007/978-3-030-92022-7_17

| Fig. 17.1A | Fig. 17.1B | Fig. 17.1C | Fig. 17.1D |

Fig. 17.1 Colorful biofortified roots, tubers, and bananas: (**a**) orange-fleshed sweetpotato. (Credit: CIP); (**b**) vitamin A cassava. (Credit: HarvestPlus); (**c**) vitamin A banana. (Credit: B. Ekesa); (**d**) iron-biofortified potato. (Credit: CIP)

the goal of accelerating the scaling readiness process for other biofortified RTB crops. Implementing gender-responsive innovation packages has been critical for reaching key nutrition and income goals. Diverse partnerships with public and private sector players and investing in advocacy for an adequate enabling environment were critical for achieving use at scale. Future scaling will depend on more nutritious sustainable food systems being at the forefront, supported by continued improvement in breeding methodologies to adapt to climate change and enhance multiple nutrient targets more quickly and to increase investment in the input and marketing infrastructure that vegetatively propagated crops require.

17.1 Introduction

Poor households dispense 60–70% of their total income on food (Bouis et al. 2020). Just rice, wheat, and maize provide at least 40% of global calories (FAO 2016). Roots, tubers, and bananas (RT&B) are associated with more localized supply chains than grains and in sub-Saharan Africa (SSA) provide more than 50% of calories in several countries (Petsakos et al. 2019).

Biofortification is the innovative concept to enhance the micronutrient content of food staples as a cost-effective, sustainable way to deliver key micronutrients, especially to the poor. This can be achieved through conventional breeding, genetic engineering, or fertilization practice (Bouis et al. 2020). A highly diversified diet is the best way to get micronutrients, but cost and access frequently undermine this goal. Biofortification, industrial food fortification, and nutrient supplementation programs are complementary strategies used to target those most at risk of micronutrient deficiencies. Whereas capsule supplementation and premix used for fortification depend on importation and passive reception by target populations, biofortified crops are produced within a country. Scaling of biofortified crops depends on proactive uptake of new varieties by producers and consumers and the generation of additional livelihood opportunities.

An estimated 1.5–2 billion people suffer from micronutrient deficiencies or "hidden hunger" (FAO et al. 2018). Iron, zinc, vitamin A, and iodine are the most widespread and severe deficiencies. Young children and women of reproductive age are most vulnerable to micronutrient deficiencies because they have greater micronutrient needs due to rapid growth and/or reproductive functions (Bailey et al. 2015). The World Bank estimated that macro- and micronutrient deficiencies underpin a 2–3% loss in annual global economic productivity (World Bank 2006). As sources of key macro- and micronutrients, biofortified staples are clear remedies to these challenges. In 2008, the Copenhagen Consensus Center ranked biofortification fifth among cost-effective solutions for global world problems (Meenakshi 2008).

Work on biofortification began in the 1990s, led by researchers in the CGIAR international agricultural centers (Bouis and Saltzman 2017; Low et al. 2017). Highlights during the past two decades include:

- From 2000 to 2009, significant progress was made in breeding and efficacy studies, convincingly demonstrating that biofortified crops could impact human health cost-effectively.
- In 2010, efforts began to intensify scaling of released biofortified varieties.
- By the end of 2019, 340 varieties of 12 biofortified crops had been released in 40 lower- and middle-income countries (LMICs).
- HarvestPlus-led delivery efforts for iron beans, orange-fleshed sweetpotato (OFSP), vitamin A orange maize, zinc rice, zinc wheat, and iron pearl millet successfully reached 8.5 million farming households, while 6.8 million farming households received OFSP through partners participating in the Sweetpotato Profit and Health Initiative (SPHI) (Fig. 17.2), co-led by the International Potato Center (CIP) and the Forum for Agricultural Research in Africa (Bouis et al. 2020).

Fig. 17.2 Participants at the 2018 SPHI Annual Meeting, held in Nairobi, Kenya. (Credit: F. Njung'e/CIP)

- By 2019, 1.7 million Nigerians were growing vitamin A cassava varieties in 26 states (Foley et al. 2021).
- In addition, other nongovernmental and governmental bodies were distributing biofortified crops whose reach was not captured by monitoring organizations.

Released varieties to date have all been conventionally bred, building on naturally occurring variations in target micronutrients in these staples. Recognition of multi-sectoral biofortification efforts included the 2016 World Food Prize to Howarth Bouis, the Director of HarvestPlus, and CIP scientists Maria Andrade, Robert Mwanga, and Jan Low for their OFSP work.

With respect to RT&B, breeding has focused on sweetpotato, cassava, potato, and banana. Increasing levels of provitamin A carotenoids, which the body converts into vitamin A, has been the priority for sweetpotato, cassava, and banana, while enhancing iron and zinc content has been the focus for potato. Since 2014, a major breeding effort is underway to increase iron content in OFSP varieties. Although a critical staple in West Africa, yam (*Dioscorea* spp.) lacked sufficient variation in any of these three micronutrients to warrant any conventional breeding investment.

The objective of this chapter is to reflect critically on where biofortified RT&B crops are concerning their development and utilization at scale. First, we will examine progress made in meeting breeding targets and review expected impacts based on ex ante studies. Second, we will present and apply the scaling readiness approach (Sartas et al. 2020) to the most advanced of RT&B crops in use: orange-fleshed sweetpotato (OFSP), yellow-fleshed vitamin A cassava (VAC), and vitamin A banana (VAB), and highlight lessons learned through a gender lens because RT&B crops are widely grown by women in SSA, but women often face constraints in benefitting from new technologies. We expect these lessons to facilitate the nascent scaling efforts of VAB and iron-biofortified potato (IP). Finally, we discuss the ways forward in light of global realities driven by climate change, the current state of the food system, and advances in breeding methods.

17.2   Status of Biofortified Crop Variety Development

17.2.1   Achieving Target Levels and Bioavailability Evidence

During the concept development of biofortification, target levels were set by a multidisciplinary working group in 2005 with the intention of achieving a measurable impact on health for children and women of childbearing age (Hotz and McClafferty 2007). Target levels combine a context-specific micronutrient baseline level measured in commercial crops, with target increments to be added to achieve a required contribution to the estimated average requirement (EAR) from the biofortified crop. Target increments are adjusted for per capita intake, retention (losses during processing, storage, and cooking), and bioavailability (Bouis et al. 2020). Hence, target increments can be achieved by breeding for higher micronutrient concentration, increasing bioavailability and increasing retention. In addition to exploiting geno-typic differences in retention in breeding, food technology has a significant role in increasing provitamin A retention. A negative effect of climate change on minerals and protein may require gradually increasing target increments for minerals in particular.

Cumulatively, more than 175 biofortified varieties of four RT&B crops have been released in more than 30 countries with a heavy emphasis on SSA (Bouis et al. 2020). Candidate biofortified varieties across the RT&B crops are being evaluated for release in an additional 20 countries (Table 17.1).

To date, OFSP is the biofortified RTB most in use, followed by VAC. VAB is still at the pilot stage and IP under development.

Orange-fleshed sweetpotato (OFSP) In OFSP the intensity of the orange color reflects the amount of beta-carotene present, and 80% of the carotenoids present are beta-carotene (Fig. 17.1). Average beta-carotene values among OFSP clones at CIP-Peru was 144 µg/g dwb, with a maximum value of 1220 µg/g dwb, meaning many clones that exceeded beta-carotene target levels were available to draw from. The breeding challenge has been to combine the beta-carotene trait with other traits relevant for adoption by adult farmers, namely, high dry matter and high starch contents, resistance to viruses, and, where needed, tolerance to drought. The devel-opment and deployment of the accelerated breeding scheme, implemented by CIP and 14 national programs, enabled the breeding cycle to be reduced from 8–10 to 4–5 years (Andrade et al. 2017). Between 2009 and 2020, 62 OFSP varieties bred in Africa were released.

Starting in 2014, CIP began breeding for a "doubly biofortified sweetpotato" with the goal of having high-iron, high-beta-carotene varieties. The positive genetic association between iron, zinc, and beta-carotene supports this effort, but genetic variation in iron and zinc content within the germplasm is much less available, requiring more cycles of breeding to reach target levels. A recent study found that bioavailability of Fe in OFSP was 8.1% in women with low Fe status and just 4.0% in women with adequate Fe status (Jongstra et al. 2020). Additional breeding cycles will be required to reach target levels for Fe biofortification.

Vitamin A cassava (VAC) The nutritional quality of cassava roots is low as roots contain mainly carbohydrates and trace elements of other micronutrients (Ceballos et al. 2017). Screening of germplasm accessions (2003–2008) found ranges of 0–19 ppm (fwb) provitamin A in roots of existing cassava varieties but good herita-bility of carotenoid content in roots, which encouraged breeders to proceed with biofortification (Ilona et al. 2017). Around 62% of the total carotenoids on average were *all-trans* beta-carotene, the most bioaccessible form of provitamin A (Ceballos et al. 2017). Two CGIAR centers collaborate in the VAC breeding effort: CIAT to generate high provitamin A sources via rapid cycling in pre-breeding and to provide in vitro clones and seed populations to the International Institute for Tropical Agriculture (IITA) for use in their breeding programs targeting African countries. As with OFSP, the goal is to have high-yielding varieties with high dry matter and

Table 17.1 Summary of countries where biofortified RT&B crops are released or under testing, target levels set, and bioavailability, bioaccessibility, and efficacy studies completed as of 2020

Crop	Micronutrient	Target level	Bioaccessibility	Bioavailability/ efficacy	# Countries in Africa		# Countries in Asia		# Countries in Latin America/ Caribbean	
					Released (R)	Under testing (T)	R	T	R	T
OFSP	Vitamin A	70.0 µg/g fwb	Bengtsson et al. 2009); (Tumuhimbise et al. (2009)	Haskell et al. (2017)	16	9	6	0	6	2
OFSP	Iron	61 µg/g dwb[b]	Andre et al. (2018)	Jongstra et al. (2020)	0	1	0	0	0	1
Cassava	Vitamin A	15.0 µg/g fwb	La Frano et al. (2013)	Haskell et al. (2017)	5	18	0	0	1	3
Dessert banana	Vitamin A	76 µg/g fwb	Not yet	Not yet	2	7	0	0	0	0
Plantain	Vitamin A	76 µg/g fwb	Ekesa et al. (2012b)	Not yet	2	3	0	0	0	0
Potato (yellow-fleshed)	Iron	7.5 µg/g fwb[b]	Andre et al. (2015)	Jongstra et al. (2020)	0	5	0	4	0	1

[a]fwb fresh weight basis, dwb dry weight basis. Underlying assumptions (Bouis et al. 2017): (1) for cassava (fresh weight to cassava meal equivalent), intake levels, 948 g/day for women; 348 g/day for children 4–6 years old; bioavailability, 20%; (2) for OFSP vitamin A, intake levels, 167 g/day for women; 101 g/day for children 4–6 years old; bioavailability, 8%; (3) for Fe, intake levels for young children, 150 g/day

[b]Target reset in 2020 by CIP breeders drawing on Jongstra et al. (2020)

high-beta-carotene content. In VAC, carotenoid concentrations are much higher in the leaves than in roots, while the opposite is true for OFSP.

Three first-wave VAC with 6–8 ppm provitamin A were released in 2011. Three second-wave varieties with up to 11 ppm were released in 2014. More than 50 VAC varieties are now under evaluation in several countries to identify those that are agronomically competitive for third-wave release (Fig. 17.1). The top five leads have more than 15 ppm, the target increment (Ilona et al. 2017).

Iron potato (IP) Potato biofortification efforts at CIP for the last 17 years have focused on iron and zinc. Breeding diploids can accelerate genetic gain. Evaluation of three cycles of recurrent selection from crosses at the diploid level revealed high heritability (0.81 for both iron and zinc), and genetic gains above 29% for iron and 26% for zinc have been demonstrated (Amoros et al. 2020). However, diploids expressed lower yield compared to local varieties in Africa (Rwanda and Ethiopia) and Asia (Nepal, Bhutan, and India). These results prompted a new series of trials in Ethiopia, Kenya, and Rwanda in 2019/2020 with 50 biofortified tetraploid clones with consumer-preferred skin and flesh colors that are also late blight- and virus-resistant. Promising results from multilocation trials indicate the feasibility of a release of tetraploid potatoes in East Africa by 2022 (Fig. 17.3).

Fig. 17.3 Farmers are always involved in assessing the performance and taste of potential new varieties (clones) during multilocational trials. (Credit: P. Demo/CIP)

Results from a human iron bioavailability study (Jongstra et al. 2020) reveal remarkably high iron absorption from yellow-fleshed potatoes (28.4%) in women from the Peruvian Andes, highlighting the potential of biofortified potatoes to contribute to increased iron intakes. Lower iron bioavailability in the purple-fleshed potato (13%) is attributed to the high levels of phenolics, important inhibitors of iron absorption (Fig. 17.1). Given typical consumption levels of 500 g daily of potato by women in highland areas of Peru, the yellow-fleshed or purple-fleshed potato studied cover about 33% of the daily absorbed iron requirement for women of reproductive age. The bioaccessibility and bioavailability of zinc has not yet been determined, but zinc levels increase as iron is selected for.

Vitamin A banana Conventional breeding of banana (*Musa* spp.) is difficult and expensive due to the long crop cycle as most commercial varieties are sterile triploids and have high cross incompatibility among fertile groups (Amah et al. 2019). Values of four boiled local cultivars in DR Congo provided vitamin A levels of 22.3–173 retinol activity equivalent (RAE) µg/100 g fwb (Ekesa et al. 2012a). Hence, the 14-year effort (2006–2019) in four East and Central African countries (Tanzania, Uganda, Burundi, DR Congo) has focused on selecting promising, more carotenoid-rich cultivars from 400 pre-screened cultivars from other countries (Fig. 17.1). About half of their carotenoid content is beta-carotene (retinol equivalence of 12:1) and the other half alpha-carotenoids (retinol equivalence of 24:1). As of March 2020, seven cultivars, from Ghana, Papua New Guinea, the Philippines, Hawaii, and Indonesia have demonstrated potential to perform well within East and Central Africa (Fongar et al. 2020) (Fig. 17.4). Sensory testing with local farmers revealed that 5 of the 15 tested cultivars have acceptable taste (Ekesa et al. 2017). Hence, pilot dissemination efforts have focused on six of the cultivars (Apantu, Bira, Lahi, Pelipita, Muracho, and Pisang Papan) in Burundi and DR Congo. By the end of 2019, nearly 13,000 farmers had been reached, of whom 61% were women (Fongar et al. 2020).

Given the challenges in conventional breeding, a breeding effort started in 2007 is using genetic modification techniques to biofortify cooking bananas. Implemented by Queensland University of Technology in Australia and the National Agricultural Research Organisation of Uganda, the Banana21 project has incorporated a gene effective at increasing provitamin A content obtained from high provitamin A carotenoid Fei banana "Asupina" (from Papua New Guinea) into M9 hybrid and East African Highland bananas (Amah et al. 2019). Activities include laboratory work and field trials in Uganda and Australia and a nutrition study in the USA. These bananas may be ready for use in 2021. IITA has also recently incorporated high provitamin A carotenoid diploids from Papua New Guinea into their plantain breeding program (Amah et al. 2019).

Fig. 17.4 Banana trial for selecting promising varieties in Burundi. (Credit: A. Simbare/Alliance Bioversity-CIAT)

17.2.2 Influence of Micronutrient Retention During Processing

For RT&B, dominant forms of consumption are driven by the perishability of the crop postharvest. Both sweetpotato and cassava can be "stored" in ground for considerable periods of time and harvested piecemeal. In contrast, potato tubers need to be harvested when they reach maturity. Banana (including East African highland bananas) plantains and dessert bananas are often harvested and used while green but also used when partially or fully ripened. In eight VAB cultivars, mean total provitamin A carotenoids increased substantially from 560–4680 mg/100 g fwb in unripe fruit to 1680–10,630 mg/100 g fwb in ripe fruit (Ekesa et al. 2015).

Once harvested, cassava roots have very limited shelf life and must be processed into a dried, storable form (Fig. 17.5). Fresh sweetpotato roots, without curing, can last 1–3 weeks, and the dominant form of consumption is boiled or steamed roots, with roasted and fried roots consumed to a lesser extent. Non-diseased potato tubers can store for months under dark and cool conditions; their dominant form of consumption in SSA is also boiled or steamed, with fried consumption concentrated in

Fig. 17.5 Preparing vitamin A cassava roots for processing in Nigeria. (Credit: HarvestPlus)

urban areas. Green plantains are boiled, while ripe plantains are typically fried. Ripened sweet bananas are eaten raw as a fruit.

In the human intestine, carotenoids like beta-carotene and alpha-carotene are cleaved to retinol (vitamin A). Beta-carotene has two times higher vitamin A activity (12:1 retinol conversion) than other carotenoids (24:1) (Ishiguro 2019). Releasing nutrients from the food matrix specific to each crop during digestion makes them *bioaccessible.* Then a person's health status and presence of other substances, like fat, determines the amount of nutrient actually absorbed by the intestines, reflecting the product's *bioavailability.*

Exposure to light, air, and heat can all contribute to the degradation of provitamin A carotenoids with levels varying considerably by genotype. De Moura et al. (2015) reported that for VAC and OFSP, boiling and steaming had much higher retention rates of vitamin A (80–98%) compared to roasting or baking (30–70%) and frying (18–54%). A significant concern for VAC in West Africa is that most cassava is consumed as *gari*, a fermented granular flour, pressed and roasted into granules – called *fufu* – fermented roots that are boiled or steamed and then pounded (Fig. 17.6). While apparent retention in *fufu* range from 44 to over 100%, *gari* had the lowest levels of retention (10–30%). Taleon et al. (2018) determined that true total carotenoid in *fufu* was only 0.8–3.1%. Reaching the biofortification vitamin A target with VAC *gari* only works because average per capita consumption levels of cassava by women in rural West Africa is high – 900 g per capita daily (fwb) (De Moura et al. 2015). By contrast, just 100–125 g of any OFSP root, regardless of how it is prepared, will meet 100% of the vitamin A EAR for young children.

Cooking bananas enhances the release of carotenoids from plastids, with amounts concentrated through water loss (Amah et al. 2019). The bioaccessibility

Fig. 17.6 Women in Nigeria use vitamin A cassava to make fufu. (Credit: HarvestPlus)

of carotenoids from boiled bananas varies by cultivar and ranged from 10% to 32% among three cultivars examined in DR Congo (Ekesa et al. 2012b). Fat does enhance bioaccessibility in OFSP (Tumuhimbise et al. 2009) and banana (Ekesa et al. 2012b).

Retention loss during storage is also of interest, as ability to store helps to address seasonal food insecurity. For VAC, drying in the shade demonstrated superior carotenoid retention (59%) than drying in the direct sun (27–56%). For OFSP, retention levels (66–96%) did not vary significantly by drying method. At issue is the substantial loss of carotenoids that can occur during subsequent storage of dried VAC or OFSP under tropical ambient conditions (Bechoff et al. 2011; Chávez et al. 2007).

Given this challenge CIP has focused on fresh sweetpotato root storage. Under commercial storage conditions in the USA, the beta-carotene content of the orange-fleshed variety Covington increased from 253 µg/g (dwb) to 260 µg/g after 4 months of storage and 291 µg/g by the end of 8 months (Grace et al. 2014). In the African context, Tumuhimbise et al. (2010) cured roots "naturally" by spreading them on the ground under the sun for 4 days (26–29 °C; 80–95% RH) prior to storage. Roots stored in pits retained higher beta-carotene content compared to roots stored in

Fig. 17.7 Making OFSP puree at Organi Ltd in Kenya – a partial substitute for wheat flour in baking. (Credit: J. Low/CIP)

sawdust, dark rooms, or ambient conditions. In all methods, OFSP varieties maintained more than 100 µg/g dwb after 4 months of storage.

Shelf-stable OFSP purée (steamed and mashed sweetpotato) has been an integral part of the innovation package designed to enhance incomes while improving the vitamin A content of the processed product (Fig. 17.7). CIP developed a vacuum-packed shelf-stable purée that is safe for storage up to 3 months at temperatures ≤ 25 °C using locally available preservatives (Musyoka 2017) and retains sufficient beta-carotene. New markets for OFSP roots would drive expansion of production, concurrently increasing levels of consumption.

17.3 Expected Impact Based on Ex Ante Studies

Ex ante simulation models have been used to estimate the potential cost-effectiveness of biofortification, based on disability-adjusted life years (DALYs) saved, or reduced prevalence of inadequate micronutrient intake. Lividini et al. (2018) reviewed 30 ex ante studies on biofortified crops from 2002 through 2015, describing 4 different categories of analysis. Since 2006, there has been increasing consideration of both supply and demand for biofortified crops, utilizing widely available household expenditure and consumption surveys. Several studies indicate that biofortification has greater impact among women and children in rural than in urban areas and benefits lower income groups more than higher income groups.

Biofortification emerges as a highly cost-effective micronutrient intervention, based on the World Bank's (2020) threshold of $270 for cost-effectiveness, *when the crop being biofortified is widely consumed and the amount of bioavailable target micronutrient is sufficient to address the deficiency* (Bouis et al. 2020). In the case of OFSP and VAB, the levels of carotenoids in orange types are quite high, making the key issues the extent of their adoption by farmers and the frequency and amount of consumption. In the case of VAC, IP and iron-biofortified OFSP, reaching target levels through repeated breeding cycles is requisite, in addition to considering and consumption levels and micronutrient bioavailability.

For example, one ex ante study for Nigeria assumed 30% of replacement of existing cassava with VAC (varieties with 8 ppm vitamin A content) and found the cost per DALY saved of $1.01, driven by the large amounts of cassava consumed by adult women. This was much lower than $50 per DALY saved for sugar fortification with vitamin A and $52 per DALY saved with supplementation (Ilona et al. 2017). Thus, these results indicate that VAC is particularly appropriate for reaching rural populations.

17.4 Innovation Package Design and the Scaling Readiness Framework

17.4.1 Concept of Scaling Readiness of Innovation Packages

"Innovation readiness" refers to the demonstrated capacity of an innovation to fulfill its contribution to development outcomes in specific locations, presented in nine stages showing progress from an untested idea (score 0) to a fully mature proven innovation (score 9). "Innovation use" indicates the level of use of the innovation or innovation package by the project members, partners, and society. Progressively broader levels of use begin with the intervention team who develop the innovation (score 0) until reaching widespread use by users who are completely unconnected with the team or their partners (score 9).

"Scaling readiness (SR)" of an innovation is a function of innovation readiness level (from 0 to 9) and innovation use (from 0 to 9). Table 17.2 provides summary definitions for each level of readiness and use (adapted from Sartas et al. 2020). The final SR score is a combination of the lowest score from two distinct components. Hence, the maximum possible SR score is 81 (9x9).

Table 17.2 Summary definition of levels of innovation readiness and use (Sartas et al. 2020)

Stage	Innovation readiness	Innovation use
1	Idea	Intervention team
2	Basic model (testing)	Direct partners (rare)
3	Basic model (proven)	Direct partners (common)
4	Application model (testing)	Secondary partners (rare)
5	Application model (proven)	Secondary partners (common)
6	Application (testing)	Unconnected developers (rare)
7	Application (proven)	Unconnected developers (common)
8	Innovation (testing)	Unconnected users (rare)
9	Innovation (proven)	Unconnected users (common)

17.4.2   Innovation Package Design Targeting Specific Livelihood Outcomes

Sartas et al. (2020) stress that scaling of any specific *core* innovation, such as a bio-fortified variety (BV), requires a set of *complementary* systems that typically entail additional innovations to enable uptake. This collection of innovations is described as an *innovation package*. Package composition is driven by the desired ultimate outcomes and the specific scaling environment.

BVs can contribute to multiple nutrition and livelihoods outcomes. These are combined in theory of change and implementation plans of research for development (R4D) programs to maximize overall benefits. For ease of interpretation, we distill this diverse set of scaling efforts into separate innovation packages that address three major development outcomes: (1) improved nutrition, (2) improved food and nutrition security, and (3) improved incomes. Efforts to improve nutritional status have been focused on those most at risk of vitamin A deficiency – women and young children. In contrast, improved food and nutrition security efforts typically target rural households. Targeting for improved incomes varies, and increased commercialization efforts for fresh products focus on rural households, or, specifically on women to assure they benefit from commercialization as farmers, traders, or processors.

Common to all three innovation packages is the core innovation of developing BVs that are acceptable to the target group(s) of each outcome. Strong evidence has demonstrated that any BV used must yield on average at least as much as the dominant local variety to be permanently adopted (Low et al. 2017). To get uptake of these varieties, positive field performance must be demonstrated and linked, typically, to on-farm trials associated with varietal testing or post-release demonstration plots. The innovation packages shown in Tables 17.3, 17.4, and 17.5 are based on actual implementation experience over the past decade. The co-authors of this article, in consultation with the colleagues within their organizations, have reviewed package components and applied the SR scales for country specific settings.

Table 17.3 Innovation packages and their scaling readiness assessments for biofortified vitamin A-rich orange-fleshed sweetpotato (maximum possible score for each readiness or use component is 9)

Innovation package number		1		2		3	
Principle outcome for innovation package		Improved nutrition		Improved food and nutrition security		Improved incomes (processed product)	
Key target group(s)		Young children; pregnant and lactating women)		Rural households		Farmers, traders, processors (focus on women)	
Countries where deployed		Uganda, Ethiopia, Kenya, Tanzania, Mozambique		Mozambique, Ethiopia, Malawi, Tanzania, Rwanda, Zambia		Rwanda, Kenya, Malawi	
Number of innovations within package		8		7		11	
Gender-responsive innovation package components	Core or complementary	Readiness	Use	Readiness	Use	Readiness	Use
1.0 New OFSP adapted varieties acceptable to consumers, bioavailability confirmed	Core	9	9	9	9	9	8
2.0 Pre-basic planting material multiplication	Complementary	9	7	9	7	9	7
3.0 Network of basic and tertiary planting material multiplication	Complementary	8	8	8	8	8	8
4.0 Demonstration of varietal performance	Core	9	9	9	9	9	9
5.0 Training package on good agronomic and postharvest practice	Core	8	8	8	8	8	8
6.0 Appropriate system for monitoring seed quality	Complementary					7	3
7.0 Product for commercialization	Core					8	4

(continued)

Table 17.3 (continued)

8.0 Laboratory services for confirming vitamin A levels of products	Core					9	8
9.0 Nutrition awareness/demand creation campaign	Complementary	8	9	8	9	8	9
10.0 Community-based nutrition-focused behavioral change model	Complementary	9	7				
11.0 Business-oriented vine and root enterprises	Complementary					6	3
12.0 Government policy supports biofortification	Complementary	9	9	9	9	7	4
Scaling readiness score (lowest readiness score x lowest use score)		**56**		**56**		**18**	

A significant complementary innovation is the pre-basic seed system, which most often falls under the domain of national research programs for RT&B crops. High-quality, disease-free starter stock (tissue culture plantlets and screen house protected cuttings) is essential for breeding programs and maximizing yields (Fig. 17.8).

Further multiplications of planting materials to increase quantities available for distribution to farmers are also complementary innovations that vary in design. Productivity and crop quality will be enhanced if complementary training on agronomic, harvesting, vine conservation, and/or other postharvest techniques is provided. Because crops biofortified for vitamin A have a distinct orange or yellow color, another complementary innovation is a demand creation campaign to build awareness about the nutritional value of the BV among end users (Low et al. 2015). Considerable investment has been made in developing and testing approaches about how to advocate both at the community level and policy level to ensure that biofortified crops are integrated into relevant national government and regional policies of agriculture, food security, and nutrition (Covic et al. 2017). The development of a strong enabling environment is critical for scaling, as government support facilitates expansion of the innovation package(s) and supplementary donor investment.

In describing innovation package components for the three distinct outcomes described, the most widely scaled BV, OFSP, will be used as an example. Packages for VAC and VAB will be presented in the subsequent section where the scaling readiness and use scores are applied and explained.

Table 17.4 Scaling readiness assessment for biofortified vitamin A cassava

Innovation package number		2		3	
Principle outcome for innovation package		Improved food and nutrition security		Improved incomes (processed product)	
Key target group(s)		Rural households in poor communities		Farmers, traders, processors	
Countries where deployed		Nigeria		Nigeria	
Number of innovations within package		10		15	
	Core or	Readiness	Use	Readiness	Use
Gender-responsive innovation package components	Complementary	Maximum 9	Maximum 9	Maximum 9	Maximum 9
1.0 New VAC-adapted varieties acceptable to consumers, bioavailability confirmed	Core	6	7	6	7
2.0 Pre-basic planting material multiplication	Complementary	9	6	9	6
3.0 Sustainable network of basic and tertiary planting material multiplication	Complementary	7	6	8	6
4.0 Appropriate agronomic and postharvest training packages	Complementary	8	6	8	6
5.0 Advocacy, based on nutrition evidence, at federal and state levels, as a result of net-mapping of key partnerships for delivery	Complementary	8	7	8	6
6.0 Community Advocacy and Sensitization using local leaders for demonstration/ dissemination of seed and products (incl. payback)	Complementary	8	6	8	6

(continued)

Table 17.4 (continued)

7.0 Seed pack concept with 50 cuttings each of new varieties in labeled polyethylene packaging (varietal info and agronomic practice)	Complementary	9	8	9	5
8.0 Commercialized production of cassava stems for seed, supported by business case development	Complementary			8	7
9.0 Commercial food enterprises using VAC-processed products, supported by business case development	Core			7	6
10.0 Appropriate protocols, quality, standards and guides established for seed, food quality, and nutrient retention	Complementary			7	6
11.0 Market stimulation, including marketer and buyer incentives and contests	Complementary			9	7
12.0 Database to track and link all value chain actors	Complementary			7	5
13.0 Nutrition awareness creation and accessibility strategies targeting women (Smart Mother Initiative), youth, and school children (NutriQuiz and farming clubs)	Complementary	8	8	8	8
14.0 Public relations including radio call-ins, interviews, press conferences, social media, Nollywood films, and celebrity support	Complementary	8	7	8	6

(continued)

Table 17.4 (continued)

15.0 Multi-stakeholder alliances for scaling innovations (e.g., Nutritious Food Alliance) and linking to other programs	Complementary	8	8	8	8
Scaling readinginess overall score		**36**		**30**	

Table 17.5 Scaling readiness assessment for vitamin A-rich banana/plantain in East and Central Africa

Innovation package number		2	
Principle outcome for innovation package		**Improved food and nutrition security**	
Key target group(s)		**Rural households with children under 5 years of age**	
Countries where deployed		**Burundi and DR Congo**	
Number of innovations within package		**6**	
		Readiness	**Use**
Gender-responsive innovation package components	**Core or complementary**	**Maximum 9**	**Maximum 9**
1. VAB varieties agronomically competitive with local non-vitamin A-rich varieties and acceptable to consumers, with bioavailability confirmed	Core	4	3
2. Macropropagation and tertiary planting material multiplication using diverse channels in the communities	Complementary	7	4
3. Community-based nutrition awareness creation and good agronomic practice model (3 sessions)	Complementary	9	4
4. Dissemination of planting materials to target farmer households	Complementary	6	4
5. Market and demand creation campaigns to increase consumption by urban consumers	Complementary	2	1
6. Government policies and donors recognize pVAC bananas as biofortified and nutritious foods	Complementary	8	2
Scaling readiness score		2	

Fig. 17.8 A Rwanda Agriculture Board tissue culture laboratory in Rubona produces quality sweetpotato plantlets. (Credit: J. Low/CIP)

Fig. 17.9 OFSP is appreciated by all household members but especially by young children. (Credit: Helen Keller International)

The OFSP innovation packages for each outcome are shown in Table 17.3. **Innovation Package #1**, with eight components, focuses on improving nutrition among children under 5 years of age and pregnant and lactating women (Fig. 17.9). Essential to this package are two complementary nutrition components. The first is a *nutrition awareness campaign* built on generating awareness locally and nationally about the importance of vitamin A for good health and the high vitamin A

Fig. 17.10 Shown at a promotional event for OFSP in the Volta Region: Nan and Kofi Annan are key advocates for using OFSP to improve diet quality in Ghana. (Credit: Tessa Smit/CIP)

content of OFSP, as well as other good sources of vitamin A available in the country. These campaigns have used radio, market-based promotions (signs, billboards), messages on orange-colored promotion materials (cloth, hats, t-shirts, vehicles), videos, social media, brochures, and television spots to inform the public and policy makers. Several prominent policy makers in SSA have become advocates themselves (Fig. 17.10). The second is a *community-based nutrition-focused behavioral change model*. Research has established that facilitated group sessions of 25–30 households meeting monthly for 6–12 months to share knowledge about better dietary and health practices at the household level and feeding practices for young children resulted in improved vitamin A intakes and vitamin A statuses among young children and their mothers (Girard et al. 2017; Hotz et al. 2012; Low et al. 2007). The use of trained community-based health workers or volunteers has been integral to the success of this approach (Girard et al. 2021) in four SSA countries (Kenya, Ethiopia, Mozambique, Tanzania).

In **Innovation Package #2**, where the focus is on improvement of food and nutrition security at the *household* level, a broad nutrition awareness campaign is used, but the activities at the community level are limited to community-sensitization meetings and one-off cooking demonstrations on how to prepare OFSP for young children and how to incorporate this food household diets (Fig. 17.11). With seven components, this approach succeeds in promoting adoption of OFSP and limited

Fig. 17.11 Demonstration of how to incorporate OFSP into a range of existing dishes in Tigray, Ethiopia. (Credit: F. Asfaw/CIP)

amounts of OFSP into the young child diet. The level of impact on young children's vitamin A levels is less than we see in Innovation Package #1 due to the lack of more intensive community-based nutrition training. The focus of many of these efforts is to strengthen food security at the household level, especially in respond to climate change and/or emergencies such as drought or flood.

OFSP **Innovation Package #3** is the most complex with 11 components and focuses on building a value chain for using OFSP in processed form to promote diversified use among urban consumers and provide a source of nutritious food and incomes for rural farming households. This package has taken 4–5 years to implement, compared to 2–3 years for Package #2, and 3–4 years for Package #1.

In Package #3, the nutrition awareness component is part of a *marketing and demand creation campaign* with an emphasized focus on building demand among urban consumers (Fig. 17.12). Two other core innovations are (1) the need to develop an economically viable product that uses the BV as a major ingredient *and* is well-liked by target consumers and (2) testing products to ensure they have retained enough beta-carotene to be marketed as a good source of vitamin A (which requires the presence of high-quality laboratory services). Additional training is required for market-oriented farmers to obtain sufficient skills to consistently provide quality roots to the processors in sufficient quantities. Associated with the development of such value chains is the development of more standardized systems for monitoring the quality of the seed provided to growers, which requires engagement with government regulatory authorities. With commercialization of the roots, the willingness of farmers to invest in more expensive and higher-quality planting

Fig. 17.12 Entrepreneur Hassana is promoting her Rahama vitamin A *gari* at a nutritious food fair in Nigeria in 2019. (Credit: HarvestPlus)

material increases along with the desire for that quality to be guaranteed. To date, scaling efforts for OFSP Package #3 have been concentrated in Kenya, Malawi, and Rwanda, building on proof-of-concept projects using OFSP purée in Kenya and Rwanda.

One critical aspect insufficiently highlighted in the SR framework by Sartas et al. (2020) is the need to be aware of gender roles and power dynamics around crop production and sale and to design packages that are aware of potential differential impacts of innovation packages on women and men. Agricultural innovations must consider the different roles that men and women take in the adoption process – be it accessing seed, crop production, marketing, processing, or household consumption. Asare-Marfo et al. (2019) systematically consider the various factors to be considered in understanding gender differences along the impact pathway, which influence the success of BVs as an innovation, but do not focus on RT&B crops. They note that men and women may receive their information through different channels or sources, and often men have more access to extension and other services. Moreover, if a BV is bringing a higher price, men may be more inclined than women to sell

Fig. 17.13 Men dominate the potato wholesaling in Kenya, reminding us of the need to consider gender appropriately in relation to innovation packages. (Credit: CIP)

rather than consume the biofortified food, a result that would have nutritional implications for the household.

Gender dynamics, of course, are context-specific, requiring adjustments by and within countries to develop effective innovation packages. For example, in East Africa male control of labor and production is typically higher for potato and banana, as they are considered cash crops, compared to sweetpotato and cassava, for which home consumption dominates (Okonya et al. 2019) (Fig. 17.13). Women participating in a commercialized OFSP value chain in Rwanda required more training sessions then men to meet quality requirements (Sindi and Low 2015). Particular attention is needed as commercialization of BVs increases to ensure women are not excluded from the benefits, nor are the nutritional goals of BV introduction unduly compromised. Monitoring is requisite. To avoid repetition, we have captured this need for almost every component through labeling the innovation package as *gender-responsive*.

17.5 Assessment of Scaling Readiness and Level of Use According to the Scaling Readiness Framework

Measurement of readiness and use levels reflects the status of biofortification breeding progress, the strategies used to scale, and available resources for different RT&B programs. The measurement of innovation use is more complex than readiness, due to the difficulty in drawing clear lines between the defined categories of actors and the extent of use, both of which vary widely by location. Sartas et al. (2020) distinguishes between:

1. The *intervention team* (typically a research organization initiating the innovation)
2. *Effective partners* (those collaborating directly with the intervention team)
3. *Innovation network stakeholders* (who influence the R4D intervention, but are not involved directly in its testing)
4. *Other stakeholders in the innovation system* (defined as other R4D teams working on similar innovations)
5. *Stakeholders or beneficiaries in the livelihood system* (who were not linked directly to the R4D innovation development)

For example, CIP, an international research organization specialized in potato and sweetpotato research, and HarvestPlus, a program (led by the International Food Policy Research Institute) dedicated to developing and promoting biofortified staples, at times have been on the same intervention team, while in other projects they have served as effective partners or innovation network stakeholders, depending on the innovation package and country. The Alliance of Bioversity International and CIAT has led the VAB effort. Only OFSP and VAC have received major donor support for scaling their BV efforts.

The assessment for each crop is summarized below. For more details, the development and scaling of OFSP have been described in Low et al. (2017) and Low and Thiele (2020). HarvestPlus' experience in scaling VAC in Nigeria and OFSP in Uganda has recently been described in Foley et al. (2021). Ilona et al. (2017) highlights key aspects of the first phase of the VAC scaling process. The pilot experience with VAB in East Africa is explored in Fongar et al. (2020).

As shown in Table 17.3 (for OFSP innovation packages), out of a maximum possible score of 81, the Food and Nutrition Security and the Improved Nutrition Packages scored 56, and the Improved Income Package scored 18. For VAC, the approach focused on Improved Nutrition and Food Security Package and scored 36, while the Income Package for processed VAC products earned 30 points (Table 17.4). As the lowest score found in any component drives the overall SR score, VAB SR rated only 2 points (Table 17.5), which reflects its pilot level and resource limitations to date (Fig. 17.14).

Given that OFSP was the first BV crop to achieve breeding targets, considerable investment was made in delivery system research using OFSP as a model biofortified crop (Low and Thiele 2020). This work produced an excellent evidence base for the improved nutrition outcomes and food and nutrition security packages. Open access investment and implementation guides (Stathers et al. 2015a, b) are available on the Sweetpotato Knowledge Portal (www.sweetpotatoknowledge.org) and provide detailed instructions on how to design, set up, and implement OFSP-focused nutrition and food security interventions. In addition, 13 modules for a *Training of Trainers* (ToT) course entitled *Everything you ever wanted to know about sweetpotato* (Stathers et al. 2012), each with activities addressing gender, are available on the Portal in 5 major languages.

Fig. 17.14 Vitamin A banana scaling efforts can learn from the OFSP and VAC experiences. (Credit: A. Simbare/Alliance Bioversity-CIAT)

As a strategy for scaling OFSP innovation packages by non-research organizations, while assuring that such organizations have access to the latest research knowledge, CIP launched the *Sweetpotato for Profit and Health Initiative* (SPHI) in 2009 with the goal of reaching 10 million households in 16 SSA countries by 2020 with improved varieties of sweetpotato and promoting their diversified use (Low 2011). During the 10-year period, NGOs were effective partners in proof-of-concept delivery projects initially but then raised independent funding and integrated OFSP varieties into their own programs, which often had many elements of the innovation packages described above, but in some cases could be entirely different – for example, the enhanced homestead garden program led by Helen Keller International (Haselow et al. 2016). Partners in the SPHI submitted annual updates on the number of beneficiary households reached directly (as program participants) or indirectly (spillover spread of varieties). By 2019, Fig. 17.15 clearly shows that the extent of scaling (use) varied enormously by country, reflecting differences in how and when adapted BVs were developed and released, and the levels of government interest and donor country prioritization. Hence, only eight of the 16 targeted SPHI

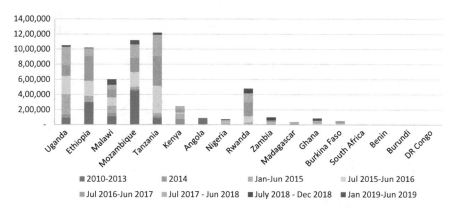

Fig. 17.15 Number of households reached by country from June 2010 to June 2019 under the Sweetpotato for Profit and Health Initiative. (Source: Okello et al. 2019)

countries had reached "scaling" levels. Two of these countries had major breeding programs (Mozambique and Uganda) to accelerate development and promotion of superior BVs.

VAC focused its breeding efforts in Nigeria and DR Congo, due to the dominance of cassava as a major staple in the diets in these countries. Scaling efforts have focused on Nigeria, a country that is home to 18% of the entire population of sub-Saharan Africa (Table 17.4). The innovation packages for VAC are more complex than those presented for OFSP, reflecting a concerted effort to develop commercialized cassava "seed" production and a diverse set of interventions to promote nutrition awareness and stimulate demand for VAC versions of a diverse array of processed products common in Nigeria. On the seed front, HarvestPlus developed a distribution system using labeled packaging with 50 stems (planting material) that were distributed for free but with recipient households agreeing to "pay back" by providing the 50 VAC stems to two households in the following season. Since 2015, emphasis has been placed on developing links between growers and stem multipliers with the goal of shifting to more commercialized seed and marketing systems. By 2018, 8% of VAC stems were being purchased (Foley et al. 2021). On demand creation, extensive use of radio, television, and social and print media and the establishment of an annual nutritious food fairs with a broad range of policy makers and celebrities helped promote VAC products as preferred choices over non-VAC options and facilitated the establishment of roadside VAC selling points to generate interest. As a result, by 2018, 50,000 ha were under VAC cultivation (Foley et al. 2021) (Fig. 17.16).

In the assessment tables (Tables 17.3–17.5), we can see the following scores for BV readiness: 9 for OFSP, 6 for VAC, and 4 for VAB. Breeding is a continual process, but a 9 indicates that appropriate varieties are available that meet the target beta-carotene levels, have been adapted to local growing conditions, meet adult consumer preferences, and have documented evidence of bioavailability. Some

Fig. 17.16 Farmers in Nigeria display vitamin A cassava. (Credit: HarvestPlus)

OFSP and VAC varieties have an additional advantage of being higher-yielding than dominant varieties on the market (Low and Thiele 2020; Foley et al. 2021), which helps drive uptake. Bioaccessibility in VAB has been confirmed, but resources to confirm bioavailability have not yet been raised.

Scores vary by and within countries, but the value of SR is in its ability to pinpoint where bottlenecks may be occurring. The challenges facing growers' access to BVs of OFSP, VAC, and VAB are no different from access issues for non-BV crops. Because RT&B planting material is easily retained, reused, and shared, private sector seed companies have not been interested to invest in these RT&B systems. CIP, HarvestPlus, and partners invested in developing networks of trained multipliers to provide greater access for growers to quality seed (Fig. 17.17). Initially, these systems were subsidized by project funding, providing free or subsidized material to growers to achieve desired nutrition and/or food security outcomes. The ability for these multiplication systems to evolve into more self-sustaining commercial entities has been highly dependent on the development of markets for BVs in fresh and/or processed form. Building on a long history of cassava agro-processing in Nigeria, VAC-processed product development has been supported among both small-scale and larger processors, with use scores reaching 7. In contrast, since most sweetpotato roots are consumed boiled or steamed, processed product development is a new phenomenon; therefore its use level is 4 in the indicated countries, showing that support is still needed from the original research for development partners. The SPHI annually monitored whether OFSP vine multipliers continued to produce planting material during and post-project. In 2019, 503 of 741 trained multipliers contacted in 11 countries were selling vines (Makokha et al. 2019). Recognizing the important of root markets to drive willingness-to-pay for quality planting materials,

Fig. 17.17 A trained OFSP multiplier harvesting planting material in Mwanza, Tanzania. (Credit: K. Ogero/CIP)

efforts are focused on validating vine-root enterprises as an integral part of the improved incomes package (readiness at 6; use at 3).

Concurrent with processed product development and its commercialization, an adoption of standards and services to validate that standardization was needed. In Kenya, the Food Analysis and Nutrition Evaluation Laboratory (FANEL) was established in 2014 as a reference service for vitamin A and other nutrient assessments, including iron and zinc. This lab facilitates labeling of biofortified products to assure consumers about vitamin A content (Muzhingi et al. 2019). For VAC, standards and guides for nutrient retention have been key, and these efforts are being strengthened through investment in tools that can more rapidly determine whether a product meets quality standards (Foley et al. 2021). Technical support in the use of such protocols and guidelines is required for validation and uptake by regulatory bodies.

On the advocacy and policy side, biofortification has been at the forefront for developing demand creation strategies and for understanding how to train and support local and regional influencers as policy advocates. Consequently, there has been widespread integration of biofortified crops as part of nutrition and agricultural policies in 24 SSA countries (Covic et al. 2017). This kind of integration into national and regional policies also facilitates enhanced government and outside donor investment in government-prioritized interventions. Readiness scores for OFSP and VAC in this area are 8 or 9 and the use of different approaches varies from 4 to 9, depending on the outcome model and country context. As VAB promotion is more recent, recognition of its potential by both governments and donors is lagging (Table 17.5), but VAB will be able to draw on the groundwork and lessons learned by from the OFSP and VAC experience.

17.6 Reaching Scale in SSA: Lessons Learned

The SR assessment provides a framework for reflecting critically on the scaling process, helping to identify bottlenecks and essential factors to support scaling. As noted by Fongar et al. (2020), "lessons for scaling regionwide adoption of VABs can be drawn from the introduction and scaling of OFSPs in SSA." The same holds true for IP, although in the latter case since iron is not a visible trait, the high-iron bean scaling effort in Rwanda is probably more relevant (Foley et al. 2021).

In their review of the 25-year OFSP development and implementation experience, Low and Thiele (2020) examine the requisite technical, organizational, leadership, and enabling environment associated with each phase examined. In this section, we will review major lessons learned concerning VAC and OFSP.

17.6.1 Technical Considerations

BVs must be agronomically competitive with dominant non-BVs and meet the taste and key quality preferences of adult consumers to achieve widespread uptake. Several cycles of breeding were required in Uganda, for instance, to produce OFSP varieties that had the desired texture. Meeting the biofortification target level is desirable for uptake but not requisite. In Nigeria, VAC varieties with <12 µg g-1 of beta-carotene have been accepted and widely cultivated by farmers. Thus, the platform is already in place for incorporating varieties with the desired levels of beta-carotene above 15 µg g-1 once released. However, caution is warranted in serving size recommendations and labeling of processed products to avoid unsubstantiated claims that may lead to a violation of trust. The establishment of protocols and laboratory services to measure nutrient retention has been a complementary component of packages focused on commercial product development. To date, research support to private sector enterprises has encouraged testing and labeling, given that the regulatory structure for managing biofortified products is still nascent.

Taste and consumer preferences vary by target groups and locations, so investment in acceptability studies among different consumer segments and across multiple locations is warranted. For example, VAB researchers found significant variation in the ranking of the same varieties in different communities within the same country (Ekesa et al. 2017). The importance of *quality* traits that capture sensory characteristics which vary by end use and postharvest considerations, such as storability, has been underemphasized in RT&B public sector breeding programs to date, but are the focus of growing interest due to their critical importance in driving adoption (Thiele et al. 2021).

In comparison to grain crops, RT&B seed systems have tended to be more informal and less commercialized. A large share of smallholder farmers retain their own planting material from year to year, only seeking new material if there is significant yield decline, loss due to drought or theft, or a new variety demanded by the market. Farmer-to-farmer sharing of seed is common. This context required significant investment to set up innovative systems for delivering high-quality planting material of biofortified varieties while also convincing farmers of the yield value in using quality seed (Fig. 17.18). For scaling, it is requisite that seed system barriers be overcome to ensure varieties are available that meet consumer demands and that quality seed is accessible to more farmers at planting time. Since seed and root supply is critical for scaling, larger multipliers and growers had to be recruited to complement smallholder farmers.

The experiences of OFSP and VAC demonstrate that is possible to develop sustainable enterprises which bring the vegetatively propagated seed closer to farmers through networks of decentralized multipliers. These small- and medium-sized businesses become critical for meeting RT&B seed demand, even more so during the 2020 pandemic. Work is still underway to improve linkages between

Fig. 17.18 Distribution of banana plantlets to members of farmers' association in Burundi. (Credit: A. Simbare/Alliance Bioversity-CIAT)

decentralized multipliers and early generation seed producers, to ensure timely renewal of quality multiplication stock. The use of digital tools and other innovations to help policy makers regulate and practitioners to monitor the seed system are included in the RTB Seed System Toolbox (https://www.rtb.cgiar.org/seed-system-toolbox), which is currently being evaluated in several SSA countries. Concerning VAC in Nigeria, development of the market for seed was done concurrently with promotion of VAC-processed products. To support private sector participation, clear business cases should be developed and tested, with the return on investment (ROI) being higher than bank interest rates.

SR assessments pinpoint the seed system as a key bottleneck. Few private sector companies in SSA are engaged in early generation seed (EGS) production (tissue culture, pre-basic seed production), and CGIAR emphasis has focused on strengthening public sector national program management capacity in this regard. A detailed study of one private company that has invested in EGS and basic seed production found that such a business requires significant upfront financial support due to high initial investment costs. The payback period required for such an investment is 3–7 years, with an average annual return of 34–70% (Rajendran and McEwan 2019). Having open-field basic seed production linked to EGS did increase economies of scale for this company.

Using disease-free planting material can have significant yield benefits, and there has been increasing interest to develop a regulatory framework for certifying the quality of RT&B planting material. In SSA, potato is the only root and tuber crop where several countries have seed certification schemes, reflecting its highly commercialized nature. Even so, less than 5% of potato seed sold outside of South Africa is certified. Such schemes typically require public sector support and investment. Cassava and sweetpotato programs have focused on developing less costly quality declared seed protocols and decentralized inspection systems. The introduction of greater regulation of seed quality correlates highly with increased commercialization of the crop. The use scores for these types of regulatory systems are among the lowest among the innovation package components due to the need for end users to be convinced of the value.

To lower risk for private sector participation, investment in demand creation campaigns and provision of technical support for production and utilization of BVs by the research for development partners has been central to jumpstarting private sector engagement. HarvestPlus has used existing platforms in Nigeria, like the Nutritious Food Fair/Alliance, Smart-Mother, and NutriQuiz to reach millions of Nigerians. Doing so increased the use scores of VAC innovation packages.

Learning how to address gender has been critical for scaling packages. For nutrition outcomes, recognizing that men have primary decision-making authority on what crops to plant, what foods are purchased, and how different foods are allocated among household members has led to the development of community-based nutrition interventions that integrate men, women, and local leaders to address household dietary quality and young child care practices (Girard et al. 2021) (Fig. 17.19). Regarding income, efforts have been made to ensure that women are not sidelined when developing or improving market interventions. This effect is achieved by

Fig. 17.19 Encouraging men to feed their young children is often part of behavioral change programs for OFSP. (Credit: CIP)

setting explicit targets for female participation and addressing specific service and capacity needs of women who may be underserved due to longstanding social and/ or economic barriers to participation. As expected, the impact of biofortified crops has been greatest for women and children in nutrition-focused interventions and on poorer rural households for broader household food security and dietary diversity (Bouis et al. 2020; de Brauw et al. 2018).

17.6.2 Organizational and Leadership Considerations

Given the complexity of the science and the need for sustained investment to achieve impact at scale, organizations such as HarvestPlus and the International Potato Center have been essential for building the evidence base for nutrition-sensitive agricultural interventions and expansion at scale (Bouis and Saltzman 2017; Bouis et al. 2020; Low and Thiele 2020). These organizations have also developed and managed cross-country, cross-project, and cross-partner monitoring systems to capture progress. This work has enabled groups to obtain timely feedback on varietal and package performance and to respond to scientific and policy queries with evidence.

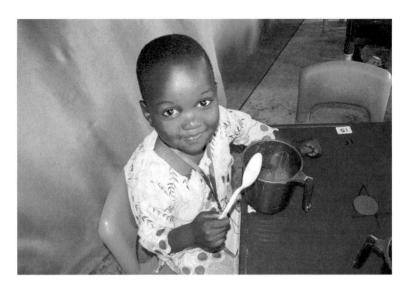

Fig. 17.20 A girl in preschool in Nigeria enjoys custard made from vitamin A cassava. (Credit: HarvestPlus)

As government and NGO partners have become increasingly involved, it is clear that the food and nutrition innovation package – the simplest of the three – is the easiest to adapt and use at scale. In this context, BVs of roots and tubers are often a part of a much broader package of crop and management practice interventions. Expanding partnerships with the World Food Program and international NGOs should continue and be strengthened to make BVs available for food assistance and resilience programming in fragile environments where nutritionally vulnerable populations are the norm.

The scalability of the improved nutrition innovation package is most likely in countries where there is strong support for community-based health workers by the government (e.g., Ethiopia, Ghana, and Malawi). Lacking that support, the higher cost of this approach compared to focusing on food and nutrition security at the household level means that a more cost-effective but longer-term solution may be to integrate nutrition education into primary and secondary school curricula and antenatal and postnatal care counseling and existing school feeding programs (Fig. 17.20).

Certainly, innovation packages focused on income enhancement are the most attractive to private sector partners. Given that many of the processing partners are small- and medium-scale enterprises, the need for significant technical support should not be underestimated, especially in developing a value proposition for investment. There is an increasing effort to focus on youth opportunities for employment associated with these efforts. While many successful examples of profitable, commercially oriented value chains exist, action research is needed to design and test ways to adapt and scale such programs to widen the scope of impact.

17.6.3   The Enabling Environment

For any innovation to take root, flourish, and scale, an enabling environment is needed for sustainability. Bouis and Saltzman (2017) have identified building blocks necessary for biofortified crops to scale:

1. Globally, biofortification must be integrated into global standards and regulatory guidelines such as Codex Alimentarius[1]. As multilateral institutions (World Bank, African Development Bank, World Food Programme, World Health Organization) integrate biofortified crops into their policies and programming, governmental and nongovernmental organizations (NGOs) can more easily include them in national policies, plans, and programs.
2. Within Africa, the endorsement of the African Union and the Regional Economic Communities can facilitate and act as an encouragement for individual countries to include biofortification as a *nutrition-sensitive* approach. HarvestPlus and CIP have jointly and individually advocated for biofortification to be included in regional and national government policies.
3. Translating policy into action involves incorporating biofortified crops into programs and plans that are then implemented through governmental and nongovernmental bodies (Covic et al. 2017). Both organizations have funded efforts to develop and test strategies to identify, recruit, and train advocates at national and regional levels.

NGOs are important because they are crucial partners for delivering innovations to more vulnerable households. To create sustainable markets, private sector participation is essential – from seed to food delivery. Private sector seed companies have the power to shorten time to market of biofortified seed varieties, although for RT&B crops this may be more problematic. Nonetheless, private sector involvement in processing is critical and facilitates the inclusion of biofortified crops as ingredients in food product value chains. As noted above, demand for the final product is critical to drive willingness-to-pay for inputs such as quality seed.

Initiatives like SPHI that bring donors and multi-sectoral stakeholders from different countries enhance the speed of innovation spread and stakeholder buy-in. Training programs that emphasize agriculture-nutrition-marketing approaches and efforts to produce qualified extension personnel are critical for within-country expansion. While integrating biofortification within national policies of agriculture, food security, and nutrition may constitute "readiness," encouraging governments and NGOs to allocate their own funds for BV dissemination is essential for achieving use at scale. Certainly, our scaling experience has shown that countries with better agricultural extension systems (Ethiopia, Kenya, Malawi) have greater capacity and more willingness to engage in diffusing new innovations than those countries where public sector extension has been downgraded (Uganda, Nigeria). As the number and type of partners grow, lead organizations play more of a facilitation and knowledge role and provide essential monitoring of progress. For example, we have seen some commercially oriented firms make unsubstantiated health claims.

Scientists and advocates need to be proactive in setting the record straight and developing outreach strategies that will prevent such claims in the first place.

Sometimes luck plays a part. Low and Thiele (2020) noted that a major OFSP-based study that demonstrated the nutritional impact of the integrated agriculture-nutrition approach coincided with a shift in the institutional environment that placed agriculture and nutrition at the forefront of the development agenda. This coincidental happenstance created an inflection point that led to increased investment in research and diffusion of biofortified crops.

17.7   Final Considerations

RT&B crops have always been among the most affordable sources of calories for rural and urban poor. Biofortified RT&Bs can serve as major affordable sources of key micronutrients in an emerging and more resilient food system in SSA. RTB crops often have shorter supply chains and are less expensive than grains; hence, market disruptions are less likely to affect availability and access, an advantage made more evident during the 2020 pandemic.

As scaling efforts expand, managing the perishability and seasonality of RT&B crops at larger scales will require greater investment in physical market chain infrastructure, storage, information systems, and enterprise development. These objectives can be achieved by working across multiple nutritious commodities to strengthen informal and formal food market systems. The value added from RT&B BVs will be to secure affordable nutrition for low-income consumers, while also providing economically attractive and nutritious ingredients in food processing and new market opportunities for producers. Stronger smallholder market participation will underpin the ability to develop commercial input supply chains for RT&B crops in SSA.

While scaling efforts utilize existing varieties, breeding must be a continuous effort, particularly in the context of climate change where new, improved materials will be essential for assuring a nutritious food supply. A major feature of climate change is the rapidly increasing carbon dioxide (CO_2) levels, predicted to rise from 400 ppm to over 550 ppm by 2050. With increasing CO_2 levels, sweetpotato, banana, and cassava yields are expected to increase (Jarvis et al. 2012; Varma and Bebber 2019), but that increase will be channeled into carbohydrate accumulation. Potato is sensitive to temperature and drought and thus likely to see decreased yields (Fleisher et al. 2017). Hence, R&TB crops, along with wheat and rice, are expected to show significant declines in nutrient density, including many nutrients critical to human health such as zinc, iron, and protein (Smith et al. 2018). Nelson et al. (2018) predict that the "greatest food security challenge in 2050 will be providing nutritious diets rather than adequate calories." Clearly, given the heavy dependence on staple foods by the poor, increased commitment to breeding for enhanced micronutrient and protein content is warranted.

All R&TB varieties released to date were developed through a targeted approach with a focus on specific crop/country combinations and tightly linked to key traits that trigger adoption. Banana is one R&TB crop where major breeding investments are using transgenic approaches to tackle disease resistance and vitamin A enhancement (Amah et al. 2019). In the future, prospects are excellent for genetic engineering to integrate full target increments in micronutrient content for several nutrients in one go, not only in next wave but also in existing commercial varieties. This potential would shortcut mainstreaming time enormously and accelerate the reach and impact of the intervention (Van Der Straeten et al. 2020). However, the regulatory environment and social acceptance of transgenic materials must improve for the value of such innovative approaches to be realized.

In some contexts, R&TB crops have an image problem to address, which reflects a lack of understanding of their role in poor people's diets in low-income countries. In their analysis of healthy and sustainable diets, the EAT Lancet Commission recommended low daily intakes of potato and cassava as staples relative to grains, in spite of the fact that these crops have much lower environmental impacts. Sweetpotatoes and *Musa* spp. were not specifically mentioned. Given the Western orientation of the article, sweetpotato was probably classified as an orange and red vegetable and *Musa* spp. designated as fruits (Willett et al. 2019). The highly negative image of potato as a junk food in the Western diet is associated with its high glycemic index and its frequent consumption as a fried product. Most potato in SSA, however, is consumed boiled. Moreover, insufficient attention has been paid to enhancing the use of micronutrient-rich leaves of cassava and sweetpotato for human consumption.

Clearly, the lessons learned from the OFSP and VAC scaling experiences can inform VAB and IP development and dissemination efforts and avoid the tendency to "reinvent the wheel." RT&B crops are well-positioned to move forward in the context of emergency recovery and gender-responsive food system transformation for more climate-resilient and nutritious foods. The SR Tool has pinpointed the degree to which different components of OFSP and VAC innovation packages were validated through evidence-based assessment. The innovation packages can be easily adapted for different country contexts, and the SR Tool is recommended for use in monitoring RTB scaling efforts over time to pinpoint bottlenecks.

Inclusion of biofortification by the Scaling Up Nutrition country programs and the potential recognition of biofortification by the African Union are two examples of policy engagement that are needed to keep biofortification and nutrition at the forefront of food policy and investment planning.

Acknowledgments This research was largely undertaken as part of, and funded by, the CGIAR Research Program on Roots, Tubers, and Bananas (RTB) and supported by CGIAR Trust Fund contributors (https://www.cgiar.org/funders/). The authors would also like to acknowledge donor agencies including the Bill & Melinda Gates Foundation, USAID, and FDCO for the financial support toward the accomplishment of this work. The authors thank Graham Thiele, Gordon Prain, Jill Lenne, and Chris Butler who provided input and feedback on earlier drafts of this chapter. We greatly appreciate Zandra Vasquez's assistance with references and formatting.

References

Amah D, van Biljon A, Brown A, Perkins-Veazie P, Swennen R, Labuschagne M (2019) Recent advances in banana (*Musa spp.*) biofortification to alleviate vitamin A deficiency. Crit Rev Food Sci Nutr 59:3498–3510

Amoros W, Salas E, Hualla V, Burgos G, De Boeck B, Eyzaguirre R, Felde T, Bonierbale M (2020) Heritability and genetic gains for iron and zinc concentration in diploid potato. Crop Sci 60:1884–1896

Andrade MI, Ricardo J, Naico A, Alvaro A, Makunde GS, Low J, Ortiz R, Grüneberg WJ (2017) Release of orange-fleshed sweetpotato (*Ipomoea batatas [l.] Lam.*) cultivars in Mozambique through an accelerated breeding scheme. J Agric Sci 155:919–929

Andre CM, Burgos G, Ziebel J, Guignard C, Hausman J-F, Zum Felde T (2018) In vitro iron bioaccessibility and uptake from orange-fleshed sweet potato (Ipomoea batatas (L.) Lam.) clones grown in Peru. J Food Compos Anal 6879–6886. https://doi.org/10.1016/j.jfca.2017.07.035

Andre CM, Evers D, Ziebel J, Guignard C, Hausman J-F, Bonierbale M, Zum Felde T, Burgos G (2015) In vitro bioaccessibility and bioavailability of iron from potatoes with varying vitamin C carotenoid and phenolic concentrations. J Agri Food Chem 63(41):9012–9021. https://doi.org/10.1021/acs.jafc.5b02904

Asare-Marfo D, Lodin JB, Birol E, Mudyahoto B (2019) Case study 4: developing gender-inclusive products and programs: the role of gender in adoption and consumption of biofortified crops. In: Quisumbing AR, Meinzen-Dick RS, Njuki J (eds) 2019 Annual trends and outlook report: gender equality in rural Africa: From commitments to outcomes. ReSAKSS Annual Trends and Outlook Report 2019. International Food Policy Research Institute (IFPRI), Washington, DC, pp 79–161

Bailey RL, West KP Jr, Black RE (2015) The epidemiology of global micronutrient deficiencies. Ann Nutr Metab 66(suppl 2):22–33

Bechoff A, Tomlins K, Dhuique-Mayer C, Dove R, Westby A (2011) On-farm evaluation of the impact of drying and storage on the carotenoid content of orange-fleshed sweet potato (*Ipomea batata Lam.*). Int J Food Sci Technol 46:52–60

Bengtsson A, Larsson Alminger M, Svanberg U (2009) In vitro bioaccessibility of β-carotene from heat-processed orange-fleshed sweet potato. J Agri Food Chem 57(20):9693–9698 https://doi.org/10.1021/jf901692r

Bouis HE, Saltzman A (2017) Improving nutrition through biofortification: a review of evidence from HarvestPlus, 2003 through 2016. Glob Food Sec 12:49–58

Bouis H, Saltzman A, Low JW, Ball A-M, Covic N (2017) Chapter 1: An overview of the landscape and approach for biofortification in Africa. Afr J Food Agri Nutr Dev 17, Special Issue on Biofortification, no. 2:11848–11864. https://doi.org/10.18697/ajfand.78.Harvestplus01

Bouis H, Birok E, Boy E, Gannon B, Haas J, Low JW, Mehta S, Michaux K, Mudyahoto B, Pfeiffer W, Qaim M, Reinberg C, Rocheford T, Stein AJ, Strobbe S, van der Straeten D, Verbeecke V, Welch R (2020) Food Biofortification—reaping the benefits of science to overcome hidden hunger. In: Council for Agricultural Science and Technology (CAST) (ed) A paper in the series on The Need for Agricultural Innovation to Sustainably Feed the World by 2050. Issue paper No. 69, Ames, Iowa (USA), 40 p

Ceballos H, Davrieux F, Talsma EF, Belalcazar J, Chavarriaga P, Andersson MS (2017) Carotenoids in cassava roots. In: Cvetkovic DJ, Nikolic GS (eds) . Carotenoids, IntechOpen, pp 189–221

Chávez A, Sánchez T, Ceballos H, Rodriguez-Amaya D, Nestel P, Tohme J, Ishitani M (2007) Retention of carotenoids in cassava roots submitted to different processing methods. J Sci Food Agric 87:388–393

Covic N, Low JW, MacKenzie A, Ball A-M (2017) Chapter 16. Advocacy for Biofortification: Building Stakeholder Support, Integration into Regional and National Policies and Sustaining Momentum. Afr J Food Agri Nutr Dev 17, Special issue on Biofortification, no. 2: 12116–12129. https://doi.org/10.18697/ajfand.78.HarvestPlus16

de Brauw A, Eozenou P, Gilligan DO, Hotz C, Kumar N, Meenakshi JV (2018) Biofortification, crop adoption and health information: impact pathways in Mozambique and Uganda. Am J Agric Econ 100:906–930

De Moura FF, Miloff A, Boy E (2015) Retention of provitamin A carotenoids in staple crops targeted for Biofortification in Africa: cassava, maize and sweet potato. Crit Rev Food Sci Nutr 55:1246–1269

Ekesa B, Kimiywe J, Van den Bergh I, Blomme G, Dhuique-Mayer C, Davey M (2012a) Content and retention of provitamin A carotenoids following ripening and local processing of four popular *Musa* cultivars from Eastern Democratic Republic of Congo. Sustain Agric Res 2

Ekesa B, Poulaert M, Davey MW, Kimiywe J, Van den Bergh I, Blomme G, Dhuique-Mayer C (2012b) Bioaccessibility of provitamin A carotenoids in bananas (*Musa spp.*) and derived dishes in African countries. Food Chem 133:1471–1477

Ekesa B, Nabuuma D, Blomme G, Van den Bergh I (2015) Provitamin A carotenoid content of unripe and ripe banana cultivars for potential adoption in eastern Africa. J Food Compos Anal 43:1–6

Ekesa B, Nabuuma D, Kennedy G, Van den Bergh I (2017) Sensory evaluation of provitamin A carotenoid-rich banana cultivars on trial for potential adoption in Burundi and Eastern Democratic Republic of Congo. Fruits 72:261–272

FAO (2016) Save and grow in practice: maize, rice, wheat-- a guide to sustainable cereal production. FAO, Rome

FAO, IFAD, UNICEF, World Food Program (WFP), WHO (2018) The State of Food Security and Nutrition in the World 2018: building climate resilience for food security and nutrition. FAO, Rome, p 183

Fleisher DH, Condori B, Quiroz R, Alva A, Asseng S, Barreda C, Bindi M, Boote KJ, Ferrise R, Franke AC, Govindakrishnan PM, Harahagazwe D, Hoogenboom G, Naresh Kumar S, Merante P, Nendel C, Olesen JE, Parker PS, Raes D, Raymundo R, Ruane AC, Stockle C, Supit I, Vanuytrecht E, Wolf J, Woli P (2017) A potato model intercomparison across varying climates and productivity levels. Glob Chang Biol 23:1258–1281

Foley J, Michaux K, Mudyahoto B, Kyazike L, Cherian B, Kalejaiye O, Ifeoma O, Ilona P, Reinberg C, Mavindidze D, Boy E (2021) Scaling up delivery of Biofortified staple food crops globally: paths to nourishing millions. Food Nutr Bull 42(1):116–132

Fongar A, Nabuuma D, Ekesa B (2020) Promoting (pro) vitamin A-rich bananas: a chronology. In: The Alliance of Bioversity and CIAT, Kampala, 33 p. https://hdl.handle.net/10568/108080

Girard AW, Grant F, Watkinson M, Okuku HS, Wanjala R, Cole D, Levin C, Low J (2017) Promotion of orange-fleshed sweet potato increased vitamin A intakes and reduced the odds of low retinol-binding protein among postpartum Kenyan women. J Nutr 147(5):955–963. https://doi.org/10.3945/jn.116.236406

Girard AW, Brouwer A, Faerber E, Grant FK, Low JW (2021) Orange-fleshed sweetpotato: Strategies and lessons learned for achieving food security and health at scale in Sub-Saharan Africa. Open Agri 6(1):511–536. https://doi.org/10.1515/opag-2021-0034

Grace MH, Yousef GG, Gustafson SJ, Truong V-D, Yencho GC, Lila MA (2014) Phytochemical changes in phenolics, anthocyanins, ascorbic acid, and carotenoids associated with sweetpotato storage and impacts on bioactive properties. Food Chem 145:717–724

Haselow NJ, Stormer A, Pries A (2016) Evidence-based evolution of an integrated nutrition-focused agriculture approach to address the underlying determinants of stunting. Matern Child Nutr 12:155–168

Haskell M, Tanumihardjo SA, Palmer A, Melse-Boonstra A, Talsma EF, Burri B (2017) Effect of regular consumption of provitamin A biofortified staple crops on vitamin A status in population in low-income countries. Afr J Food Agric Nutr Dev 17:11865–11878

Hotz C, McClafferty B (2007) From harvest to health: challenges for developing Biofortified staple foods and determining their impact on micronutrient status. Food Nutr Bull 28:S271–S279

Hotz C, Loechl C, Lubowa A, Tumwine JK, Ndeezi G, Nandutu Masawi A, Baingana R, Carriquiry A, de Brauw A, Meenakshi JV, Gilligan DO (2012) Introduction of beta-carotene-rich orange sweet potato in rural Uganda resulted in increased vitamin A intakes among children and women and improved vitamin A status among children. J Nutr 142:1871–1880

Ilona P, Bouis H, Palenberg M, Moursi M, Oparinde A (2017) Chapter 9. Vitamin A cassava in Nigeria: crop development and delivery. Afr J Food Agric Nutr Dev Special issue on Biofortification 17(2):12000–12025

Ishiguro, K., 2019. Sweet potato carotenoids, in: Mu, T.-H., Singh, J. (Eds.), Sweet potato: chemistry, processing, and nutrition. Elsevier Inc., London, pp. 223–242

Jarvis A, Ramirez-Villegas J, Herrera Campo BV, Navarro-Racines C (2012) Is cassava the answer to African climate change adaptation? Trop Plant Biol 5:9–29

Jongstra R, Mwangi MN, Burgos G, Zeder C, Low JW, Mzembe G, Liria R, Penny M, Andrade MI, Fairweather-Tait S, Zum Felde T, Campos H, Phiri KS, Zimmermann MB, Wegmüller R (2020) Iron absorption from iron-biofortified sweetpotato is higher than regular sweetpotato in Malawian women while iron absorption from regular and iron-biofortified potatoes is high in Peruvian women. J Nutr 150:3094–3102

La Frano MR, Woodhouse LR, Burnett DJ, Burri BJ (2013) Biofortified cassava increases β-carotene and vitamin A concentrations in the TAG-rich plasma layer of American women. Br J Nutr 110(2):310–320. https://doi.org/10.1017/S0007114512005004

Lividini K, Fiedler J, De Moura F, Moursi M, Zeller M (2018) Biofortification: a review of ex-ante models. Glob Food Sec 17:186–195

Low JW (2011) Unleashing the potential of sweet potato to combat poverty and malnutrition in Sub-Saharan Africa through a comprehensive initiative. Acta Hort 921:171–179

Low JW, Thiele G (2020) Understanding innovation: the development and scaling of orange-fleshed sweetpotato in major African food systems. Agr Syst 179:1–17

Low JW, Arimond M, Osman N, Cunguara B, Zano F, Tschirley D (2007) A food-based approach introducing orange-fleshed sweet potatoes increased vitamin A intake and serum retinol concentrations in young children in rural Mozambique. J Nutr 137:1320–1327

Low J, Ball A, van Jaarsveld PJ, Namutebi A, Faber M, Grant FK (2015) Assessing nutritional value and changing behaviours regarding orange-fleshed sweetpotato use in Sub-Saharan Africa. In: Low J, Nyongesa M, Quinn S, Parker M (eds) Potato and sweetpotato in Africa: transforming the value chains for food and nutrition security. CAB International, Oxfordshire, UK, pp 551–579

Low JW, Mwanga ROM, Andrade MI, Carey E, Ball A-M (2017) Tackling vitamin A deficiency with biofortified sweetpotato in sub-Saharan Africa. Glob Food Sec 14:23–30

Makokha P, Wanjohi L, Okuku H, Kwikiriza N, Okello J, Low JW (2019) 2019 update of the status of sweetpotato decentralized vine multipliers in sub-Saharan Africa. International Potato Center, Nairobi, 33 p

Meenakshi JV (2008) Cost-effectiveness of Biofortification, best practice paper: new advice from the Copenhagen consensus of 2008. Copenhagen Consensus Center, Copenhagen, 21 p

Musyoka JN (2017) Effect of selected preservatives on microbial growth and stability of β-carotene during storage of orange fleshed sweetpotato puree, food science, nutrition and technology. University of Nairobi, Nairobi, 85 p

Muzhingi T, Malavi D, Moyo M (2019) Food and nutritional evaluation laboratory (FANEL): growth towards a sustainable service unit. Sweetpotato Action for Security and Health in Africa Project (SASHA). SASHA Brief 23. International Potato Center, Nairobi. https://hdl.handle.net/10568/105950

Nelson G, Bogard J, Lividini K, Arsenault J, Riley M, Sulser TB, Mason-D'Croz D, Power B, Gustafson D, Herrero M, Wiebe K, Cooper K, Remans R, Rosegrant M (2018) Income growth and climate change effects on global nutrition security to mid-century. Nat Sustain 1:773–781

Okello J, Wanjohi L, Makokha P, Low JW, Kwikiriza N (2019) Sweetpotato for profit and health initiative: status of sweetpotato in sub-Saharan Africa in 2019. International Potato Center, Nairobi, 29 p

Okonya JS, Petsakos A, Suarez V, Nduwayezu A, Kantungeko D, Blomme G, Legg JP, Kroschel J (2019) Pesticide use practices in root, tuber, and banana crops by smallholder farmers in Rwanda and Burundi. Int J Environ Res Public Health 16:400

Petsakos A, Prager SD, Gonzalez CE, Gama AC, Sulser TB, Gbegbelegbe S, Kikulwe EM, Hareau G (2019) Understanding the consequences of changes in the production frontiers for roots, tubers and bananas. Glob Food Sec 20:180–188

Rajendran S, McEwan M (2019) Early generation seed production for roots, tubers and bananas is financially viable for private sector seed companies in East Africa. SASHA Brief 13. International Potato Center, Nairobi. https://hdl.handle.net/10568/105960

Sartas M, Schut M, Proietti C, Thiele G, Leeuwis C (2020) Scaling Readiness: science and practice of an approach to enhance impact of research for development. Agr Syst 183:102874

Sindi K, Low JW (2015) Rwanda sweetpotato super foods: market chains that work for women and for the poor, SASHA Brief. International Potato Center (CIP), Nairobi, 4 p. https://hdl.handle.net/10568/69200

Smith MR, Thornton PK, Myers SS (2018) The impact of rising carbon dioxide levels on crop nutrients and human health, gender, climate change and nutrition integration initiative (GCAN). International Food Policy Research Institute (IFPRI), Washington DC

Stathers T, Low JW, Mwanga R, Carey T, David S, Gibson R, Andrade M, Namanda S, McEwan M, Bechoff A, Malinga J, Katcher H, Blakenship J, Agili S, Abidin E (2012) Everything you ever wanted to know about sweetpotato: reaching agents of change ToT manual. International Potato Center (CIP), Nairobi

Stathers T, Mkumbira J, Low JW, Tagwireyi J, Munyua H, Mbabu A, Mulongo G (2015a) Orange-fleshed sweetpotato investment guide. International Potato Center, Nairobi, 39 p

Stathers T, Mkumbira J, Low JW, Tagwireyi J, Munyua H, Mbabu A, Mulongo G (2015b) Orange-fleshed sweetpotato investment implementation guide. International Potato Center, Nairobi, 57 p

Taleon V, Sumbu D, Muzhingi T, Bidiaka S (2018) Carotenoids retention in biofortified yellow cassava processed with traditional African methods. J Sci Food Agric 99(3):1434–1441, 8 p. https://doi.org/10.1002/jsfa.9347

Thiele G, Dufour D, Vernier P, Mwanga ROM, Parker ML, Schulte Geldermann E, Teeken B, Wossen T, Gotor E, Kikulwe E, Tufan H, Sinelle S, Kouakou AM, Friedmann M, Polar V, Hershey C (2021) A review of varietal change in roots, tubers and bananas: consumer preferences and other drivers of adoption and implications for breeding. Int J Food Sci Technol 56:1076–1092

Tumuhimbise GA, Namutebi A, Muyonga JH (2009) Microstructure and in vitro beta carotene bioaccessibility of heat processed orange fleshed sweet potato. Plant Foods Hum Nutr 64:312–318

Tumuhimbise GA, Namutebi A, Muyonga JH (2010) Changes in microstructure, beta-carotene content and in vitro bioaccessibility of orange-fleshed sweet potato roots stored under different conditions. Afr J Food Agric Nutr Dev 10:3015–3028

Van Der Straeten D, Bhullar NK, Steur HD, Gruissem W, MacKenzie D, Pfeiffer W, Qaim M, Slamet-Loedin I, Strobbe S, Tohme J, Trijatmiko KR, Vanderschuren H, Montagu MV, Zhang C, Bouis H (2020) Multiplying the efficiency and impact of biofortification through metabolic engineering. Nat Commun 11:5203

Varma V, Bebber DP (2019) Climate change impacts on banana yields around the world. Nat Clim Chang 9:752–757

Willett W, Rockström J, Loken B, Springmann M, Lang T, Vermeulen S, Garnett T, Tilman D, DeClerck F, Wood A, Jonell M, Clark M, Gordon LJ, Fanzo J, Hawkes C, Zurayk R, Rivera JA, De Vries W, Majele Sibanda L, Afshin A, Chaudhary A, Herrero M, Agustina R, Branca F, Lartey A, Fan S, Crona B, Fox E, Bignet V, Troell M, Lindahl T, Singh S, Cornell SE, Srinath Reddy K, Narain S, Nishtar S, Murray CJL (2019) Food in the anthropocene: the EAT Lancet Commission on healthy diets from sustainable food systems. Lancet 393:447–492

World Bank (2006) Repositioning nutrition as central to development: a strategy for large scale action, directions in development. World Bank, Washington, DC, 246 p

World Bank (2020) GDP per Capita (Current US$). World development indicators. https://data.worldbank.org/indicator/NY.GDP.PCAP.CD

Index

© The Author(s) 2022
G. Thiele et al. (eds.), *Root, Tuber and Banana Food System Innovations*,
https://doi.org/10.1007/978-3-030-92022-7

Printed in the United States
by Baker & Taylor Publisher Services